T0343159

Process Intensification

Process Intensification
Engineering for Efficiency, Sustainability and Flexibility

Second edition

David Reay

Colin Ramshaw

Adam Harvey

AMSTERDAM • BOSTON • HEIDELBERG • LONDON
NEW YORK • OXFORD • PARIS • SAN DIEGO
SAN FRANCISCO • SINGAPORE • SYDNEY • TOKYO
Butterworth-Heinemann is an imprint of Elsevier

Butterworth-Heinemann is an imprint of Elsevier
The Boulevard, Langford Lane, Kidlington, Oxford, OX5 1GB, UK
225 Wyman Street, Waltham, MA 02451, USA

First edition 2008
Second edition 2013

Notice

Knowledge and best practice in this field are constantly changing. As new research and experience
broaden our understanding, changes in research methods, professionals practices, or medical treatment
may become necessary.

Practitioners and researchers must always rely on their own experience and knowledge in evaluating and
using any information, methods, compounds, or experiments described herein. In using such information
or methods they should be mindful of their own safety and the safety of others, including parties for
whom they have a professional responsibility

To the fullest extent of the law, neither the Publisher nor the authors, contributors, or editors, assume any
liability for any injury and/or damage to persons or property as a matter of products liability, negligence
or otherwise, or from any use or operation of any methods, products, instructions, or ideas contained in
the material herein.

British Library Cataloguing-in-Publication Data
A catalogue record for this book is available from the British Library

Library of Congress Cataloging-in-Publication Data
A catalog record for this book is available from the Library of Congress

ISBN: 978-0-08-098304-2

For information on all Butterworth-Heinemann publications
visit our website at www.elsevierdirect.com

Typeset by MPS Ltd., Chennai, India
www.adi-mps.com

Printed in Great Britain

13 14 15 16 17   10 9 8 7 6 5 4 3 2 1

Contents

Foreword

In the early 1990s my research team at Dow Chemical was challenged to overcome the technical barriers to create an economically viable process for making hypochlorous acid (HOCl). A number of chemical routes were documented in the literature, but no one had successfully commercialized any of the proposed routes. We selected a reactive distillation approach as the most promising. However, the conventional equipment and process technology did not meet the project objectives. We had not heard of 'process intensification' at the time, but the work of Colin Ramshaw on the rotating packed bed (Higee or RPB) was known. Believing that the Higee could solve the technical issues, we undertook its application. In fact, the RPB exceeded expectations, becoming the enabler to bring the HOCl process to full commercial status in 1999. Solving the technical challenges of the process development was only half the problem; the other half was convincing business managers, project managers, and plant personnel to take the risk to implement new technology. Not only did we have a new chemical process which no one else had been able to commercialize, but the new process was based on new equipment technology which had never been scaled up beyond the pilot scale. Though eventually successful, what we lacked in the 1990s was a broad-based understanding of process intensification principles and successful commercial examples to facilitate the discussion on risk management.

What was lacking a decade ago in terms of process principles and examples has now been supplied by David Reay, Colin Ramshaw, and Adam Harvey in this book on *Process Intensification* (PI). The authors chronicle the history of PI with emphasis on heat and mass transfer. For the business manager and project manager the PI Overview presents the value proposition for PI including capital reduction (smaller, cheaper), safety (reduced volume), environmental impact, and energy reduction. In addition, PI offers the promise of improved raw material yields. The authors deal with the obstacles to implementing PI, chief of which is risk management.

For the researcher and technology manager the authors provide an analysis of the mechanisms involved in PI. Active methods (energy added) to enhance heat and mass transfer are emphasized. A thorough look at intensified unit operations of heat transfer, reaction, separation, and mixing allows the reader to assess the application of PI to existing or new process technologies. The examples of commercial practice in the chemical industry, oil and gas (offshore), nuclear, food, aerospace, biotechnology, and consumer products show the depth and breadth of opportunities for the innovative application of PI to advance technology and to create wealth.

The final chapter provides a methodology to assess whether PI provides opportunities to improve existing or new processes. The step-by-step approach reviews both business and technical drivers and tests, including detailed questions to answer, to determine the potential value of applying PI. Not to be overlooked in this assessment process are the helpful tables in Chapters 2, 5, and 11. Table 2.5 lists the equipment types involved in PI and the sections of the book where additional information is located. Table 5.5 provides a list of the types of reactors employed in PI. Table 11.2 reviews the applications of PI.

This book on process intensification would have helped my research team to accelerate its study of the RPB (Higee) for production of HOCl, but would have also exposed us to much broader application of PI principles to other opportunities. The content would have been useful in the process of convincing the business and project managers to undertake the risk of implementing the new process and equipment. The book comes on the scene at an opportune time to influence and impact the chemical and petroleum industries as they face increasing global competition, government oversight, and social accountability. Business as usual will not meet these demands on the industry; the discipline of process intensification provides a valuable set of tools to aid the industry as we advance into the twenty-first century.

David Trent
Retired Scientist of Dow Chemical

Preface

As with all relatively new technologies, change can be rapid and research can quickly come to fruition in applications. On the other hand, some aspects of new technologies may not live up to expectations, or take much longer to become reality. Process intensification is no exception, and this second edition has examples of rapid successes, slow successes and some failures. A scan of the appendix listing organisations involved in process intensification gives a vivid picture of the changes, when compared to the earlier edition.

It is also interesting to observe the geographical variations over a relatively short period (the first edition was published in 2008) in the extent to which process intensification has moved from the laboratory to the field (metaphorically speaking!). The growth in PI research in the Far East (Taiwan, Japan and particularly, China) is reflected to a certain extent in increased application activities there. In Europe, much of the research remains in academic and industrial laboratories, while in the USA, we now see large-scale applications of micro-reactors becoming reality.

Areas of PI activity that have moved forward at a rapid rate are the development of carbon capture technologies, an increasing interest in green chemistry, and the start of serious research into linking renewable energies (in particular solar energy) to intensified processes such as chemical reactions.

Beyond the chemical engineering field, the interest in thermal control of micro-electronics continues to challenge intensified heat and mass transfer engineers. Electric fields, in particular microwaves and ultrasound, are finding wider uses, and the use of electrokinetic forces at the micro- and nanoscale continue to interest.

Large studies involving many specialists, such as those participating in the Delft Skyline Debates, are pushing some towards concepts such as the perfect reactor. Process intensification is but one concept in helping us to achieve a sustainable society – how important its role will be is still open to debate! The authors hope that this new edition will help to guide those, be they new to PI or established in the field, in their research; product development and design; production and application, in a constructive and stimulating way.

D. A. Reay
C. Ramshaw
A. P. Harvey

Acknowledgements

The authors are indebted to a number of organisations and individuals for providing data, including case studies, for use in this book and the previous edition. They include:

David Trent, recently retired from Dow Chemical, Texas, for the Foreword and data included in the text.

Chart Energy and Chemicals, for data on the compact heat exchangers and micro-reactors made by his company, including illustrations in Chapters 4 and 5 and the cover reactor photograph used on the cover.

Glen Harbold and latterly Matthew Brinn of GasTran Systems, USA, for case studies on HiGee systems in Chapters 8 and 9.

Anita Buxton and TWI, Cambridge, for case studies and illustrations of Surfisculpt. (See Figure 3.2.)

Robert Ashe and Gilda Gasparini of AM Technology, and Mayank Patel of Imperial College, University of London, for the case study in Chapter 5 on the innovative reactors produced by the company, and data in Chapter 8

Prof. Asterios Gavriilidis of University College, London, for data on the mesh reactor.

Dr. Carmen Torres-Sanchez of Heriot-Watt University, for the case study on ultrasound in baking (Chapter 10).

Hugh Epsom of Twister BV, The Netherlands, for data on Twister technology in Chapter 9.

Sarmad Ahmed, Daniel Fraser, James Hendry and Kieron Hopper, graduate chemical engineering students at Heriot-Watt University, Edinburgh, for the four case studies in the Appendix to Chapter 12.

Robert MacGregor, FLAME postgraduate student at Heriot-Watt University, for preparing the equations associated with the SDR in Chapter 5.

The MEng/MSc students on the process intensification module at Heriot-Watt University, for compiling much of the data in Appendices 4 and 5.

Figures 1.7 and 1.8 courtesy of Jim Wem, ex-ICI and consulting engineer, Teesside, UK.

Figure 1.14 reprinted from Ozkan, L., Backx, T., Van Gerven, T. and Stankiewicz, A.I. Towards perfect reactors: gaining full control of chemical transformations at molecular level. *Chemical Engineering and Processing*, 51, pp. 109–116, 2012, with permission from Elsevier.

Figure 3.5 reprinted from Lu, W., Zhao, C.Y. and Tassou, S.A. Thermal analysis on metal-foam filled heat exchangers. Part I: Metal-foam filled pipes. *International Journal of Heat and Mass Transfer*, Vol. 49, Issues 15–16, pp. 2751–2761, July 2006, with permission from Elsevier.

Figure 3.6 reprinted from Wang, L. and Sunden, B. Performance comparison of some tube inserts. *Int. Comm. Heat Mass Transfer*, Vol. 29, No. 1, pp. 45–56, 2002, with permission from Elsevier.

Figure 3.8 reprinted from Janicke, M.T., Kestenbaum, H., Hagendorf, U., Schüth, F. Maximilian Fichtner and Schubert, K. The controlled oxidation of hydrogen from an explosive mixture of gases using a microstructured reactor/heat exchanger and Pt/Al$_2$O$_3$ Catalyst. *Journal of Catalysis*, Vol. 191, pp. 282–293, April 2000, with permission from Elsevier.

Figure 3.9 reprinted from Legay, M., Le Person, S., Gondrexon, N., Boldo, P. and Bontemps, A. Performances of two heat exchangers assisted by ultrasound. *Applied Thermal Engineering* 37, pp. 60–66, 2012, with permission from Elsevier.

Figure 3.10 reprinted from Karayiannis, T.G. EHD boiling heat transfer enhancement of R123 and R11 on a tube bundle. *Applied Thermal Engineering*, Vol. 18, Issues 9–10, pp. 809–817, September 1998, with permission from Elsevier.

Figure 3.13 reprinted from Chen, M., Yuan, L. and Liu, S. Research on low temperature anodic bonding using induction heating. *Sensors and Actuators A*, 133, pp. 266–269, 2007, with permission from Elsevier.

Figure 3.14 reprinted from Takei, G., Kitamori, T. and Kim, H.B. Photocatalytic redox-combined synthesis of L-pipecolinic acid with a titania-modified microchannel chip. *Catalysis Communications*, Vol. 6, pp. 357–360, 2005, with permission from Elsevier.

Figure 3.15 reprinted from Bolshakov, A.P., Konov, V.I., Prokhorov, A.M., Uglov, S.A. and Dausinger, F. Laser plasma CVD diamond reactor. *Diamond and Related Materials*, Vol. 10, pp. 1559–1564, 2001, with permission from Elsevier.

Figure 3.16 reprinted from Butrymowicz, D., Trela, M. and Karwacki, J. Enhancement of condensation heat transfer by means of passive and active condensate drainage techniques. *Int. J. Refrigeration*, Vol. 26, pp. 473–484, 2003, with permission from Elsevier.

Figure 3.17 reproduced from Qian, S. and Bau, H.H. Magneto-hydrodynamic stirrer for stationary and moving fluids. *Sensors and Actuators B*, Vol. 106, pp. 859–870, 2005, with permission from Elsevier.

Figure 3.18 reproduced from Garnier, N., Grigoriev, R.O. and Schatz, M.F. Optical manipulation of microscale fluid flow. *Physics Review Letters*, Vol. 91, Paper 054501, 2005, with permission from Elsevier.

Figure 3.19 reproduced from Shah, P., Vedarethinam, I., Kwasny, D. Andresen, L., Dimaki, M., Skov, S. and Svenden, W.R. Microfluidic bioreactors for culture of non-adherent cells. *Sensors and Actuators B: Chemical*, 156, pp. 1002–1008, 2011, with permission from Elsevier.

Figures 4.2 (a) and (b) reproduced with permission of Within – www.within-labs.com

Figure 4.9 reproduced from Tsuzuki, N., Kato, Y. and Ishiduka, T. High performance printed circuit heat exchanger. *Applied Thermal Engineering*, Vol. 27, pp. 1702–1707, 2007, with permission from Elsevier.

Figure 4.10 reproduced from Tsuzuki, N., Utamura, M. and Ngo, T.L. Nusselt number correlations for a microchannel heat exchanger hot water supplier with S-shaped fins. *Applied Thermal Engineering*, 29, 3299–3308, 2009, with permission from Elsevier.

Figure 4.14 reproduced from Boomsma, K., Poulikakos, D. and Zwick, F. Metal foams as compact high performance heat exchangers. *Mechanics of Materials*, Vol. 35, pp. 1161–1176, 2003, with permission from Elsevier.

Figure 4.15 and Figure 4.16 reproduced from Zhao, C.Y., Lu, W. and Tassou, S.A. Thermal analysis on metal-foam filled heat exchangers. Part II: Tube heat exchangers. *International Journal of Heat and Mass Transfer*, Vol. 49, pp. 2762–2770, 2006, with permission from Elsevier.

Figure 4.17 and Figure 4.18 reproduced from Tian, J., Lu, T.J., Hodson, H.P., Queheillalt, D.T. and Wadley, H.N.G. Cross flow heat exchange of textile cellular metal core sandwich panels. *International Journal of Heat and Mass Transfer*, Vol. 50, pp. 2521–2536, 2007, with permission from Elsevier.

Figure 4.20 reproduced with permission from Drummond Hislop, SES Ltd.

Figure 4.22 reproduced from Alm, B., Imke, U., Knitter, R., Schygulla, U. and Zimmermann, S. Testing and simulation of ceramic micro heat exchangers. *Chemical Engineering Journal*, Vol. 135, Supplement 1, pp. S179–S184, 2007, with permission from Elsevier.

Figure 4.23 reproduced from Mala, G.M. and Li, D. Flow characteristics of water in microtubes. *Int. J. Heat and Fluid Flow*, Vol. 20, pp. 142–148, 1999, with permission from Elsevier.

Figure 4.24 reproduced from Jeng, T-M., Tzeng, S-C. and Lin, C-H. Heat transfer enhancement of Taylor–Couette–Poiseuille flow in an annulus by mounting longitudinal ribs on the rotating inner cylinder. *International Journal of Heat and Mass Transfer*, Vol. 50, Issues 1–2, pp. 381–390, 2007, with permission from Elsevier.

Figure 4.25 reproduced from Lockerby, D.A. and Reese, J.M. High-resolution Burnett simulations of micro-Couette flow and heat transfer. *J. Computational Physics*, Vol. 188, pp. 333–347, 2003, with permission from Elsevier.

Figure 4.26 reproduced from Qin, F., Chen, J., Lu, M., Chen, Z., Zhou, Y. and Yang, K. Development of a metal hydride refrigeration system as an exhaust-gas driven automobile air conditioner. *Renewable Energy*, Vol. 32, pp. 2034–2052, 2007, with permission from Elsevier.

Figure 5.19 reproduced from Mackley, M.R. and Stonestreet, P. Heat transfer and associated energy dissipation for oscillatory flow in baffled tubes. *Chem. Eng. Sci.*, Vol. 50, pp. 2211–2224, 1995, with permission from Elsevier.

Figure 5.13 reproduced from Dutta, P.K. and Ray, A.K. Experimental investigation of Taylor vortex photocatalytic reactor for water purification. *Chemical Engineering Science*, Vol. 59, pp. 5249–5259, 2004, with permission from Elsevier.

Figure 5.30, Figure 5.31 and Figure 5.32 reproduced from Haugwitz, S., Hagander, P. and Noren, T. Modelling and control of a novel heat exchanger reactor, the Open Plate Reactor. *Control Engineering Practice*, Vol. 15, pp. 779–792, 2007, with permission from Elsevier.

Figure 5.43 reproduced from Wasewar, K.L., Pangarkar, V.G., Heesink, A.B.M., and Versteeg, G.F. Intensification of enzymatic conversion of glucose to lactic acid by reactive extraction. *Chemical Engineering Science*, Vol. 58, pp. 3385–3393, 2003, with permission from Elsevier.

Figure 5.44 reproduced from Centi, G., Dittmeyer, R., Perathoner, S. and Reif, M. Tubular inorganic catalytic membrane reactors: advantages and performance in multiphase hydrogenation reactions. *Catalysis Today*, Vol. 79–80, pp. 139–149, 2003, with permission from Elsevier.

Figure 5.48 reproduced from Zhang, H. and Zhuang, J. Research, development and industrial application of heat pipe technology in China. *Applied Thermal Engineering*, Vol. 23, Issue 9, pp. 1067–1083, 2003, with permission from Elsevier.

Figure 6.1 and Figure 6.2 reproduced from Kaibel, B. Distillation – dividing wall columns. *Encyclopedia of Separation Science*, pp. 1–9, Elsevier, Oxford, 2007, with permission from Elsevier.

Figure 6.3 reproduced from Bruinsma, O.S.L., Krikken, T., Cot, J., Saric, M., Tromp, S.A., Olujic, Z. and Stankiewicz, A.I. The structured heat integrated distillation column. *Chemical Engineering Research and Design*, 90, pp. 458–470, 2012, with permission from Elsevier.

Figure 6.4 reproduced from Maleta, V.N., Kiss, A.A., Taran, V.M. and Maleta, B.V. Understanding process intensification in cyclic distillation systems. *Chemical Engineering and Processing: Process Intensification*, 50, pp. 655–664, 2011, with permission from Elsevier.

Figure 6.11 reproduced from Wang, G.Q., Xu, Z.C., Yu, Y.L. and Ji, J.B. Performance of a rotating zigzag bed – a new Higee. *Chemical Engineering and Processing*, doi:10.1016/j.cep.2007.11.001, 2007, with permission from Elsevier.

Figure 6.14 reproduced from Day, N. Why centrifuges play an important role in the production of sugar. *Filtration and Separation*, Vol. 41, Issue 8, pp. 28–30, October 2004, with permission from Elsevier.

Figure 6.15 reproduced from Caputo, G., Felici, C., Tarquini, P., Giaconia, A. and Sau, S. Membrane distillation of HI/H_2O and H_2SO_4/H_2O mixtures for the sulphur-iodine thermochemical process. *Int. J. Hydrogen Energy*, Vol. 32, pp. 4736–4743, 2007, with permission from Elsevier.

Figure 6.16 reproduced from Belyaev, A.A. et al. Membrane air separation for intensification of coal gasification process. *Fuel Processing Technology*, Vol. 80, pp. 119–141, 2003, with permission from Elsevier.

Figure 6.22 reproduced from Matsushima, H., Nishida, T., Konishi, Y., Fukunaka, Y., Ito, Y. and Kuribayashi, K. Water electrolysis under microgravity – Part 1. Experimental technique. *Electrochim. Acta,* Vol 48, pp. 4119–4125, 2003, with permission from Elsevier.

Figure 7.1 reprinted from Hessel, V., Lowe, H., and Schoenfeld, F. Micromixers – a review on passive and active mixing principles. *Chemical Engineering Science*, Vol. 60, Issues 8–9, pp. 2479–2501, 2005, with permission from Elsevier.

Figure 7.2 reprinted from Ferrouillat, S., Tochon, P., Garnier, C. and Peeerhossaini, H. Intensification of heat transfer and mixing in multifunctional heat exchangers by artificially generated streamwise vorticity. *Applied Thermal Engineering*, Vol. 26, pp. 1820–1829, 2006, with permission from Elsevier.

Figure 8.1 reprinted from Neelis, M., Patel. M., Bach, P. and Blok, K. Analysis of energy use and carbon losses in the chemical industry. *Applied Energy*, Vol. 84, pp. 853–862, 2007, with permission from Elsevier.

Figure 8.4 reprinted from Pedernera, M.N., Pina, J., Borio, D.O. and Bucala, V. Use of a heterogeneous two-dimensional model to improve the primary steam reformer performance. *Chemical Engineering Journal*, Vol. 94, pp. 29–40, 2003, with permission from Elsevier.

Figure 8.5 reprinted from Perez-Ramirez, J. and Vigeland, B. Lanthanum ferrite membranes in ammonia oxidation. Opportunities for 'pocket-sized' nitric acid plants. *Catalysis Today*, 105, pp. 436–442, 2005, with permission from Elsevier.

Figure 8.6 Calvar, N., Gonzalez, B. and Dominguez, A. Esterification of acetic acid with ethanol: Reaction kinetics and operation in a packed bed reactive distillation column. *Chemical Engineering and Processing*, Vol. 46, pp. 1317–1323, 2007.

Figure 8.7 reprinted from Cao, Enhong and Gavriilidis, A. Oxidative dehydrogenation of methanol in a microstructured reactor. *Catalysis Today*, Vol. 110, pp. 154–163, 2005, with permission from Elsevier.

Figure 8.8 reprinted from Enache, D.I., Thiam, W., Dumas, D., Ellwood, S., Hutchings, G.J., Taylor, S.H., Hawker, S. and Stitt, E.H. Intensification of the solvent-free catalytic hydroformylation of cyclododecatriene: comparison of a stirred batch reactor and a heat-exchanger reactor. *Catalysis Today*, Vol. 128, pp. 18–25, 2007, with permission from Elsevier.

Figure 8.14 reprinted from Cornelissen, R., Tober, E., Kok, J. and van de Meer, T. Generation of synthesis gas by partial oxidation of natural gas in a gas turbine. *Energy*, Vol, 31, pp. 3199–3207, 2006, with permission from Elsevier.

Figure 8.15 reprinted from Hugill, J.A., Tillemans, F.W.A., Dijkstra, J.W. and Spoelstra, S. Feasibility study on the co-generation of ethylene and electricity through oxidative coupling of methane. *Applied Thermal Engineering*, Vol. 25, pp. 1259–1571, 2005, with permission from Elsevier.

Figure 8.16 reprinted from Weatherley, L.R. Electrically enhanced mass transfer. *Heat Recovery Systems & CHP*, Vol. 13, No. 6, pp. 515–537, 1993, with permission from Elsevier.

Figure 8.17 and Figure 8.18 reprinted from Tai, C.Y., Tai, C-t, and Liu, H-s. Synthesis of submicron barium carbonate using a high-gravity technique. *Chemical Engineering Science*, Vol. 61, pp. 7479–7486, 2006, with permission from Elsevier.

Figure 8.23 reprinted from Dalmoro, A., Barba, A.A., Lamberti, G. and d'Amore, M. Intensifying the microencapsulation process: Ultrasonic atomisation as an innovative approach. *European Journal of Pharmaceutics and Biopharmaceutics*, 80, pp. 471–477, 2012, with permission from Elsevier.

Figure 8.26 reprinted from Chen, G., Li, S., Jiao, F. and Yuan, Q. Catalytic dehydration of bioethanol to ethylene over TiO_2/γ-Al_2O_3 catalysts in microchannel reactors. *Catalysis Today*, Vol. 125, pp. 111–119, 2007, with permission from Elsevier.

Figure 8.32 (a) and (b) reprinted from Moldenhauer, P., Ryden, M., Mattisson, T. and Lyngfelt, A. Chemical looping combustion and chemical looping reforming of kerosene in a circulating fluidised-bed 300 W laboratory reactor. *International Journal of Greenhouse Gas Control*, 9, 1–9, with permission from Elsevier.

Figure 8.33 reprinted (adapted) with permission from Halonsen, S.F. and Blom, R. Chemical looping combustion in a rotating bed reactor – finding optimal process conditions for prototype reactor. *Environmental Science and Technology*, 45 (22), pp. 9619–9626, 2011 American Chemical Society.

Figure 9.4 reprinted from Zhang, D., Zhang, P-Y., Zou, H-K., Chu, G-W., Wu, W., Zhu, Z-w., Shao, L. and Chen, J-F. Application of HIGEE process intensification technology in synthesis of petroleum sulfonate surfactant. *Chemical Engineering and Processing: Process Intensification*, 49, 508–513, 2010, with permission from Elsevier.

Figure 9.7 reprinted from Petty, C.A. and Parks, S.M. Flow structures within miniature hydrocyclones. *Minerals Engineering*, Vol. 17, pp. 615–624, 2004, with permission from Elsevier.

Figure 9.12 reprinted from Tonkovich, A.L., Jarosch, K., Arora, R., Silva, L., Perry, S., McDaniel, J., Daly, F. and Litt, R. Methanol production FPSO plant concept using multiple microchannel unit operations. *Chemical Engineering Journal*, Vol. 135S, pp. S2–S8, 2008, with permission from Elsevier.

Figure 9.13 Courtesy of Oxford Catalysts Group

Figure 10.14 Courtesy of Torftech Ltd.

Figure 10.15 and Figure 10.16 reprinted from Van der Bruggen, B., Curcio, E. and Drioli, E. Process intensification in the textile industry: the role of membrane technology. *J. Environmental Management*, Vol. 73, pp. 267–274, 2004, with permission from Elsevier.

Figure 10.17 reprinted from Warmoeskerken, M.M.C.G., van der Vlist, P., Moholkar, V.S. and Nierstrasz, V.A. Laundry process intensification by ultrasound. *Colloids and Surfaces A: Physicochem. Eng. Aspects*, Vol. 210, pp. 277–285, 2002, with permission from Elsevier.

Figure 11.2 and Figure 11.3 reprinted from Gilchrist, K., Lorton, R. and Green, R.J. Process intensification applied to an aqueous LiBr rotating absorption chiller with dry heat rejection. *Applied Thermal Engineering*, Vol. 22, pp. 847–854, 2002, with permission from Elsevier.

Figure 11.4 reprinted from Izquierdo, M., Lizarte, R., Marcos, J.D., and Gutierrez, G. Air conditioning using an air-cooled single effect lithium bromide absorption chiller: results of a trial conducted in Madrid in August 2005. *Applied Thermal Engineering*, Vol. 28, pp. 1074–1081, 2008, with permission from Elsevier.

Figure 11.7 reprinted from Determan, M.D. and Garimella, S. Design, Fabrication, and Experimental Demonstration of a Microscale Monolithic Modular Absorption Heat Pump, *Applied Thermal Engineering*, 47, pp. 119–125, 2012, with permission from Elsevier.

Figure 11.8 reprinted from Determan, M.D. and Garimella, S. Design, Fabrication, and Experimental Demonstration of a Microscale Monolithic Modular Absorption Heat Pump, *Applied Thermal Engineering*, 47, pp. 119–125, 2012, with permission from Elsevier.

Figures 11.9(a) and 11.9(b) reprinted from Goodman, C., Fronk, B.M. and Garimella, S. Transcritical carbon dioxide microchannel heat pump water heaters: Part 1 – validated component simulation modules. *International Journal of Refrigeration*, 34, pp. 859–869, 2011, with permission from Elsevier.

Figure 11.10 reprinted from Heppner, J.D., Walther, D.C. and Pisano, A.P. The design of ARCTIC: a rotary compressor thermally insulated micro-cooler. *Sensors and Actuators A*, Vol. 134, pp. 47–56, 2007, with permission from Elsevier.

Figure 11.11 reprinted from Bradshaw, C.R., Groll, E.A., and Garimella, S.A. A comprehensive model of a miniature-scale linear compressor for electronics cooling. *International Journal of Refrigeration*, 34, pp. 63–73, 2011, with permission from Elsevier.

Figure 11.12 reprinted from Marcinichen, J.B., Olivier, J.A. and Thome, J.R. On-chip two-phase cooling of datacenters: Cooling system and energy recovery evaluation. *Applied Thermal Engineering*, 41, pp. 36–51, 2012, with permission from Elsevier.

Figure 11.13 reprinted from Critoph, R.E. and Metcalf, S.J. Specific cooling power intensification limits in ammonia-carbon adsorption refrigeration systems. *Applied Thermal Engineering*, Vol. 24, pp. 661–678, 2004, with permission from Elsevier.

Figure 11.14 reprinted from Tamainot-Telto, Z., Metcalf, S.J. and Critoph, R.E. Novel compact sorption generators for car air conditioning. *International Journal of Refrigeration*, 32, pp. 727–733, 2009, with permission from Elsevier.

Figure 11.15 reprinted from Munkejord, S.T., Maehlum, H.S., Zakeri, G.R., Neksa, P. and Pettersen, J. Micro technology in heat pumping systems. *Int. J. Refrigeration*, Vol. 25, pp. 471–478, 2002, with permission from Elsevier.

Figure 11.16 reprinted from Garimella, S., Determan, M.D., Meacham, J.M., Lee, S. and Ernst, T.C. Microchannel component technology for system-wide application in ammonia/water absorption heat pumps. *International Journal of Refrigeration*, 34, pp. 1184–1196, 2011, with permission from Elsevier.

Figure 11.17 reprinted from Yu, H., Chen, H., Pan, M., Tang, Y., Zeng, K. Peng, F. and Wang, H. Effect of the metal foam materials on the performance of methanol-steam micro-reformer for fuel cells. *Applied Catalysis A: General*. Vol. 327, pp. 106–113, 2007, with permission from Elsevier.

Figures 11.18 and 11.19 reprinted from Kundu, A., Jang, J.H., Gil, J.H., Jung, C.R., Lee, H.R., Kim, S.-H., Ku, B. and Oh, Y.S. Review Paper. Micro-fuel cells – Current development and applications. *Journal of Power Sources*, Vol. 170, pp. 67–78, 2007, with permission from Elsevier.

Figure 11.20 reprinted from Ribaud, Y. La micro turbine: L'example du MIT. *Mec. Ind.*, Vol. 2, pp. 411–420, 2001, (in French), with permission from Elsevier.

Figure 11.21 reprinted from Weiss, L. Power production from phase change in MEMS and micro devices, a review. *International Journal of Thermal Sciences*, 50, pp. 639–647, 2011, under the 'Attribution No Derivatives' License. This work may be used under the Creative Commons 'Attribution No Derivates' License. http://creativecommons.org/licenses/by-nd/2.0/uk.

Figure 11.22 reprinted from Cheng, Hsu-Hsiang and Tan, Chung-Sung. Reduction of CO_2 concentration in a zinc/air battery by absorption in a rotating packed bed. *Journal of Power Sources*, Vol. 162, pp. 1431–1436, 2006, with permission from Elsevier.

Figure 11.25 reprinted from Figus, C. et al. Capillary fluid loop developments in Astrium. *Applied Thermal Engineering*, Vol. 23, pp. 1085–1098, 2003, with permission from Elsevier.

Figure 11.26 reprinted from Moon, Seok Hwan, et al. Improving thermal performance of miniature heat pipe for notebook PC cooling. *Microelectronics Reliability*, Vol. 44, pp. 315–321, 2004, with permission from Elsevier.

Figure 11.29 reprinted from Hu. X. and Tang, D. Experimental investigation on flow and thermal characteristics of a micro phase-change cooling system with a microgroove evaporator. *Int. J. Thermal Sciences*, Vol. 46, pp. 1163–1171, 2007, with permission from Elsevier.

Figure 12.1 reprinted from Kothare, M.V. Dynamics and control of integrated microchemical systems with application to micro-scale fuel processing. *Computers and Chemical Engineering*, Vol. 30, pp. 1725–1734, 2006, with permission from Elsevier.

Introduction

Process intensification (PI) may be defined in a number of ways. The chemist or chemical engineer will appreciate the two-part definition used by one of the major manufacturers of PI equipment:

- PI significantly enhances transport rates.
- It gives every molecule the same processing experience.

This definition can be usefully interpreted as being a process development involving dramatically smaller equipment which leads to:

1. Improved control of reactor kinetics giving higher selectivity/reduced waste products.
2. Higher energy efficiency.
3. Reduced capital costs.
4. Reduced inventory/improved intrinsic safety/fast response times.

The heat transfer engineer will note that 'intensification' is analogous to 'enhancement', and intensification is based to a substantial degree on active, and to a lesser extent, passive enhancement methods, that are used widely in heat and mass transfer, as will be illustrated regularly throughout the book.

Readers will be well placed to appreciate and implement the PI strategy once they are aware of the many technologies which can be used to intensify unit operations and also, of some successful applications.

Perhaps the most commonly recognisable feature of an intensified process is that it is smaller – perhaps by orders of magnitude – than that it supersedes. The phraseology unique to intensified processes – the pocket-sized' nitric acid plant being an example – manages to bring out in a most dramatic way the reduction in scale possible, using what we might describe as extreme heat and mass transfer enhancement (although one is unlikely to put a nitric acid plant in one's pocket!). Cleanliness and energy-efficiency tend to result from this compactness of plant, particularly in chemical processes and unit operations, but increasingly in other application areas, as will be seen in the applications chapters of this book. To this may be added safety, brought about by the implicit smaller inventories of what may be hazardous chemicals that are passing through the intensified unit operations. So it is perhaps entirely appropriate to regard PI as a green technology – making minimum demand on our resources – compatible with the well-known statement from the UN Bruntland Commission for '......*a form of sustainable development which meets the needs of the present without compromising the ability of future generations to meet their own needs*'.

In the UK, the Institution of Chemical Engineers' (IChemE), *A Roadmap for 21st Century Chemical Engineering*, celebrating 50 years since it was awarded its

Royal Charter, set the scene for process intensification in the context of sustainable technology, (Anon, 2007):

> *As chemical engineers we have readily accepted the principle of the economy of scale, and as a result have designed and built ever larger production units, increasing plant efficiency and reducing per unit costs of production. The downsides of this policy include increased safety and environmental risks arising from higher inventories of hazardous material, the economic risk of overcapacity from simultaneous multiple world-scale plant expansions, and the legacy effects of written down plant impeding the introduction of new products and technology.*
>
> *New concepts such as process intensification, flexible, miniaturised plants, localised production and industrial ecology must become mainstream and we must continually reassess our approach to plant design and the acceptance of innovative concepts to render the chemical industry sustainable.*
>
> *IChemE believes that the necessary change in business strategy to speed the introduction of innovative and sustainable technologies should be led from the boardroom, facilitated and encouraged by chemical engineers at all levels in industry, commerce and academia.*

The compact heat exchanger, one of the first technologies addressed in this book (in Chapter 4), is a good example of an *evolutionary* process technology which now forms the basis of very small chemical reactors (and possibly new generations of nuclear reactors), as well as being routinely used for its primary purpose, heat transfer, in many demanding applications. The rotating distillation unit, known as HiGee, invented over 30 years ago by co-author Professor Colin Ramshaw when at ICI, represented a *revolutionary* change (in more ways than one) in process plant size reduction – in the words of Bart Drinkenberg of the major chemical company, DSM, reducing distillation columns '...*the size of Big Ben, to a few metres in height*'. This Second Edition contains a photo of the original rotor of the HiGee rotating packed bed (see Chapter 1).

As well as building awareness of what remains, to many, an obscure technology, a further aim of the book is to show that process intensification, whether its technology has evolved over the years or involves a step change in thinking, is not limited to chemical processes. The electronics industry, first with the transistor and then with the chip, has achieved amazing performance enhancements in modern microelectronic systems – and these enhancements have necessitated parallel increases in heat removal rates, typified by intensified heat exchangers and even micro-refrigerators. Note that 'intensification' has a slightly different connotation here – the micro-refrigerator used to cool the electronics chip does not have the cooling capacity of its large counterparts, whereas the HiGee separator or the plate reactor, as will be demonstrated later, do retain the capability of their ubiquitous, but now obsolescent, large predecessors.

It is highly relevant to note that some of the most compact intensified process plants are fabricated using methodologies developed within the electronics sector – micro-technology and MEMS, (micro-electro mechanical systems) are synonymous with modern manufacturing technology and also with intensification. The printed circuit heat exchanger (Chapter 4), as its name implies, bears not a small relationship to electronics. In this edition, it will be noted that there is reference to 3D printing, a technique that some believe will revolutionise manufacturing over the next two decades. The technique, also called additive manufacturing, allows one to construct 3D solid objects from a digital image, commonly using laser sintering. Highly compact heat exchangers and micro-reactors are examples where this technique is being, or could be, applied.

Biological and biochemical systems can also be intensified – food production and effluent treatment are examples. In its roadmap, the IChemE extends its comments to the food industry, again citing PI as an important contributor:

Innovation within the food industry bridges a spectrum from far market and blue sky, usually supported by the larger organisations, to incremental development, often the preserve of small companies. Chemical engineering has an essential role in areas such as the scale-up of emerging technologies, e.g. ultra high pressure, electrical technologies, pulsed light; the control of processes both in terms of QA (Quality Assurance) approaches (e.g. HACCP/HAZOP/HAZAN) and process engineering control approaches; the validation and verification of the effectiveness of processing systems; the optimisation of manufacturing operations; increasing flexibility in plant and process intensification; and the application of nanotechnology concepts to food ingredients and products. Commercial viability of innovative technologies is key, as is the consumer perception of the risks and benefits of new technologies. Education is vital in informing such perceptions. The environmental impact of the new approach will be one of the key factors.

Considering the range of these topics, it is clear that some are far from application in the manufacturing sector of today and require fundamental research to develop the knowledge of the science that underpins the area, together with the engineering approaches necessary to implement the new technology in the manufacturing arena. This is clearly a role for strategic research funding within the academic community. It is important to encourage the blue sky development of science on a broad front compatible with the key challenges for the industry. Sustainability is vital and must be an active consideration for all involved in the food sector.

While those processes involving enzymes tend to progress at rather leisurely paces, some fermentation processes may be limited by oxygen availability and therefore susceptible to mass transfer intensification. The ability to intensify such reactions remains attractive in food production, some pharmaceutics production and waste disposal – in fact, reactors such as those based upon oscillatory baffle movement are becoming increasingly a commercial reality. (As an aside, a literature search of process intensification inevitably encompasses intensive agriculture – PI on a grander scale!)

(At this stage, it is useful to point out that whilst a knowledge of chemistry, biochemistry and/or chemical engineering helps in the detailed appreciation of some of the arguments for process intensification in the chemicals and related sectors, particularly when discussing reaction kinetics, it is not essential – other texts such as that by Stankiewicz and Moulijn (2004) deal in greater depth with the chemistry and chemical engineering aspects. Most engineering or science graduates will have no difficulty in following the logic of the arguments presented. Where theory is necessary to appreciate concepts, or to emphasise arguments, equations are included.)

Where a concept is used, albeit in different forms, across a range of industries, there is opportunity for technology transfer, and it is hoped that this book will stimulate this by demonstrating the broad application of PI. For a good example of technology transfer in the longer term, the Delft Skyline Debates take us towards 2050 with a broad vision of PI across a wide range of end uses (Web 1, 2011).

The benefits of PI are several, but readers from industry or research laboratories will identify their own priorities when contemplating whether PI will be beneficial to their own activities. However, environmental considerations will inevitably weigh increasingly heavily when considering investment in new processes within a context of global climate change. Data towards the end of the book should help potential users of PI technologies to make the case for an investment. Giving guidance on how to incorporate them in the plant design process and to use them effectively is an essential part of confidence building in supporting new investment arguments. Although many PI technologies are still under development, considerable thought has been given by most research teams to ways for ensuring that they are effective in practice, as well as in the laboratory. In fact, as pointed out by Professor Ramshaw in his many papers on PI, the dominant feature of PI plant – its small size coupled to high throughput – can in many instances make the laboratory plant the production unit as well!

This book should help the reader, if a student or academic researcher, to obtain a good appreciation of what PI is, and, if working in industry, to make a judgement as to whether PI is relevant to his/her business (be it a global player or a small company) and, if positive, provide sufficient information to allow him/her to make a first assessment of potential applications. Where the topic is of particular relevance, the reader should be able to initiate steps towards implementation of the technology.

In order to be able to fulfil the above, this book should help the reader:

- To obtain an understanding of the concept of process intensification, an appreciation of its development history and its relationship to conventional technologies.
- To gain an appreciation of the contribution process intensification can make to improving energy use and the environment, safety, and, most importantly, the realisation of business opportunities.
- To gain a knowledge of the perceived limitations of process intensification technologies and ways of overcoming them.

- To gain a detailed knowledge of a range of techniques which can be used for intensifying processes and unit operations.
- To obtain a knowledge of a wide range of applications, both existing and potential, for PI technologies.
- To gain a basic appreciation of the steps necessary to assess opportunities for PI, and to apply PI technology.

References

Anon, A, 2007. Roadmap for the Twentyfirst Century. Inst. Chem. Eng. May.

Stankiewicz, A., Moulijn, J.A., 2004. Re-engineering the Chemical Processing Plant: Process Intensification. Marcel Dekker, New York.

Web 1, 2011. Research Agenda for Process Intensification towards a Sustainable World. Editors: Andrzej Gorak and Andrzej Stankiewicz. <http://www.3me.tudelft.nl/fileadmin/Faculteit/3mE/Actueel/Nieuws/2011/docs/DSD_Research_Agenda.pdf> (accessed 22.10.12.).

A Brief History of Process Intensification

OBJECTIVES IN THIS CHAPTER

The objectives of this chapter are to summarise the historical development of process intensification, chronologically and in terms of the sectors and unit operations to which it has been applied.

1.1 Introduction

Those undertaking a literature search using the phrase 'process intensification' will find a substantial database covering the process industries, enhanced heat transfer and, not surprisingly, agriculture. For those outside specialist engineering fields, 'intensification' is commonly associated with the increases in productivity in the farming of poultry, cattle and crops where, of course, massive increases in yield for a given area of land can be achieved. The types of intensification being discussed in this book are implemented in a different manner, but have the same outcome.

The historical aspects of heat and mass transfer enhancement, or intensification, are of interest for many reasons. We can examine some processes that were intensified some decades before the phrase 'process intensification' became common in process engineering (particularly chemical) literature. Some used electric fields, others employed centrifugal forces. The use of rotation to intensify heat and mass transfer has, as we will see, become one of the most spectacular tools in the armoury of the plant engineer in several unit operations, ranging from reactors to separators. However, it was in the area of heat transfer – in particular, two-phase operation – that rotation was first exploited in industrial plants. The rotating boiler is an interesting starting point, and rotation forms the essence of process intensification (PI) within this chapter, with interesting newly revealed data on the HiGee test plant installed at ICI, (Web 3, 2012).

It is, however, worth highlighting one or two early references to intensification that have interesting connections with current developments. One of the earliest references to intensification of processes was in a paper published in the US in 1925 (Wightman et al., 1925). The research carried out by Eastman Kodak in the US was directed at image intensification: increasing the 'developability' of latent images on plates by a substantial amount. This was implemented using a small addition of hydrogen peroxide to the developing solution.

Process Intensification.
© 2013 Elsevier Ltd. All rights reserved.

T.L. Winnington (1999), in a review of rotating process systems, reported work at Eastman Kodak by Hickman (1936) on the use of spinning discs to generate thin films as the basis of high-grade plastic films (UK Patent, 1936). The later Hickman still, alluded to in the discussion on separators later in this chapter, was another invention of his. Application of PI in the image reproduction area was brought right into the 21st century by the activity reported in 2007 at Fujifilm Imaging Colorants Ltd in Grangemouth, Scotland, where a small continuous three-reactor intensified process has replaced a very large 'stirred pot' in the production of an inkjet colorant used in inkjet printer cartridges. The outcome was production of 1 kg/h from a lab-scale unit costing £15,000, whereas a conventional reactor for this application would have required a 60 m^3 vessel costing £millions (Web 1, 2007).

In the period between the first and second editions of this book, much of the emphasis on PI has become concentrated in Continental Europe, China and Taiwan. In particular, active intensification methods, particularly using rotating packed beds (see Chapters 5 and 6), have blossomed in the Far East, with organisations such as the Beijing University of Chemical Technology continuing to lead the way with its dedicated 'Research Center for High Gravity Engineering & Technology' (Chen, 2009) – see also section 1.4.

Within Europe, the Delft Skyline Debates represent to date the most ambitious attempt to focus PI on meeting the challenges facing humanity. In a special issue of a journal reporting on the outcomes, Professors Andrzej Górak and Andrzej Stankiewicz described how the debates have attempted to define a 'set of key technological achievements' that will contribute to sustainability in 2050 and beyond. The areas that the 75 experts within the team addressed were health, transport, living, and food and agriculture (Górak and Stankiewicz, 2012). Aspects of this are discussed in Chapter 2.

1.2 Rotating boilers

One of the earliest uses of HiGee forces in modern day engineering plant was in boilers. There are obvious advantages of using rotation plant in spacecraft, as they create an artificial gravity field where none existed before, see for example, Reay and Kew (2006). However, one of the first references to rotating boilers arises in German documentation cited as a result of post-Second World War interrogations of German gas turbine engineers, where the design was used in conjunction with gas and steam turbines (Anon, 1932, 1946).

1.2.1 The rotating boiler/turbine concept

The advantages claimed by the German researchers on behalf of the rotating boiler were that it offered the possibility of constructing an economic power plant of *compact dimensions* and *low weight*. No feed pump or feed water regulator were required, the centrifugal action of the water automatically taking care of the feed water supply. Potential applications cited for the boiler were small electric generators, peak load generating plant (linked to a small steam turbine), and as a starting motor for gas turbines, etc. A rotating boiler/gas turbine assembly using H_2 and O_2 combustion was also studied for use in torpedoes. The system in this latter role is illustrated in Figure 1.1. The boiler tubes are located at the outer periphery of the unit, and a contra-rotating integral steam turbine drives both the boiler and the power shaft (it was suggested that start-up needed an electric motor.)

The greatest problem affecting the design was the necessity to maintain dynamic balance of the rotor assembly while the tubes were subject to combined stress and temperature deformations. Even achieving a static 'cold' balance with such a tubular arrangement was difficult, if not impossible, at the time.

During the Second World War, new rotating boiler projects did not use tubes, but instead went for heating surfaces in two areas: a rotating cylindrical surface which formed the inner part of the furnace, and the rotating blades themselves – rather like the NASA concept described below. In fact, the stator blades were also used as heat sources, superheating the steam after it had been generated in the rotating boiler. One of the later variants of the gas turbine design is shown in Figure 1.2.

Steam pressures reached about 100 bar, and among the practical aspects appreciated at the time was fouling of the passages inside the blades (2 mm diameter) due to deposits left by evaporating feed water. It was even suggested that a high temperature organic fluid (diphenyl/diphenyl oxide – UK trade name, Thermex) be used instead of water. An alternative was to use uncooled porcelain blades, with the steam being raised only in the rotating boiler.

1.2.2 **NASA work on rotating boilers**

As with the German design above, the first work on rotating boilers by NASA in the US concentrated on cylindrical units, as illustrated in Figure 1.3. The context in which these developments were initiated was the US space programme. In spacecraft it is necessary to overcome the effect of zero gravity in a number of areas that it adversely affects, and these include heat and mass transfer. The rotating boiler is often discussed in papers dealing with heat pipes, which also have a role to play in spacecraft, in particular rotating heat pipes (Gray, 1969; Gray et al., 1968; Reay et al., 2006).

The tests by NASA showed that high centrifugal accelerations produced smooth, stable interfaces between liquid and vapour during boiling of water at one bar, with heat fluxes up to 2570 kW/m^2 (257 W/cm^2) and accelerations up to 400 g and beyond. Boiler exit vapour quality was over 99% in all the experiments. The boiling heat transfer coefficients at 'high g' were found to be about the same as those at 1 g, but the critical heat flux did increase, the above figure being well below the critical value. Gray calculated that a 5 cm diameter rotating boiler, generating 1,000 g, could sustain a heat flux of 1.8 million Btu/h (6372 kW/m^2 or 637.2 W/cm^2).

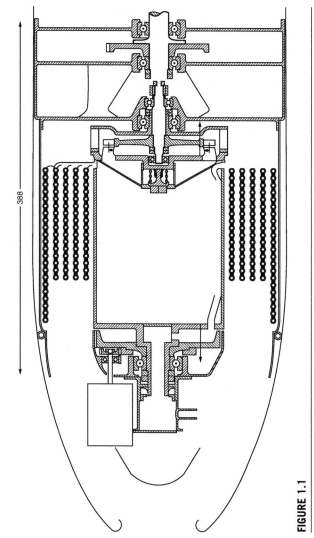

FIGURE 1.1

The German 20h.p. starting motor concept with a rotating tubular boiler (tubes are shown in cross section).

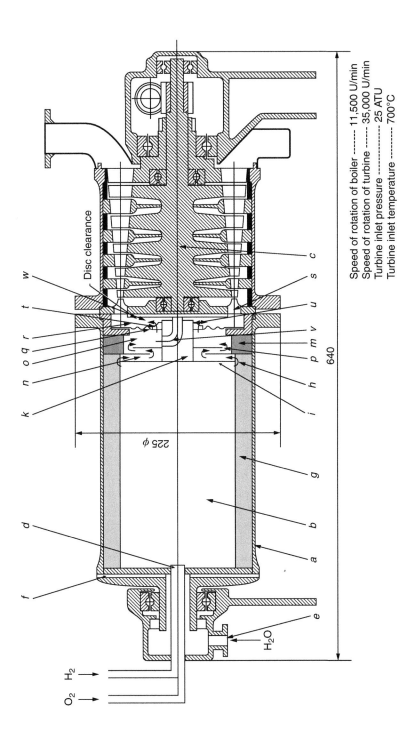

Speed of rotation of boiler --------- 11,500 U/min
Speed of rotation of turbine -------- 35,000 U/min
Turbine inlet pressure -------------- 25 ATU
Turbine inlet temperature ---------- 700°C

FIGURE 1.2

One of the final designs of the gas turbine with rotary boiler (located at the outer periphery of the straight cylindrical section).

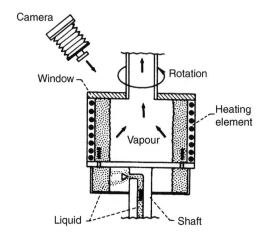

FIGURE 1.3

Schematic diagram of the experimental NASA rotating boiler.

1.3 The rotating heat pipe

The rotating heat pipe is a two-phase closed thermosyphon in which the condensate is returned to the evaporator by centrifugal force. The device consists, in its basic form, of a sealed hollow shaft, having a slight internal taper along its axial length[1] and containing a fixed amount of working fluid (typically up to 10% of the void space). As shown in Figure 1.4, the rotating heat pipe, like the conventional capillary-driven unit, is divided into three sections: the evaporator region (essentially the rotating boiler part of the heat pipe), an adiabatic section, and the condenser. The rotational forces generated cause the condensate, resulting from heat removal in the condenser section, to flow back to the evaporator, where it is again boiled.

There is a suggestion that peak heat fluxes in the evaporators of rotating heat pipes increase at one-fourth the power of acceleration (Costello and Adams, 1960). While the condenser performance has been less well documented, high G forces allow very thin film thicknesses and continuous irrigation of the surface, reducing the thermal resistance across it. Because of the sealed nature of heat pipes and other rotating devices, even further enhancement of condenser performance could be achieved by promoting drop-wise condensation.

There is an interesting observation made in a rotating heat pipe with a stepped wall. Work supervised in China by a highly renowned heat pipe laboratory (IKE,

---

[1] The taper has since been shown to be unnecessary – as the liquid is being removed from the evaporator, the rotation of an axi symmetrical tube will ensure that condensate takes up the space on the surface thus released. However, it has been calculated that a shaft with an internal taper of 1/10 degrees would need 600 G to just pump against gravity (see Gray, 1969, for more data on this).

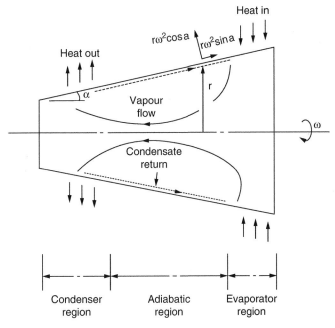

FIGURE 1.4

The basic rotating heat pipe concept (Daniels and Al-Jumaily, 1975).

Stuttgart) indicates the existence of 'hygrocysts', or walls of water in rapidly rotating partially filled horizontal cylinders, which can lead to increased thermal resistance due to thicker films. The particular system studied had a stepped wall, either in the condenser or evaporator section, which suggests that the hygrocyst may be created by such a discontinuity. In this case, it may affect the performance, under certain conditions, of rotating discs with circumferential surface discontinuities (Balmer, 1970). The reader may wish to examine this in the context of spinning disc reactors, etc., as discussed in Chapter 6. There are numerous applications cited of rotating heat pipes, some conceptual, others actual. An interesting one which bears some relationship to the Rotex chiller/heat pump (see Chapter 11) is the NASA concept for a rotating air conditioning unit.

1.3.1 **Rotating air conditioning unit**

An application of a rotating boiler, and all other components in the rotating heat pipe described above, is in a rotating air conditioning unit. Illustrated in Figure 1.5, the motivation behind the design of this vapour compression unit was principally compactness. The heat pipe forms the central core of the unit, but rotation is employed in several other ways with the intention of enhancing performance. As shown, the air conditioning unit spans the wall of a building, requiring a relatively

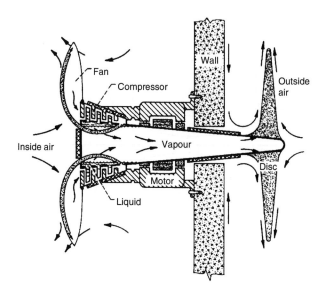

FIGURE 1.5

The rotating air conditioning unit, based upon heat pipes.

small hole to connect the condenser section to the inside of the room. The reject heat from the cycle is dissipated by convection induced in the outside air by a rotating conductive fin, or by a fan (not shown)[2]. In the space to be air conditioned the liquid refrigerant flows into the hollow fan blades, where it expands through orifices near the blade tips to fill them with cold vapour which extracts heat from the room air. The warmed vapour enters the compressor and then flows to the condenser (data given in Gray, 1969).

Other rotating air conditioning unit concepts are discussed later, but chronologically, it is now appropriate to introduce the work at ICI, the major UK chemical company, that some 35 years ago established the foundation of the majority of the concepts that are presented in this book.

1.4 The chemical process industry – the process intensification breakthrough at ICI

The use of rotation for separations and reactions has been the subject of debate for many years and, particularly in the case of separations, the literature cites examples dating back 65 years or so. The Podbielniak extractor was one of the earliest references, cited in a Science and Engineering Research Council (SERC, now EPSRC) document reviewing centrifugal fields in separation processes (Ramshaw, 1986). However, it was the developments by Colin Ramshaw and his colleagues at

---

[2]One could envisage the rotating fin as being hollow but not connected to the main vapour space. This could then act as another rotating heat pipe, in series with the main unit, to aid dissipation.

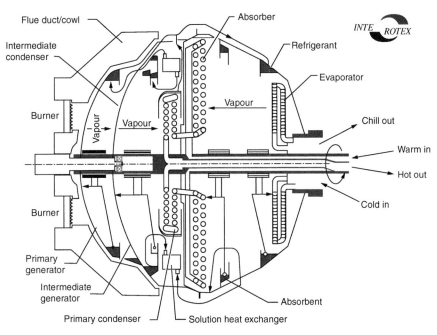

FIGURE 1.6

The Rotex absorption cycle heat pump.

ICI in the 1970s that really demonstrated the enormous potential of PI in the chemical process industries, where 'big is beautiful' had been the order of the day.

The original process intensification thinking at ICI in the 1970s and early 1980s was lent substance by several technical developments by Ramshaw and his co-workers (see also Chapter 2). These comprised:

- The HiGee rotating packed bed gas/liquid contactor
- The Rotex absorption cycle heat pump (see Figure 1.6)
- The mop fan deduster/absorber
- The rotating chlorine cell
- The printed circuit heat exchanger, or PCHE (This was independent of parallel developments in Australia by Johnson)
- The polymer film compact heat exchanger
- The catalytic plate reactor (see section 1.6)

Most of these are discussed in later chapters. The last three are static pieces of plant based upon compact and/or micro-heat exchanger technologies, which may be assigned a reactor capability by introducing catalysts (see Chapters 4 and 5). The first patent on the Rotex concept makes interesting reading (Cross and Ramshaw, 1985).

Having shown that a laboratory HiGee (or high gravity) unit was very effective for distillation and absorption, a substantial pilot-scale distillation facility was built at Billingham in the UK, in 1981, to study ethanol/propanol separation.

New data of considerable historical interest concerning HiGee was revealed recently by Jim Wem in 2012, the original project engineer associated with the design, construction and testing of the very first HiGee machine at ICI. Figure 1.7 shows the rotor of HiGee at ICI, while Figure 1.8 illustrates the whole test facility ready for unloading at site.

FIGURE 1.7

The rotor used in the first HiGee unit tested at ICI Billingham.

(Courtesy: Jim Wem)

FIGURE 1.8

The HiGee test facility prior to unloading at site. The HiGee unit is in the vertical cylindrical section to the upper left in the photograph.

(Courtesy: Jim Wem)

Jim Wem started working on HiGee in 1978, the original suggestion being that Broadbent, a centrifuge manufacturer, would make the HiGee machine. The original concept considered a number of stacked stages, but this was found to be difficult to implement. The design of the rotor led to a diameter of 800 mm o.d. and 150 mm in width. However, in order to progress, an end-use was needed. The group examined the distillation columns being used in ICI Petrochemicals Division, and found that, of the 60 being considered, 39 could use HiGee, therefore they could fund the building of a unit. Detailed design was handed to Broadbent and the go-ahead was given to construct the unit in June 1980; 18 months later it was constructed at a cost of £1.2 million. The unit weighed 40 t and was made from stainless steel. The operating rotational speed was 1800 rpm. A smaller rotor was also constructed that could run at higher speed.

The unit was operated with a vertical shaft and the machine was designed for 22 bar, as no columns operated at a pressure greater than 21 bar. Jenkins of Rotherham made the high pressure unit.

The design performance was achieved: initially, 18 stages in 300 mm height. Retimet (a Dunlop metal foam) was used. This had a surface area density of $2,000 \, m^2/m^3$ and was preferable to the other packing considered (Knitmesh) that was found to be difficult to pack. Alternative casing designs were also examined.

The concept found a use for cleaning the air in submarines which included CO_2 removal, and had to withstand a 35 g acceleration test – the machine was exhibited in the Science Museum in London. A full selection of early drawings, photographs and other sketches relating to the first HiGee are given in Web 3 (2012).

With regard to distillation applications of HiGee, ICI concluded that the best place to use it was offshore, but they first required 10 years experience onshore. One idea was to try it at the BP onshore plant at Wytch Farm in Dorset, but this came to nought.

With interest waning at ICI, the technology was licensed to Glitsch Inc (Dallas) who specialised in the manufacture of packed tower absorption systems. Glitsch initiated several projects which included natural gas sweetening and groundwater remediation (Travers City). While the machines proved to be mechanically reliable and appeared to meet their design specifications, Glitsch withdrew from the market around 1990. Much later, David Trent (Dow Chemicals) pioneered the application of HiGee (see Sections 5.4.3 and 8.3.1.2) to the manufacture of hypochlorous acid. Following the successful operation of the pilot unit, several full-scale machines were installed and have achieved their design specification.

A parallel development of micro-reactors for chemical manufacture and analysis gained considerable popularity in the mid-1990s. The spirit of the work falls under the PI heading and has been spearheaded by Mainz University, though many other groups are now involved. The design is based on the use of arrays of very fine channels (1–10 microns) which have been etched or engraved into a range of substrates. As with the catalytic plate reactor, which has channels in the 1–2 mm range, the reactor performance relies on the short diffusion/conduction path lengths associated with small passage diameters. However, they must be regarded with some reservations for realistic chemical processing in view of their extreme susceptibility to fouling.

Some limited results from the ICI HiGee tests were reported in a patent during a 1983 Gordon Research Conference in the US. Professor Nelson Gardiner of Case Western University attended the conference and subsequently set up a HiGee research programme, which involved one of his mature students, Chong Zheng, who had recently arrived from Beijing. Zheng later returned to China where he was able to persuade the Chinese Government to support a five-year HiGee development programme centred in the newly created Higrav Research Institute at the Beijing University of Chemical Technology (BUCT). This has resulted in China being responsible for dozens of the full-scale industrial applications of HiGee, notably for water deaeration in oil recovery and for precipitation duties. Following the retirement of Chong Zheng, the Institute is now applying HiGee for the manufacture of nano-particles under the direction of Professor J Chen (2009). In 1991, Ramshaw left ICI and was appointed to the chair of Chemical Engineering at Newcastle University, where he assembled a team to further develop the various aspects of PI he had been working on at ICI.

With modest initial funding from the UK's Engineering and Physical Science Research Council, the Process Intensification Network (PIN) was set up in 1998 and run by the Chemical Engineering Department at Newcastle University, in order to promote awareness of PI and to stimulate further developments. PIN membership has now reached 450. A sister organisation was later established at Delft University in The Netherlands and remains active, with around 60 members. More recently, the European Federation of Chemical Engineering launched a website on PI, reporting on the efforts of the EFCE Working Party on Process Intensification. (See Appendix 5.)

1.5 Separators

1.5.1 The Podbielniak extractor

This was designed specifically for liquid–liquid extraction, and was the subject of a US patent (Podbielniak, 1935). The rotor consists of a perforated spiral strip (a design adopted by others). Heavy liquid enters at the centre and then moves out towards the periphery on the inner face of the spiral, see Figure 1.9. The perforations generate droplets of heavy phase while the light continuous phase moves radially inwards.

Ramshaw pointed out that the device probably operated as a cross- or counter-flow spray column, and he felt that, because the perforations represented only a small proportion of the area of the spiral strip, they would impose a severe restriction on the flooding performance of the rotor. The height of a transfer unit (htu) was about 10cm.

Other units cited in the SERC review by Ramshaw included a rotary demister by Smith, and the Hickman rotary still, illustrated in Figure 1.10. (Hickman is also responsible for developing the method for thin film production during the 1930s at Eastman Kodak, which is referenced in the majority of subsequent patents dealing with rotating separation equipment.)

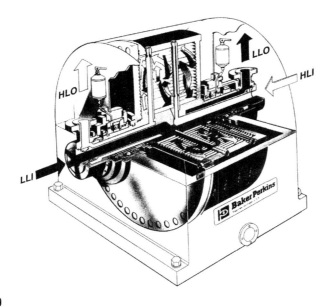

FIGURE 1.9

The Podbielniak extractor. HLO 5 heavy liquid out; HLI 5 heavy liquid in; LLO 5 light liquid out; LLI 5 light liquid in (Ramshaw, 1986).

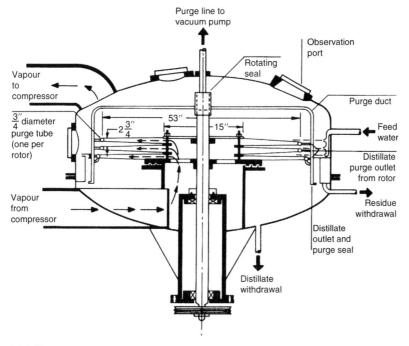

FIGURE 1.10

The Hickman rotary still (Hickman, 1936).

1.5.2 **Centrifugal evaporators**

Thin film evaporators, often operating under vacuum, have been routinely manufactured for many years. They can be used for evaporation, concentration, distillation, stripping, deodorising and degassing. Often used for heat-sensitive products, the main advantages include compactness, controllability and uniformity of concentrated product. It is interesting to note that many of the comparisons between intensified separation processes and 'conventional' plant, neglect this well-established piece of equipment. With regard to PI, the existence of such reliable precedents augurs well for the development of PI technology based on rotary equipment.

Another centrifugal evaporator concept was proposed by Porter and Ramshaw (1988), but unlike the unit described above, it uses a large number of plates located normal to the rotating shaft, and is designed to accommodate a wider range of separations than the above apparatus. As described in Chapter 4 and in the context of spinning disc reactors in Chapter 5, the enhanced evaporation (and condensation) heat transfer coefficients on 'spinning disc' surfaces are of considerable benefit.

1.5.3 **The still of John Moss**

This UK resident filed a patent, (published in the US) on a rotating still (Moss, 1986; Figure 1.11a and b). Although specified as a separator for light and heavy fractions (multi-stage, countercurrent distillation) the so-called lamellar bodies normal to the radial flow path are perforated and are shown in the detail in Figure 1.11b.

It is also interesting to note the patent of Pilo and Dahlbeck (1960). Here, in a device for countercurrent contacting of two fluids, a variety of internal structures are covered ranging from spherical packing to blades.

1.5.4 **Extraction research in Bulgaria**

In the early 1980s, Ballinov at the Central Institute of the Chemical Industry in Sofia, Bulgaria, reported on the intensification of extraction processes based on vibrating plates (Ballinov, 1982). Stating that a five to fifteen times increase in column productivity could be achieved (depending upon the form the intensification takes), the main aim of Ballinov's study was to identify the mechanism of intensification in his column.

It was concluded that the mass transfer resistance was concentrated mainly in the continuous aqueous phase (using a water–iodine–carbon tetrachloride mixture), and enhancement was due both to an increase in the interfacial area (caused by the vibration) and the increase in mass transfer coefficients due to increased turbulence.

1.6 **Reactors**

In the area of reactors, intensification has been brought about in a number of ways. Two stand out in the history of PI reactor development – the catalytic plate reactor

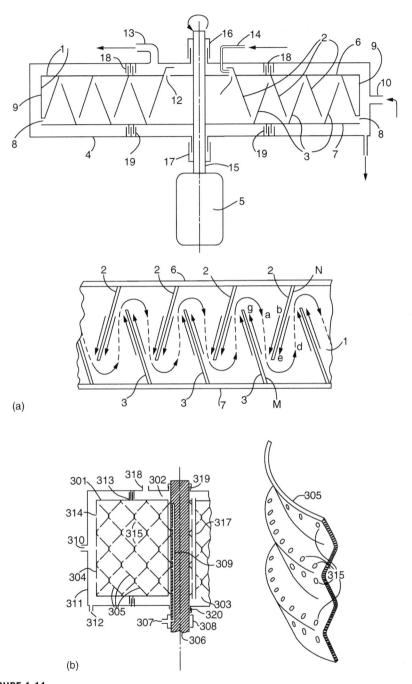

FIGURE 1.11

(a) The structure inside the Moss still, and (b) the perforated plates used in the Moss still.

and the spinning disc reactor, both of which came out of the ICI stable. Both concepts are discussed at length in Chapter 5.

1.6.1 Catalytic plate reactors

The work at ICI on PI on the laminar flow compact plate heat exchanger in achieving very high volumetric heat transfer coefficients was soon recognised as offering benefits to chemical reactor design. The heat transfer matrix could be the basis of a very intense catalytic reactor, provided thin layers of highly active catalyst could be bonded to one or both sides of the plates. The inherent attraction of this approach is that it effectively short-circuited the heat and mass transfer resistances between the reaction site and the heating or cooling medium. When the process reaction is endothermic, the heat needed to drive the reaction can, in principle, be provided by catalytic combustion on the other side of the plate.

It is well known that thin films of combustion catalyst supported on ceramic or metal surfaces are capable of stimulating high heat fluxes when the surface is in contact with an appropriate gas mixture. This notion was used many years ago as the basis of a town gas igniter (before natural gas became available in the UK from the North Sea). It employed thin filaments of a platinum group metal. When placed in a hydrogen–air mixture the wire glowed brightly and ignited the surrounding gas. Methane is much less reactive than hydrogen, so it would be expected that methane–air mixtures would require higher reaction temperatures. This was confirmed by some unpublished ICI research that showed that heat fluxes of about $10 \, kW/m^2$ could be generated by 10 micron layers of Pd/Al_2O_3 in stoichiometric methane/air at about 600°C.

The intimate linking of the combustion heat source with the endothermic process reaction virtually eliminates the overall heat transfer resistance. The long radiation path lengths needed for conventional furnaces are replaced by channel dimensions of 1–2 mm in plate matrices, with an obvious impact on the size of reactor needed for a given production rate. A comparison of a conventional and a plate catalytic reformer is shown in Figure 1.12. In Figure 1.12 (a) the catalyst pellets are shown inside a tube, typically of 10 cm diameter, only one wall of which is shown in the cross-section.

1.6.2 Polymerisation reactors

Rotation has been proposed by several organisations to enhance polymerisation reactions. An early reference was made by Ramshaw (1993) to a US patent taken out in 1964 by DuPont Company which highlighted the benefits of polymerising in thin films at up to 400°C with a residence time of seconds. Not all subsequent inventors have jumped in at the deep end in producing rotating reactors.

Other polymerisation reactor designs include agitation without rotation in order to enhance heat transfer into and out of the reactants (Goebel, 1977). Agitation/rapid mixing was used in the continuous polymerisation reactor proposed by Phillips Petroleum (Witt, 1986). The use of an agitator is claimed to have a significant positive effect upon mixing, reducing residue formation.

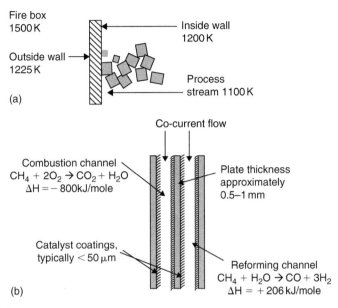

FIGURE 1.12

Comparison of conventional, (a); and catalytic plate, (b) reformers.

1.6.3 **Rotating fluidised bed reactor**

The use of centrifugal forces in conjunction with fluidised beds is relatively well known in reactors – the Torbed unit (see Figure 1.13) manufactured by Torftech Ltd and described at the November 1999 PIN meeting, is an example of this (see Chapter 5 and Web 2 (1999) for a full description). The fluidisation takes place using a process gas stream, as in a conventional fluidised-bed, but in the Torbed, the angled slots through which the gas passes, (above which are the particles) impart a velocity component in a circumferential direction, causing the particles to move around the bed, as shown by the arrows.

1.6.4 **Reactors for space experiments**

As with the rotating heat pipe which generates its own gravity field, a rotating reactor also does this when conceived for experiments in space. Most frequently these were designed for biological reactions. One of the concepts involved a rotating tubular membrane (Schwarz and Wolf, 1991; Schwarz et al., 1991).

1.6.5 **Towards perfect reactors**

Arising out of the Delft Skyline Debate, one of the areas to be targeted for around 2050 is the 'perfect reactor'. The authors (Ozkan et al., 2012) argue that if we are to achieve sustainable and efficient chemical processes, it is at the molecular level

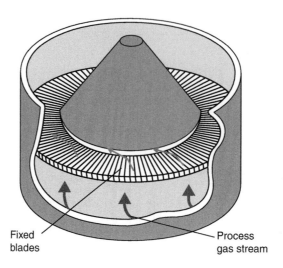

Fixed blades

Process gas stream

FIGURE 1.13

The Torbed compact bed reactor.

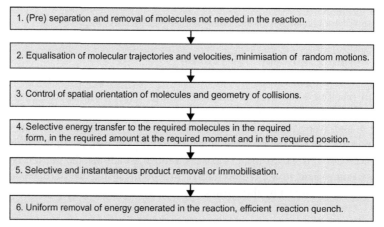

1. (Pre) separation and removal of molecules not needed in the reaction.

2. Equalisation of molecular trajectories and velocities, minimisation of random motions.

3. Control of spatial orientation of molecules and geometry of collisions.

4. Selective energy transfer to the required molecules in the required form, in the required amount at the required moment and in the required position.

5. Selective and instantaneous product removal or immobilisation.

6. Uniform removal of energy generated in the reaction, efficient reaction quench.

FIGURE 1.14

Schematic representation of the steps in the 'perfect reactor' (Ozkan et al., 2012).

that control has to be achieved. One can see an analogy between this and the phrase of the company Protensive Ltd, that PI equipment can give 'each molecule the same experience...', in that the Universities of Delft, Eindhoven and Leuven see that '...control of molecular alignment and geometry of collisions...selective and efficient activation of the...molecules' are critical areas to be addressed.

The basic functional steps to be addressed in such a 'perfect reactor' are illustrated schematically in Figure 1.14.

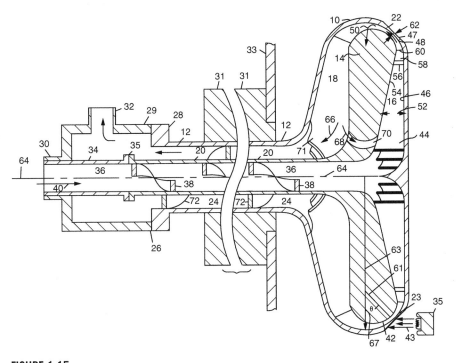

FIGURE 1.15

Liquid cooled rotating anode (hot side on the right).

1.7 Non-chemical industry-related applications of rotating heat and mass transfer

As is obvious from Section 1.3, the chemical industry is not unique in studying rotating heat and mass transfer devices. Technology transfer can work both ways and some of the concepts studied outside the industry may have strong relevance to the needs to today's rotating intensified unit operation.

1.7.1 Rotating heat transfer devices

1.7.1.1 Liquid cooled rotating anodes

One of the more recent technological developments which necessitates enhanced heat transfer is the target for energy beams, such as lasers. Targets absorb large amounts of thermal energy over a comparatively small surface area, and a number of innovative methods for effectively removing the heat from them have been studied. One is the rotating anode (Iversen and Whitaker, 1991), illustrated in Figure 1.15. This system uses rotation and a system of vanes on the inside surface of the hot face (the right-hand side of the drawing) in order to enhance the flow of coolant radially across the inside of the target face. By judicious design of the surface, the inventors also claim to generate multiple independent centrifugal force pressure gradients on the heat transfer surface, thereby increasing the heat flux removal.

An additional feature is the use of nucleate boiling sites to promote bubble generation and rapid removal on the inner surface of the target, although the initial concept is more concerned with single-phase cooling. As with any rotating heat transfer enhancement method, the concept may be relevant to exothermic reactions, which could take place on the opposite side of the wall.

1.7.1.2 The Audiffren Singrun (AS) machine

The first hermetic compressor was invented by a French abbot, Abbe Audiffren, in 1905 and manufactured by Singrun at Epinal, also in France. Later manufactured by major refrigeration companies, the machine, shown schematically in Figure 1.16, had many innovative features (Cooper, 1990). A sphere formed the condenser and an oval-shaped cylindrical vessel the evaporator, these were connected by a hollow shaft and the whole assembly was rotated by a belt drive. SO_2 was the working fluid.

When the unit was rotated, SO_2 gas was drawn from the evaporator through the hollow shaft into the compressor. This was then discharged onto the inner wall of the spherical condenser. Here, the refrigerant–oil mixture was collected by a stationary scoop (those interested in the Rotex machine mentioned earlier may recognise this concept) and taken to the separator, where oil is taken over a weir onto the moving components. The liquid refrigerant then goes via the high side float regulator to expand through the small pipe in the hollow shaft into the evaporator. Here again, centrifugal force is used to wet the heat transfer surfaces. The author of an article in which a recent write-up appeared likened the technology to the Rotex machine, indicating that it might reappear in this form (Cooper, 1990).

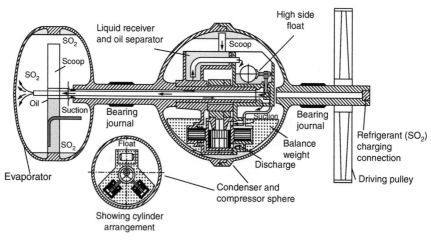

FIGURE 1.16

The AS rotating refrigeration unit.

1.7.1.3 John Coney rotating unit

Dr Coney (1971) researched Taylor vortex flow with particular interest in rotary heat exchangers, which he later used in a rotary vapour compression cycle heat pump. These days, Taylor-Couette flows are used in commercial intensified reactors (see Chapter 5).

1.8 **Where are we today?**

1.8.1 **Clean technologies**

John Glaser of the US Environmental Protection Agency, writing in *Clean Technologies and Environmental Policy* (Glaser, 2012) presented a selection of abbreviated journal articles relating to PI, particularly in the area of green chemistry. The emphasis was on conservation of materials and energy, as well as reducing cost. PI was seen as having a significant role to play here, although it is conceded that some aspects of PI are still in their infancy. Glaser cites examples in modern flow chemistry, where micro-reactors attract attention because of their versatility in the laboratory, and goes further down in scale to discuss nano-reactors derived from a surfactant that forms nano-micelles. These facilitated a range of organic reactions and permitted easy recycling of the surfactant. A further contribution related to the recognition of Velocys, Inc (see Chapter 5) with the Kirkpatrick Award for Chemical Engineering Achievement from *Chemical Engineering*. This award recognises 'the most noteworthy chemical engineering technology that was commercialised globally during 2009 and 2010'. Velocys received it for their micro-channel reactor that has applications in gas-to-liquids, biomass-to-liquids and coal-to-liquids conversions.

In a press release on 23 May 2012, Oxford Catalysts Group, owners of Velocys, said: 'The first commercial scale microchannel Fischer-Tropsch (FT) reactor has been successfully started-up and is operating as expected at a client's facility. The reactor, with a nominal capacity of over 25 barrels per day (bpd), is expected to be run continuously for up to six months at the client's facility. The reactor was transported to the client's site during the first quarter of 2012 with the catalyst already loaded. Catalyst activation/reduction was performed at the client's site under commercially relevant conditions and synthesis gas operations began last week. The client, an integrated energy company, is operating the reactor to provide detailed engineering information for the design of commercial medium-scale modular synthetic fuels plants that it plans to build on sites around the world' (Velocys, 2012).

1.8.2 **Integration of process intensification and renewable energies**

PI may have much to offer in the renewable energy field. While we tend to think of many renewable energy sources as diffuse – the antithesis of PI – the International Energy Agency (IEA), via AEE INTEC, the Institute for Sustainable Energy in Austria, co-ordinates a number of activities relating to solar energy applications (Anon, 2012).

Of particular interest is the IEA solar heating & cooling task 49/IV subtask B, and the section below is taken from subtask B on the IEA website (the italics are those of the authors):

Using the foreseen topics as a foundation, subtask B will be based upon the following structure:

- *Combining process intensification technologies and solar process heat:*
    - *For existing processes.*
    - *In new applications (not limited to thermal processes).*

In general, subtask B will focus on the following main objectives:

- Improved solar thermal system integration for production processes by advanced heat integration and storage management; advanced methodology for decision on integration place and integration types.
- *Increase of the solar process heat potential by combining process intensification and solar thermal systems and fostering new applications for solar (thermal/ UV) technologies.*

AEE INTEC (see Appendix 4) has stated that in the EU27, thermal energy needs are 49% of total energy demand, of which industry needed 30% of the total at temperatures $>250°C$ and 14% of the total at $<250°C$. In the EU25, the potential from solar energy is 70 TWh/year at less than $250°C$, or 3–4% of the total thermal energy demand in Europe.

1.8.3 **PI and carbon capture**

There has been a substantial growth in the literature reporting research and near-market activities in the area of capture of CO_2, generally called carbon capture (CC). The logic behind this is that most current CC plants use old chemical engineering technologies such as static absorption/desorption towers for the most common form of carbon capture – post-combustion capture using absorption of the CO_2. If PI can be used to reduce the size and capital cost of such plants, and the associated pumping losses, which are a substantial drain on the power station electricity output, several benefits can result. The rotating packed bed (RPB) has been examined in Europe and in China in this respect – see for example Cheng and Tan (2009) and Yi et al., (2009).

Discussed in more detail in later chapters, CC, which can also involve membranes and intensified adsorption reactions, could become one of the most important applications of PI within the next two decades.

As pointed out in the first edition, several authors have emphasised that process intensification has, or will have, a major role to play in the future of chemical engineering. Charpentier (2007) used the phrase 'molecules into money' in proposing that chemical process engineering drives today's economic development and wealth creation, the process engineering being, of course, based on PI. This is not far removed from the Protensive Ltd phrase – 'making every molecule count' – used in the introduction to this book, and while we may argue as to whether biologists, physicists, chemists, engineers (of all disciplines) or economists and accountants drive our economic development and wealth creation, there is no doubt that PI is likely to be an important weapon in supporting a sustainable future.

That PI is, and will remain a business with substantial technical and financial risks associated with it is exemplified by the fact that Protensive Ltd failed as a company, (although its technologies, in particular the spinning disc reactor, live on).

1.9 Summary

Process intensification first attracted serious attention during the 1970s, in the chemicals sector (ICI in the UK), where it is most widely known to this day. As a result of the research there the path has been eased for some other companies developing and using PI techniques and equipment. Now, thanks to initiatives such as the Delft Skyline Debate in Europe and the active research groups in the Far East, PI is seen as having a long-term and highly important future.

We should not forget, however, that PI did not begin with ICI in the 70s. Many PI technologies had been developed before the phrase itself even existed. For instance, any continuous process represents an intensification compared to the batch process, and the Podbielniak centrifugal contactor was developed in the 1930s!

It is also important to realise that the advantages of PI are not applicable only to the chemical industry. PI could have significant benefits in most process industries, with the food and pharmaceutical industries standing out as obvious candidates, due to their scale and changing economics (the increase in generic pharmaceuticals for instance).

References

Anon, 1932. Dampferzeuger mit Turbine z.V.D.I Bd. 76, No. 41.

Anon, 1946. Vorkauf rotating boiler (Drehkessel) and rotating boiler gas turbine (Drehkessel Turbine). British intelligence objectives sub-committee final report No. 931, item no. 29, London, HMSO.

Anon, 2012. Solar Process Heat for Production and Advanced Applications. IEA Solar Heating and Cooling Programme. <http://www.iea-shc.org/task49/> (accessed 16.06.12.).

Ballinov, Y., 1982. On the mechanism of process intensification in vibrating-plate extraction. Chem. Eng. J. 25, 219–221.

Balmer, R.T., 1970. The hygrocyst – a stability phenomenon in continuum mechanics. Nature 227, 600–601.

Charpentier, J.-C., 2007. In the frame of globalisation and sustainability, Process Intensification, a path to the future of chemical and process engineering (molecules into money). Chem. Eng. J. 134, 84–92.

Chen, J.F., 2009. The recent developments in the HiGee technology. Proc. Green Process Engineering-European Process Intensification Conference, Venice, June, 14–17.

Cheng, H.H., Tan, C.S., 2009. Carbon dioxide capture by blended alkanolamines in rotating packed bed. Energy Procedia 1, 925–932.

Coney, J.E.R., 1971. Taylor vortex flow with special reference to rotary heat exchangers. PhD thesis, Dept. Mechanical Eng., Leeds University.

Cooper, A., 1990. The world below zero: a history of refrigeration. Part 25. ACR News March, 50–52.

Costello, C.P., Adams, J.M., 1960. Burn-out fluxes in pool boiling at high accelerations. Report of Mech. Eng. Dept., University of Washington, Washington DC.

Cross, W.T., Ramshaw, C., 1985. Centrifugal heat pump. US Patent 4553408, filed 9 March 1984, issued 19 November 1985.

Daniels, T.C., Al-Jumaily, F.K., 1975. Investigations of the factors affecting the performance of a rotating heat pipe. Int. J. Heat Mass Transf. 18, 961–973.

Glaser, J.A., 2012. Process intensification. Clean Technol. Environ. Policy 14, 155–160.

Goebel, P., 1977. Polymerisation reactor with gilled-tube radiator and axial agitator. US Patent 4029143, 14 June.

Gorak, A., Stankiewicz, A., 2012. Editorial on the delft skyline debate. Chem. Eng. Process. 51, 1.

Gray, V.H., 1969. The rotating heat pipe – a wickless, hollow shaft for transferring high heat fluxes. ASME Paper 69-HT-19, ASME, New York.

Gray, V.H., Marto, P.J., Joslyn, A.W., 1968. Boiling heat transfer coefficients, interface behaviour, and vapour quality in rotating boilers operating to 475 Gs. NASA TN D-4136, March.

Hickman, K.C.D., 1936. UK Patent 482880, July.

Iversen, A.H., Whitaker, S., 1991. Liquid cooled rotating anode. US Patent 5018181, May 21.

Moss, J., 1986. Still. US Patent 4597835, 1 July.

Ozkan, L., Backx, T., Van Gerven, T., Stankiewicz, A.I., 2012. Towards perfect reactors: gaining full control of chemical transformations at molecular level. Chem. Eng. Process. 51, 109–116.

Pilo, C.W. Dahlbeck, S.W., 1960. Apparatus for intimate contacting of two fluid media having different specific weight. US Patent 2941872, 21 June.

Podbielniak, 1935. US Patent 2044996.

Porter, J.E. Ramshaw, C., 1988. Evaporator. US Patent 4731159, 15 March.

Ramshaw, C., 1986. Separation processes: the opportunities for exploiting centrifugal fields. Report for the Science and Engineering Research Council (now EPSRC), Oct.

Ramshaw, C., 1993. Opportunities for exploiting centrifugal fields. Heat Recovery Syst. CHP 13 (6), 493–513.

Reay, D.A., Kew, P.A., 2006. Heat Pipes: Theory, Design and Applications, fifth ed. Elsevier, Oxford.

Schwarz, R.P., Wolf, D.A., 1991. Rotating bioreactor cell culture apparatus. US Patent 4988623, 29 Jan.

Schwarz, R.P. et al., 1991. Horizontally rotated cell culture system with a coaxial tubular oxygenator. US Patent 5026650, 25 June.

Velocys, 2012. Microchannel Fischer-Tropsch reactor: Successful start-up under commercial conditions. Technical Press Release, 23 May 2012. <http://www.velocys.com/press/pr/ocgtpr20120523.php>.

Web 1, 2007. Proceedings of the Fourteenth Process Intensification Network Meeting, Grangemouth, April 2007. See <www.pinetwork.org>.

Web 2, 1999. Minutes of the 2nd Meeting of Pin, DTI Conference Centre, 16 November. See <www.pinetwork.org>.

Web 3, 2012. Proceedings of the Twentieth Process Intensification Network Meeting, Newcastle University, 2 May 2012. See <www.pinetwork.org>.

Wem, J., 2012. HiGee – Developments a ICI. Proceedings of the Twentieth Process Intensification Network Meeting, Newcastle University, 2 May 2012. <www.pinetwork.org>.

Witt, M.S., 1986. Continuous polymerisation reactor. US Patent 4587314, 6 May.

Wightman, E.P., Trivelli, A.P.H., Sheppard, S.E., 1925. Intensification of the latent image on photographic plates. J. Franklin Inst. 200, 335.

Winnington, T.L., 1999. The evolution of rotating process systems. Proceedings of the Third BHRG Conference on Process Intensification, Antwerp, Belgium, 25–27 October.

Yi, F., Zou, H.K., Chu, G.W., Shao, L., Chen, J.F., 2009. Modelling and experimental studies on absorption of CO_2 by Benfield solution in rotating packed bed. Chem. Eng. J. 145, 377–384.

Process Intensification – An Overview

OBJECTIVES IN THIS CHAPTER

The objectives in this chapter are to build upon the earlier definition of process intensification given in the introduction, using examples and to discuss the advantages of, and obstacles to PI. The chapter also looks at the principal unit operations, the PI type(s) used to improve them, and their potential applications.

2.1 Introduction

In this chapter we give an overview of process intensification, with brief examples and, at the end of the chapter, the first of our three Key Tables giving data on the unit operations that can be intensified and the applications where they might be used (other key tables can be found at the end of Chapters 5 and 11). This chapter can also be used as a 'signpost' to help readers find sections of the book that may be of direct interest to them.

After defining process intensification in greater detail than in the introduction, and explaining its principal *raison d'être* and subsequent development at ICI, the main advantages of PI are described. These include allowing safer plant, reducing environmental impact and reductions in carbon emissions – this last feature being a key motivator in several national and international PI R&D programmes. Most importantly to business, the opportunities afforded by PI to companies who wish to develop new and/or improved products in relatively short times are discussed.

In this second edition of the book we see no reason to change the definition of PI, although some organisations have attempted to broaden its scope in directions that are more speculative. A number of players, such as Bayer (Buchholz, 2010) usefully identify more specific perceived barriers to PI and other process improvements, directed ultimately at the European Commission (EC)-funded FP7 project 'F^3 Factory'. Discussed later, in the context of increased EC participation in PI, this project (Web 2, 2012) is directed at combining process innovation with 'modularisation and equipment standardisation…to allow widespread implementation of continuous, cost beneficial equipment and manufacturing platforms'. So Bayer, in this discussion, considers continuous manufacturing as the principal benefit of PI.

There will be discussion in Chapter 12 of a number of systematic approaches to PI, which have been updated since 2007. However, we do not see the need to classify PI as something new in relation to unit intensification and plant intensification, as proposed by Ponce-Ortega et al., (2012) as an adjunct to their systematic approach. It has always, since its conception at ICI, been the belief among the main groups involved in PI development in the UK that intensification of the whole plant is superior to PI applied to individual unit operations. Unfortunately, and the authors are maybe lamenting this fact, whole plant intensification is much more complex than PI applied to individual unit operations!

2.2 What is process intensification?

Writing in *Chemical Engineering Progress* Keller and Bryan (2000) highlighted the fact, one with which most directors of process companies will agree, that growing worldwide competition will necessitate major changes in the way plants are designed. The authors, both leading scientists in industry and academia in the US, pinpointed seven key themes that would mould developments in industry. These key themes are as follows:

- Capital investment reduction.
- Energy use reduction.
- Raw material cost reduction.
- Increased process flexibility and inventory reduction.
- Ever greater emphasis on process safety.
- Increased attention to quality.
- Better environmental performance.

Later, the reader will be given opportunities to see if his or her own activities or business can use PI concepts in order to benefit from any of the key themes. The reader may, of course, recognise that there are other ways of achieving these benefits, it is not suggested that PI is a panacea for all problems relating to business and commerce. Consider the following:

- Size reduction for its own sake, however, is not the be-all and end-all of PI. There are other intensified processes which offer us the opportunity to create new or better products *with properties that are better controlled*. Pharmaceutical products, which cannot be made to such a tight specification in any other way, are a case in point.
- Increasing the speed of some processes (compatible with knowledge of reaction kinetics, where appropriate) can also be a strong incentive.
- Bayer appears to see continuous production as the most beneficial outcome of PI, as identified as a role for the F3 Factory, but this is not of course appropriate for all processes, particularly distributed manufacture/processing (another benefit offered by PI).

One of several definitions of PI sets out a selection of these themes, all of which have already been identified in the introduction:

Any chemical engineering development that leads to a substantially smaller, cleaner, safer and more energy efficient technology is process intensification.

Stankiewicz and Moulijn (2000), who first used this definition, missed safety out of their original statement. It has been added in this book because it is considered by some to be an important driver in spurring businesses to seriously consider PI technologies – particularly as we become more risk averse. This is particularly the case when dealing with reactions or dangerous substances.

The most impressive examples of PI, when viewed from almost any vantage point, are those that reveal non-incremental reductions in process plant size. Some of these have already been highlighted in Chapter 1. These can be unit operations – the HiGee distillation unit of Colin Ramshaw (1983) is an obvious and very early example. Figure 2.1 shows a remarkable reduction in visual impact of a PI technology, while concepts such as the 'desktop process plant', the pocket-sized nitric acid plant, and the lab-on-a-chip stimulate our imaginations today. Very small chemical reactors as power sources in our mobile phones are being prepared for the market by companies such as Toshiba (see Figure 2.2; Anon, 2005a). This shows the direct methanol fuel cell (DMFC) with an output of 100 mW – which was awarded the title of 'the world's smallest DMFC' by Guinness World Records in 2005! Other research on micro-fuel cells for portable and transportation uses is described by Chen and Tsao (2006). It remains in the sphere of micro-electronics that perhaps the most impressive claims for size reductions of massively functional equipment reigns – IBM claims that supercomputers 'will fit in a sugar cube' in the future, (Palmer, 2010). Interestingly, the principal feature outside the manufacture of micro-circuits,is the cooling system used for cooling stacks of computer processors – water flowing between each stack.

2.3 **The original ICI PI strategy**

The PI strategy was conceived by the Process Technology Group at ICIs New Science Laboratory in the late 1970s, as discussed in Chapter 1. At the time, the company was reviewing its capital development philosophy, having concluded that major capital cost savings in future plant were essential in order to secure its future. These savings had to be made without compromising plant output. The review stimulated a fundamental reappraisal of the capital cost structure of the company's process plants, and was based on the Lang Factor approach to capital cost estimation. The Lang Factor is the ratio of the total cost of installing a process in a plant to the cost of its major technical components. This procedure recognises that the total delivered cost of the principal equipment needed to perform a process is much less than that of the eventual operating system. The difference is accounted

FIGURE 2.1

The volume of HiGee compared to a conventional distillation column, the HiGee unit is on the lower left-hand side (Fishlock, 1982).

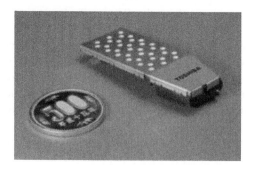

FIGURE 2.2

A Toshiba mobile phone – an early home for the adjacent micro-reactor fuel cell.

for by the installation costs. Installation factors – the ratio of the cost of installed equipment to its delivered cost on site – may be used on either a whole plant basis or as the sum of individual items, depending upon the required accuracy of the estimate.

Table 2.1 shows a cost breakdown for a relatively simple plant for ammonium sulphate manufacture. The figures in the table correspond to the proportion of the total capital cost (%) with the 100% total at the foot of the last column. It can be seen that, in order to convert the main plant items (MPI) to a working system, many other cost components are involved, such as piping, support structure, civil engineering and foundations, etc. It is both surprising and significant that piping costs in this case (last column) are almost equivalent to the total MPI cost. In the case of Table 2.1 the global Lang Factor is around 4, though for some other plants it may be as high as 8.

The PI concept arose from a rather provocative question: '*If* we could make a dramatic reduction in the size and/or volume of *all* the process plant components, without compromising output, would there be a significant impact on the total plant capital cost?' The size reduction envisaged corresponded to a volume reduction of 100–1,000 or about an order of magnitude in the linear dimension. Such a target abandons the incremental or apple polishing philosophy of plant development in favour of the radical or step-out approach, in order to reduce Lang Factors across the board. Even if an individual intensified process component is more expensive than its conventional equivalent, (although hopefully it would be cheaper) it must generate substantial overall cost savings for the plant *system*.

PI is not a strategy for the faint-hearted. Bearing in mind the size reduction sought, novel or unusual approaches to plant design are essential. Many searching questions will be posed, such as the need for turbulent flow, the use of batch rather than continuous operation and the extensive application of merely terrestrial acceleration to multi-phase systems. Some PI examples are described in the following chapters. It is a sobering thought that if chemical engineers were to design the human digestive and metabolic system on a conventional basis, our bodies would be much larger and require a great deal more than the basic metabolic rate of around 150 W to operate. On the other hand, nature operates unobtrusively with laminar flow in capillary matrices having a high surface density, for example in kidneys and lungs, while coping with some fouling problems by coughing or sneezing. As scientists and engineers we must not be too arrogant to learn some lessons from the natural world.

PI conflicts directly with the two thirds power rule which has traditionally been used to consider capital costs at various production scales. The rule is based on the notion that plant component cost (C) varies in proportion to the equipment's surface area, whereas production capacity is proportional to its volume. For a given plant item having a characteristic dimension D, the cost varies in proportion to area, i.e.:

$$C \sim D^2$$

Table 2.1 Cost Breakdown of a 90,000 Tonne/Annum Ammonium Sulphate Plant (The figures Represent the Percentage of the Total Plant Cost).

	Unit	Spares	Electronics	Insulation/painting	Design	Contingencies	TOTAL
Mechanical	26.1	1.2	8.1	2.5	1.3	2.6	42.0
Civil	7.4	–	–	0.9	1.3	1.3	10.9
Structure	8.3	–	–	1.0	1.3	1.3	12.9
Pipework	13.8	–	4.3	1.5	0.2	2.0	21.8
Instruments	4.5	0.1	0.8	–	1.0	0.9	7.4
Electrical	1.6	–	0.1	–	0.3	0.2	2.3
Control room	2.4	–	–	–	0.4	0.4	3.1
TOTAL	64.1	1.3	13.3	6.8	5.8	8.6	100.0

Whereas, production capacity (P) is proportional to D^3. i.e.:

$$P \sim D^3$$

Hence, $C \sim P^{2/3}$ and $C/P \sim 1/D$.

This is the basis of the big-is-beautiful philosophy of plant design. The reader will recognise that if all the production of a larger plant cannot be sold, the reasoning breaks down. In addition, for an expanding market, the two thirds power rule tends to encourage capital investment to be made in large increments, which may destabilise a market as scarcity is abruptly transformed into a glut. A technology which allows production capacity to be adjusted economically in smaller increments may prove to be attractive.

2.4 The advantages of PI

While capital cost reduction was the original target for PI, it quickly became apparent that there were other benefits, some of which have become even more important since PI was conceived.

2.4.1 Safety

Given the anticipated reductions in plant volume, the toxic and flammable inventories are correspondingly reduced, thereby making a major contribution to intrinsic plant safety, bearing in mind that large distillation and reactor systems can easily contain several hundred tonnes of hazardous material. The best way to deal with the problem is not to have it in the first place and this point has been well made by Trevor Kletz (1991), the renowned safety expert, who pointed out that 'what you don't have can't leak'! His work has been extended by Dennis Hendershot in the US at Rohm and Haas Company, who coined the phrase 'inherently safer plant' (Hendershot, 1997). Hendershot said that approaches to the design of inherently safer plant could be grouped into four major strategies:

- Minimise
- Substitute
- Moderate
- Simplify

The first of these mirrors Trevor Kletz's point – use small quantities of hazardous substances. PI can also allow one to *moderate* conditions to minimise risk of explosions and to *simplify* processes by having fewer unit operations and less complex plant. Illustrations of the impact of PI on safety are neatly described by Hendershot in a later publication, (Hendershot, 2004).

The recent history of the process industries is littered with cases where disastrous releases have occurred. These have done nothing to improve the industry's poor public image. For example, in the 1974 Flixborough disaster at the Nypro (UK) site the cyclohexane content (about $200\,m^3$ of liquid cyclohexane) of a train of five pressurised (8 bar) oxidation reactors escaped, giving rise to a large vapour cloud which caught fire from an ignition source. Unfortunately, the vapour cloud was of such a size as to allow the flame front to accelerate and become a shock wave with horrendous consequences. A 60 acre site was destroyed, 28 people were killed and 89 were injured.

The court of enquiry concluded that the discharge of a vapour cloud of cyclohexane resulted in an explosion which was equivalent to the detonation of 15–45 tonnes of TNT, (Health and Safety Executive, 1975). While the mechanism by which the deflagration accelerated to become a detonation has not been made clear, it seems reasonable to expect that if the cyclohexane inventory had been reduced by (say) two orders of magnitude, a discharge would only have involved a deflagration.

There is some evidence to support this contention in the field of rocket engine combustion. In this case, it is possible to generate an interaction between the vaporising liquid fuel–oxidant whereby a flame front accelerates to a detonation or shock wave within the combustion chamber, with disastrous consequences. The coupling mechanism is believed to result from the step change in gas velocity as the shock wave passes through an array of fuel–oxidant droplets. The resulting 'emulsification' enhances the local combustion rate and augments the intensity of the shock wave. If this mechanism was involved at Flixborough then it is reasonable to expect that a minimum flame path length would be needed to amplify a deflagration to a shock wave. This implies that the discharge of a volatile flammable liquid should be kept below some critical size in order to avoid the damage potential of a detonation. Obviously, PI is a considerable help in this regard.

Other disasters at chemical plants followed: in 1976 at Seveso (Italy) there was a dioxin escape which polluted over 4,000 acres of farmland, killed 100,000 grazing animals and led to the evacuation of 1,000 people. In India in 1984, a release of 40 tonnes of methyl isocyanate from a batch operation at Bhopal resulted in 40,000 deaths and 100,000 injuries. Incidents such as these would have either been avoided or extensively mitigated had the processes been intensified. In the Flixborough case, a very much smaller vapour cloud would probably have been incapable of developing the shock wave which proved to be so damaging. A continuous intensified version of the Bhopal reactor may still have caused fatalities, had the contents been released, but there would have been far fewer.

It is perhaps not generally realised that a switch from batch to continuous reactor design has intrinsic beneficial PI and safety implications when exothermic reactions and their associated runaway risks are involved. For batch operation, the time during which the reaction exotherm is generated is only a fraction of the batch cycle time. In order to control the reactions, it is imperative that provision is made to cope with the maximum likely heat evolution load so as to inhibit runaway. On the other hand, the heat exchanger provision for a continuous process operating at the same production rate needs to be considerably less than that for the batch equivalent, because the heat load is uniformly time-distributed rather than being concentrated in a fraction of the batch residence time. Hence, continuous versions of batch processes have both safety and intensification benefits.

Since the first edition of this book was published, there have been many assessments of the safety of PI plant, and many more examples where safety has been put forward as a principal reason for looking at PI. In Finland, Ebrahimi et al., (2012). Finland have developed a checklist for assessing the safety of PI and other plants, based to some extent upon the important work of Hendershot (2004), who divided the safety of chemical processes into four layers of protection. The application of this to hydrogen peroxide production will be briefly examined in Chapter 12, together with other new case studies, such as alkylation and MEA production.

2.4.2 **The environment**

On a relatively superficial level, an intensified plant will be much less obtrusive, with the absorption and/or distillation towers of our present chemical complexes being replaced by much more compact and correspondingly inconspicuous equipment. In addition, the cost of any necessary gaseous and aqueous effluent treatment systems should be reduced, thereby allowing tighter emission standards to be achieved economically. In some cases this should result in us questioning the accepted wisdom of using large centralised operations for chemical manufacture. The potentially hazardous products from these plants must often be delivered using the public transport network, with corresponding environmental risk implications. An alternative concept is for an intensified plant to be operated at the customer's boundary fence, with the product being piped over. This is similar to the existing practice for tonnage oxygen supply, where the plant is installed within or adjacent to the customer's site, as described below.

Prior to 1970, all commercial scale oxygen and nitrogen demands, provided by companies such as Air Liquide, were met by cryogenic distillation and distribution from a centralised production facility serving many customers. Distribution methods varied, depending upon the demand of specific customers – large users may had a dedicated pipeline. For needs of less than 20 tonnes/day, liquid oxygen or nitrogen could be delivered by road or rail tankers, or as a gas in pressurised bottles.

At the beginning of the 1990s, a revolution in the sector occurred – the introduction of locally produced liquid gases, on the users' sites. The reasons for this included: close match of the needs of the oxygen user (high purity, reactive capacity and pressure) and cost reduction (in terms of energy use and capital cost). This trend in the production and distribution of industrial gases had the following benefits:

- The standardisation of cryogenic air separation units.
- The adaptation of small air separation units to produce low purity oxygen (95–98 mol %).
- The development and commercialisation of non-cryogenic air separation technology, based upon N_2 adsorption, to produce impure O_2 (90–95 mol %).

A more radical idea in the above context, being actively pursued by the major oil companies, is to exploit the time taken in the seaborne transport of crude oil in order to convert the oil on board ship to a range of fuels and petrochemicals. This would have a profound and beneficial impact upon the storage and distribution systems associated with our existing refinery complexes. 'Mini-refinery' intensified technology is required, which allows a high operating performance to be maintained in a moving platform. These FPSOs (floating production storage and offloading) vessels are now made by many specialist shipbuilders around the world. The technologies could also be used at an offshore well head. Companies such as Velocys have designed plant for use on such FPSOs, and we now see the extension of the concepts first studied by the oil companies to other areas, where devices such as rotating packed beds can be used to carry out chemical processes *en route*.

One of the most interesting and challenging areas of technology (that has absorbed vast sums of expenditure within the last few years, and one that was barely visible on the horizon in the PI context when the first edition was written), is carbon capture (CC), normally combined with storage (CCS) in most literature. Although PI has been examined in the context of a variety of carbon capture techniques, including in both pre- and post-combustion regions around the plant combustors (normally power plant, but other CO_2 emitting processes can be considered), it is post-combustion capture that has attracted most interest to date. Ramshaw stated recently (Wang et al., 2011) that:

It should be recognised that the flue gas volumetric flow rate involved for a typical coal-fired power plant is extremely large. In a typical subcritical coal-fired power plant, an electricity output of 500 MWe corresponds to a combustion rate of roughly 1500 MWth. Assuming that coal's calorific value is based on the heat of formation of CO_2 and with 10% excess air, the flue gas flow rate is roughly 550 m^3/s. Hence, with a flue duct velocity of 15 m/s, in order to limit flue pressure drops to reasonable values, the duct cross-sectional area must be substantial (around 65 m^2). The CO_2 chemical absorption system need only be a fairly crude affair which would provide, say, the equivalent of 2–3 transfer units which corresponds roughly to a 10-fold CO_2 reduction. This is based on the assumption of a low CO_2 partial pressure over the CO_2- rich solvent flowing to the stripper system. With respect to the absorption stage, as shown above, large gas ducts are needed whether or not carbon capture is involved. It is, therefore, worth considering the installation of a simple liquid spray system to implement the absorption stage. This leaves us with the issue of the stripping duty and the lean/rich solvent cross heat exchanger where PI may be employed to good effect.

The application of PI here is directed at reducing the sizes of the absorption and desorption columns, for example, by carrying out the processes in rotating packed beds and by using compact heat exchangers in other parts of the plant. Other methods, discussed more fully in Chapter 6, include intensified adsorption.

Environmentally, however, the most telling impact of PI is likely to be in the development of reactor designs for truly green technology. It is well understood that the reactor is the heart of any chemical process, as it dictates both the product quality and the extent of the downstream separation and treatment equipment. Rather than accept mere end-of-pipe remediation of pollution problems, we must generate reactor fluid dynamics which allow us to fully exploit the *intrinsic* chemical process kinetics, however rapid these are. There is then a much better chance of designing reactors which operate intensively and which give high conversion and selectivity with minimal by-product formation. This will permit us to approach the green ideal of delivering a high quality product without an extensive downstream purification sequence. This observation leads us neatly into the discussion of the ways in which PI can benefit energy use, which, of course, impacts on the environment in a number of ways, in particular, in carbon emission mitigation.

2.4.3 **Energy**

The effectiveness of any PI strategy is ultimately dependent upon success in identifying techniques for dramatically increasing the intensity of the fluid dynamic environment, so as to accelerate the transfer of heat, mass and momentum within a process or operation. A high fluid dynamic intensity may be used to minimise the concentration, temperature and velocity differentials involved in transferring mass heat and momentum. When a process is considered from a thermodynamic standpoint these differentials may be regarded as irreversibilities that impair the ideal process energy efficiency. For example, a heat transfer operation performed with very high heat transfer coefficients will generate less entropy and preserve most of the thermodynamic potential of the heat transferred. An analogous situation is provided by an electrochemical operation for which the cell voltage attains the thermodynamic ideal value at very low current densities. However, the delivered voltage can fall substantially when realistic currents are involved. This voltage loss is incurred by mass transfer, conductivity and catalytic factors, which can be alleviated by an intensification approach.

Based upon the above, innovative applications of PI thinking will improve the process industry's capacity to meet the energy and global warming targets agreed at Kyoto and by the European Union and national governments since Kyoto was signed.

As an example of the changes taking place in the environment, the data in Figure 2.3 shows the variation in the length of the thermal growing season in the UK from 1772 to date (DECC, 2012).

The length of a thermal growing season is the longest period within a year that meets the following requirements:

- Begins at the start of a period of five successive days where the daily-average temperature is greater than 5.0°C.
- Ends on the day before of a period of five successive days when the daily-average temperature is less than 5.0°C.

The increase in growing season length since 1980 is largely due to the earlier onset of spring. The earliest start of the thermal growing season was in 2002 when

Departures in temperature in °C (from the 1990 value)

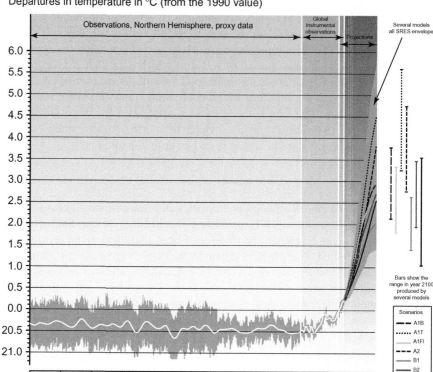

FIGURE 2.3

Variation in the thermal growing season in the UK (1772–2010). DECC data.

it began on 13 January. The longest growing season in the 239-year series was 330 days, in 2000. The shortest growing season was 181 days and occurred in both 1782 and 1859. In 2010, the thermal growing season was 255 days, down from 298 days in 2009 and above the 1961–1990 average of 252 days.

The smoothed line gives an indication of the trend but there is greater uncertainty for the first and last decade of the series.

Between 1900 and 1955, the average rate of global energy use rose from about 1 TW to 2 TW. Between 1955 and 1999, energy use rose from 2 TW to about 12 TW, and in 2006, a further 16% growth in primary energy use was recorded worldwide.

There are recommendations by the UK Royal Commission on Environmental Pollution, subsequently supported by others in the UK, that we need to reduce CO_2 emissions by more than 50% in order to stabilise their impact on global warming (CO_2 is believed to be the principal gas contributing to this phenomenon).

The following are the current estimates of the impact of climate change on our temperatures given by DECC (2012):

- The Earth's surface has warmed by about 0.8°C since around 1900 and by around 0.5°C since the 1970s.
- The average rate of global warming over the period from 1901 to 2010 was about 0.07°C per decade.
- More than 30 billion tonnes of CO_2 are emitted globally each year by burning fossil fuels.
- Average global temperatures may rise between 1.1°C and 6.4°C above 1990 levels by the end of this century.

2.4.3.1 Early UK energy assessments

The UK carried out an assessment of the potential for PI energy savings some years ago. In the 1990s, the then UK Energy Efficiency Office supported the development of strategies in three areas – compact heat exchangers; heat and mass transfer enhancement; and PI – all related to saving energy. There is logic and synergy in the linking of these three strategies – compact heat exchangers are used as the basis of several intensified unit operations (e.g. heat exchanger reactors) and, as we have seen in Chapter 1 (and in more detail in Chapter 3), enhancement is the principal mechanism behind most intensified plant (e.g. rotation, electric fields). As part of these strategies, studies were carried out by Linnhoff March, of the energy savings made, and data are given in Table 2.2.

Most of these data relate to unit operations within the chemicals and related sectors, and the technical potential for whole plant intensification, recognised as the most effective way of gaining the benefits of PI, was about 1,000 ktoe/annum[1], but it is estimated that a further 1,000 ktoe/annum would be saved if PI was extended to other process industries sectors, including glass and metals (Reay, 2007).

The above data do not take into account the increased knowledge of the potential (and actual) applications of PI since the strategies were formulated. In some instances, the opportunities will have increased – in others, they may have been

Table 2.2 Potential Energy Savings Due to Investment in PI in a Range of Process Unit Operations (Chemicals Sector Only).

Compact heat exchangers	–	16 PJ/a
Separators	–	6.2 PJ/a
Reactors	–	11 PJ/a
Overall plant intensification	–	40 PJ/a (technical potential)
Effluent treatment	–	1 PJ/a

[1] ktoe – thousand tonnes of oil equivalent.

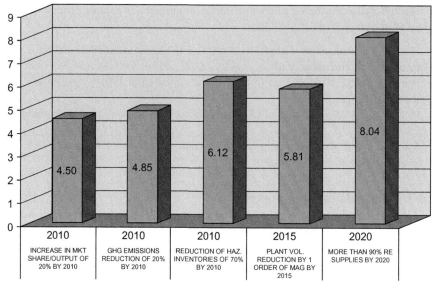

FIGURE 2.4

PI business scenario observations.

superseded by other process improvements. An important factor is that major process changes/plant replacement would be needed to realise the savings, and a second critical observation is that integration is necessary in all except minor unit operation substitution. Effective integration can maximise emission reductions.

Interestingly, a survey on attitudes to PI (Nikoleris et al., 2002) revealed, (as shown in Figure 2.4), that a change to a non-carbon based economy would be a major stimulus for PI. This was borne out by other data that revealed that 60% of those surveyed felt that a switch to a hydrogen or bioethanol type economy would greatly benefit PI uptake. The work of Dr Adam Harvey at Newcastle University (Harvey, 2006) hinted at a trend towards a portable intensified bio-diesel production unit.

This would give a substantial stimulus to integration methodologies which have yet, with a few exceptions, to fully incorporate the novel PI technologies emerging in the marketplace. One example, was the work on fluidised-bed reactors in the Netherlands and Germany (Deshmukh et al., 2007). Since then, many PI projects, such as those supported by the European Commission within the 7th Framework Programme, mention integration as a 'key word', but do not obviously implement it. (The EU scene is discussed below.)

2.4.3.2 Current UK strategy on PI

Several years ago, the Carbon Trust (the UK body at the time overseeing implementation of the UK Government's strategy for reducing carbon emissions), had process intensification as one of its top priorities for action and support. In 2005, however, a study by Arthur D. Little (ADL) for the Carbon Trust (CT) effectively downgraded support for industry at the expense of renewables and buildings (2005b). (Another ADL report carried out on behalf of a number of European countries and organisations, however, [see Anon, 2006] lent substantial support to PI). The study for the CT considered 'process substitution' as the generic area in which PI would fall. This was described as: '...the modification or replacement of existing processes to result in less energy being consumed'. The areas chosen for further investigation were 'the potential to exploit micro-reaction technology...' (see Chapter 5); '...and advances in heat exchanger technology' (see Chapter 4). It is interesting to read the context in which these two technologies were discussed, summarised in the next paragraph.

With regard to advanced heat exchangers, the scope was extended to include electric enhancement methods (see Chapter 3), as follows:

> '...in some cases the combination of heat exchangers with alternative energy sources (e.g. microwaves or radio frequency) has energy efficiency benefits.'

However, there were no specific recommendations in the document (Anon, 2005b) to the types of advanced heat exchanger that should be supported. The micro-reaction technology category was listed under 'micro-fluidic processes' (see Chapters 3 and 11), and the comments reported were slightly more positive:

> 'Micro-fluidic processes (reactions which occur at a micro scale [sic]) help to improve energy efficiency, mixing and product yield. There are many practical challenges to overcome in applying these processes to industrial applications, particularly in scaling-up from small volume to bulk manufacture....Substantial UK funding over a prolonged period would be required to compete effectively with the best research available outside the UK.'

The authors most likely had Germany and Japan in mind in coming to this conclusion. The overall conclusion, contrary to what many involved in the application of PI believe, is that the areas of micro-fluidics for chemical processes and advanced heat exchangers should be reviewed periodically to reassess whether support is necessary.

2.4.3.3 A major European initiative

Arthur D. Little, in 2005–6 carried out a study with SenterNovem of The Netherlands. The authors of this book were involved in giving information during the study (Anon, 2006). The outcome of the study was the identification of a plan for implementing PI in a number of sectors, using specified PI technologies. The effort was costed and potential stakeholders identified, and the work is being followed by a current study as part of the European Roadmap of Process Intensification, principally involving The Netherlands and Germany. With regard to energy use, there were many interesting and encouraging applications data, which are summarised in Table 2.3.

Table 2.3 Energy Saving Potential of PI Technologies as Identified by the ADL-SenterNovem Study.

	Bulk Chemicals	Fine Chemicals	Food
Multifunctional equipment (advanced distillation)	50–80% energy savings in 15% of processes. 9–18 PJ	Limited to separation processes, i.e. 10% of sector. Increase efficiency by 50%, saving <1 PJ	Drying and crystallisation. 10% total energy saving, worth 3–5 PJ
Micro/milli-reactors	A study by ECN in Holland suggests 20 PJ savings using heat exchanger reactors. Micro-reactors extend this to 25 PJ.	Applications in 20% of processes in the sector saving 20% of energy – 1 PJ. Reduce feedstock and additives by 30% in 10% of processes saving 5–7 PJ	Spill-over from fine chemicals: <1 PJ
Microwaves (electrical enhancement)	*	Reduce feedstock and additives by 20–40% in 5% of processes: 2–3 PJ	20–50% saving in 10% of drying market: 1–1.5 PJ; 10% energy reduction in product processing: 1–1.5 PJ
High gravity fields (e.g., spinning disc reactor, HiGee)	*	Reduce feedstock, solvents, etc., by 50% in 5% of processes: 1–3 PJ	Assuming 20% of electricity in food production goes to emulsification, mixing, etc., 10–20% saving worth 0.5 PJ

Note: energy saving is in PJ/annum
*ECN suggested that too little was known of the effects of microwaves and HiGee in the bulk chemicals sector.

The overall energy savings across the three sectors were estimated to be of the order of 50–100 PJ per annum by 2050. The energy savings were largely due to better selectivity and reduced energy use in separation processes, as well as improved control. Although not specifically highlighted in the summary of the ADL data above, any chemical reactions which are currently carried out as batch processes in stirred tanks could be carried out in continuously operated, intensified reactors. This can give reductions in energy due to, for example:

- Less unwanted byproducts – reducing downstream processing.
- Moving from batch to continuous processing will reduce the energy need for cleaning the plant.

- More scope for process heat recovery – and higher grade heat may be available.
- Reaction rates may be speeded up, hence there will be reduced energy losses, due to shorter processing times.
- Reduced system losses.

The opportunities for distributed processing should not, of course, be neglected in considering the carbon footprint. The food sector is particularly amenable to distributed processing – see spinning disc reactor applications later.

The UK pioneered advanced compact heat exchangers, such as the printed circuit (PCHEs) and Marbond units in the process industries, and the success of initiatives in this area – which led to the emphasis of enhancement and PI – was seen by many as the first official recognition of process intensification technologies as major tools in aiding more efficient use of energy. This has now extended internationally, and to reactors (Haugwitz et al., 2007), as illustrated in Chapter 5.

2.4.3.4 Support for PI in the US

The US Department of Energy has also initiated programmes supporting PI in the process industries (Web, 2007). Relevant to PI are the activities on hybrid distillation and novel reactors, together with advanced water removal (drying/evaporation) methods. The Industrial Technologies Programme (ITP) highlights the following points:

- Multifunctional reactors – e.g. Sandia slurry bubble-column reactor.
- Fuel and electricity savings of >50%.
- Reduced waste – e.g. less acid used in alkylation.
- Potential savings of >70 PJ/a by 2020.

Process intensification, although yet to fully emerge as an established technology in the process industries, offers significant opportunities for carbon reduction in sectors ranging from chemicals to food and glass manufacture. The possibility of increased 'local' production and the growing use of biological 'renewable' feedstocks open up new challenges and opportunities for those active in integration as well as intensification.

2.4.4 The business process

In the years since PI was first mooted it has become evident that its potential business impact is by far the most important factor in any decision to adopt it. It has already been pointed out above that 'boom and bust' in capital investment may be avoided by smaller up-rating steps in production. However, the full business implications of PI are rather more subtle.

Since the essential idea of PI is a large increase in production per unit plant volume, it follows that residence times are correspondingly decreased – typically from hours to seconds for some operations. This has profound implications for the plant's ability to respond quickly to desired process changes in general and for the control philosophy in particular. Indeed, a legitimate question is to ask whether we need conventional feedback controllers when the reaction time constant is around one second. In any event, it will become possible to switch product grades rapidly with little intermediate off-specification product being generated. This facilitates a just-in-time manufacturing business policy which could result in a substantial reduction in product warehousing costs.

A further aspect of PI is relevant to the pharmaceutical and fine chemical sectors which are attempting to bring their products to the market much more rapidly as they strive to combat generic competition. However, they have hitherto been culturally obsessed with batch production in stirred vessels, a manufacturing approach which has held sway for well over 100 years. The batch/stirred-pot mindset stems from the moment that the process development chemist reaches for a beaker or flask to produce trial quantities of a new molecule. Once valuable therapeutic properties have been demonstrated, the molecule is patented and the patent 'clock' starts to tick – typically for 20 years. While clinical trials proceed, the new process is assigned to the engineers for scale-up in order for them to design a bigger beaker. Unfortunately, the reaction performance of a stirred vessel may be profoundly influenced by its size (see Chapter 5), so the regulatory authorities insist that the process is assessed at the laboratory, pilot and full-scale, each step taking perhaps two years. However, an intensified process at laboratory scale can probably generate product at the typical rates needed – say 500 tonnes/year (equivalent to continuous liquid flows of only 50 ml/s). It will be recognised that PI may give the opportunity for several extra years of production, while under patent cover, provided that clinical trials can be accelerated. This leads to the concept of a 'desktop' process which will help the drug companies to commercialise their inventions more quickly. Scale-up delays can be avoided once the laboratory scale becomes full-scale. Being continuous, the equipment will not be subject to the labour costs or potential contamination problems associated with inter-batch cleaning. If the batch culture is to be overcome, it is imperative that development chemists have access to, and appreciation of, intensified continuous reactors.

2.4.4.1 The PI 'Project House' at Degussa

One example of business-driven PI investment was announced in 2005 and took place at Degussa (Hahn, 2005). The project-house concept had already been applied at Degussa in areas such as nano-materials and catalysis, and in this instance, an interdisciplinary project team of 15 with a budget of € 15 million was established in Germany to operate from the beginning of 2005 until the end of 2007. In collaboration with seven Degussa business units, ranging from speciality acrylics to coatings and colourants, the aim was to:

> '...develop two to three new products, to make two to three new key technologies available for Degussa, and significantly improve two to three production processes'.

Specific areas of investigation are highly active catalysts – which will involve micro-reactors, and functional materials – where Degussa has coined the term 'nano-droplet reactors' as part of the route to products such as 'activatable additives' for adhesives. Two other areas of research are disperse systems, directed at reducing process times while improving throughput – thus increasing efficiency in, for example, fermentation processes, and what Degussa calls Chemical ExplorENG. This is a modular plant concept aimed at rapid assembly and commissioning on site from a number of components. This should reduce the 'time-to-market' for products. An example of this is given in Chapter 8.

2.4.4.2 Benefits of PI for Rhodia

An earlier insight into the motivation behind a major company's interest in PI was given during a presentation by Sylvaine Neveu of Rhodia at a meeting at Cranfield University in 2002 (Neveu, 2002). The main processes at Rhodia that might be of interest to a PI approach were polymerisation and organic or inorganic synthesis. Rhodia saw the multiple reactions in organic synthesis as being potential beneficiaries of PI. Impinging jet mixer technology was one opportunity and in the Bourne reaction one could improve selectivity and diminish secondary products, thus removing the need for separation (or at least making it easier). In exothermic reactions, the introduction rates of reactants are limited by the speed of heat removal – PI could improve this.

For inorganic processes, a typical product would undergo precipitation; separation; and/or washing; drying; calcination and then presentation to the customer. All steps must be intensified to improve productivity, but it was recognised that this was not easy. The precipitation process is a quality-critical step, and nucleation, growth, aggregation and agglomeration all need controlling. Mixing has an effect on these, as do hydrodynamics. Using agitated batch or semi-batch reactors, one gets a large particle size distribution and a problem of scale-up, due to limited heat and mass transfer. Scale-up criteria are also difficult to assess, for example the relationship between hydrodynamics and nucleation rate. Using PI would accelerate heat and mass transfer, give better contact between reactants, better hydrodynamic control, and make the system more easily scaleable.

In 1994, Rhodia was granted a patent for the production of an Al/OHx/Cly polyaluminiumchloride compound for water treatment. The production was, however, associated with the formation of an undesirable insoluble secondary product. Use of a $14 \, m^3$ semi-batch reactor did not allow separation, and the solution was to install a 'quick mixer' which was Y-shaped and mixed the reactants under good stoichiometric conditions. This allowed the desired composition to be achieved with no byproduct production.

For polymerisation, Rhodia is proposing to switch from batch/semi-batch to continuous processing. But there is a need to change the product each day in the same plant (change in demand) and in continuous configurations daily changeover can give rise to problems with intermediate products. However, PI decreases the plant size, raises production and decreases intermediates. In devolatisation, thin films also assist mass transfer. In summary, Rhodia saw PI as giving greater production, lower investment cost, greater speed of product development and high product quality.

An interesting aspect of the meetings held by the PI Network is the discussion that follows talks. The company was already using PI techniques for heat and mass transfer intensification and believed the best way to adopt a proactive approach to PI was to use it from the beginning of a product development. One can show the advantages at the laboratory scale and then one has the choice in production between, for example, a standard reactor or its intensified equivalent. With regard to investment, one can add PI equipment without necessarily changing or replacing everything.

With regard to developing criteria for testing processes for PI relevance, the company had adopted a systematic approach, starting with precipitation and extending to other fields. Researchers carefully studied the efficiency of a range of PI devices, heat and mass transfer, hydrodynamics, etc., and then made laboratory scale devices. Only then would a study be made of process parameters for the larger scale. It was suggested at the time that one difficulty of PI was the range of modules and lack of hydrodynamic data – therefore one goes for the safe bet – the batch reactor. Rhodia, (confirming the approach recommended in Chapter 12) stated that one needs a good knowledge of the plant to convince the plant managers that PI is a worthwhile investment.

2.5 **Some obstacles to PI**

Despite the compelling benefits of PI for the process industry it has to be admitted that there have been, and still are, serious obstacles which have been responsible for its relatively slow adoption since its inception around two decades ago. In the context of an existing fully established and depreciated plant, it is extremely difficult to introduce unproven intensified equipment. On one hand there are concerns that any malfunction will compromise the output of the whole plant, while on the other, the full benefits of PI are unlikely to be realised unless the whole plant is intensified. A policy of 'scrap and start again' is unlikely to be accepted until a whole range of intensified equipment has established its reliability credentials. Consequently, a very wide range of PI technologies must be developed which covers the major reactor types and separation duties (in the chemical process sector for example) in order to ensure that PI is readily accepted. Most plant managers are understandably conservative and 'rush to be second'. (In other words, they require full-scale evidence of successful operation before they are prepared to take any risk). A good opportunity for the adoption of the new technology is when a new plant is being considered and where there are significant space or weight problems associated with conventional technology.

A mature technology in the process industry is usually associated with the existence of design codes and packages. Thus, for example, a shell and tube exchanger will be designed according to the Tubular Exchanger Manufacturers Association (TEMA) code, which lends confidence regarding the reliability and safe operation of these units. A developing technology such as PI is not yet embodied in such design codes. The provision of this information is an important next step for the relevant equipment vendors.

A very effective technique for PI involves the use of laminar flow in narrow channels (see Chapter 4) so as to improve both heat and mass transfer coefficients and consequently the transfer area available per unit equipment volume. Unfortunately, small passages are notoriously prone to blockage if the system is liable to deposit scale or contains solids having a size greater than about 20% of the channel width. In some cases, such as with crystal scale or carbon deposition, the improved heat and mass transfer associated with PI may alleviate the problem by reducing surface supersaturations or surface temperature. Nevertheless, in general, a severe fouling environment may seriously limit the intensification potential for a given unit operation – this is discussed in the context of crystallisation fouling in micro-channels in Chapter 4 (Section 4.3).

Even in the twenty-first century, rotating machinery can still be regarded by companies, or the engineers within them, as unreliable. Perhaps the best way to counter this is to use a quote from URENCO (the leader in centrifuge uranium enrichment [see Chapter]) 10 when the company was addressing a meeting of the UK Process Intensification Network:

> *'URENCO's existence depends upon the reliability of high speed rotating equipment'.*

In this case, the reference was to centrifuges rotating at over 1,000 revs/second for about 25 years with a failure rate of less than 1% per annum.

2.6 A way forward

The extensive consolidation which has occurred in the chemical industry in the last two decades has generally distracted management attention away from technical development. Long-term engineering research has often been sacrificed in order to cut overhead costs, while the positive business process opportunities presented by PI have largely gone unrecognised. Now that the consolidation has hopefully run its course, the pendulum should swing back, and interest in manufacturing excellence may revive, particularly under pressure from the increasingly strident environmental movement.

It has already been noted that there are cultural issues regarding the adoption of PI. A long-term approach to overcoming these is to ensure that the next generation of chemists and engineers is fully aware of the benefits of PI technology. Therefore,

Table 2.4 Comparison of the Principal Parameters of Two Water Deaeration Units – One Conventional (Vacuum) and One Using HiGee Technology.

Parameter	Existing Technology	Rotating Bed Technology
Investment	1.0	0.6
Power consumption	1.0	<1
Ground space	1.0	0.4
Land requirement	1.0	0.6
Weight	1.0	0.2
Plant height	1.0	0.25
Oxygen in water supply (ppm)	6–12	6–12
Oxygen in water after treatment (ppb)	200–800 (1 tower)	50
	50–200 (2 towers in series)	

(New Data, Chen, 2009)

PI modules should be incorporated into relevant university courses and students must be encouraged to question conventional thinking and be given the chance to experiment with intensified equipment. This is already being done at a small number of universities in the UK and The Netherlands.

Another important factor involves the lack of demonstration facilities in which clients' processes may be performed on intensified equipment, such as those at the Centre for Process Innovation on Teesside in the UK (see Appendix 4). A successful outcome to such trials is a very powerful motivator for the adoption of the new technology. This is an area where the national and European funding agencies can play a vital role, by providing funding and encouraging collaboration between industrial organisations with similar technical interests. It is worth observing in this context that European exploitation of compact (i.e. intensified) heat exchangers is much higher than in the US, largely because of the raised awareness generated by the Heat Exchanger Action Group (HEXAG). This UK network has been managed and directed by Professor Reay over the last twenty years and its success provides encouragement regarding the future impact of the PI networks.

2.7 To whet the reader's appetite

The data in Table 2.4 from China on a HiGee vacuum deoxygenation unit with a capacity of 250 t/h give some comparison of cost/performance data on an existing commercial PI unit, showing up to fivefold reductions in some of the prime

FIGURE 2.5

The Chinese use of process intensification for deaeration offshore.

parameters[2]. The plant is shown in Figure 2.5, the intensified plant being on the right-hand side of the photograph. Further data are given in Chapter 9 where offshore applications of PI are discussed.

Most of the themes listed earlier can arise out of 'compact' plant, where physical volume reductions may not be described in terms of orders of magnitude. However, without wanting to be prescriptive in defining PI, the *real benefits* will arise when the size of the plant can be produced, or, in the words of the then chief engineer of BP, Terry Lazenby, (at a meeting of HEXAG in 1998): '...honey, I shrank the plant!'

2.8 Equipment summary – finding your way around this book

The range of unit operations employing process intensification is rising rapidly, and is based to a large extent on R&D outputs in Europe, North America and, increasingly, China and Japan. Table 2.5 is intended to assist the reader in navigating the text by summarising the principal types of equipment that have been intensified, the mechanism(s) used to do this, and the main applications. The chapters where the topics are discussed in more detail are listed in the right-hand column.

---

[2]Note that China has exploited rotating intensified separation plant offshore and onshore rather rapidly – see the applications chapters. In Europe and North America, exploitation of these particular technologies has been slower.

Table 2.5 Summary of Principal Intensified Equipment Types and Applications Discussed in this Book.

Equipment Type	Techniques Used to Intensify	Typical Applications	See Chapter
Heat exchangers			
Compact heat exchangers (CHEs)	Channel size reduction	Process industries, aerospace, power generation, etc.	4
Printed circuit heat exchanger (PCHE)	Chemical etching of channels in sheets, then diffusion bonding together	Offshore, HEX-reactors, process heat transfer, power generation	4
Chart-flo	Chemical etching of channels in sheets, then diffusion bonding together	Heat exchange, HEX-reactors (Chart-kote)	4
Polymer film heat exchanger	Cross corrugation of thin polymer sheets	Heat recovery, hex-reactors, membrane technology	4
Foam heat exchanger	Foam produced from a metal-coated polymer foam	Aerospace, micro-power generation, compact and micro-reactors	3, 4
Mesh heat exchanger	Bonded layers of woven wire mesh	As for foams, also giving structural contribution	4
Micro-heat exchangers	Chemical etching, spark erosion, UV LIGA techniques	'Lab-on-a-chip', micro-electronics thermal control	4
Passive enhancement methods	Several – involving geometrical and surface changes or additional materials	Many heat (and mass) transfer applications	3, 4
Active enhancement methods	Several – rotation, vibration, electric fields, etc.	Many heat (and mass) transfer applications, including spinning disc reactors, heat pumps/chillers, drying, heat exchangers	3 and throughout the book
Reactors			
Units based on compact heat exchangers	Coating brazed or diffusion bonded heat exchanger surfaces with a catalyst. Metal foam-based variants	Combining endothermic and exothermic gas–gas and liquid–gas reaction. Thermal control of reactions	5

(Continued)

Table 2.5 (Continued)

Equipment Type	Techniques Used to Intensify	Typical Applications	See Chapter
Helix reactor	Twisted tubes provide intense mixing of the flow	Initially for replacing highly exothermic batch reactions	5
Packed-bed types	Insert catalyst particles into a compact heat exchanger passage structure	Oxidation, hydrogenation and reforming reactions	5
Micro-reactors	As for CHEs	Drug and fine chemical manufacture. Fuel cell reformers for mobile phones	5
Spinning disc	Rotation of catalysed discs	Several, including polymerisation	5
Rotating fluidised-bed	Suspended particles in a fluid bed are given a tangential velocity	Combustion, calcining, catalyst reactivation, gasification/pyrolysis, pasteurisation (clinical waste, herbs and spices), and desorption	5
Oscillatory baffle	Either baffles or the fluid is 'oscillated' to enhance performance	Suspension polymerisation (e.g.)	5
Electrically-enhanced reactions	Use of electric fields for induction heating, or ultrasonic agitation	Bioreactions (e.g.)	5
Membrane reactors	Membrane materials are coated with a reaction catalyst	Dehydrogenation, dehydration. Gas–liquid–solid reactions	5
Supercritical processes (aid to reactions)	CO_2 used as solvent to help catalyst recovery. Overcome mass transfer limitations	Hydrogenation, acid catalyst recovery	5
Adsorption reactors	Use a solid/gas system, e.g., ammonia/activated carbon	Refrigeration, heat pumping, long distance heat transport	4, 10, 11
Separators Distillation			
Dividing wall	Combining columns into one	Most common separations	6

(Continued)

Table 2.5 (Continued)

Equipment Type	Techniques Used to Intensify	Typical Applications	See Chapter
Columns			
Rotating discs/ packing (HiGee)	Porous matrix of polymer or metal packed into disc form and then rotated	Offshore, deaeration, distillation	6
Compact heat exchanger inside column	CHE replaces conventional rectification 'internals'	Air separation, ethylene recovery, CO2 purification, etc.	6
Mechanical vapour recompression plus CHE	Integration of a compressor and a compact heat exchanger.	Evaporators, e.g., food, drinks, chemicals, effluent	8
Centrifuges			
Conventional types	Assembly of closely-spaced conical discs, rotated at high speed	Slurry concentrating, dewatering (e.g.)	6
Ultra-centrifuge	High gravity forces	Nuclear processing sector	6, 10
Membranes	Use of permeability characteristics	Solvent/solute separation; dehydration	6
Emulsion liquid membranes	Use of permeability characteristics	Copper extraction; water treatment	10
Electrically-enhanced separations			
Biological	Electrostatic fields to enhance droplet motion, for example	Fermentation broths	6
Drying	Use of microwave energy (or other wavelengths)	Powders	6
Extraction			6
Mop fan/deduster	A pack of flexible spines are rotated at high speed. These may be coated to absorb impurities in a stream	Odour control, effluent clean-up	6

(Continued)

Table 2.5 (Continued)

Equipment Type	Techniques Used to Intensify	Typical Applications	See Chapter
Cleaning	Ultrasound	Textiles	10
Crystallisation/ precipitation	Increase contact surface, high gravity forces	Sodium chloride isolation, thin film crystallisation (electronics)	6
Mixers			7
In-line mixers	Specific 'inserts' located in tubes in fluid stream	Polymer processing, food processing, etc.	7
Spinning disc as mixer	High gravity forces	Mixing prior to crystallisation	7
Ejectors	Intense turbulence formation or shock gives mixing	Gas-liquid contacting; reactors	7
Mixer-heat exchangers	Mixers with vortex generators or other enhancement devices to aid heat transfer	HEX-reactors, polymer processing	7
Miscellaneous			
Intensified mass transfer units	Rotation, vibration, mixing, electric fields	Many processes throughout text	3
Micro-fluidic processes	Electrokinetics	Many processes	3
High pressure systems	Pressure effects	Cooking, etc.	3
Plasma electrolytic oxidation	Pulsed plasmas	Coating of metals	10
Sintering	Use of microwave energy	Sintering of metals	10
Whole plant intensification	Combination of any of the above	Organic and inorganic chemical plants, pharmaceuticals, offshore processing, food and drink	12
Rotex/Rotartica	Rotation and compact heat exchangers	Air conditioning/heat pumping	1, 11

2.9 Summary

Process intensification is frequently defined in terms of its benefits to the process industries, in particular the chemicals sector. An appreciation of the benefits that PI can give in terms of energy saving, safety, company profitability and new opportunities will help readers to understand the motivation behind PI use when studying the applications chapters. There are also limitations to PI, and an awareness of these is essential if correct application is to be ensured. This chapter has started to raise the awareness of the reader to all these aspects, as well as directing him/her to other chapters where the technologies and uses of PI are elaborated upon in more depth.

References

Anon, 2005a. Toshiba's DMFC officially certified as world's smallest by Guinness World Records. Toshiba Press Release. <www.toshiba.co.jp/about/press/2005_02/pr2801.htm>.

Anon, 2005b. Carbon Trust research landscape study. <www.thecarbontrust.co.uk> CTC511.

Anon, 2006. Building a business case on Process Intensification. Unpublished report, Arthur D. Little and SenterNovem, July.

Buchholz, S., 2010. Future manufacturing approaches in the chemical and pharmaceutical industry. Chem. Eng. Process.: Process Intensification 49, 993–995.

Chen, J.-F., 2009. The recent developments in the HiGee technology. Presentation to the GPE-EPIC Conference, Venice, Italy, June 14–17, 2009.

Chen, C.Y., Tsao, C.S., 2006. Characterisation of electrode structures and the related performance of direct methanol fuel cells. J. Hydrogen Energy 31, 391–398.

DECC – Department of Energy and Climate Change, 2012. <http://www.decc.gov.uk/en/content/cms/tackling/> (accessed 25.06.12.).

Deshmukh, S.A.R.K., Heinrich, S., Moerl, L., van Sint Annaland, M., Kuipers, J.A.M., 2007. Membrane assisted fluidized-bed reactors: potentials and hurdles. Chem. Eng. Sci. 62, 416–436.

Ebrahimi, F., Virkki-Hatakka, T., Turunen, I., 2012. Safety analysis of intensified processes. Chem. Eng. Proces.: Process Intensification 52, 28–33.

Fishlock, D., 1982. Gravity: new essence of distillation. Financ. Times, 33. Thursday 11 November.

Hahn, H., 2005. Market pull meets technology push. Degussa Science Newsletter 10, 4–6.

Harvey, A.P., 2006. Biodiesel process intensification projects at Newcastle University. Presentation to PIN, 16 November. See <www.pinetwork.org>.

Haugwitz, S., Hagander, P., Noren, T., 2007. Modelling and control of a novel heat exchanger reactor, the open plate reactor. Control Eng. Pract. 15, 779–792.

Health and Safety Executive, 1975. The Flixborough Disaster: Report of the Court of Inquiry. HMSO.

Hendershot, D.C., 1997. Inherently safer chemical process design. J. Loss Prev. Process. Ind. 10, 51–157.

Hendershot, D.C., 2004. Process intensification for safety. In: Stankiewicz, A., Moulijn., J.A. (Eds.), Re-Engineering the Chemical Process Plant: Process Intensification Dekker, New York. (Chapter 13).

Keller II, G.E., Bryan, P.F., 2000. Process engineering: moving in new directions. Chem. Eng. Progress January, 41–50.

Kletz, T.A., 1991. Plant Design for Safety. The Institution of Chemical Engineers, Rugby, UK.

Neveu, S., 2002. Potential benefits of process intensification for Rhodia. Proceedings of the Third meeting of the Process Intensification Network, Cranfield University, 14 November. See <www.pinetwork.org> for overheads and minutes.

Nikoleris, D., Arias, R., O'Conner, M., 2002. Delphi study on process intensification technologies MSc Project. Heriot-Watt University, Edinburgh, Available on <www.pinetwork.org>.

Palmer, J., 2010. Supercomputers will fit in a sugar cube, IBM says. BBC News Technology. <www.bbc.co.uk/news/technology-11734909>.

Ponce-Ortega, J.M., Al-Thubaiti, M.M., El-Halwagi, M.M., 2012. Process intensification: new understanding and systematic approach. Chem. Eng. Process.: Process Intensification 53, 63–75.

Ramshaw, C., 1983. Process intensification by miniature mass transfer. Process Eng. January.

Reay, D.A., 2007. The role of process intensification in cutting greenhouse gas emissions. Plenary paper, Proceedings of PRES 07, Ischia, June.

Stankiewicz, A.I., Moulijn, J.A., 2000. Process intensification: transforming chemical engineering. Chem. Eng. Progress January, 22–34.

Wang, M., Lawal, A., Stephenson, P., Sidders, J., Ramshaw, C., 2011. Post-combustion CO_2 capture with chemical absorption: a state-of-the-art review. Chem. Eng. Res. Des. 89, 1609–1624.

Web 2, 2012. The future factory project. <www.f3factory.com> (accessed 23.06.12.).

The Mechanisms Involved in Process Intensification

OBJECTIVES IN THIS CHAPTER

A wide range of unit operations are capable of intensification. In this chapter, some of the more important intensification techniques are briefly discussed, in order to prepare the reader for the more detailed treatment provided in subsequent chapters.

Where appropriate, developments taking place after the writing of the first edition are included. It is particularly noticeable that there has been continuing and increasing activity, in particular in the use of electric fields of all types, and in further scale reductions (micro- and now nano-scales).

3.1 Introduction

A general characteristic of the process industry is that it deals with a wide range of multi-phase systems. Indeed, about two thirds of the unit operations fall into this category. Thus we have gas and/or liquid operations:

- Boiling
- Condensation
- Absorption
- Distillation
- Mist disengagement
- Electrolysis/fuel cells (where a gas phase is generated or consumed)

    And solid and/or fluid operations:

- Fluidisation
- Dust disengagement
- Filtration

The fluid dynamics of all these operations is controlled by the buoyancy term $\Delta \rho \, g$. If either ρ or g is zero, the system is controlled by surface forces. Under these conditions counter-current flow of the phases cannot occur, coalescence is suppressed and the droplet or bubble size becomes very large. In the absence of interfacial shear, heat and mass transfer coefficients are severely reduced. A good demonstration of this state of affairs is given by the astro lamp (or lava lamp), shown in Figure 3.1.

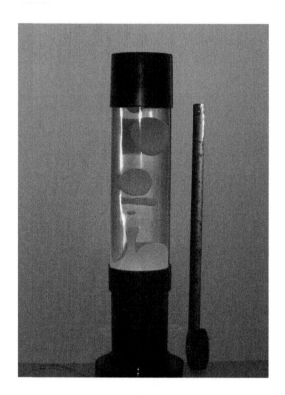

FIGURE 3.1

The astro lamp, illustrating milli-g behaviour. The clear section is 20 cm in length.

Here, a molten wax and an immiscible aqueous phase are heated from below by a low power electric bulb, thereby generating an axial temperature gradient. Wax which is slightly less dense at the base is marginally denser in the cooler upper region of the column. Typical density differences are of the order of 1kg/m^3 so the system can be considered to simulate milli-g behaviour. All the characteristics noted above are evident. The wax globules are about 5 cm in diameter, their rise and fall velocity is only a few millimeters per second and coalescence never occurs. Clearly the intensity of operation is very slow. An obvious corollary is that multi-phase systems in a high acceleration environment will behave much more intensely, with high shear stresses, fine bubbles and high counter-current velocities. The exploitation of centrifugal acceleration has proved to be an effective strategy for PI, since it influences so many process unit operations.

This is one of several active enhancement methods that have found favour in PI, and these and their main features are discussed in this chapter. Active, and to a lesser extent passive, enhancement techniques are the bedrock of intensified unit operations, as initially shown in Table 2.5. The use of fine channels, and the conversion from batch to continuous operation, have already been mentioned and are covered in more detail in Chapters 4 and 5.

It is necessary to have knowledge of the techniques and technologies that form the basis of intensified unit operations, so that factors affecting their design, performance and, in some cases, inferior behaviour or failure can be understood. In this chapter, data on the enhancement methods, both active and passive, that can be used to intensify heat and mass transfer are given. Included is electric field enhancement, a generic area that is becoming increasingly popular in PI. Towards the end of the chapter, the area of fluid dynamics now known as micro-fluidics is introduced. It is believed by many that micro-fluidics will play an increasing role in very small unit operations such as micro-reactors and the lab-on-a-chip.

3.2 Intensified heat transfer – the mechanisms involved

Intensified heat transfer is also known as enhanced or augmented heat transfer. As highlighted by Webb and Nae-Hyun (2005), intensified heat transfer is used routinely in a range of everyday consumer systems. The car radiator has, within a relatively small volume, a substantial amount of 'finning' or extended surface to ensure that sufficient heat removal from the engine coolant takes place, even on hot days. Most air conditioning units have similar heat exchangers, the aim of the enhancement being to reduce the heat exchanger size and cost, without compromising performance – part of the PI philosophy.

The mechanism of intensification is a strong function of the nature of the fluid stream (gas, liquid, in some cases a solid, or mixture of all three), and the mode of heat transfer. For example, with single-phase gas streams, the fins (as on the above radiator) are ideal for enhancement. If we are boiling a fluid in a narrow tube, fins are likely to be impractical and a surface treatment on the micro-scale may be more appropriate.

Since it is well accepted (and theoretically underpinned) that heat, mass and momentum transfer are qualitatively governed by the same general rules, they will be considered together in the context of PI. It is generally recognised that high degrees of turbulence, generated at high fluid Reynolds numbers, give rise to intense surface shear stresses and high transfer coefficients. For fully developed flow in pipes this idea is encapsulated in the familiar Dittus-Boelter equations:

$$Nu = \varnothing_1\, Re^{0.8}\, Pr^{0.33} \tag{3.1}$$

$$Sh = \varnothing_2\, Re^{0.8}\, Sc^{0.33} \tag{3.2}$$

By manipulating these equations it can be shown that the film transfer coefficients h, h_D vary as follows:

$$h \text{ or } h_D \propto d^{-0.2}\, u^{0.8} \tag{3.3}$$

Thus, there is a weak dependence on d and a fairly strong influence of enhanced velocities. Unfortunately, the pipe pressure drop varies as u^2 so, while higher intensities may be obtained at greater fluid velocities, this is achieved at an increasingly severe pumping energy penalty. These observations suggest that it may be worth exploring the alternative strategy of using *lower* Reynolds numbers such that the flow becomes laminar, which can be done by using fine tubes or narrow channels. In this case, the Nusselt number for fully developed flow is about 3.8 and is *independent* of velocity. Thus $h \propto d^{-1}$ and narrow channels/pipes with their short diffusion/conduction path lengths are clearly advantageous.

This benefit is compounded when it is recognised that the specific transfer area (a) for close-packed tubes or channels is inversely proportional to diameter. The volumetric performance intensity is therefore h a $\propto 1/d^2$, implying that narrow capillaries are beneficial. The surface shear stress is given by:

$$R = \frac{8u\mu}{d}$$

Showing that low velocities will minimise pumping energy. It is worth noting that the strategy of exploiting laminar flow in fine capillaries has been used in nature for millions of years, notably in lungs, kidneys, etc., where high performance at low pumping energy is paramount (see also Cross and Ramshaw, 1986). As a current illustration of this in an engineering application, there is a Swedish heat exchanger, used as an oil cooler in large vehicles, produced by Laminova AB (see Appendix 3 and Anon, 2012b). This uses rows of fins as flow passages, with gaps between each fin to prevent the boundary layer changing from laminar to turbulent flow. So, as one proceeds along the (highly compact) heat exchanger, new laminar boundary layers are being formed on the heat transfer surface every few mm. Thus the criterion above is satisfied.

Another technique for heat and/or mass transfer enhancement relies simply upon the geometry of the surface in contact with the fluid. It is relevant at very low fluid velocities, when diffusion/conduction can be considered to operate in a stagnant situation. Under these conditions, the Nusselt or Sherwood number for an isolated sphere in an infinite medium is given by:

$$Nu = Sh = 2$$
$$\text{i.e.:} \, h, h_D \propto 1/d$$

Clearly, in the immediate neighbourhood of a surface with a high curvature, the concentration/temperature field is very intense and this can be obtained with no power input. This approach is analogous to the pointed lightning conductor which relies upon sharp points to precipitate local electrical breakdown. It is directly relevant to the design of electrodes in electrochemistry where asperities can shoulder a disproportionate share of the current density.

Substantial energy cost savings across the whole of the process industries can result if the performance of heat exchangers is improved. One technique which is already used to improve heat exchanger performance (and indeed process heat transfer in general) is enhancement. In fact, heat transfer enhancement[1] is the most

---

[1] Sometimes called heat transfer augmentation or heat transfer intensification.

common way of increasing the heat transfer in equipment, typified by fins on most air-cooled equipment. Of particular relevance is the fact that enhancement, be it of heat and/or mass transfer, is critical to many other PI items of plant.

Enhancement is normally concerned with increasing the heat transfer coefficient on one or both sides of a heat exchanger. An enhanced surface, for example, one physically treated to increase turbulence, will yield higher heat transfer coefficients than a plain surface (Reay, 1991). The goal of enhancement techniques may be to reduce the size of the heat exchanger for a given duty; to increase the capacity of an existing heat exchanger; or to reduce the approach temperature difference. A combination of these goals is of course feasible (Bergles, 1999).

Enhancement of heat transfer can lead to an extra energy need. This may be due to an increased resistance to flow, giving a greater pressure drop. Some techniques do not incur an obvious penalty of this type while others, more specifically active enhancement methods, by definition, require an external energy input. Both are considered here.

3.2.1 Classification of enhancement techniques

Enhancement techniques may be conveniently divided into two classes – *passive* methods and *active* techniques. Extended surfaces which require no direct application of external power fall within the passive category. Rotation (as used in the Rotex heat pump initially mentioned in Chapter 1) is an active technique, as obviously energy is needed to rotate the device. Additionally, each technique may be applicable to one or more modes of heat transfer (e.g. forced convection, boiling, condensation, etc.) and compound enhancement – the use of more than one technique on a single heat exchanger – is practised.

Passive enhancement techniques include the following:

- Treated surfaces (coatings and promoters).
- Rough surfaces.
- Extended surfaces.
- Displaced enhancement devices.
- Swirl flow devices.
- Surface tension devices.
- Porous structures.
- Additives (for liquids and gases).
- Coiled tubes.
- Surface catalysis.

Active methods include:

- Mechanical aids.
- Surface vibration.
- Fluid vibration (including ultrasonics).
- Electrostatic fields.
- Other electrical methods.
- Suction or injection.

- Jet impingement.
- Rotation.
- Induced flow instabilities (e.g. pulses).
- Grooves and rivulets.

The demarcation line between techniques can be imprecise, particularly in the case of physical changes to heat transfer surfaces which form the basis of a variety of passive methods. The techniques may also be categorised in terms of the mode of heat transfer (e.g. single-phase convection, boiling, etc.) to which they are applied.

It should be noted that many of the passive techniques are used to make shell and tube heat exchangers more compact, and some are less well known in the compact heat exchanger or other unit operation area. Tube inserts are particularly useful in shell and tube heat exchangers for debottlenecking. A selection of the enhancement techniques most relevant to current PI technologies is discussed below.

Noteworthy in the intervening period between the two editions of this book are developments in manufacturing procedures that will have an increasing influence on what we can fabricate and how rapidly it can be made. Rapid prototyping has a new meaning now we can 3D-print equipment!

3.2.2 Passive enhancement techniques

3.2.2.1 Treated surfaces

These are surfaces which have a relatively fine scale alteration to the surface fins, or have a coating applied. The changes are generally used to promote enhancement of boiling or condensation. In the case of condensation, data (Cuthbertson et al., 1999) have shown that a power station condenser using tubes capable of sustaining drop-wise condensation could have double the overall heat transfer coefficient of the conventional unit, with positive benefits for unit volume. (Drop-wise condensation gives much higher heat transfer coefficients than film-wise condensation.) The use of this condensation mechanism may be, for example, a way of sustaining heat input to an endothermic reaction.

Relevant to intensified plant of many types are the implications of treated surfaces, in this case ion implantation methods, for fouling prevention. The catalyst expert may see opportunities for applying catalysts using some of these methods. The drawback of some of these coatings is the difficulty in guaranteeing a long life.

The Surfi-Sculpt process developed at TWI, Cambridge (see Appendix 4), is a recent development that has substantial implications for PI where surface modifications can intensify heat and/or mass transfer, as well as a range of other uses, highlighted in a paper devoted to the aerospace sector (Buxton and Dance, 2010). The process derives from electron beam welding with full or partial penetration of the material being processed. During the drilling process, a reduction in beam power allows surface texturing to take place, instead allowing the expulsion of material around the outer depression in the surface. If this is followed by the application of magnetic deflection coils re-entrant features can be made. Surfi-Sculpt gives greater feature control by allowing protrusions to be built and shaped as separate

FIGURE 3.2

Examples of the surface topography possible using Surfi-Sculpt, the process developed at TWI.
(Courtesy: TWI Ltd., Cambridge, UK)

operations, allowing a wide range of features to be formed on flat or curved surfaces. Even honeycomb structures can be built up on a plate. Typical sizes of the protrusions are hundreds of microns to millimetres, and the smallest holes that can be created using the beam are around a few hundred microns in diameter. Tens of microns may be possible in the future. Examples are shown in Figure 3.2.

With regard to applications, these include mixing; aero- and hydrodynamics; heat dissipation; and chemical reactions. One could use the features to, for example, change the location of transition from laminar to turbulent flow, or change the drag characteristics of a surface. Curved protrusions could be used to create swirl, and the holes could function as micro-injectors – even for chemicals – or to increase cooling. Surfi-Sculpt could also function as a mechanical interlock. Shape memory alloys could also be improved in their functionality. Many materials could be processed in this manner: metals, polymers, ceramics and glass are all feasible. The time to process $5\,cm^2$ of material is a few seconds, and the equipment needed includes an electron beam machine and a vacuum chamber.

The example illustrated in Figure 3.3, arising out of research at KTH in Stockholm (Furberg and Palm, 2007), is a surface modification caused by electrochemical deposition. As well as a porous surface with regular cavities, the structure forming each cavity has protruding features with characteristic dimensions of one micron or less. This results in enhanced heat transfer in pool and flow boiling, with improvements in heat transfer coefficient of 18 times at low ($1\,W/cm^2$) heat fluxes, to 7 times at higher heat fluxes of approximately $9\,W/cm^2$. The performance is illustrated in Figure 3.4 for varying surface characteristics in terms of cavity density (the higher the better).

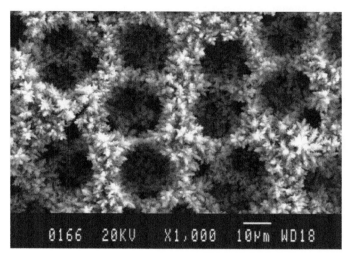

FIGURE 3.3

The surface finish caused by electrodeposition, used to enhance boiling.

(Courtesy: KTH, Stockholm)

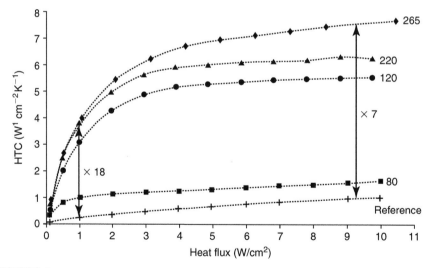

FIGURE 3.4

The heat transfer enhancement, above the reference surface, as a function of heat flux.

(Courtesy: KTH, Stockholm)

3.2.2.2 Extended surfaces

The fin is the most common form of extended surface, associated with both tubular and plate-type heat exchangers – the car radiator is a highly compact finned unit. Most heat exchangers used in gas streams have fins to improve the gas-side heat transfer. Compact units such as the plate-fin heat exchanger have fins, or secondary surfaces, between each pair of plates. Fins are used in both natural and forced convection. They are also used, (sometimes called ribs) to aid boiling or to provide drainage for condensate. The plate-fin heat exchanger (see Chapter 4) uses a variety of extended surfaces to enhance performance. The finned heat exchangers are often used as the basis of heat exchanger reactors (see Chapter 5) where there is frequently a need to use enhanced heat transfer to control, for example, an exotherm.

In the UK, a company called Porvair introduced a metal foam which can double as an extended surface (although it is marketed principally as a filter). One example of its use is in a gas–liquid heat exchanger, the tubes penetrating the bulk of the foam, which replace conventional extended surfaces. The tubes are sintered into the foam, giving good contact. The pore size of the foam is 0.1–5mm and it could be put inside tubes or between plates, configuring a gas–gas heat exchanger. The foam can be described as an *anisotropic three-dimensional extended surface*. Work was carried out at Brunel University (see Figure 3.5 and Lu et al., 2006) and the University of Warwick to characterise the heat transfer and pressure drop behaviour of the foam in tubular heat exchanger form. This could lead to designs for tube-in-foam heat exchangers of co/counter or cross-flow configuration. More recently, Mancin et al., (2011) carried out measurements on the heat transfer and pressure drop of 20ppi (pores per inch) aluminium foam utilising air as the fluid. See Chapter 4 for more information on compact heat exchangers made using metal foams and meshes.

Advantages of the foam include a choice of metals resistant to temperatures up to 1,000°C, low density, compact size and the ability to be formed into complex shapes. The foam can be catalysed, a feature that could be highly relevant to HEX-reactors. Extended surfaces and other features of compact/intensified heat exchangers are covered in full by Hesselgreaves (2001).

3.2.2.3 Swirl flow devices

These normally take the form of tube inserts which assist mixing and forced convection by creating rotating or secondary flows, or a combination of both. Twisted tapes, wire coils, coiled tubes and vortex generators are common forms. The principal application has been in forced convection in liquids, but the use in gases has been investigated. They have been examined for enhancing condensation, where up to 60% improvements in heat transfer were recorded, at a penalty in terms of pressure drop (Briggs et al., 1999).

Figure 3.6 shows other data for overall enhancement factors (taking into account both Nusselt number improvement and possible degradation in friction factor) for laminar flow in tubes using various twisted tape and wire coil tube inserts. The data are compared with that of a plain tube (Wang and Sunden, 2002). The authors found that tube insert technology was more effective in the laminar region than the turbulent region.

> ... *In the laminar region, the heat transfer enhancement ratio and the overall enhancement ratio can be up to 30 and 16, respectively. But, in the turbulent region, they can be only up to 3.5 and 2.0, respectively. This is probably*

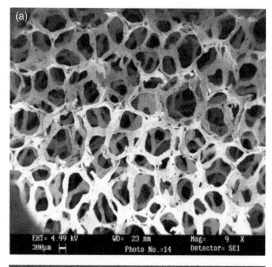

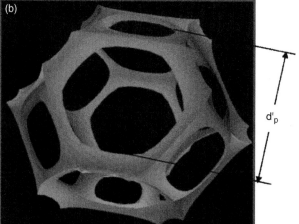

FIGURE 3.5

An example of a tube filled with metal foam, tested at Brunel University, further data given in Chapter 4 (Lu et al., 2006).

the reason that in practice, tube insert technology is often used in the laminar region, not in the turbulent region, where other enhancement technologies are common, such as ribs, low fins, etc.

These single-phase flow data represent only one application area of inserts. They are being used to enhance two-phase boiling and condensation and can reduce fouling.

Tube inserts, as their name suggests, are used in tubular heat exchangers, such as shell and tube or plate and fin heat exchangers. They can permit size reductions,

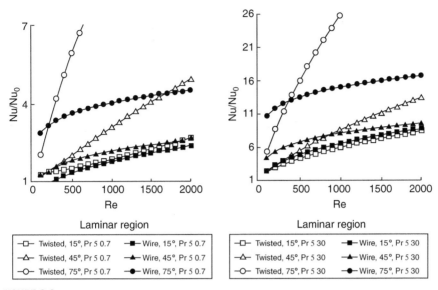

FIGURE 3.6

Comparison of the performance of tube inserts in the laminar flow region, Pr 5 0.7 (left-hand side) and 30 (right-hand side) (Wang and Sunden, 2002).

but are frequently used to improve the performance of an existing heat exchanger, e.g. to carry out debottlenecking. Tube inserts are also capable of reducing fouling in some fluids, such as heavy oils, and have recently been examined for minimising cooling water fouling. They have been studied for use in small diameter (3 mm) tubes, and have also been coated with catalysts to induce in-tube reactions – work at the University of Bath in the UK in this area was carried out some years ago. HiTRAN elements, used in this example to enhance mixing in a falling film evaporator, are illustrated in Figure 3.7 (Anon, 2012a).

3.2.2.4 Additives (for liquids and gases)

A variety of additives can be used to enhance heat transfer in both single- and two-phase flow. Convective heat transfer in gases can be aided by the addition of fine mists. In liquids, the addition of bubbles or solid particles can also be beneficial. There is evidence that the latter can reduce fouling – the 'Microfloss' concept, using polymer fibres introduced into a liquid where they create an abrasive effect at the heat exchanger wall, has been shown to reduce biofouling.

The use of nano-particles to enhance heat transfer is a phenomenon that has received a massive amount of attention since the first edition, although arguments remain as to how this is manifest, particularly in boiling heat transfer. Wen et al., (2009) suggest that a number of benefits arise from using nano-particles as solid additives, compared to other solids. However, they do point out that positives such

FIGURE 3.7

HiTRAN elements of Cal-Gavin Ltd, under investigation at the University of Manchester to enhance the operation of a falling film evaporator.

as an increase in liquid thermal conductivity may be countered by an increase in viscosity, or adverse changes in specific heat (for single-phase flow). For boiling applications, Wen presets data that in some cases show benefits and on others, a decrease in heat transfer coefficient. In conclusion, he rightly calls for 'more systematic and strategic work' to be carried out.

More recently, Chandrasekar et al., (2012), in a most comprehensive review of thermophysical properties of nano-fluids and their convective heat transfer characteristics, presented generally positive data, but did suggest as did Wen et al., that further examination is needed, in particular to be able to predict performances of such fluids.

Of course, nano-particles may need to be removed from the stream at some point, necessitating good filtration. In the applications chapters (8, 9 and 11), there is discussion of nano-fibres for the relatively new energy recovery process – energy scavanging.

3.2.2.5 Surface catalysis

Twenty-five years ago, surface catalysis would not have featured in a list of heat transfer enhancement methods. Now, heat exchanger reactors, frequently based on CHEs such as the plate-fin or printed circuit types, routinely have catalyst coatings on surfaces where heat transfer takes place. Applications range from fuel-cell reformers to chemical processes, and extend to the more complex surfaces of the metal foam in Figure 3.5. A catalyst may be applied to one or both sides of the heat exchanger; in the former, the cooling gas stream is used to remove heat from an exothermic reaction. (See Chapter 5 and applications Chapters 8, 9 and 11.). An example here, Janicke et al., (2000), shown in Figure 3.8, illustrates the

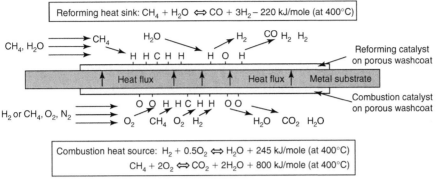

FIGURE 3.8

A combustion catalyst, used to take heat through a wall to a reforming catalyst in a heat exchanger-reactor.

mechanisms involved, including heat transfer from a combustion catalyst located in the lower gas stream.

A study of two different heat exchanger-reactor configurations, involving catalysts for both exo- and endothermic reactions is reported by Bayat et al., (2012). This is typical of the application of heat exchanger (or HEX-reactors). At the micro-scale, we have companies such as Velocys (Oxford Catalysts) at the forefront of HEX-reactor technology.

3.2.3 Active enhancement methods

3.2.3.1 Mechanical aids

Mechanical aids are normally applied to a liquid, often in a region close to the heat exchanger wall. Stirrers (e.g. the stirrer in a stirred pot) and mixers (not static mixers – which should strictly be regarded as passive enhancement devices) would be included, while scraped surface heat exchangers would also feature in the active category. (Rotation of the heat transfer surface, also a mechanical aid, is discussed separately below.)

3.2.3.2 Fluid vibration

It is much easier to excite the fluid than to create vibrations in heat exchangers themselves. A range of frequencies has been used to excite fluids, ranging from 100 Hz to ultrasound. Ultrasound has received attention recently because of its ability to enhance reactions and mixing. The potential for heat exchangers is currently less certain, although it could be used to minimise fouling. The sonic horn used to free large process regenerators of fouling will be familiar to heat transfer engineers with experience of such units. More recently, ultrasound has been used in an anti-fouling application for frost suppression on a finned tube evaporator, but the authors

(Wang et al., 2012) concluded that it would be better to combine this with a hydrophobic coating on the fins (not yet tested).

Instead of vibrating a surface directly, ultrasound can be used to create rapid movement in a fluid adjacent to a heat exchanger surface. The effects of ultrasound are commonly witnessed in ultrasonic cleaning baths, for removing contamination from the surfaces of metal objects, for example. Ultrasound, (as discussed later) is used to enhance reactions and mixing – see for example, Prosonix in Appendix 3.

With regard to heat transfer enhancement, ultrasound use has progressed considerably since the publication of the first edition of this book. Fundamental work at the University of Pisa in Italy on heat transfer enhancement in single-phase free convection and in saturated pool boiling has produced encouraging results (Baffigi and Bartoli, 2012). This examined the effect on distilled water around a heated cylinder and the authors correlated the effect of ultrasound on performance. They concluded that the benefits of ultrasound could be useful in the latest generation of electronic component cooling systems. The enhancement of the heat transfer coefficient achieved under specific single-phase conditions is shown in Figure 3.9. In France, a double-tube heat exchanger had its performance successfully enhanced using ultrasound at 35 Hz. The enhancement factor, in terms of improved overall heat transfer coefficient, was between 1.5 and 2.3 (Legay et al., 2012).

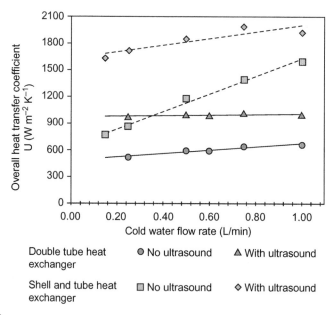

FIGURE 3.9

Overall heat transfer coefficient versus cold water flow rate: constant hot water flow rate $Q_h = 0.8\,\mathrm{L\,min^{-1}}$ (double-tube heat exchanger: $Re_h = 1986$; shell-and-tube heat exchanger: $Re_h = 5673$).

The main thrust of fluid vibration in recent years has been to enhance reactions – discussed in Section 3.4 in the context of electrical enhancement.

3.2.3.3 Electrostatic fields

Electrostatic fields, the application of which is sometimes known as electrohydrodynamics (EHD), are of growing interest in the inventory of tools for heat and mass transfer enhancement. Normally limited in their application to dielectric fluids, such as refrigerants and transformer oil, the high voltages (and low currents) associated with such fields are reflected in increased activity close to the heat transfer surface. The technique has been successfully used to enhance boiling and condensation. (See Allen and Karayiannis (1995) for an historical review.)

The way in which electrodes can be sited to give strong EHD fields close to the surface of a heat exchanger tube is an essential component in maximising the effect of the electric field. By suitable electrode choice and location, the engineer interested in intensifying a process may be able to 'map' field strength to give optimum benefits. Figure 3.10 illustrates the impact of EHD on the performance of a two-phase system involving a bundle of tubes with R123 as the working fluid (Karayiannis, 1998). The decrease in enhancement as the heat flux increases,

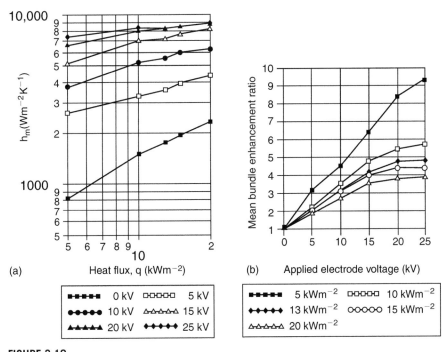

FIGURE 3.10

Data including heat transfer enhancement as a function of applied voltage, for a tube bundle with R123 as the fluid.

perhaps contrary to what might be expected, follows trends noted in other research in this area. Karayiannis points out that it is due to the attenuation of the effective local field strength caused by the increasing number of departed bubbles as the heat flux increases.

There are several other electrical methods for enhancing heat and mass transfer, some of which are finding application in intensified processes, (see also Section 3.5) where it can be seen that magnetohydrodynamics (MHD) is also seeing a revival of interest. Laohalertdecha et al., (2007) have reviewed the impact of EHD on boiling and condensation heat transfer.

3.2.3.4 Rotation

Rotation can be used to aid both heat and mass transfer. By passing a fluid across a rotating disc, increases in evaporation heat transfer coefficients can be achieved. Disc speeds are relatively low to implement this, being typically 500–1,000 rpm. The spinning disc reactor, which uses enhanced heat and mass transfer, is one of the most important PI developments of the last decade and is described in Chapter 5. Rotation in another form is used inside rotating heat pipes, enabling higher heat fluxes to be handled, as well as overcoming any difficulties in transporting liquid from the condenser to the evaporator. Several historical examples of rotation to aid heat (and in some cases mass) transfer are given in Chapter 1.

The earliest significant work on rotating discs as heat transfer devices was done initially at Newcastle University and then at Protensive Ltd, until recently a PI equipment manufacturer. The data given below from Protensive are important in showing how spinning discs can enhance heat transfer. The heat transfer aspect is associated with the Spinning Disc Reactor (SDR) in the discussion below, but standalone spinning disc heat exchangers are feasible.

Core to the heat transfer performance of an SDR is the characteristic of the (liquid) film as it moves across the disc. Waves tend to form in the film that significantly enhance heat transfer (as well as mass transfer), giving film coefficients of around 20–50 kW/m^2 K for low viscosity fluids. However, the process film coefficient is only part of the story; the film coefficient for the heating/cooling fluid and the thermal resistance of the disc itself are critical.

Before the company ceased trading, Protensive developed a disc that incorporates special channels for the service fluids, see Figure 3.11. This design gives enhanced heat transfer thanks to the fin effect of the channels and, when used with suitable heat transfer fluids at high flow rates, can give an effective overall film coefficient of around 15–20 kW/m^2 K. The disc is fabricated from copper because of its excellent thermal conductivity, with a thin layer of chrome plating for chemical resistance. Taking into account heat transfer on the other side of the disc, an SDR has an overall heat transfer coefficient of approximately 10 kW/m^2/K even for organic liquids. This is typically five to ten times that achieved by most heat transfer devices and enables small discs with low process fluid inventory to handle significant thermal duties. As an example, a water-like fluid such as milk could be heated on a disc from 25°C to 85°C using a temperature difference driving force of 20°C. Despite the high specific heat, a 1 m diameter disc could process approximately 2 t/h.

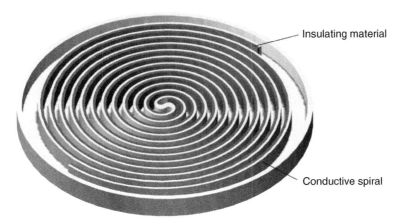

Insulating material

Conductive spiral

FIGURE 3.11

The structure of the spinning disc heat transfer surface. Overall heat transfer coefficient of $10\,kW/m^2\,K$ for low viscosity process fluids and water or glycol as heat/cool fluid. A 0.5 m diameter disc with delta T 20 K will transfer 39 kW, thanks to patented double spiral disc design, copper construction (with chrome plating for corrosion resistance).

The evaporation of an organic solvent is a common commercial process. For example, using an SDR to evaporate toluene, a 1 m diameter disc could evaporate approximately 1.5 t/h of toluene, again with a temperature different of 20°C. Although the use of a 20°C overall temperature difference is typical of industry, the very high process film heat transfer coefficient of an SDR means that the temperature difference between the process fluid and the disc surface is very low. For an overall temperature difference of 20°C the film to process temperature difference is only about 5°C. This can be very important when processing heat sensitive materials where low delta-Ts can reduce product degradation.

Domestic equipment should not be overlooked for examples of active enhancement. The domestic washing machine spins at high speed to enhance water removal, while the novel Dyson machine also agitates the load, to enhance contact between the clothes and cleaning medium.

Much of the work on rotation has shifted to China, in particular to Beijing University of Chemical Technology, (see for example Chen (2009)), although centres of excellence exist in Europe, the USA and the UK.

3.2.3.5 Induced flow instabilities

Otherwise called pulses, induced flow instabilities at relatively low frequencies (<5 Hz) have been shown to improve heat transfer and have a positive effect on fouling. An analogy can be seen in pulse combustion systems, where the effect is to break down steady-state conditions and to create instantaneous changes in the fluid velocity and direction. The effect on fouling has been likened to 'coughing'.

In an interesting application in catalysis, Eindhoven University of Technology in The Netherlands has studied pulses as a way of affecting the temperature of reactions, creating temperature pulses of the order of 500 K in under 20 µs (Stolte et al., 2011). They call this 'pulsed activation'.

3.2.4 System impact of enhancement/intensification

Heat transfer enhancement can have benefits beyond the heat exchanger itself, affecting the whole system of which the heat exchanger is a component. Three examples, relating to evaporation and condensation heat transfer enhancement, are given below. It will be noticed that the improvements brought about by heat transfer enhancement are the same as those attributable to compact heat exchangers.

It will also be recognised that many of the types of equipment benefiting from enhancement can, with a little more thought, be taken a stage further to become 'radically improved PI systems'. Intensification of more than just the heat transfer surfaces may then be considered.

In order to illustrate the systems benefit that can be achieved, it is helpful to classify the main types of equipment using two-phase heat exchangers as follows:

- Power-producing systems, such as those based on the Rankine cycle.
- Power-consuming systems, such as vapour compression refrigeration and vacuum distillation plant.
- Heat-actuated systems such as absorption refrigeration plant and conventional distillation operations.

For power-producing systems heat transfer enhancement can reduce the boiler and/or condenser surface for a given duty, i.e. for a specific turbine output. Alternatively, by adding enhancement to the existing boiler and condenser, for example using tube inserts, the turbine output might be increased – a form of debottlenecking – increasing the power output by relatively low cost improvements to one component that was limiting the system output.

In the case of power-consuming systems, three possible benefits can be realised:

- The heat transfer area can be reduced for a given compressor power.
- The evaporator duty can be increased for a given compressor lift.
- The compressor power can be reduced for a given evaporator duty (due to closer approach temperatures).

In heat-actuated systems, again three possible benefits exist:

- The heat transfer area can be reduced for fixed operating temperatures.
- The heat exchanger capacity can be increased, while keeping the surface area constant.
- The log mean temperature difference (LMTD) can be reduced for a given surface area, and in this case the thermodynamic efficiency of the process can be improved.

The above examples show that enhancement can benefit processes involving heat exchangers in many ways. Each route chosen has its own trade-offs, and the comparison of benefits can be complex. Nevertheless, their potential is considerable.

When one indulges in debottlenecking using one of the above enhancement methods, one is of course taking a first step towards process intensification. The 'active' methods, which require more sophisticated changes to the plant, are the ones which potentially offer the 'step change' improvements associated with the best PI techniques.

The selection of an appropriate enhancement technique must take into account the nature of the fluid stream(s) involved. For example, some enhancement methods, particularly those involving fine surface features, are susceptible to contaminants which could reduce the effectiveness. Heavily fouled streams, or those containing oil, such as a refrigerant circuit using an ineffective oil separator, would not be ideal candidates. Some incur a pressure drop penalty, and, as highlighted earlier, active methods invariably involve an extra power input.

The reader who has an interest in power generation plant efficiency may like to consider how drop-wise condensation on a large power station condenser might improve the electricity output, and by how much. Bear in mind that only one side of the condenser is being enhanced in this way.

3.3 **Intensified mass transfer – the mechanisms involved**

It is not surprising that some methods for heat transfer enhancement can be used to improve mass transfer. There are a number of ways of intensifying mass transfer, including:

- Rotation – in a cyclone or on a rotor.
- Vibration – high frequency ultrasound, for example.
- Mixing – the newer designs of in-line mixers are highly effective.

3.3.1 **Rotation**

A major mass transfer intensification opportunity centres upon the use of enhanced acceleration in a rotating system, either within a cyclone/vortex or a rotor. Since the fluid behaviour of all multi-phase systems is acutely dependent upon the buoyancy term $\Delta \rho g$, higher applied accelerations can produce thinner films, smaller bubbles/droplets and increase flooding velocities (for counter current systems). About two thirds of the unit operations in process engineering involve multiple phases and are therefore susceptible to this intensification approach. Vortex fluidic devices and HiGee rotating packed-bed developments are typical examples of this technology (see, e.g. Peel et al., 1998). The latter can give many counter-current stages in one unit while needing a rotor to create the field. The vortex units are simple co-current contactors, but need a pump or other source of pressure to drive them.

FIGURE 3.12

Enhanced bubble removal from molten glass using ultrasound.

An excellent review of rotation as a way of intensifying mass transfer is given by Zhao et al., (2010), based upon activities in China. Taiwan is also active in research in this area.

3.3.2 Vibration

While vibration is often created in the first instance by an electric field, it is effectively a mechanical phenomenon. One of the first applications of intensive vibration forces using a sonic horn was to remove fouling from the surfaces of massive boiler heat exchangers. On a laboratory scale, the ultrasonic cleaning bath is an intensive process. Any surface mass transfer enhancement need would be worthy of investigation with ultrasound effects.

As will be described in Chapter 10, ultrasound can be used to improve the quality of cast aluminium. The effect of ultrasound seen in Figure 3.12, in this case for glass refining, is to enhance bubble removal from glass melts, allowing it to take place in minutes rather than hours. (Again, see Chapter 10 for further data and a fuller explanation. In Figure 3.12, the ultrasound probe is visible top-centre, penetrating into the molten glass. This locally encourages the bubbles to agglomerate and rise to the top of the molten glass bed.

There is considerable interest in ultrasonic enhancement of reactions, as introduced in Section 3.2.3.2. This is discussed further in Chapter 5.

3.3.3 Mixing

Depending upon the viscosity of the components being mixed, the mechanisms used in mixers are similar to those for heat transfer when fluid paths are disturbed to create turbulence. The design of many types of mixers has improved in leaps and bounds recently, and mixing combined with reactions, together with 'induction-heated mixers', are of particular interest to PI engineers (see below and Chapter 5).

In the context of mixing, whether it is associated with heat exchangers or reactors (or other unit operations), the first approach is to reduce the mixing path length. In the context of the tubular heat exchanger, the inserts described above do just this. For reactors, instead of using a large pot with a mixing path length equivalent to its radius or thereabouts, we should bring thin films or fine jets into contact – another nail in the coffin for the conventional pot.

3.4 **Electrically enhanced processes – the mechanisms**

Electrical enhancement has been with us for many years, and although the tranche of mechanisms involved have not in the past been recognised by all as process intensification, most are highly effective in this role and increasingly are making appearances in the PI literature across a wide range of intensified unit operations. There are nine or more electrical processes which could be used as intensification tools. All of these are available, and several have already been used in practical applications. The more common are listed in Table 3.1.

The advantage of dielectric heating arises from the fact that heat is generated directly within the work material itself. Many of the common barriers to heat transfer by 'external' heat sources are broken down. Thus, it is much more rapid than, say, external firing with hot gases, and is excellent for processing material of low thermal conductivity. Critical features of the material include its dielectric constant and dissipation factor, and materials with polar molecules are those most attractive to dielectric heating (see data in Guyer and Brownell, 1989).

The use of microwaves is well-recognised in the food processing sector and in other process industries (see Chapter 10). There have been attempts to use microwaves in catalytic micro-reactors, following on from success in using this energy form in organic synthesis, but results to date have been mixed. The authors of a research paper (Cecilia et al., 2007) have suggested that modifying the properties of composite material in which the palladium catalyst was loaded could lead to a positive outcome. Some positive outcomes were reported by Patil et al., (2011) in the context of microwaves used in a continuous flow heat exchanger-reactor. Desorption is a thermally demanding process in many applications, and work in Poland on microwave irradiation to intensify desorption suggests that microwave-assisted desorption is better applied to fluidised bed adsorbers than fixed bed units (Cherbanski and Molga, 2009). This paper also has a useful review of applications within it.

Induction heating is resistance heating, the alternating current generating electric resistance losses in the metal conductor, hence heating the work piece located between the coils as seen in Figure 3.13 (Chen et al., 2007). The induction heating bonding system consisted of a high-frequency power supply operating at 400 kHz (left hand side) with an output power of 1 kW, an induction coil made from copper tubing

Table 3.1 Common Electric Enhancement Methods.

Method	Frequency (where applicable)	Applications (selection)
Dielectric heating		
Radio frequency (r.f.)	<100 MHz	Plastics
Microwave	>500 MHz	Food, organic synthesis reactors, cracking, pyrolysis (Metaxas and Meredith, 1983)
Induction heating		Heating metals, e.g., reactor body, paddle
Plasma heating		Calcining, producing metal spheres
Ultrasound	16 kHz to 100 MHz	Glass processing, metallurgy; (see Chapter 10), adsorption (Ji et al., 2006), reactions, including wastewater treatment.
Electromagnetic irradiation		Catalytic reactions
Electrostatics	e.g. 40 kV applied voltage	e.g. enzymatic hydrolysis, liquid–liquid contacting in general
Electrohydrodynamics		
Photocatalysis (light via electricity or solar energy)	UV wavelengths	Enhancement of reactions in several reactor types
Plasma chemical reactions		Methane upgrading (Kim et al., 2007) Hydrocarbon reforming (Lindner and Besser, 2012)
Lasers	e.g. infra-red, 3×10^{13} Hz	Reactions

Note: Sound and light can be available from natural sources (e.g. solar energy in the latter case), but for the purposes of the categorisation here it is assumed that electricity is used to provide the energy input.

and cooled with water (right-hand side), a quartz rod, thermocouple, two graphite blocks and the bonding pair, as shown in Figure 3.13. The graphite blocks were held in the centre of the three-turn helical coils. Pre-cleaned silicon and glass pieces were sandwiched between the graphite blocks which were connected to the high-voltage power supply, ensuring a positive electrode potential on the silicon side with respect to the glass. The quartz rod was used to press the bonding pair through the graphite blocks and the thermocouple was used to measure the process temperature.

When the power is applied, the magnetic field created in the coil traversed perpendicular to the surface of the graphite and bonding pair, so the graphite blocks are heated by induction heating. This results in heating the bonding pair, via conduction,

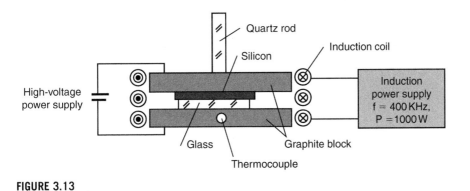

FIGURE 3.13

The concept of induction heating, in this case used for anodic bonding, that can be used to heat a reactor.

to a predetermined temperature. Next, a high-voltage direct current is applied to the graphite electrodes, just as in conventional silicon–glass anodic bonding. The rapid heating possible using electric fields can help to achieve catalyst light-off in combustion systems, as well as helping to sustain other catalytic reactions.

Ultrasound and lower frequency sound are becoming routine in PI applications – see also Section 3.2.3.2. Gomaa et al., (2004) cite several electrochemical processes where sonic vibration of the electrodes or the fluid adjacent to them can enhance mass transfer. Electroplating and bio-leaching are mentioned, as well as controlled electro-organic synthesis. Gomaa and colleagues found that electrode vibration could increase mass transfer by up to 23 times the static value. Ultrasound is used in a range of sonochemical reactors (see Gogate et al., 2004) for phenol destruction. See Chakinala et al., (2007) for hydroxyl radical production.

Similarly, photocatalysis is growing in importance as a PI tool, having been demonstrated on a variety of reactors, including the spinning disc reactor. The review by Van Gerven et al., (2007) is a useful starting point for anyone interested in this enhancement method. One example is the micro-reactor for photo-catalytic reactions constructed by Takei et al., (2005). The unit was composed of two Pyrex glass substrates (0.7 mm thick). To get stable flow and sufficient amounts of reaction products, branched micro-channels were fabricated by photolithography (wet etching) techniques. The micro-channels were 770 µm wide and 3.5 µm deep. A TiO_2 thin film was prepared on another substrate with a sol–gel method using titanium tetra n-butoxide ethanol solution as a starting material. The two substrates were thermally bonded at 650°C for four hours. Cross-sectional scanning electron microscope images of the channel showed that the film was 300 nm in thickness and was composed of approximately 100 nm diameter TiO_2 particles. Tests revealed the successful integration of photoactive TiO_2 into the micro-channel system.

Illustrated in Figure 3.14, under operation a high-pressure mercury lamp was used to irradiate the photo-sensitive titanium dioxide thin film in the reactor, loaded

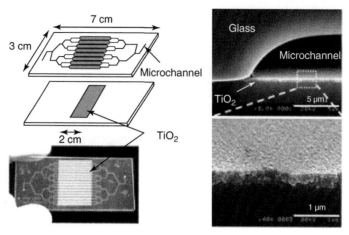

FIGURE 3.14

TiO$_2$ modified micro-channel chip for photocatalysis reaction, and cross-sectional images of the channel (Takei et al., 2005).

with platinum catalyst. Conversion of the feedstock at 87% was achieved in 52 seconds, compared to 60 minutes for the non-photocatalytic reaction. For a broader review that includes electromagnetic fields in the area of heterogeneous catalytic reactions, see Toukoniitty et al., (2005).

Plasma torches produce the highest industrial temperatures, locally 4,000°C in the gas around the arc column. Niche applications in reactors, ceramics and metals processing exist. The laser, which itself is becoming increasingly 'intensified' with small solid state models in consumer electronic equipment, has yet to find an important role in PI in the context of the process industries. However, as hinted at below, the potential for exploitation of laser energy, which can be created at a variety of wavelengths, certainly exists. With beam power densities in excess of 10^5 W/cm^2 and the ability to pulse at high frequency, as well as deliver the energy continuously, it merits more attention. The patent literature reveals Russian activity in optically activated chemical reactions (Anon, 2003). Here irradiation of chemicals with the energy supplied via fibre optics and concentrated in a single vessel allows enhancement of chemical reactions. Work is also reported on a laser plasma chemical vapour deposition diamond reactor, see Figure 3.15 and Bolshakov et al., (2001). More recently, a micro-plasma reactor has been successfully operated in the USA for hydrocarbon reforming (Lindner and Besser, 2012).

Here the beam of a 2.5 kW multimode CO$_2$ laser passes through a sealed window into the reaction chamber and is focused by a lens near the nozzle exit where a subsonic gas stream was formed. The gas mixture consisted of the plasma forming gas (argon or xenon) and feed gases (in most experiments H$_2$/CH$_4$), and could be prepared either before or after the nozzle. In the latter approach, H$_2$ and CH$_4$ could be convectively admixed in the volume between the plasma and tungsten or

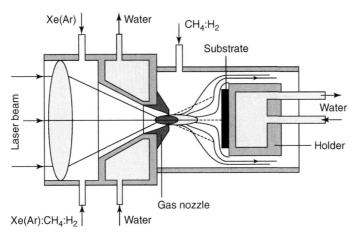

FIGURE 3.15

Layout of a laser plasma chemical vapour deposition (CVD) reactor, the gases being introduced to deliver reactants to the substrate surface where plasma was maintained using a CO_2 laser.

molybdenum substrates fixed on water-cooled substrates. For plasma ignition the high voltage electrical discharge or thin tungsten wire were used. Substrate temperature was monitored by an optical pyrometer and/or thermocouple. The US Army Missile Command has examined lasers for inducing the reaction of nitrogen dioxide with gases such as propane and butane (Stanley and Godbey, 1989).

Two specific uses of electric fields, for heat transfer and extractions (e.g. biological), are discussed in Allen and Karayiannis (1999), and by Weatherley (1993), respectively. A review of EHD was made by Laohalertdecha et al., (2007), while Weatherley and Rooney (2007) have discussed the use of electrostatics for processing natural oils (sunflower oil).

- Electrohydrodynamics – an enhancement method for boiling/condensation and single phase (corona wind) effects (see also magnetohydrodynamics below).
- Electrically-enhanced extractions – charged droplets, electrostatic spraying, etc.

In the example shown in Figure 3.16, EHD is used to assist the draining of condensate from between fins on a heat exchanger tube (Butrymowicz et al., 2003). The main feature is the arrangement of the tube–electrode system. In this case the electric field is generated between the rod-type electrode and the lower part of the finned tube. This method involves a DC potential. Due to the condensate flow induced by the EHD force, a significant area of the tube surface remains unflooded, thus making favourable conditions for condensate heat transfer. Therefore, the proper application of the EHD condensate drainage shell leads to an increase of the flooding angle.

The advantages of electricity are its flexibility, ease of control, point-of-use delivery and the fact that it can add value – giving better quality and faster product throughput. The processes include microwave-assisted firing; ceramic firing assisted

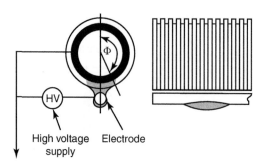

FIGURE 3.16

EHD-enhancement to remove condensate trapped between exchanger fins.

by microwaves overcomes temperature differentials and allows quicker firing and reduced HF emissions. This is generally twice as fast as a conventional oven, but 3–4 times faster in some cases. Other electric processes include magnetic bed reactors, venturi aeration, enhanced membranes and ohmic heating. One example is the magnetic bed reactor – an enhanced fluid-bed unit, for uses in, e.g. the biotechnology sector. Magnetic particles excited in a fluid bed can enhance throughput, (by up to 4 in one trial) and give better yields. Electric processes could allow faster, smaller, higher output per unit of energy, and lower inventory processes to be designed and operated.

Some electrical phenomena arise 'naturally', for example electrokinetics. These can be artificially induced by applying an external electric field. Such mechanisms have come to attention through the area of micro-fluidics – flow at the micro-scale – and are briefly discussed in Section 3.5.

A more esoteric area, but one which modern laser technology is in sight of allowing considerable progress to be made, is that of laser/plasma reaction systems, briefly discussed above. Specifically, infra-red multi-photon dissociation (IRMPD) allows the targeting of radicals, e.g. CH_3, within, for example, methane. This can assist combustion or upgrading to higher added value chemicals – a future challenge for PI! For a discussion of IRMPD in the more common context of mass spectrometers, and how it works, see Newsome and Glish (2009).

Of increasing interest in a wide variety of applications is the practice of 'harvesting' energy (mechanical or thermal). The energy can then be used for powering micro-devices (using current technology), and possibly larger systems in the future. As an example, Wang (2012) has reviewed piezoelectric nano-generators that can take energy from 'ambient' mechanical resources – even respiration motion in the human nostril!

3.5 Micro fluidics

Also discussed later in the context of micro-electronic thermal control in Chapter 11, and expanded upon when we start discussing heat exchangers with very small channels Chapter 4 (Section 4.4), micro-fluidics involves a number of interesting enhancement phenomena that will increasingly play a role in PI. Three are introduced below.

The reader is also advised to consult a short review (Poulikakos, 2009) on micro- and nano-scale thermofluidics, based upon work being carried out at ETH Zürich.

3.5.1 Electrokinetics

Electrokinetics is the name given to the electrical phenomena that accompany the relative movement of a liquid and a solid. These phenomena are ascribed to the presence of a potential difference at the interface between any two phases where movements occur. Thus, if the potential is supposed to result from the existence of electrically-charged layers of opposite sign at the interface, then the application of an electric field must result in the displacement of one layer with respect to the other. If the solid phase is fixed while the liquid is free to move, as in a porous material, the liquid will tend to flow through the pores as a consequence of the applied field. This movement is known as electro-osmosis.

Electro-osmosis using an alternating current (ac) has been used as the basis of micro-fluidic pumps, allowing normal batteries to be used at modest applied voltages. This helps to integrate such a pumping method with portable 'lab-on-a-chip' devices. Workers in Zürich (Cahill et al., 2005) have shown that a velocity of up to 100 µm/s could be achieved for applied potentials of less than 1 V_{rms}. The impact of the chemical state of the surfaces of the channel on electro-osmotic flow has also been investigated. Recently, the pumps have been investigated in the context of nano-channels, as well as micro ones, (Geri et al., 2011). However, of significance to reactions and other unit operations carried out in micro-fluidic devices is the ability to enhance mixing (Wu and Liu, 2005). Wu and Liu found that it was relatively easy to create complex flow fields in a straight micro-channel without the need for complex micro-fabrication procedures or actuators (internal or external). The software package CFD-ACE+ was used to simulate the effect of the herringbone-shaped electrodes deposited on the channel wall. The mixing is fully implemented within an axial distance of about 3 mm, and experiments showed over 90% mixing within 5 mm.

In nature, electrokinetics is used, in the form of electro-osmosis, by earthworms to allow them to move over soil. The flow within a micro-thin liquid layer near the worm's body surface is induced by what is known as the electric double layer (EDL) interaction. This is essentially electro-osmosis at the micro-scale providing lubrication between its body and moist soil, thus reducing surface adhesion (Yan et al., 2007). The reader will find similar research using the term 'biomimetics' in his or her search engine. Yan and colleagues (Li and Yan, 2011) has transferred this technology to chemical engineering by examining how the phenomenon might be applied in solid desiccant dehumidification.

3.5.2 Magnetohydrodynamics (MHD)

For those willing to contemplate more complex systems, the use of MHD (or possibly EHD – electrohydrodynamics – see Jones, 1973) can be used for fluid propulsion,

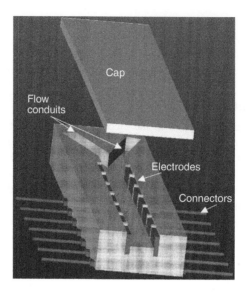

FIGURE 3.17

The Y-shaped micro-channels lead in to a section with electrodes on both walls. The gap between adjacent electrodes is 0.5 mm (Qian and Bau, 2005).

mixing and separations. Qian and Bau (2005) have tested an MHD stirrer, illustrated in Figure 3.17. When a potential difference (PD) is applied across one or more pairs of electrodes, the current that results interacts with the magnetic field to induce Lorentz forces and fluid motion. The alternating application of the PD results in chaotic advection and mixing, but the authors point out that the system needs perfecting.

During construction, a polycarbonate template in the shape of the stirrer cavity was milled and positioned on a glass substrate. The closed stirrer's template consisted of a rectangular slab (L × W × H = 85 mm × 8 mm × 2 mm) while the flow-through stirrer's template was shaped like a Y (L × W × H = 85 mm × 4 mm × 2 mm). The leading edge of the first electrode was 10 mm downstream from the straight conduit's entrance (the point where the two legs of the Y connect with the third). Two plastic tubes (1.75 mm o.d. and 1.2 mm i.d.) were connected to the two legs of the Y. A third tube was connected to the chamber's exit. Finally, a polymeric solution was cast around the template. After curing the solution, the template and frame were removed, leaving behind a cavity with patterned electrodes along its sidewalls. The cavity was capped with a glass slide as shown in Figure 3.17. The device was positioned on top of a neodymium permanent magnet that provided a nearly uniform intensity magnetic field of approximately 0.4 T.

Ibanez and Cuevas (2010) have examined MHD pumps at the micro-scale, in particular highlighting the irreversibilities in a micro-channel due to friction, heat flow and the presence of the electric field. By assessing the entropy generation, they are able to optimise the micro-fluidic pumping system and its associated irreversibilities.

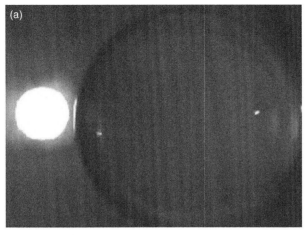

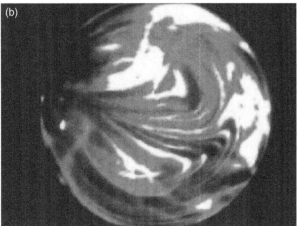

FIGURE 3.18

(a) Top and (b) lower showing the effect of a light beam on mixing (Garnier et al., 2005).

3.5.3 Opto-micro-fluidics

Research at the Georgia Institute of Technology (Garnier et al., 2003) is directed at using a light beam as an energy source for liquid manipulation. As shown in Figure 3.18(a) and (b), a dye injected into a larger droplet (the dyed droplet being the white spot in Figure 3.18(a)) is mixed in the bulk liquid drop by, as with MHD, chaotic advection, as shown in the lower figure. In other experiments, the modulated light field can be used to drive liquids and droplets using thermocapillary forces thus generated. (Of course, it may be ultimately shown that thermocapillary driving forces could be dominant in a system where previously conventional capillary forces had, perhaps to the detriment of performance, governed flow characteristics.) Even optical tweezers are proposed in the patent literature for micro-manipulation.

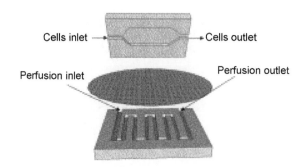

FIGURE 3.19

Schematic of a micro-fluidic bioreactor. (Top) culture chamber, (bottom) perfusion chamber and (middle) the sandwiched PC membrane (Shah et al., 2011).

For those wishing to follow up the numerous techniques for influencing capillary forces, the paper by Le Berre et al., (2005) on generating electrocapillary forces in micro-fluidic elements is worthy of study. At the small scale where particles might be involved in the fluid stream, the impact of electrophoresis and thermophoresis – the motion of a particle induced by a temperature gradient (from a hot to a cold region) should not be neglected. For the aficionado of particle motion, the Magnus effect (forces created by an object rotating in a fluid stream) may well have an influence at the micro-scale. Of course, this assumes that the object will fit into the appropriate passage.

It is in the area of biological systems that micro-systems, including reactors and analysis units, are receiving a large amount of attention, as in the example of the earthworm earlier in this chapter. Micro-bioreactors and similar small-scale units are described by Jensen (2007), who has reported on research at the Massachusetts Institute of Technology in this area. More recently the literature has revealed interest in these reactors for cell culture, as an example. Illustrated in Figure 3.19, the cell culture micro-bioreactor (Shah et al., 2011) has channels of 30–50 microns micro-milled in polymer.

3.6 Pressure

Supercritical processing using CO_2 is discussed elsewhere in this book but it is, for the sake of completeness, useful to include pressure as an enhancement procedure. Those working in the food processing industries will be most familiar with high pressure processing, but the use of high pressure reactors for chemical processing has been discussed, see for example Elliott and Sealock, (2005). The range of pressures used in high pressure processing, according to Virginia Tech in the US, is 100–1,000 MPa (Anon, 2007), and in the area of food it can accelerate the destruction of micro-organisms, speed up cooking (the pressure cooker, of course) and increase the extraction rate of desirable compounds. Pressure can also increase bubble collapse in cavitation (Toukoniitty et al., 2005).

It will be evident from the discussions in this chapter that the mechanisms involved in intensification may involve an additional energy input. In building up the 'case for support' the project manager will, of course, include the additional energy inputs in the cost equations. In most instances, the extra cost of the energy input will be much less significant than the financial rewards gained by investment in the intensified process and/or plant.

3.7 Summary

The basis of most intensified unit operations is related to an enhancement mechanism that improved heat and/or mass transfer. The most important methods of enhancement, as far as separations and reactions are concerned, are those that require an external energy input – active methods. More recently, reductions in scale as typified by micro- and nano-systems have opened up many new opportunities for intensification at scales previously not considered, and it is here that electrokinetics has a major role to play.

References

Allen, P.H.G., Karayiannis, T.G., 1995. Electrohydrodynamic enhancement of heat transfer and fluid flow. Heat Recovery Syst. CHP 15 (5), 389–423.

Anon, 2003. Method and apparatus for optically activating chemical reactions. Patent RU2210022, published 8 October. Source: European patent database.

Anon, 2007. Virginia Tech high pressure processing web site. <www.hpp.vt.edu>.

Anon, 2012a. Company web site of Cal-Gavin Ltd., <www.calgavin.com> (accessed July 2012).

Anon, 2012b. Company web site of Laminova AB., <www.laminova.se> (accessed July 2012).

Baffigi, F., Bartoli, C., 2012. Influence of the ultrasounds on the heat transfer in single phase free convection and in saturated pool boiling. Exp. Therm. Fluid Sci. 36, 12–21.

Bayat, M., Rahimpour, M.R., Taheri, M., Pashaei, M., Sharifzadeh, S., 2012. A comparative study of two different configurations for exothermic–endothermic heat exchanger-reactor. Chem. Eng. Process.: Process Intensification 52, 63–73.

Bergles, A.E., 1999. Enhanced heat transfer: endless frontier, or mature and routine. Paper C565/082/99. Proceedings of the UK National Heat Transfer Conference, Heriot-Watt University. IMechE, Bury St Edmunds, September.

Bolshakov, A.P., Konov, V.I., Prokhorov, A.M., Uglov, S.A., Dausinger, F., 2001. Laser plasma CVD diamond reactor. Diamond Related Materials 10, 1559–1564.

Briggs, A., Kelemenis, C., Rose, J.W., 1999. Augmentation of in-tube condensation in the presence of non-condensing gas using wire inserts. Paper C565/065/99. Proceedings of the UK National Heat Transfer Conference, Heriot-Watt University. IMechE, Bury St Edmunds, September.

Butrymowicz, D., Trela, M., Karwacki, J., 2003. Enhancement of condensation heat transfer by means of passive and active condensate drainage techniques. Int. J. Refrig. 26, 473–484.

Buxton, A.L., Dance, B.G.I., 2010. The potential of EB surface processing within the aerospace industry. Proceedings on International Conference on Power Beam Processing Technologies (ICPBPT2010), Beijing, 25–29 October, 2010.

Cahill, B.P., Heyderman, L.J., Gobrecht, J., Stemmer, A., 2005. Electro-osmotic pumping on application of phase-shifter signals to interdigitated electrodes. Sens. Actuators B: Chemical 110, 157–163. (September).

Cecilia, R., Kunz, U., Turek, T., 2007. Possibilities of process intensification using microwaves applied to catalytic microreactors. Chem. Eng. Process. 46, 870–881.

Chakinala, A.G., Gogate, P.R., Burgess, A.E., Bremner, D.H., 2007. Intensification of hydroxyl radical production in sonochemical reactors. Ultrason. Sonochem. 14, 509–514.

Chandrasekar, M., Suresh, S., Senthilkumar, T., 2012. Mechanisms proposed through experimental investigations on thermophysical properties and forced convective heat transfer characteristics of various nanofluids – a review. Renewable Sustainable Energy Rev. 16, 3917–3938.

Chen, J.-F., 2009. The recent developments in the HiGee technology. Presentation to the GPE-EPIC Conference, Venice, Italy, June 14–17, 2009.

Chen, M., Yuan, L., Liu, S., 2007. Research on low temperature anodic bonding using induction heating. Sens. Actuators A 133, 266–269.

Cherbanski, R., Molga, E., 2009. Intensification of desorption processes by use of microwaves – an overview of possible applications and industrial perspectives. Chem. Eng. Process. 48, 48–58.

Cross, W.T., Ramshaw, C., 1986. Process intensification: laminar flow heat transfer. Chem. Eng. Res. Dev. 64, 293–301. (July).

Cuthbertson, G., McNeil, D.A., Burnside, B., 1999. Dropwise condensation of steam over a bundle of tubes at utility turbine condenser pressure. Paper C565/051/99. Proceedings of the UK National Heat Transfer Conference, Heriot-Watt University. IMechE, Bury St Edmunds, September.

Elliott, D.C., Sealock, L.J. Jr., 2005. Chemical processing in high-pressure aqueous environments. Proceedings of the first International Conference on Science, Engineering and Technology of Intensive Processing, University of Nottingham, 18–20, September.

Furberg, R., Palm, B., 2007. A novel porous surface for enhancing heat transfer in pool and flow boiling. Proceedings of the Annex 33 Heat Pump Meeting, Stockholm. See <http://www.heatpumpcentre.org/en/projects/completedprojects/annex%2033/Sidor/default.aspx>.

Garnier, N., Grigoriev, R.O., Schatz, M.F., 2003. Optical manipulation of microscale fluid flow. Phys. Rev. Lett. 91 (Paper 054501).

Geri, M., Lorenzini, M., Morini, G.L., 2011. Proceeding of the Third Micro and Nano Flows Conference, Thessaloniki, Greece, 22–24, August.

Gogate, P.R., Mujumdar, S., Thampi, J., Wilhelm, A.M., Pandit, A.B., 2004. Destruction of phenol using sonochemical reactors: scale up aspects and comparison of novel configuration with conventional reactors. Sep. Purif. Technol. 34, 25–34.

Gomaa, H., Al Taweel, A.M., Landau, J., 2004. Mass transfer enhancement at vibrating electrodes. Chem. Eng. J. 97, 141–149.

Guyer, E.C., Brownell, D.L. (Eds.), 1989. Handbook of Applied Thermal Design McGraw-Hill, New York.

Hesselgreaves, J.E., 2001. Compact Heat Exchangers – Selection, Design and Operation. Pergamon, Oxford.

Ibanez, G., Cuevas, S., 2010. Entropy generation minimisation of a MHD (magnetohydrody-namic) flow in a microchannel. Energy 35, 4149–4155.

Janicke, M.T., Kestenbaum, H., Hagendorf, U., Schüth, F., Maximilian Fichtner, Schubert, K., 2000. The controlled oxidation of hydrogen from an explosive mixture of gases using a microstructured reactor/heat exchanger and Pt/Al_2O_3 Catalyst. J. Catal. 191, 282–293. (April).

Jensen, K.F., 2007. Chemical and biological microsystems for discovery and scaling to production. Proceedings of the European Congress of Chemical Engineering (ECCE-6), Copenhagen, pp. 16–20, September.

Ji, J.-B., Lu, X.-H., Xu, Z.-C., 2006. Effect of ultrasound on adsorption of geniposide on polymeric resin. Ultrason. Sonochem. 13, 463–470.

Jones, T.B., 1973. Electrohydrodynamic heat pipe. Int. J. Heat Mass Transfer 16, 1045–1048.

Karayiannis, T.G., 1998. EHD boiling heat transfer enhancement of R123 and R11 on a tube bundle. Appl. Therm. Eng. 18 (9–10), 809–817. (September).

Kim, S.-S., Lee, H., Choi, J.-W., Na, B.-K., Song, H.K., 2007. Methane conversion to higher hydrocarbons in a dielectric-barrier discharge reactor with $Pt/\gamma-Al_2O_3$ catalyst. Catal. Commun. 8, 1438–1442.

Laohalertdecha, S., Naphon, P., Wongwises, S., 2007. A review of electrohydrodynamic enhancement of heat transfer. Renewable Sustainable Energy Rev. 11, 858–876.

Le Berre, M., Chen, Y., Crozatier, C., Zhang, Z.L., 2005. Electrocapillary force actuation of microfluidic elements. Microelectron. Eng. 78–79, 93–99.

Legay, M., Le Person, S., Gondrexon, N., Boldo, P., Bontemps, A., 2012. Performances of two heat exchangers assisted by ultrasound. Appl. Therm. Eng. 37, 60–66.

Li, B., Yan, Y.Y., 2011. Solid desiccant dehumidification techniques inspired from natural electro-osmosis phenomena. J. Bionic. Eng. 8, 90–97.

Lindner, P.J., Besser, R.S., 2012. A microplasma reactor for chemical process intensification. Chem. Eng. Technol. 35 (7), 1249–1256.

Lu, W., Zhao, C.Y., Tassou, S.A., 2006. Thermal analysis on metal-foam filled heat exchangers. Part I: metal-foam filled pipes. Int. J. Heat Mass Transfer 49 (15–16), 2751–2761. (July).

Mancin, S., Zilio, C., Rossetto, L., Cavallini, A., 2011. Foam height effects on heat transfer performance of 20 ppi aluminium foam. Appl. Therm. Eng. doi: 10.1016/j.applthermaleng.2011.05.015.

Metaxas, A.C., Meredith, R.J., 1983. Industrial Microwave Heating. Peter Peregrinus Ltd. For: Institution of Electrical Engineers, Power Engineering Series 4, Stevenage.

Newsome, G.A., Glish, G.L., 2009. Improving IRMPD in a quadrupole ion trap. J. Am. Soc. Mass Spectrom. 20, 1127–1131.

Patil, N.-G., Hermans, A.I.G., Rebrov, E.V., Meuldijk, J., Hulshof, L.A., Hessel, V., et al., Optimisation of energy use for flow processing under microwave heating by using integrated reactor-heat exchanger. IChemE Symposium Series No. 157, EPIC 2011, IChemE, Rugby.

Peel, J., Howarth, C.R., Ramshaw, C., 1998. Process intensification: HiGee seawater deaeration. Chem. Eng. Res. Des. 76 (A5), 585–593. (July).

Poulikakos, D., 2009. On emerging micro- and nanoscale thermofluidic technologies. Proceedings of the Second Micro and Nano Flows Conference, West London, UK, 1–2, September 2009.

Qian, S., Bau, H.H., 2005. Magneto-hydrodynamic stirrer for stationary and moving fluids. Sens. Actuators B 106, 859–870.

Reay, D.A., 1991. Heat transfer enhancement – a review of techniques and their possible impact on energy efficiency in the UK. Heat Recovery Syst. CHP 11 (1), 1–40.

Shah, P., Vedarethinam, I., Kwasny, D., Andresen, L., Dimaki, M., Skov, S., et al., 2011. Microfluidic bioreactors for culture of non-adherent cells. Sens. Actuators B: Chemical 156, 1002–1008.

Stanley, A.E., Godbey, S.E., 1989. The laser-induced nitrations of several hydrocarbons. Appl. Spectrosc. 43, 674–681. (May–June).

Stolte, J., Ozkan, L., Backx, A.C.P.M., 2011. Pulsed activation in heterogeneous catalysis. IChemE Symposium Series No. 157, EPIC 2011, Manchester, UK. IChemE, Rugby.

Takei, G., Kitamori, T., Kim, H.B., 2005. Photocatalytic redox-combined synthesis of L-pipecolinic acid with a titania-modified microchannel chip. Catal. Commun. 6, 357–360.

Toukoniitty, B., Mikkola, J.-P., Murzin, D.Y., Salmi, T., 2005. Utilisation of electromagnetic and acoustic irradiation in enhancing heterogeneous catalytic reactions. Appl. Catal. A: General 279, 1–22.

Van Gerven, T., Mul, G., Moulijn, J., Stankiewicz, A., 2007. A review of intensification of photocatalytic processes. Chem. Eng. Process. 46, 781–789.

Wang, L., Sunden, B., 2002. Performance comparison of some tube inserts. Int. Comm. Heat Mass Transfer 29 (1), 45–56.

Wang, D., Tao, T., Xu, G., Luo, A., Kang, S., 2012. Experimental study on frosting suppression for a finned-tube evaporator using ultrasonic vibration. Exp. Therm. Fluid Sci. 36, 1–11.

Wang, X., 2012. Piezoelectric nanogenerators – harvesting ambient mechanical energy at the nanometer scale. Nano Energy 1, 13–24.

Weatherley, L.R., 1993. Electrically enhanced mass transfer. Heat Recovery Syst. CHP 13 (6), 515–537.

Weatherley, L.R., Rooney, D., 2007. Enzymatic catalysis and electrostatic process intensification for processing of natural oils. Chem. Eng. J. (In Press).

Webb, R.L., Nae-Hyun, K., 2005. Principles of Enhanced Heat Transfer, second ed. Taylor & Francis, New York.

Wen, D., Lin, G., Vafaei, S., Zhang, K., 2009. Review of nanofluids for heat transfer applications. Particuology 7, 141–150.

Wu, H.-S., Liu, C.-H., 2005. A novel electrokinetic micromixer. Sens. Actuators A 118, 107–115.

Yan, Y.Y., Zu, Y.Q., Ren, L.Q., Li, J.Q., 2007. Numerical modeling of electro-osmotically driven flow within the micro-thin liquid layer near an earthworm surface – a biomimetic approach. Proc. IMechE 221 (Part C: J. Mechanical Engineering Science).

Zhao, H., Shao, L., Chen, J.-F., 2010. High-gravity process intensification technology and application. Chem. Eng. J. 156, 588–593.

Compact and Micro-heat Exchangers

OBJECTIVES IN THIS CHAPTER

The objectives of this chapter are to enlighten the reader with the current status of compact and micro-sized heat exchangers. Heat exchangers of these types were among the first process sector unit operations to be intensified, and many chemical reactors are based upon their structural features. A good knowledge of available heat exchanger concepts is a useful precursor to the introduction of more complex and combined unit operations based upon such compact heat exchangers. Out with the process sector, the challenges to heat transfer in the aerospace and electronics areas have also greatly encouraged the development of such heat exchangers.

4.1 Introduction

In this chapter, intensified heat exchangers, ranging from compact to so-called micro-heat exchangers are described, the discrimination between the two being largely a feature of flow channel diameter and/or width. Some of the heat exchangers described use one or more of the enhancement methods discussed in Chapter 3. The use of metal foams as heat exchangers and the enhancement of heat transfer by rotation are two examples that are included in this chapter. Compact heat exchangers (CHEs) and micro-heat exchangers can also be used as the basis of reactors (see Chapter 5) and enhancement methods may be employed in reactors as, for example, additional support structures for catalysts, or for fluid mixing. Towards the end of this chapter, the use of chemical reactions as a means of improving or enhancing heat transfer is briefly examined. A type of surface that forms the basis of some current CHEs, and that offers good mixing qualities, is shown in Figure 4.1. This design forms the basis of developments such as the Chart and Velocys shims in small heat exchangers and heat exchanger reactors.

The degree of intensification of compact heat exchangers as defined in terms of their area densities (see below) can be quite modest, as in the case of the plate heat exchanger, for example. However, even this simple and well-established CHE has seen its transformation into a heat exchanger reactor by Alfa Laval (see Chapter 5).

The last few years have seen a major growth in activities involving micro-fluidics – fluid flow at the micro scale (already introduced in Chapter 3). Instead of involving passages of a few millimetres in diameter, micro-fluidic systems can include features where flow passages are a few micro-metres, or microns (μm), in

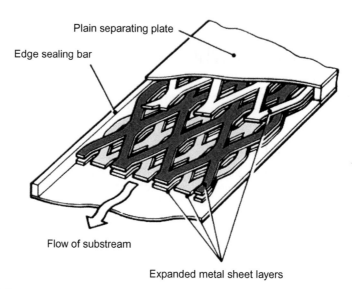

Plain separating plate

Edge sealing bar

Flow of substream

Expanded metal sheet layers

FIGURE 4.1

The highly compact heat exchanger surface invented by John Hesselgreaves at NEL (Anon, 1992a).

diameter. The design of heat exchangers, and other devices, which employ such small passages can require new approaches and a section at the end of this chapter introduces aspects of micro-fluidics relevant to micro-heat exchanger design.

As mentioned previously, it is not just design that can be affected by the scale and form of heat exchangers – 3D printing and/or rapid prototyping methods will revolutionise the way we construct unit operations, including heat exchangers, in the future, making integration of intensified units, we can confidently say, easier. Figure 4.2 (a) and (b) shows views of a heat exchanger made using 3D printing (additive manufacturing) by Within (Anon, 2012b).

In the context of integration, it is interesting to note that the widely applied process integration software – incorporating heat exchanger networks (HENs) is starting to incorporate intensified heat exchangers/heat transfer techniques. Appropriately led by the Centre for Process Integration at Manchester University (see Appendix 4), recent work has highlighted the retrofitting of enhancement techniques to HENs, incorporating many large shell-and-tube heat exchangers (Pan et al., 2012).

The number of studies on using nano-fluids (fluids – commonly liquids – dosed with nano-sized particles) has increased over the last few years. – see for example Anon (2012a). Section 4.3 will also touch upon this aspect, particularly in the context of micro-heat exchangers[1]. Nano-heat exchangers as such have yet to make their mark, except perhaps in biological systems!

---

[1]Note that the reader may be confused by the terminology in some papers. Microchannels can be used in compact heat exchangers, whereas in this chapter, most specifically in Section 4.3, a micro-heat exchanger is substantially smaller than a compact heat exchanger, in overall size and in surface/volume ratio.

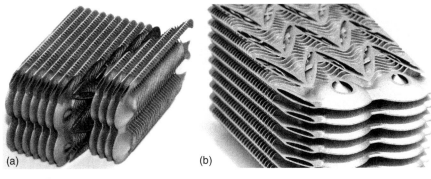

FIGURE 4.2

Views of a heat exchanger made using 3D printing.

(Courtesy Within – Anon 2012b)

4.2 Compact heat exchangers

Compact heat exchangers, while accounting for perhaps 10% of the worldwide market for heat exchangers, have in recent years seen their sales increase by about 10% per annum, compared to 1% for all heat exchangers. The majority of automotive heat exchangers are 'compacts' and they are being used increasingly where fluid inventory is an important factor, such as in refrigeration and heat pumping equipment using flammable working fluids. Aerospace and electronics are also major uses, where small size and low weight are important. The chemical process industries were relatively slow to adopt these units because of concerns about fouling, except in areas such as cryogenics, where the aluminium plate-fin heat exchanger is difficult to better for flexibility. An example is shown in Figure 4.3. However, now hundreds of examples of units such as printed circuit heat exchangers abound offshore, on gas processing duties, and in other uses where fouling can be avoided or controlled (Anon, 1992b).

There are a variety of compact heat exchangers which can be used in the process industries and elsewhere, listed in Table 4.1. While many are of interest to designers and users of intensified plant, four types in particular should be highlighted:

- The plate heat exchangers of Alfa Laval (and a range of other manufacturers).
- The printed circuit heat exchanger (PCHE) of Heatric.
- The Chart-flo unit of Chart Heat Exchangers.
- The polymer film heat exchanger (less common at present).

These can also form the basis of heat exchanger reactors. The laminar flow printed circuit heat exchanger of ICI (Cross and Ramshaw, 1986) is also discussed, as a parallel but independent development of PCHE-type structures. The degree of compactness of these heat exchangers – given in the table as 'compactness' or 'area density' and illustrated in Figure 4.4 – is indicative of the amount of heat transfer surface that can be included within the unit volume of the heat exchanger, hence its dimensions

FIGURE 4.3

A plate-fin heat exchanger – note the multi-stream capability offered only by a few other exchanger types (Anon, 1992).

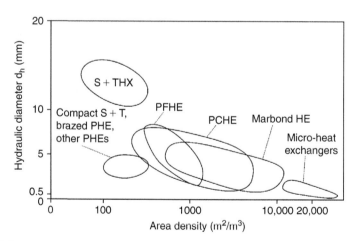

FIGURE 4.4

The area densities of a range of compact heat exchangers.

Table 4.1 Main Parameters of a Range of Compact Heat Exchangers.

Type of Heat Exchanger Features	Plate and Frame (Gaskets)	Fully Welded Plate (Alfarex)	Brazed Plate	Spiral	Brazed Plate-fin	Diffusion-bonded Plate-fin	Printed Circuit	Polymer (e.g. Channel-plate)	Chart-flo
Compactness (or area density) (m²/m³)	200	200	200	®200	800–1500	700–800	200–5000	450	®10 000
Stream types (1)	liq–liq gas–liq 2-phase	liq–liq gas–liq 2-phase	liq–liq 2-phase	liq–liq 2-phase	gases liquids 2-phase	gases liquids 2-phase	gases liquids 2-phase	gas–liq (8)	gases liquids 2-phase
Materials (2)	s/s, Ti, Incoloy Hastelloy graphite polymer	s/s Ti Ni alloys	s/s	c/s, s/s, Ti, Incoloy Hastelloy	Al, s/s Ni alloy	Ti s/s	s/s, Ni, Ni alloys Ti	PVDF (11) PP (12)	S/s, Ni, Ni alloys, Ti
Temperature range (°C)	−35 to +200	−50 to +350	−195 to +220	400	Cryogenic to +650	550	−200 to +900	150 (9)	−200 to +900
Max pressure (bar) (3)	25	40	30	25	90	>200	>600	6	>400
Cleaning methods	Mech (10)	Chem	Chem	Mech (10)	Chem	Chem	Chem	Water wash	Chem
Corrosion resistance	Good (4)	Excellent	Good (5)	Good	Good	Excellent	Excellent	Excellent	Excellent
Multi-stream capability	Yes (6)	No	No	No	Yes	Yes	Yes	No	Yes
Multi-pass capability	Yes	Yes	No (7)	No	Yes	Yes	Yes	Not usually	Yes

Liq = liquid; Mech = mechanical; Chem = chemical. Notes for Table 4.1 1. Two-phase includes boiling and condensing duties. 2. Other special alloys are frequently available. 3. The maximum pressure capability is unlikely to occur at the higher operating temperatures, and assumes no pressure/stress-related corrosion. 4. Function of gasket as well as plate material. 5. Function of braze as well as plate material. 6. Not common. 7. Not in a single unit. 8. Condensing on gas side. 9. PEEK (polyetheretherketone) can go to 250°C continuous use temperature. 10. Can be dismantled. 11. Polyvinylidene diflouride. 12. Polypropylene.

FIGURE 4.5

Regenerative burner.

of m²/m³. A perhaps more meaningful comparison, all other things being equal, is the volumetric heat transfer coefficient. Cross and Ramshaw quote, for their laminar flow matrix, a volumetric heat transfer coefficient of $7\,MW/m^3K$, compared to only $0.2\,MW/m^3K$ for a shell and tube heat exchanger – still the workhorse of the chemical process industries. Hesselgreaves (2001) provides one of the most comprehensive compilations of compact heat exchanger design information.

Although relatively rare in the chemical process industries, compact regenerators challenge many other types of CHE in terms of surface area/unit volume. This is particularly the case for rotating regenerators, but for regenerative heat exchangers as used in combustion systems – the static regenerator, is compact and very efficient – see an example in Figure 4.5. Examples are also found in cryogenics and Stirling engines. The reader is advised to consult Wnek (2012) for further data on static regenerators in efficient combustion systems.

4.2.1 The plate heat exchanger

Plate heat exchangers (PHEs) are the most common form of compact heat exchanger, but their area densities are inferior to other types discussed in this chapter. There are several forms of construction, mainly associated with the methods used to seal the edges of the plates – gaskets, braze, welding or laser welding. The PHE is widely used in the liquid food processing industry, where the use of stainless steel in its fabrication and the ability to clean effectively are great advantages. The effectiveness of these units can be as high as 95% in terms of the amount of heat recoverable from one stream that can be transferred to an adjacent stream.

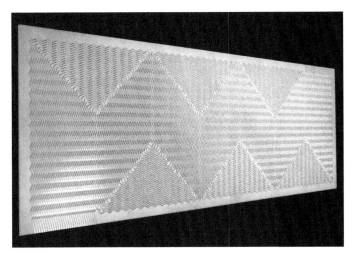

FIGURE 4.6

The chemically etched plate of a PCHE.

Also, the PHE has been developed by Alfa Laval and a number of other companies for use in refrigeration and heat pumping duties, where close approach temperatures aid cycle efficiency and the small volume allows lower fluid inventories of potentially toxic or flammable working fluids – a strong feature of PI plant.

Within common definitions of PI, in terms of size reduction, the plate heat exchanger is a relatively modest step in this direction. Units such as the PCHE have significantly greater amounts of surface within a given volume. Nevertheless, the PHE is a feature of several intensified plants, including the Rotex/Rotartica heat pump/chiller described in Chapter 11, and it is making a major contribution to reducing the inventory of refrigerants that are greenhouse gases in refrigeration systems worldwide. Alfa Laval has now extended the PHE, with different internal geometries to aid mixing, to act as a heat exchanger reactor – see Chapter 5.

4.2.2 Printed circuit heat exchangers (PCHE)

The printed circuit heat exchanger (PCHE) derives its name from the procedure used to manufacture the plates that form the core of the heat exchanger; the fluid flow passages are produced by chemical milling, a technique similar to that used to manufacture printed circuit boards in the electronics industry. One benefit of this is the flexibility it affords in terms of flow passage geometry. Channel depths are typically in the range 0.5–2.0 mm, larger for some streams. The most common variant of the PCHE is that manufactured by Heatric Ltd in the UK. Their typical assembly procedure involves diffusion bonding the stack of plates to form a compact, strong, all-metal heat exchanger core. Finally, fluid headers and nozzles are welded to the cores. A PCHE plate is illustrated in Figure 4.6.

FIGURE 4.7

Gas treatment and export coolers for the Kvitebjorn platform in the Norwegian North Sea.

(Photograph courtesy of Heatric)

Heatric states that

Diffusion-bonded heat exchangers are highly compact, highly robust exchangers that are well established in the upstream hydrocarbon processing, petrochemical and refining industries. With customers such as BP, Shell and ExxonMobil, we have supplied hundreds of heat transfer products to projects all over the world and continue to develop heat exchange solutions for a wide range of new market sectors.

Figure 4.7 shows a selection of Heatric PCHEs destined for a platform in the North Sea. There are at least two different plate designs for each block, one for each fluid stream. The channels are optimised for counterflow exchange or to encourage flow disturbances, (the zig-zag pattern shown in Figure 4.6). Fluid manifolds are normally in the form of half cylinders. Where large capacities are required, several diffusion-bonded blocks are welded together. In small capacity units the fluid headers can be created by the plates themselves, ports being etched through the thickness of the plates prior to bonding.

The PCHE is capable of being used at temperatures within the range −200°C to 900°C. Pressure differentials in excess of 600 bar can be accommodated. The design is particularly versatile in terms of the number of passes and number of streams which can be built into a PCHE block, and the type can be used for a wide variety of duties involving single-phase liquids and gases or two-phase streams. PCHEs are typically four to six times smaller and lighter than shell and tube heat

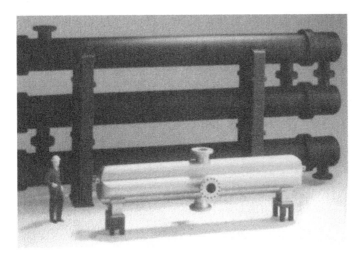

FIGURE 4.8

Comparison of the size of a PCHE gas–gas heat exchanger with its shell and tube equivalents.

exchangers of equivalent performance. This is illustrated for a gas–gas exchanger in Figure 4.8.

The PCHE is characterised by the flexibility of design and the high strength offered by the techniques used in construction. Until about ten years ago this gave it a unique place in the process industry heat exchanger field. Now, however, other manufacturers are challenging the PCHE features pioneered by Heatric, based on research in Japan, Germany, the Republic of Korea and the UK. For example, work at the Nuclear Reactor Research Laboratory in Tokyo (Tsuzuki et al., 2007, 2009) has resulted in a new flow channel configuration, for use in supercritical CO_2 cycle applications – for example residential air–water heat pumps. Illustrated, together with the conventional herringbone type pattern, in Figure 4.9, the use of the discontinuous S-shaped fins has resulted in a pressure drop only 20% of that of the conventional unit. Nusselt number correlations have been obtained and agree well with experimental data obtained using the unit in Figure 4.10. Interestingly, recent work by Heatric on supercritical CO_2 Brayton cycles using the PCHE suggests that the Heatric design is superior to the Japanese equivalent in terms of friction factor and Nusselt number – see Le Pierres et al. (2011). The conditions are somewhat more demanding than those in the heat pump application, with turbine inlet temperatures well in excess of 500°C. A PCHE was constructed in the Korea Advanced Institute of Science and Technology (Baek et al., 2012) for cryogenic applications. The problem of axial heat conduction can affect the performance of such units at low Reynolds numbers, and the researchers were able to modify the PCHE design to reduce conduction.

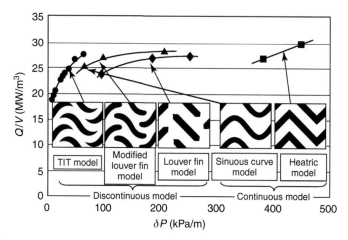

FIGURE 4.9

Comparison of the conventional and discontinuous models of the two surface configurations on printed circuit heat exchangers (Tsuzuki et al., 2007). Q/V is the heat transfer rate per unit volume.

FIGURE 4.10

Geometry of rounded Tsuzuki model, the radius of roundness is 0.2 mm (Tsuzuki et al., 2009).

The very large numbers of installations of the Heatric PCHE on offshore platforms has ably demonstrated the PI benefits which result from their use – namely the opportunities given to innovative offshore operators to cut costs through improved safety and greatly reduced topside platform size and weight. This is visible in specific areas such as compressor after coolers, gas coolers, gas dehydration trains and cryogenic processes for the removal of inerts from streams. Southall and

Dewson (2010) discuss design and application options, including their use in the nuclear sector (see Chapter 10).

Additional functions can be included in the exchanger design, such as chemical reaction, mass transfer, and mixing, optimising the process considerably. In fact, the Heatric PCHE is also marketed as a reactor – the printed circuit reactor (PCR) (see Chapter 5).

4.2.3 The Chart-flo heat exchanger

The Chart-flo heat exchanger produced some eight years ago and its current stablemates in the heat exchanger reactor field (see Chapter 5) are among the more recent truly innovative designs to enter the CHE marketplace. Produced by Chart Energy and Chemicals, the Chart-flo unit extends the option for those who are looking for high integrity, highly compact units able to operate over a range of pressures and temperatures not met with more conventional gasketed or welded CHEs. The manufacturing procedures are similar to those of the PCHE (chemical etching and diffusion bonding – although brazing is possible – see below), but the construction allows the use of small passageways, which significantly increases the porosity (free volume fraction) of the heat exchanger core. This can result, in appropriate applications, in a substantially higher area density than the PCHE. For example, a doubling of porosity, other factors being equal, results in a halving of the volume for a given surface area. An expanded view of a Chart-flo unit is shown in Figure 4.11, while some plates – the shims – are illustrated in Figure 4.12.

The heat exchanger can be fabricated in any material which can be diffusion bonded – stainless steel, titanium, higher alloy steels and nickel. As an alternative,

FIGURE 4.11

The Chart-flo compact heat exchanger – in this variant, larger passages are provided to take possible fouling into account.

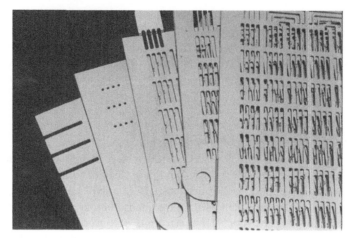

FIGURE 4.12

Plates used in the Chart-flo heat exchanger.

brazing can be used, which allows metals such as aluminium to be considered. This potentially could lead to capital cost reductions compared to exchangers fabricated using more bespoke procedures.

A particular feature of the Chart-flo is that it was developed at the outset for applications which could encompass reactions, as well as pure heat transfer duties, in its Chart-pak form, amongst others (see Chapter 5). As a heat exchanger reactor, the scope of application becomes very wide, including continuous chemical reactions, fuel cells and other reforming applications. The compactness (see below) implies that it could be an integral part of many intensified processes, and a study of its implications for reducing the size of absorption cycle refrigeration plant has illustrated the benefits of multi-stream, multi-pass and multi-functional use within a single module.

As mentioned earlier, the high porosity of the Chart-flo unit, together with the ability to use low hydraulic diameters, gives it advantages over several other CHEs in terms of compactness. For clean stream duties, the volume of a Chart-flo heat exchanger could be as low as 5% of that of the equivalent shell and tube heat exchanger.

An interesting aspect of compact heat exchanger structures such as the Chart-flo, some Heatric variants, and materials like metal foams (see Section 4.2.5), is that they may have a dual role – a heat transfer and a structural one. The nature of some compact heat exchanger surfaces, when formed into thin multiple plate assemblies, is that they are light in weight but strong and/or rigid. This allows heat transfer functions to be integrated directly with a structural member, such as the skin of an aircraft. Of course the availability of 3D printing as a way of fabricating heat exchangers opens up the multi-functional role even wider!

FIGURE 4.13

Polymer film cross-flow heat exchanger plates.

4.2.4 **Polymer film heat exchanger**

There are several polymer heat exchangers on the market, including shell and tube units with mm diameter polymer tubes, and low cost plate types for domestic applications. However, the introduction of polymers which can be resistant to temperatures well in excess of those we commonly associate with materials, such as PVC, and new joining methods, such as laser welding, has allowed their use to be extended. Polymers, such as PEEK (polyetheretherketone), which can be used continuously at temperatures in excess of 200°C, and which are also available in films of 100 micron thickness, permit CHEs of novel design to be fabricated. Such thin film units avoid the heat transfer disadvantages of using polymers where wall thicknesses of 0.5– 1.0 mm lead to high resistances. As the materials have excellent chemical stability, they can be used in potentially corrosive streams, but polymers lose strength with increasing temperature, so in general, pressure resistance is limited.

Work at Newcastle University in the UK (Zaheed and Jachuck 2004; 2005) led to the development of cross-flow and spiral configurations of PEEK CHEs. The cross-flow unit has been studied for gas–gas and liquid–liquid duties, while the spiral configuration was directed at gas–liquid uses. Areas of application include the chemicals industry, as well as domestic heating appliances (e.g. condensing boilers) and aircraft. A photograph of the plates used in the heat exchanger is shown in Figure 4.13. The greatest potential for the spiral form of PEEK heat exchanger is in applications where low gas-side pressure drop is essential, whilst maintaining high surface area densities. As well as PEEK, other polymers such as PVDF and PVC may be used and there are also applications in intensified membrane technology. T'Joen et al., (2009) suggest that heating, ventilating and air conditioning are application areas for such polymer heat exchangers (see Chapter 11), but interestingly, point out that nano-fibres can be used to increase the strength of the polymer plate. PEEK has been investigated for multiple effect distillation for seawater desalination, where corrosion resistance is demanded. To meet the thermal abilities of metallic heat transfer surfaces in falling film units, the PEEK had to be no more than around 25 μm thick, and mechanical tests revealed that this gave adequate strength (Christmann et al., 2012).

Table 4.2 lists the principal application areas of the more common compact heat exchanger types. For a full review of CHEs see Hesselgreaves (2001).

Table 4.2 Application Areas of Compact Heat Exchangers.

Heat Exchanger Type	Sector and Application

Plate and frame heat exchanger	Chemicals and petrochemicals
	Food and drink
	Paper and board
	Textiles and fabric care
	Oil and gas processing
	Prime movers
	Generic:
	   Refrigeration
	   Air compressors
	   MVR
	   Hazardous stream separation

Brazed plate heat exchanger	Chemicals and petrochemicals
	Food and drink
	Oil and gas processing
	Prime movers
	Generic applications:
	   Air compressors
	   MVR
	   Refrigeration
	   Hazardous stream separation

Welded plate heat exchangers	Chemicals and petrochemicals Food and drink Oil and gas processing
Spiral heat exchanger	Chemicals and petrochemicals Paper and board Effluent treatment Generic application: Hazardous stream separation

Plate-fin heat exchanger	Chemicals and petrochemicals
	Cryogenics
	Oil and gas processing
	Prime movers
	Generic applications:
	   Refrigeration
	   Hazardous stream separation

Printed circuit heat exchanger	Chemicals and petrochemicals
	Cryogenics
	Oil and gas processing
	Nuclear reactors
	Generic applications:
	   Refrigeration
	   Hazardous stream separation

Table 4.2 Continued

Heat Exchanger Type	Sector and Application
Chart-flo heat exchangers Heat exchangers Chart-pak heat exchanger/reactors	Chemicals and petrochemicals (1, 2) Prime movers (1, 2) Process intensification (1, 2) Absorption refrigeration (1)
Polymer film heat exchanger	Chemicals and petrochemicals Prime movers Aerospace
Compact shell and tube heat exchangers	Chemicals and petrochemicals Prime movers
Compact types retaining a 'shell'	Chemicals and petrochemicals Food and drink

The heat exchangers discussed above, with the possible exception of the polymer film unit, tend to rely upon relatively sophisticated methods of assembly – for example diffusion bonding and vacuum brazing. It is interesting to note that the use of adhesive bonding, which is likely to be highly cost-effective for uses at modest temperatures (<200°C) has benefits in terms of improving heat exchanger heat transfer performance (Paulraj et al., 2012). Working on a micro-channel heat exchanger, the team at Oregon State University in the USA found that adhesive bonding allowed a greater area to be available between plates than when diffusion bonding was employed. The plates were photo-chemically machined (PCM) to obtain surface features in all cases. In the case of PCM/diffusion bonded units, the active area was 0.00594 m^2 while an optimised PCM/adhesive bonded unit had an active micro-channel surface area of 0.00777 m^2, an increase of 33%. Higher pressure drops were noted in the latter case, attributed to manifold design that the researchers believed could be overcome in subsequent designs.

4.2.5 Foam heat exchangers

Rigid foam has benefits for heat transfer enhancement (see also Chapter 3, Section 3.2.2.2) and it is also used as the basis of heat exchangers. One example of its use is in a gas–liquid heat exchanger, the tubes penetrating the bulk of the foam which replaced conventional extended surfaces. The tubes are sintered into the foam, giving good contact. Foam pore size is 0.1–5 mm. Foam can be put inside tubes or between plates, configuring a gas–gas heat exchanger. The foam can be described as a *three-dimensional extended surface*. Work has been underway at Brunel University in the

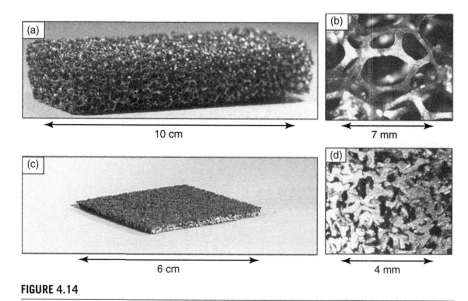

FIGURE 4.14

An open-cell aluminium foam. This can be compressed to form smaller pore structures for CHEs, as was done at ETH (c and d).

UK (Zhao et al., 2006) to characterise the heat transfer and pressure drop behaviour of the foam, involving some mathematical modelling. This could lead to designs for tube-in-foam heat exchangers of co/counter or cross-flow configuration – although fin-foam contact may inhibit performance (see below). Other research at ETH, Zürich has investigated the use of metal foams as compact heat exchangers (Boomsma et al., 2003).

Advantages of the foam include a choice of metals to 1,000°C, low weight, compaction and the ability to be formed in complex shapes. Some potential applications include gas turbine tail gas heat exchangers, hydrogen liquefaction plant and partial oxidation of hydrocarbons to hydrogen. The foam can be catalysed, of relevance to HEX-reactors. Within the area of heat exchangers, the work at ETH was based on aluminium foam of the type shown in Figure 4.14.

The work at Brunel on tubes containing foams is illustrated in Figures 4.15 and 4.16 (Zhao et al., 2006). It can be seen that the overall heat transfer coefficient of the metal foam filled heat exchanger is significantly higher than that of conventional finned tube heat exchangers. For example, the overall heat transfer coefficient of a 10 ppi (pores per inch) copper foam-filled heat exchanger is shown to be more than double that of a finned tube heat exchanger (spiral fins 5000 fins/m, height 1 mm, thickness 0.075 mm and inner tube grooves width 0.1 mm, height 1 mm). Therefore, it is clear that the use of metal foams can greatly enhance the heat transfer and have significant potential in the manufacture of compact heat exchangers. This was borne out by recent work in China – in collaboration with Carrier Corporation, where flow boiling of refrigerant-oil mixtures in tubes filled

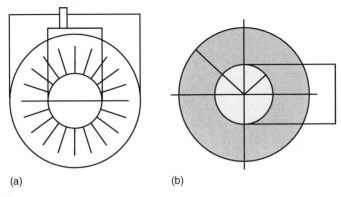

(a)   (b)

FIGURE 4.15

The two geometries tested at Brunel, a tube-in-tube exchanger with internal fins, and one of identical size with the fins replaced by foam.

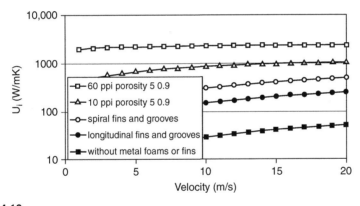

FIGURE 4.16

Performance data comparing the Brunel University foam and finned units. The superiority of the foam is evident.

with 10 ppi metal foam showed heat transfer coefficient enhances reaching a maximum of 185% of the non-foam case (Zhu et al., 2012).

Nevertheless, care must be taken in using foams in, for example, finned heat exchangers, on the air side. Sertkaya et al., (2012) report that using aluminium foam (10, 20 and 30 ppi), the advantages over a finned heat exchanger were less evident. The effectiveness of the finned unit was about twice that of the foam-based unit. This was attributed to the point contact available with the foam/tube junction, while of course with fins the contact is line or surface – significantly better. In such heat exchangers, air-side fouling and pressure drop (as well as cleanability) may well favour the conventional solution. For a numerical study on foam-wrapped tubes, see also Odabaee and Hooman (2011).

ERG Aerospace in the USA produces foams for heat exchangers in both metallic and non-metallic materials (Anon, 2011).

Other applications include aircraft wing structures for the aerospace industry, core structures for high strength panels, and containment matrices and combustors. ETH has stated that due to the high surface area density and strong mixing capability for the fluid, open cell metal foams are now regarded as one of the most promising materials for the manufacture of efficient compact heat exchangers. Again, it is worth highlighting the dual role of heat transfer and structural members that materials such as foam can satisfy.

4.2.6 Mesh heat exchangers

Structure is mentioned specifically in research reported at Cambridge University (Tian et al., 2007). Here, metal meshes (woven as textiles) have been tested as heat exchangers. Although not a new concept, the mesh heat exchanger is an alternative to metal foams, and as readers who are familiar with heat pipe technology will know, woven meshes and twills are available in a wide range of materials and pore sizes, and they can be stacked and brazed, where necessary.

The unit illustrated in Figure 4.17 may be configured as a cross-flow heat exchanger. In the case illustrated the cooling flow is flowing horizontally through the mesh, removing heat from devices convecting to the lower surface face sheet. Pressure may be applied (as it is a load-bearing structure). The measurements carried out by the research team are shown in Figure 4.18, where data are compared

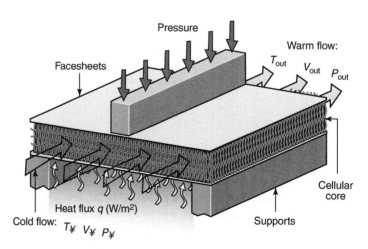

FIGURE 4.17

The mesh unit where, in sandwich construction with textile technology: (a) a transient liquid phase joins the wire mesh screen laminated at all points of contact; (b) facesheets are added to the textile core.

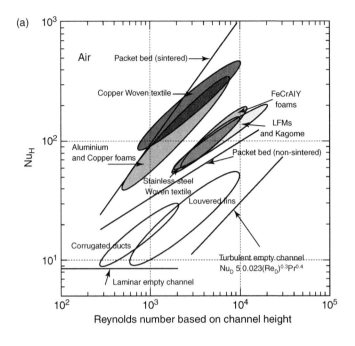

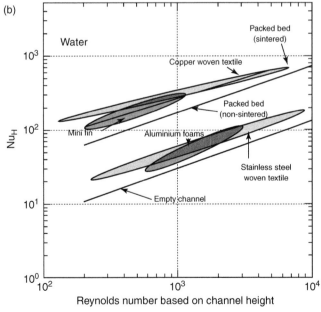

FIGURE 4.18

Comparison of the mesh textile unit with other porous heat exchangers, such as sinters and metal foams. The Nusselt number is base upon channel height (as for Re).

with other porous heat exchanger types. Note that an LFM is a lattice-frame material. The mesh is comparable in terms of heat transfer to the foam.

Pressure drop and heat transfer are strong functions of the path length through the mesh (or foam). Probably the mesh is less reproducible than foam structures, once several mesh layers are stacked, but there may be advantages in accessibility of material and flexibility.

More recent work on modelling the heat transfer in heat exchangers with wire mesh in the channels suggests that the thermal gain outweighs the pressure drop penalty in the case of single-phase gas flows, giving an energetic benefit of 40% compared to an empty channel (Dyga and Placzek, 2010). A niche application for mesh-based heat exchangers is in cryogenics (small cryo-coolers and helium liquefaction systems). Chen et al., (2009) have modelled the thermal behaviour of such units, which have a high effectiveness and are very compact.

Interestingly, a mesh reactor, albeit with the mesh fulfilling a slightly different purpose, was constructed at University College London (see Chapter 5 and www.pinetwork.org).

4.3 Micro-heat exchangers

Micro-heat exchangers are generally defined as having area densities well in excess of the best values achieved with compact heat exchangers (see Figure 4.4). Micro-heat exchangers with area densities in excess of $15,000 \, \text{m}^2/\text{m}^3$ are available. Heat transfer and pressure drop measurements have been carried out on structures with channels having a depth of 0.5 microns. The manufacture of metallic micro-heat exchangers may be carried out in a similar way to the PCHE described above, and they resemble 'compacts', except for significantly smaller passage sizes. Other methods, such as selective laser melting, can increase the design flexibility, by using a laser and metal powder to build a heat exchanger from a virtual three-dimensional computer-generated image. (3D printing mentioned at the beginning of this chapter.) Examples constructed at Liverpool University in the UK are shown in Figure 4.19. A discussion on the process and a variety of applications is in Osakada and Masanori (2006), and a unit from Sustainable Engine Systems Ltd. is illustrated in Figure 4.20.

Tubular units with micro-tubes having internal diameters of 3 to 80 microns have been studied, both for single and two-phase flows. A slightly larger variant for use for cryo-coolers is shown in Figure 4.21. (See Longsworth [1993], for a useful discussion of Joule-Thomson cryo-coolers based on such finned tube geometries.) Micro-sized shell and tube heat exchangers have been studied (as well as the compact variants, used in aircraft engine fuel-oil heat exchangers, for example). Dai et al., (2011) studied a micro-unit with 61.1 mm o.d. tubes, wall thickness 0.16 mm, with the active length of the tubes 140 mm. Used in single-phase tests with R142b in the tubes, heated by hot water on the shell-side, satisfactory data were obtained for the Nusselt numbers, but at the expense of increased friction factor, compared to conventional units.

FIGURE 4.19

Micro-heat exchangers fabricated using selective laser melting.

FIGURE 4.20

SES 3D-printed heat exchanger.

(Courtesy SES Ltd.)

The micro-heat exchanger concept has been driven in the past by the increased cooling requirements of micro-electronics systems. The principal applications include: micro-sensors, micro-machines, and the chemical plant-on-a-chip. A group led by Newcastle University is looking at the active cooling of chips using micro-refrigerators – also a challenge for micro-heat exchangers – see Chapter 11.

There are European centres of excellence in micro-heat exchangers at the Nuclear Research Centre, Karlsruhe; and at IMM, Mainz, both in Germany, (see Appendix 4). A number of laboratories around Europe are developing chip-sized processes involving chemical reactions, etc., where such technologies will be essential – in fact, much of the work in the past five years on micro-heat exchangers has been driven by the demands of micro-reactors. For example, the Forschungszentrum Karlsruhe in Germany has been developing ceramic micro-heat exchangers, ultimately as the basis for high temperature reactions where polymers and metals may

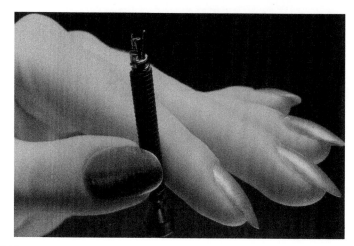

FIGURE 4.21

A finned heat exchanger used for cryogenics applications. Diameter over the fins is 1.0 mm.

(Courtesy Honeywell Hymatic Ltd)

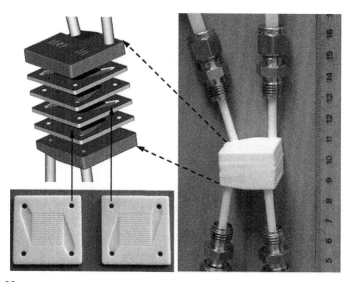

FIGURE 4.22

One variant of the ceramic micro-heat exchanger developed at Forschungszentrum Karlsruhe (Alm et al., 2007).

not be suitable (Alm et al., 2007). Illustrated in Figure 4.22, counter-flow and cross-flow examples have been made, and performance testing using water at flows to 120 kg/h has been reported. The Institute for Micro Process Engineering (IMVT) that grew out of the Forschungszentrum Karlsruhe has developed some exciting micro-structured devices for two-phase duties (Anurjew et al., 2011). This work has led to designs that, using electric heating, can produce superheated steam that was not feasible in some micro-channel devices. (Further examples of micro-fabrication can be seen in Chapter 5. For example, Corning SA micro-reactors.)

Recently, a milestone in testing of the ultimate micro-tubular heat exchanger was reached with a unit used as a compressor precooler on the SABRE engine designed to propel the Skylon space vehicle to Mach 5. The UK company, Reaction Engines (Anon, 2012c), has spent many years developing the precooler, as part of the overall cycle. The heat exchanger cools hot air from 1,000°C to −140°C, heating helium, in 0.01 s. With a duty of 400 MW, the exchanger is fabricated using 50 km of 1 mm o.d. by 20 μm wall thickness Inconel 718 tubing – giving a duty/unit volume of 1 GW/m³.

Of course, one of the main concerns about micro-heat exchangers (as well as CHEs) relates to fouling. Mayer et al., (2012), working at the Technical University of Braunschweig and the Institute for Micro-Process Engineering near Karlsruhe, studied crystallisation fouling (calcium carbonate) on a heat exchanger whose fluid heat transfer surface comprising 12 micro-channels, each 200 μm deep, 400 μm wide and 24 mm long. It was found that the behaviour mimicked what happened in macro-heat exchangers. Reductions in heat transfer performance and large increases in pressure drop were noted. At low flow rates the impact of the crystal build up was much greater than in normal heat exchangers. Cleaning procedures were recommended for further investigation (as done in larger units, of course) as was redesign of the manifolds to cater for maldistribution as channels become inhibited for fluid flow. As with heat exchangers such as the Heatric PCHE, it is logical that one either avoids streams where fouling could occur, introduces filters as appropriate and/or introduces a proven cleaning/fouling inhibition strategy.

4.4 **What about small channels?**

Before we leave the subject of micro-channel heat exchangers, it is timely to introduce a discussion regarding the fluid flow in such small systems. Although written approximately 13 years ago, an important review by Mehendale et al., (1999) threw down the gauntlet to those active in micro-heat exchanger technology, (much of which involves micro-fluidics) by asking: 'Down to what channel sizes can we confidently apply continuum theory and the conventional Navier-Stokes equations?'. The examples given, which prompted researchers to raise such a question are numerous and include:

- Large differences in reported friction factors (0.5 < (friction factor)/(friction factor)$_{\text{conventional theory}}$ < 3.5) and Nusselt numbers (0.21 < Nu/Nu$_{\text{conventional theory}}$ < 16) for single-phase flow in channels with hydraulic diameters of between 0.96 microns and 2.6 mm.
- Transition from laminar to turbulent flow occurring at Re ranging from 200 to 900.
- In refrigerant condensation, heat transfer coefficients were seen to increase as channel diameter reduced from 5 to 1 mm. What happens when surface tension forces begin to become significant?

- What is the influence of electrokinetic effects, neglected at the macro-scale, as we reduce characteristic dimensions?
- Channel roughness can affect f in laminar flow at such small scales.
- The Knudsen number (mean free path/characteristic length) observed in gas flow in micro-channels suggests that rarefied gas effects come into play – some MEMS (micro-electro-mechanical systems) devices, for example, operate in a regime normally associated with low pressure, high altitude (aerospace) devices.

Janson et al., (1999) used the Knudsen number[2], Kn, to illustrate that for air at STP in very small channels (1 µm) corresponds to Kn = 0.07, and they state that while for Kn<0.01, standard continuum fluid models are fine, as Kn rises effects such as slip flow manifest themselves, and for higher Kn values, Navier-Stokes equations do not apply. The Direct Simulation Monte Carlo (DSMC)[3] technique is one method that can then be applied. It is also pointed out that the surface conditions in MEMS can vary from atomically smooth to 'rough'. This can affect the study of viscous effects. Smith (2005) cites McAdams as setting the standard for gas regimes in 1954, suggesting:

Continuum flow: Kn < 0.001 (giving a hydraulic diameter > 0.068 mm)
Slip flow: 0.001 < Kn < 2
Free-molecule flow: Kn > 2

As an example of the above, Choi et al., (1991) had channels with very low roughnesses – expressed as a mean relative roughness (MRR) the range was 0.0001 to 0.008. As a result, $f/f_{laminar}$ was not affected by the surface roughness. However, another group (see Mala and Li, 1999), found that mean relative roughness values in the range 0.007 to 0.035 did influence the pressure drop in single-phase flow, making it higher than that predicted by classical theory. These data are shown in Figure 4.23.

Jason Reese of Strathclyde University has a simple test problem – can standard continuum equations predict the mass flow rate in a micro-fluidic system in which helium is flowing along a duct of dimensions (l) 7500 µm × (w) 53 µm × 1.33 µm? The answer to this is illustrated in Figure 4.24.

Of particular relevance to an area of process intensification that is gaining momentum at present is another observation by Reese and colleagues (Lockerby et al., 2003; 2004). The Taylor-Couette flow phenomenon is becoming increasing popular as a means of intensifying chemical reactions (Wang et al., 2005; see Chapter 5). The topic remains one of much interest, and Jeng et al. (2007) have shown that longitudinal ribs in a Taylor-Couette system (the inner cylinder rotating in this example) can enhance heat transfer. The arrangement is illustrated in Figure 4.25.

---

[2] Kn = l/H, where l is the typical mean free path of the gas and H the characteristic dimension of the channel.

[3] The DSMC method uses probabilistic simulation to solve the Boltzmann equation for fluid flows. The DSMC method was initially proposed as a method for predicting rarefied gas flows where the Navier-Stokes equations are inaccurate, and it has now been extended to near-continuum flows.

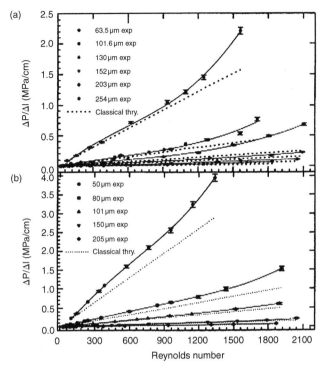

FIGURE 4.23

The data of Mala and Li, showing the deviation (an increase) from the norm in pressure drop per unit length of micro-tube as predicted by classical theory for micro-tubes of various diameters. Graph (a) is for stainless steel micro-tubes, (b) for fused silica tubes.

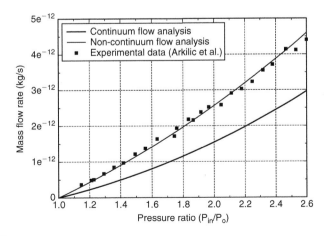

FIGURE 4.24

Can standard continuum equations predict flows at the micro-scale? Not in this case.

(Data courtesy Jason Reese)

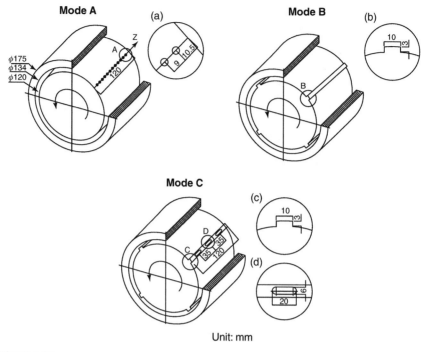

FIGURE 4.25

Longitudinal ribs used on the inner rotating cylinder in a Taylor-Couette system. The flow is essentially axial (Jeng et al., 2007).

Couette flow is a laminar circular flow occurring between a rotating (inner) cylinder and a static one, and the extension via increased speed of rotation to centrifugally-driven instabilities leads to laminar Taylor vortex flow, tending to turbulent flow as speed increases. Poiseuille flow is axial.

Illustrated in Figure 4.26, the Burnett equations have been used to predict the pressure distribution across the gap between the inner (rotating) and outer cylinders. These data are compared with the DSMC data of Nanbu (1983) and that obtained using the Navier-Stokes equation. Regardless of the value of Kn, the conventional solution shows no pressure variation across the channel, while the Burnett and DSMC solutions are essentially in agreement.

Continuum can be applied in passages with characteristic dimensions down to 50 nanometres or less but, at the micro-scale and below, the wall becomes dominant and we must ask ourselves whether the bulk fluid properties such as viscosity are appropriate in these regions. If they are not, the procedure for designing and simulating flows and heat transfer at such scales may need reassessment – a topic for the reader interested in research at such small scales to investigate more fully. A start has been made by Kandlikar (2010) in the area of flow boiling in micro-channels,

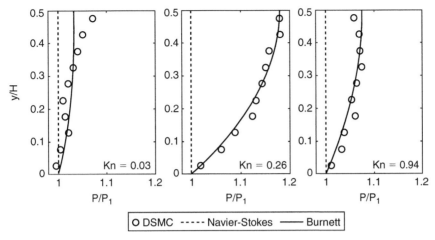

FIGURE 4.26

Distribution of non-dimensional pressure (p/p_1) across a micro-channel (distance y from wall, channel full height H), M = 3, for various values of Kn (Lockerby et al., 2003).

where he has investigated the effects of surface tension and evaporation momentum forces on FC-77 and water working fluids.

4.5 Nano-fluids

Since the first edition appeared, the volume of literature dealing with nano-fluids has massively increased. While we are not concerned here with nano-heat exchangers (yet!), the number of researchers and organisations adding nano-particles to fluids in heat exchangers at both the large and micro-scale is very great. There are several major collaborative projects, including NanoHex – Enhanced Nano-Fluid Heat Exchange (Anon, 2010), which is funded by the European Commission, and led by Thermacore in the UK, a company specialising in electronics thermal management (see Appendix 3). For a comprehensive review of nano-fluids, see Chandrasekar et al., (2012). As with others, the authors highlight disagreements between groups of researchers as to the benefits, or otherwise, of nano-fluids. Stanford University in the United States is working with support from ONR, Sandia National Laboratories to investigate nano-particle behaviour in micro-channels, and to date concludes that '... *while particle based nano-fluids show little promise for heat exchanger design, there is some indication that CNT (carbon nano-tube)-based nano-fluids may offer improvements in thermal conductivity that outpace the impact of the viscosity increase.*' (Anon, 2012d)

So the jury is still out! Also, bear in mind that nano-particles could be classed as fouling particles!

4.6 Summary

Knowledge of compact heat exchangers and the smaller (in terms of passage size) micro-heat exchangers is a necessary first step to appreciating a variety of intensified unit operations that involve heat transfer. As well as forming the basis of many intensified reactor types, (as will be seen in Chapter 5) highly compact heat exchangers in their own right are components of a range of intensified plants.

The introduction of micro-fluidics into the equation can both complicate the assessment of, and at the same time open up more opportunities for, PI in heat exchangers. This is one area where the challenges are substantial, but the rewards may be yet another step reduction in process equipment size, without compromising capability.

Nano-fluids may have a role to play, but their performance in compact and micro-heat exchangers is not clear-cut, and care is needed.

References

Alm, B., Imke, U., Knitter, R., Schygulla, U., Zimmermann, S., 2007. Testing and simulation of ceramic micro heat exchangers. Chem. Eng. J. 135 (Supplement 1), S179–S184.

Anon, 1992a. Investigation of a novel compact heat exchanger surface. Future Practice – R&D Profile 29, Best Practice Programme, UK Carbon Trust, July.

Anon, 1992b. Plant design using multi-stream heat exchangers. Future Practice – R&D Profile 30, Best Practice Programme, UK Carbon Trust, September.

Anon, 2010. The Web Site of the NanoHex European Commission-funded project. <www.nanohex.eu> (accessed 10.06.12.).

Anon, 2011. The Web Site of ERG Aerospace, USA, manufacturer of foam-based heat exchangers. <www.ergaerospace.com> (accessed 07.07.12.).

Anon, 2012a. The Web Site of the NanoHex project – Enhanced Nano-Fluid Heat Exchange. <www.nanohex.eu> (accessed 06.07.12.).

Anon, 2012b. The Web Site of Within, 3-d Printers (including heat exchangers). <www.within-lab.com> (accessed 07.07.12.).

Anon, 2012c. The Web Site of Reaction Engines, developing the SABRE engine and its recuperator. <www.reactionengines.co.uk> (accessed 11.07.12.).

Anon, 2012d. Stanford University nanofluid project. <www.nanoheat.stanford.edu/projects/nanofluids> (accessed 01.07.12.).

Anurjew, E., Hansjosten, E., Maikowske, S., Schygulla, U., Brandner, J.J., 2011. Microstructure devices for water evaporation. Appl. Therm. Eng. 31, 602–609.

Baek, S., Kim, J.-H., Jeong, S., Jung, J., 2012. Development of highly effective cryogenic printed circuit heat exchanger (PCHE) with low axial conduction. Cryogenics 52, 366–374.

Boomsma, K., Poulikakos, D., Zwick, F., 2003. Metal foams as compact high performance heat exchangers. Mech. Mater. 35, 1161–1176.

Chandrasekar, M., Suresh, S., Senthilkumar, T., 2012. Mechanisms proposed through experimental investigations on thermophysical properties and forced convective heat transfer characteristics of various nanofluids – a review. Renewable Sustainable Energy Rev. 16, 3917–3938.

Chen, S., Hou, Y., Zhao, H., Xi, L., 2009. A numerical model of thermal analysis for woven wire screen matrix heat exchanger. Cryogenics 49, 482–489.

Choi, S.B., et al., 1991. Liquid flow and heat transfer in microtubes In: Choi, Micromechanical Sensors, Actuators and Systems, 32 ASME DSC, pp. 123–134.

Christmann, J.B.P., Kratz, L.J., Bart, H-J., 2012. PEEK film heat transfer surfaces for multi-effect distillation: a mechanical investigation. Appl. Therm. Eng. 38, 175–181.

Cross, W.T., Ramshaw, C., 1986. Process intensification: laminar flow heat transfer. Chem. Eng. Res. Des. 64 (July), 293–301.

Dai, C. Wang, Q., Li, B., 2011. An experimental study on heat transfer and pressure drop of MTHE (micro-sized shell and tube heat exchanger). Proceedings of the Third Micro and Nano Flows Conference, Thessaloniki, Greece, 22–24 August.

Dyga, R., Placzek, M., 2010. Efficiency of heat transfer in heat exchangers with wire mesh packing. Int. J. Heat Mass Transfer 53, 5499–5508.

Hesselgreaves, J.E., 2001. Compact heat exchangers Selection, Design and Operation. Pergamon, Oxford.

Janson, S., et al., 1999. Micropropulsion systems for aircraft and spacecraft. In: Helvajian, H. (Ed.), Microengineering Aerospace Systems The Aerospace Press and AIAA, USA, pp. 657–696.

Jeng, T-M., Tzeng, S-C., Lin, C-H., 2007. Heat transfer enhancement of Taylor-Couette-Poiseuille flow in an annulus by mounting longitudinal ribs on the rotating inner cylinder. Int. J. Heat Mass Transfer 50 (1–2), 381–390.

Le Pierres, R., Southall, D., Osborne, S., 2011. Impact of mechanical design issues on Printed Circuit Heat Exchangers. Proceedings of SCO_2 Power Cycle Symposium 2011, University of Colorado at Boulder – University Memorial Center, CO, May 24–25. (Paper downloadable from <www.heatric.com> (accessed 06.07.12.).

Lockerby, D.A., Reese, J.M., 2003. High-resolution Burnett simulations of micro-Couette flow and heat transfer. J. Comput. Phys. 188, 333–347.

Lockerby, D.A., Reese, J.M., Emerson, D.R., Barber, R.W., 2004. Velocity boundary condition at solid wall in rarefied gas calculations. Phys. Rev. E. 70 (1) (3rd Series, Paper 017303, July).

Longsworth, R.C., 1993. Heat exchangers for Joule Thomson cryocoolers. In: Shah, R.K., Hashemi., A. (Eds.), Aerospace Heat Exchanger Technology Elsevier, The Netherlands, pp. 1993.

Mala, G.M., Li, D., 1999. Flow characteristics of water in microtubes. Int. J. Heat Fluid Flow 20, 142–148.

Mayer, M., Bucko, J., Benzinger, W., Dittmeyer, R., Augustin, W., Scholl, S., 2012. Exp. Therm. Fluid Sci. 40, 126–131.

Mehendale, S.S. et al., 1999 Heat exchangers at micro- and meso-scales. In: Compact Heat Exchangers and Enhancement Technology for the Process Industries. Proceedings of the International Conference, Banff, Canada, pp. 55–74, July 18–23.

Nanbu, K., 1983. Analysis of the Couette flow by means of the new direct-simulation method. J. Phys. Soc. Jpn. 52, 5.

Odabaee, M., Hooman, K., 2011. Metal foam heat exchangers for heat transfer augmentation from a tube bank. Appl. Therm. Eng. 36, 456–463.

Osakada, K., Masanori, S., 2006. Flexible manufacturing of metallic products by selective laser melting of powder. Int. J. Mach. Tools Manuf. 46 (11), 1188–1193.

Pan, M., Bulatov, I., Smith, R., Kim, J-K., 2012. Optimisation for the retrofit of large scale heat exchanger networks with different intensified heat transfer techniques. Appl. Therm. Eng. doi: 10.1016/j.applthermaleng.2012.04.038.

Paulraj, P., Paul, B.K., Peterson, R.B., 2012. Development of an adhesive-bonded counterflow microchannel heat exchanger. Appl. Therm. Eng. doi: 10.1016/j.applthermaleng.2012.05.001.

Smith, E.M., 2005. Advances in Thermal Design of Heat Exchangers. A Numerical Approach: Direct-sizing, Step-wise Rating and Transients. John Wiley, Chichester.

Sertkaya, A.A., Altinisik, K., Dincer, K., 2012. Experimental investigation of thermal performance of aluminium finned heat exchangers and open-cell aluminium foam heat exchangers. Exp. Therm. Fluid Sci. 36, 86–93.

Southall, D., Dewson, S.J., 2010. Innovative compact heat exchangers. Proceedings of ICAPP '10, Paper 10300, San Diego, CA, USA, June 13–17, 2010.

Tian, J., Lu, T.J., Hodson, H.P., Queheillalt, D.T., Wadley, H.N.G., 2007. Cross-flow heat exchange of textile cellular metal core sandwich panels. Int. J. Heat Mass Transfer 50, 2521–2536.

T'Joen, C., Park, Y., Wang, Q., Sommoers, A., Han, X., Jacobi, A., 2009. A review on polymer heat exchangers for HVAC&R applications. Int. J. Refrig. 32, 763–779.

Tsuzuki, N., Kato, Y., Ishiduka, T., 2007. High performance printed circuit heat exchanger. Appl. Therm. Eng. 27, 1702–1707.

Tsuzuki, N., Utamura, M., Ngo, T.L., 2009. Nusselt number correlations for a microchannel heat exchanger hot water supplier with S-shaped fins. Appl. Therm. Eng. 29, 3299–3308.

Wang, L., et al., 2005. CFD simulation of aggregation and breakage processes in laminar Taylor-Couette flow. J. Colloid Interface Sci. 282, 380–396.

Wnek, M., 2012. Ceramic or metallic? – Material aspects of compact heat regenerator energy efficiency IOP Conference Series. Mater. Sci. Eng. 35, 012022.

Zaheed, L., Jachuck, R.J.J., 2004. Review of polymer compact heat exchangers, with special emphasis on polymer film unit. Appl. Therm. Eng. 24, 2323–2358.

Zaheed, L., Jachuck, R.J.J., 2005. Performance of a square, cross-corrugated, polymer film, compact, heat-exchanger with potential application in fuel cells. J. Power Sources 140, 304–310.

Zhao, C.Y., Lu, W., Tassou, S.A., 2006. Thermal analysis on metal-foam filled heat exchangers. Part II: tube heat exchangers. Int. J. Heat Mass Transfer 49, 2762–2770.

Zhu, Y., Hu, H., Ding., G., Sun, S., Jing, Y., 2012. Influence of metal foam on heat transfer characteristics of refrigerant-oil mixture flow boiling inside circular tubes. Appl. Therm. Eng. <http://dx.doi.org/10.1010/j.applthermaleng.2012.06.045>.

Reactors

5

OBJECTIVES IN THIS CHAPTER

The reactor is often described as the heart of a chemical process. This is based on the fact that the downstream processes are a function of the reactor performance, in particular, features such as degree of control, selectivity and heat and mass transfer (which are mostly interdependent). Furthermore, the upstream processes can be viewed as a set of steps whose sole purpose is to get the reactants to the correct condition ready for inputting to the reactor.

Consequently, in the field of process intensification, the reaction step has received more attention than other unit operations. This is illustrated by the large number of intensified reactors either on the market or under development. This chapter introduces several of the most important of these, following a short tutorial on reactor theory that puts the main characteristics of reactors in the appropriate context.

5.1 Reactor engineering theory

It is generally taught that reactor engineering has two key elements: reaction kinetics and reactor design. The reactor design is based upon a mass or mole balance, simplistically of the form:

$$\text{In} + \text{Made} = \text{Out} + \text{Accumulation}$$

Where, for a simple reaction, such as A→B, 'Made' is the rate at which a species is created or lost by reaction. If the balance were on species A, this would be equal to $r_A V$, where r_A is the rate of reaction, therefore in this case, the rate of loss of species A; and V is the volume of the reactor. 'In' would be the molar flow rate of species A into the reactor, given by e.g. $Q.C_{Ai}$, where Q is the volumetric flow rate of material into the reactor, and C_{Ai} is the molar concentration of species A in this stream. 'Out' would be the flow rate of species A out of the reactor, equal to $Q.C_{Ao}$, assuming the flow rate remained the same throughout the reactor (C_{Ao} is simply the concentration of A at the outlet of the reactor). The accumulation term is the rate at which the total amount of species A in the reactor changes with time. This will be zero when operating a continuous reactor at steady state. This would usually be equal to $d(\Sigma C_A.V)/dt$.

The kinetics of the reaction depend only upon the concentrations, rate constant and order of reaction, expressed as r_A. For example, for a first order loss of A, the rate would be expressed in the following form: $r_A = k_A C_A$ (k_A is the rate constant, and C_A the concentration of species A). It should be noted that although temperature is not explicit in this expression, rate constants are usually very strong functions of temperature, often following Arrhenius kinetics, in which k is proportional to $\ln(-1/T)$, where T is absolute temperature.

The balance can only be accurate, however, if the reaction is proceeding at its inherent rate, i.e. there is perfect mixing and mass transfer within the reactor. The equation above becomes considerably more difficult to solve when there are mass transfer limitations. The observed rate of reaction will be the rate of the slowest step, which can often be the rate of mass transfer. When developing a process, if a certain rate of reaction has been observed at laboratory scale (where mass transfer limitations are easily overcome), there is no guarantee that this rate will be observed at larger scales, as mass transfer/mixing limitations become more difficult to overcome at larger scales. If reactor scale-up calculations are based upon the false assumption that a reaction will proceed at its inherent rate regardless of scale, then the reactor will be undersized and unable to provide the necessary residence time for the desired conversion.

In this section, a brief introduction to some of the reactor design issues of relevance to the rest of the chapter is given. For greater depth the authors recommend texts such as *Chemical Reactor Engineering* (Levenspiel, 2007).

5.1.1 Reaction kinetics

It is often the case in larger stirred tank design reactors that reactions do not proceed at their intrinsic rate, as it is limited by the mixing within the vessel, i.e. the reactants are not perfectly mixed together, known as 'mixing limited'. The role of PI is often to remove such limitations, so that the reaction *can* reach its intrinsic rate. Thus, the aim is to ensure that the mixing and heat/mass transfer rates will be relatively fast compared to the fundamental process kinetics.

Homogeneous reactions can usually be described by Arrhenius type expressions:

$$k = A \exp\left(\frac{-\varepsilon_A}{RT}\right)$$

where k is the rate constant, A is the pre-exponential function, a function of the reaction. (Note: the units of k and A depend upon the stoichiometry of the reaction). ε_A is the activation energy ($J\,mol^{-1}$); R is the gas constant ($J\,mol^{-1}\,K^{-1}$) and T is temperature (K). The dependence on temperature is clear. This equation is directly applicable if the reaction mixture is well-mixed and/or the phases are fully miscible.

Analysis of reactions involving multiple phases is more complex, as these assumptions do not necessarily apply. In certain cases, particularly at larger scale, there may be the spurious impression that the reaction rate is slow, when a more detailed appraisal suggests, for instance, that a critical intermediate has a very short half life.

When mixing is slow, the inter-penetration of reactants is poor, so that reaction only occurs in the immediate vicinity of the interface between them (depending upon the relative rates of reaction and diffusion). Therefore, a reduced proportion of the volume available is used and the reaction proceeds at various rates throughout the volume as local stoichiometric ratios differ greatly. Thus the space–time yield of the system is depressed, and accurate control over the selectivity and by-products is compromised. On the other hand, with a higher degree of mixing, reactants can become intimately mixed with each other before significant reaction occurs. Now reaction proceeds throughout the available volume and the desired control over stoichiometric ratio is achieved, and the reaction is said to be reaction-controlled, rather than mixing-controlled. This illustrates the point made above, that the intrinsic reaction kinetics must be allowed to proceed at a rate unimpeded by other phenomena. When this ideal is not achieved, which occurs frequently in industrial-scale batch stirred vessels, the reaction is significantly mixing-limited.

One industrial example, where there is probably a gross mismatch between kinetics and mixing rates, is the p-xylene oxidiser used to manufacture terephthalic acid in acetic acid solution. The low solubility of oxygen in the reaction mixture implies that the oxygen concentration in an element of oxygenated p-xylene will fall to zero in only a few seconds. With the air-sparged reactor, (a stirred tank unit into which air is bubbled) the circulation time of the vessel contents is of the order of 1 minute. Thus, there is at least the likelihood that after the reaction mixture has absorbed oxygen near the impeller, where aeration is greatest, the concentration (and reaction rate) in the rest of the circulation loop is minimal. If this is the case, then this reaction is an ideal candidate for equipment, such as the spinning disc reactor described in Section 5.2.

5.1.2 Residence time distributions (RTDs)

The residence time distribution of the flow through a given geometry can be thought of as a probability distribution for the time an element of fluid takes to travel through that geometry. It is often easier to envisage an RTD as the response to the input of a perfect, infinitely narrow pulse, known as a Dirac delta input.

The two extremes of behaviour in reactor engineering are the continuously stirred tank reactor (CSTR) and the plug flow reactor (PFR). The CSTR is said to be fully back mixed or perfectly mixed, and, as such, has a broad residence time distribution (in fact, an exponential decay, as shown in Figure 5.1).

PFRs, on the other hand, have an extremely narrow RTD. In fact, in an ideal PFR, the output is exactly the same as the input, as there is no axial dispersion of information along the reactor (see Figure 5.2, below), due to its flat velocity profile usually achieved by ensuring high levels of turbulence. Real reactors' RTDs lie between these two extremes: no PFR has an infinitely narrow RTD, and no CSTR is perfectly mixed.

It is often desirable to have plug flow operation rather than CSTR, as it gives very tight control of the processing history: each element of fluid spends exactly the same amount of time in each environment in the reactor. This leads to reduced

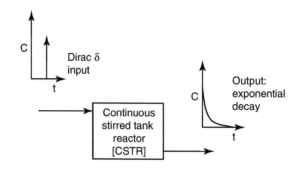

FIGURE 5.1

The CSTR and its RTD.

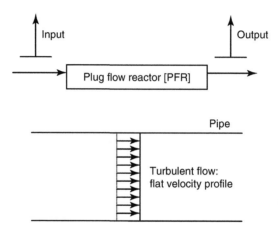

FIGURE 5.2

The PFR, its RTD and the velocity profile within the tube.

by-products, as elements of fluid are not, for example, exposed to higher temperatures for any longer than is necessary for the required conversion, and to a smaller reactor. The spinning disc reactors, oscillatory baffled reactors and Taylor-Couette reactors, covered elsewhere in this book, are examples of novel, intensified plug flow reactors.

5.1.3 Heat and mass transfer in reactors

One of the key aims in reactor design, or indeed in the design of a range of other unit operations, is to achieve 'good' mixing, but what constitutes good mixing? Generally, a vessel containing more than one substance can be described as micro-mixed if the concentrations of all substances are uniform throughout, or macro-mixed, meaning that there is a distribution of concentrations throughout the vessel.

5.1.3.1 Mass transfer

Before considering the detailed attributes of intensified reactors in the following sections, it is worth highlighting the shortcomings of conventional equipment, particularly the batch stirred tank. A key issue is the scaleability of mixing. A batch stirred vessel at laboratory scale can be perfectly mixed, as it is relatively easy to operate the stirrer at a high enough rate to ensure this, i.e. a high power density is easily achieved. However, the costs of achieving this degree of mixing soon become prohibitively expensive and/or technically difficult at larger scales, not least because the power number for mixing is proportional to the impeller diameter, d, to the inverse power 5:

$$ N_p = \frac{P}{\rho n^3 d^5} $$

where:
N_p = power number
P = power
ρ = fluid density
d = impeller diameter (all SI units)

In practice, this means that large stirred vessels are not well-mixed, which in turn often means that the reactions are mixing-limited. This is the reason for the often-reported decrease in reactor volume when converting from batch to continuous processing using intensified processes. The length scales of mixing are typically shorter in such devices, and the mixing methods are often improved, so the reaction is no longer mixing-limited and is able to proceed at a rate closer to its inherent rate, i.e. it is reaction-controlled.

If, for example, a stirred vessel were to be used for a gas–liquid reaction it would include some form of agitator, coupled with a set of baffles to inhibit the generation of a free vortex, while promoting circulation by means of two toroidal vortices, as shown in Figure 5.3. The gas is usually injected directly below the impeller via some suitable sparging arrangement. The popularity of the stirred vessel is due to its simplicity and adaptability, and because it is, in many senses, straightforward to scale-up from the laboratory beaker that was used when the process was being developed. Unfortunately, the stirred tank suffers from several serious problems, when scaling up. Its mixing capability is a strong function of its size. Scale-up usually proceeds on the basis of a constant impeller tip speed, and since the mean circulation speed in the vortices is broadly proportional to the tip speed chosen, the circulation time is proportional to the vessel diameter. Thus the turnover time of the vessel contents increases at the larger scale and the macro-mixing performance deteriorates. Once the research chemist has specified the production sequence for a target molecule using a laboratory beaker or flask, scale-up to production levels is usually the responsibility of the process engineer. In most cases the

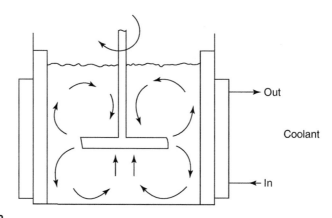

FIGURE 5.3

Circulation pattern in a stirred vessel.

approach chosen is simply to use a bigger beaker (a stirred tank reactor), resulting in a number of scaling problems.

These fundamental shortcomings of the stirred vessel have generated a considerable degree of uncertainty when fine chemical or pharmaceutical processes are being developed for full-scale operation. This has led the relevant regulating authorities, e.g. the US Food and Drug Administration, to insist on process validation at laboratory-, pilot- and full-scale. Since each validation entails significant administration and delay, the procedure can hold up the implementation of commercial production for several years. Because a new metabolically active molecule will be patented as soon as possible and certainly before clinical trials and process development, this delay significantly erodes the time available under patent cover to recoup a company's R&D expenditure, and then make a profit, from a new drug.

5.1.3.2 Heat transfer

Heat transfer in reactors is often underplayed in undergraduate reactor engineering courses, but in practice it is often critical to their design and operation. The reactor designer or operator must ensure that temperature can be controlled to the required accuracy: an important calculation for any reactor designer is to determine whether thermal runaway is possible. This refers to a situation in which the reactor and associated temperature control system are unable to remove the heat generated by an exothermic reaction, and consequently the temperature climbs exponentially due to positive feedback until some other limit, such as a safety valve or control system, comes into play.

This is particularly important when considering new reactor technologies, as the enhanced mixing will allow the reaction a closer approach to its inherent rate, causing the rate of release of heat of reaction to be increased, perhaps substantially. This may reveal the exotherm of a reaction to be substantially greater than previously thought i.e. when operated in a poorly mixed batch reactor, often at greater dilution. A standard chemical engineering textbook, such as Coulson and Richardson's *Chemical Engineering Volume 1*, can provide a sufficient level of heat transfer theory to conduct such calculations.

For a stirred tank batch reactor, in the normal case of a geometrically similar scale-up, it can be readily shown that the surface area per unit reactor volume varies inversely with the vessel diameter. Thus, larger vessels are more difficult to cool, since the heat generated by a reaction in a potential runaway situation is proportional to the vessel volume, whereas the surface available to dissipate a given heat output is decreased. Vigorous exothermic reactions may therefore require the reactor to be 'unintensified' by operating with more dilute feedstocks, in order to reduce the full-scale reaction intensity and therefore the risk. This decreases the reaction rate, and could influence the reaction temperature trajectory and compromise the yield and selectivity. Intensified reactors, on the other hand, tend to have enhanced heat transfer capabilities. This is due to increased surface-area-to-volume ratio, and/or flow structures within the reactor being more conducive to transport of heat.

It should not be assumed that because the unintensified larger scale reactor was safely operated that the intensified reactor will be safe with regard to thermal run-away. The exotherm may increase due to the mixing, but the surface-area-to-volume ratio will also increase, so the calculation to determine the relative importance of these two effects must be performed.

It should be stated that, in general, intensified reactors are safer than their conventional counterparts (see below) due to the reduced inventory of material at hazardous conditions, and because, in most (but not all) cases, the surface-area-to-volume ratio is the more significant effect.

5.2 **Spinning disc reactors**

5.2.1 **Exploitation of centrifugal fields**

As mentioned in Chapter 3, the rationale for the exploitation of centrifugal fields in process technology is based on the recognition that the fluid dynamic behaviour of multiphase systems is dictated by the magnitude of the buoyancy term $\Delta \rho g$, where $\Delta \rho$ is the density difference and g is the applied acceleration. If either factor approaches zero then the fluid dynamic intensity decreases, with surface forces tending to dominate the system. Under these conditions, bubble terminal velocities, interfacial shear stresses and flooding mass fluxes all decrease, while bubble sizes tend to increase. The lava lamp mentioned in Chapter 3 simulates the behaviour of a milli-g system since the interfacial density difference is only about $1 \, kg/m^3$. It is a convincing illustration of the trend towards large 'lazy' bubbles which imply poor interfacial heat/mass transfer and low specific surface area. Further confirmation of the importance of the applied acceleration in controlling multiphase systems is provided by some recent electrolytic experiments in a microgravity environment.

Kaneko et al. (1993) electrolysed water for up to 25 seconds during a parabolic aeroplane flight. During the micro-g period, it was shown that much larger bubbles were generated while the current density fell dramatically. A second study by Matsushima et al. (2003) used a 400 m drop shaft to generate micro-g conditions for about 8 seconds. It was shown that the electrolyte gas volume fraction and the gas coverage of the electrodes were much higher during the micro-g period compared with the situation under terrestrial conditions. It should be pointed out that much more severe gas-blinding of the system would be expected if the micro-g environment could have been maintained for longer, more realistic, periods.

Having shown that lower acceleration environments are not conducive to process intensity, the corollary is that *high* acceleration should generate immediate PI benefits, particularly when it is recognised that approximately two thirds of the process unit operations involve multiphase systems, e.g. distillation, boiling, absorption, etc. Hence any progress in this direction may have broad application. It is assumed that elevated acceleration is achieved for as long as required by operating within a rotating system. One embodiment of this approach is based on the flow of a liquid film over a rotating surface of revolution. Usually this will simply be a spinning flat disc: the spinning disc reactor (SDR).

5.2.2 **The desktop continuous process**

The predominant culture which prevails for the production of drugs and fine chemicals at an output of up to, say, 500 tonnes per year, is to operate batchwise. As already noted, this stems from the fact that the process is almost always developed from a batch-operated beaker or flask. However, it is worth observing that an output of 500 tonnes/year of active substance corresponds to a continuous process flow rate of around only 70 ml/sec. This allows various items of intensified equipment to be assembled and operated continuously, literally on a desktop, to meet the production demand. The decision to switch from batch to continuous processing immediately confers a useful intensification benefit because the peak batch process loads (e.g. heat output, liquid removal, etc.) are distributed in time, so equipment size can be reduced. Thus, with a new process, the laboratory scale becomes the full-scale when allowed to run continuously and the scale-up delays described earlier are largely avoided. This strategy is generating considerable industrial interest as the commercial pressure to bring new molecules to market rapidly continues to increase.

A further factor that favours continuous manufacture is its potential impact on the overall business process of making fine chemicals. It goes without saying that with very short process residence times, the operation can be much more responsive so that grade changes can be effected in seconds rather than hours. This facilitates a just-in-time approach to manufacture which can lead to dramatic reductions in the capital costs associated with the multiple grades of stock that may be needed rapidly to satisfy demanding customers.

5.2.3 The spinning disc reactor

When a liquid flow is supplied to, or near, the centre of a rotating surface of revolution, an outwardly flowing liquid film is generated. The film is initially accelerated tangentially by the shear stresses generated at the disc/liquid interface. Subsequently, having nearly reached the local angular velocity, the liquid moves outwards as a thinning/diverging film under the prevailing centrifugal acceleration as will be shown below. Provided the rotating surface is fully wetted, the films generated may be very thin, typically 50 microns for water-like liquids. Such short diffusion/conduction path lengths stimulate excellent heat, mass and momentum transfer between the gas phase and the liquid, and between the rotating surface and the liquid. These features make the SDR an ideal basis for performing fast exothermic reactions involving water-like-to-medium viscosity. It is, therefore, an effective enabler for the desktop strategy noted above. The ability to cope with moderate liquid viscosities also allows the SDR to function as a very effective polymeriser.

5.2.4 The Nusselt flow model

While the fluid dynamics of the actual film-flow process is dauntingly complex, a very approximate interim flow model may be based on Nusselt's (1916) treatment of the flow of a condensate film. This model assumes that there is no shear at the gas–liquid interface, that the film is ripple-free and that there is no tangential slip at the disc–liquid surface. The treatment is based on the schematic representation given in Figure 5.4. Part b represents the local film at a radius, r. The shear stress on the annular plane at a distance y from the disc provides the radial acceleration for the fluid lying between $y = y$ and $y = s$. Thus, a force balance on the film lying between $r = r$, and $r = r + dr$, with zero shear stress at the gas-liquid interface, gives:

$$\omega^2 r\rho(s - y) = \mu\frac{du}{dy} \tag{5.1}$$

The boundary conditions are:

1. $u = 0$ at $y = 0$ since there is no fluid slip at the disc/liquid interface.
2. $du/dy = 0$ at $y = s$ since there is no shear stress at the gas/liquid interface.

Hence:

$$u = \frac{\omega^2 r\rho}{\mu}\left(sy - \frac{y^2}{2}\right) \tag{5.2}$$

The average film velocity is given by:

$$U_{av} = \frac{1}{s}\int_0^s u\,dy = \frac{\rho\omega^2 r s^2}{3\mu} \tag{5.3}$$

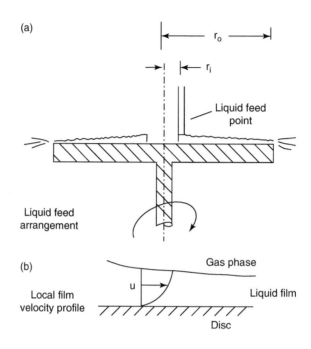

(a)

Liquid feed point

Liquid feed arrangement

(b)

Gas phase

Local film velocity profile

Liquid film

Disc

FIGURE 5.4

(a) Sketch of a liquid film on a rotating disc. (b) Detail of a liquid film on a rotating disc.

The maximum film velocity (at $y = s$) is:

$$U_{max} = \frac{\rho \omega^2 r s^2}{2\mu} = 1.5 U_{av} \tag{5.4}$$

Referring to Figure 5.4, the liquid is supplied to the disc at a radius r_i and a mass flow rate M. It is deemed to instantaneously acquire and maintain the disc angular velocity as it moves over the disc to be discharged at its periphery. At a radius r the mass flow rate is given by:

$$M = \rho U_{av} s 2 \pi r \tag{5.5}$$

Eliminating U_{av} from Eqns. 5.4 and 5.5 gives:

$$s = \left(\frac{3 \mu M}{2 \pi \omega^2 \rho^2} \right)^{1/3} r^{-2/3} \tag{5.6}$$

Inserting Eqn. 5.6 into Eqn. 5.5 gives:

$$U_{av} = \frac{\rho \omega^2 r}{3\mu} \left(\frac{3\mu M}{2\pi \rho^2 \omega^2 \rho^2} \right)^{2/3}$$

$$= \left(\frac{M^2 \omega^2}{12\pi^2 \rho \mu} \right)^{1/3} r^{-1/3}$$

(5.7)

Hence the average time required for the liquid to travel from r_i to r_o is:

$$t = \int_{R_i}^{R_o} \frac{dr}{U_0} = \frac{3}{4} \left(\frac{12\pi^2 \rho \mu}{M^2 \omega^2} \right)^{1/3} (r_0^{4/3} - r_i^{4/3})$$

(5.8)

If we consider a typical example of water flowing over a disc under the following conditions:

$$M = 3 \times 10^{-2} \, \text{kg/s}; \mu = 10^{-3} \, \text{N s/m}^2$$
$$\rho = 10^3 \, \text{kg/m}^3;$$
$$r_i = 5 \times 10^{-2} \, \text{m}; r_o = 0.25 \, \text{m}; \omega = 100 \, \text{s}^{-1} \, (955 \, \text{rpm});$$

...then from Equation 5.8 the average liquid transit time on the disc is 0.25 s, and from Equation 5.6, the film thickness at the disc edge (provided that the film does not break up into rivulets) is 28 microns. A more viscous liquid, such as a polymer (say, $\mu = 10 \, \text{N s/m}^2$), would have a thickness at the disc periphery of 600 microns and a transit time of about 5 seconds.

As already noted, the foregoing calculations must be regarded as a guide only, since the films are intrinsically unstable, with waves being amplified as the liquid proceeds to the edge of the disc. It will be appreciated that this process proceeds more rapidly with relatively inviscid liquids.

5.2.5 Mass transfer

A conservative estimate of the disc's mass transfer performance may be obtained from the Nusselt model, assuming that there is no film mixing as it proceeds to the edge of the disc. For unsteady diffusion into a finite stagnant slab, having a thickness 2l, the plot shown in Figure 5.5 from Carslaw and Jaeger (1959) gives the relative concentration distribution within the slab at various times, with a zero initial concentration and a surface concentration C_o imposed at time $t = 0$. The parameter on the curves is the Fourier number, F_o, where:

$$F_o = \frac{Dt_e}{s^2}$$

(5.9)

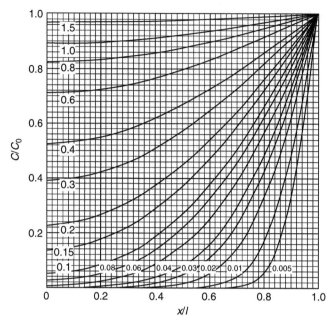

FIGURE 5.5

Concentration distribution at various times in a slab −l<X<l for zero initial concentration and surface concentration C_o.

D = solute diffusivity within the film
t_e = exposure time of the surface film
s = film thickness

As can be seen from Figure 5.5, if $F_o < 0.02$, the concentration changes within the film are confined largely to the surface layer and the local mass transfer coefficient is given by the Higbie penetration theory (Higbie, 1935) as:

$$k_L = \left(\frac{D}{\pi t_e}\right)^{1/2} \tag{5.10}$$

For the previous example of a polymer flowing over the disc, a typical Fourier number may be calculated from:

$$D = 10^{-9}\,\mathrm{m^2/s}$$
$$t_e = 5\,\mathrm{s}$$
$$s = 6 \times 10^{-4}\,\mathrm{m}$$

Thus $F_o = 0.014$

Equation 5.4 shows that the film surface velocity is given by:

$$U_{max} = 1.5U_{av} = \left(\frac{9}{32} \cdot \frac{M^2\omega^2}{\pi^2\rho\mu}\right)^{1/3} r^{-1/3}$$

Hence from Equation 5.8:

$$t_e = \left(\frac{32}{9}\frac{\pi^2\rho\mu}{M^2\omega^2}\right)^{1/3}\frac{3}{4}\left(r^{4/3} - r_1^{4/3}\right) \tag{5.11}$$

Inserting this into Equation 5.10 gives:

$$k_L = \left(\frac{D}{\pi}\right)^{1/2}\left(\frac{2M^2\omega^2}{3\pi^2\rho\mu}\right)^{1/6}\frac{1}{\left(r^{4/3} - r_1^{4/3}\right)^{1/2}} \tag{5.12}$$

However, it must be noted that as the film flows over the disc, the film thickness progressively decreases, provided the liquid fully wets the disc. As this occurs, the concentration profiles normal to the disc plane are compressed, thereby causing a proportionate enhancement of the solute diffusion rate beyond that predicted by penetration theory. Thus, the local value of k_L can be corrected to account approximately for the steepened concentration gradients by multiplying by a factor s_1/s, where s_1 is the film thickness at radius r_1 as given by Equation 5.6. The corrected local value of k_L is then:

$$k_L = \left(\frac{D}{\pi}\right)^{1/2}\left(\frac{2M^2\omega^2}{3\pi^2\rho\mu}\right)^{1/6}\left(\frac{r}{r_1}\right)^{2/3}\frac{1}{\left(r^{4/3} - r_1^{4/3}\right)^{1/2}} \tag{5.13}$$

At the point of film formation, where $r = r_1$, Equation 5.13 shows that $k_L=\infty$. However, the average value of k_L over the disc surface is given by:

$$k_{L_{av}} = \frac{1}{\pi\left(r_2^2 - r_1^2\right)}\int_{r_1}^{r_2} 2\pi k_L r\, dr \tag{5.14}$$

This requires numerical integration. As pointed out at the outset, these estimates of the mass transfer performance are likely to be conservative, as the disturbance of the film by ripples has been neglected. This will reduce the exposure time significantly, particularly with inviscid liquids.

5.2.6 Heat transfer

The Nusselt model was originally developed to correlate the performance of vapour condensers. In this case, the latent heat of condensation is discharged at the gas–liquid interface and subsequently conducted through the draining condensate film, the conduction path length being the local film thickness. When a liquid film is heated or cooled on a spinning disc, the conduction path length is less (about 50% of the thickness) because all of the sensible heat does not have to be conducted through the entire film. Since the thermal diffusivity of most liquids is typically of the order of $10^{27}\,m^2/s$, compared with a mass diffusivity of around $10^{29}\,m^2/s$, the Fourier numbers involved in the heat transfer version of Figure 5.5 are approximately 100 times their mass transfer equivalent. This implies that the heat transfer process involves the whole liquid film rather than merely a thin layer near the disc surface. The Higbie model for heat transfer is therefore inappropriate.

For the larger Fourier numbers involved in the heat transfer it is reasonable to represent the film temperature profile approximately by a quadratic expression:

$$T = A + By + Cy^2 \tag{5.15}$$

A, B and C are constants determined by the following boundary conditions:

$$1.\, T = T_w \text{ at } y = 0$$
$$2.\, T = T_s \text{ at } y = s$$
$$3.\, dT/dy = 0 \text{ at } y = s$$

It can be shown that:

$$T = T_w - 2(T_w - T_s)\frac{y}{s} + (T_w - T_s)\frac{y^2}{s^2} \tag{5.16}$$

and:

$$\frac{dT}{dy} = \frac{2(T_w - T_s)}{s}\left(\frac{y}{s} - 1\right) \tag{5.17}$$

Since the film temperature gradient perpendicular to the disc will be much greater than that in the radial direction, the local heat flux Q into the film will be controlled by the value of dT/dy at the disc surface. Hence:

$$Q = k\left(\frac{dT}{dy}\right)_{y=0} = \frac{2k}{s}(T_w - T_s)$$

Thus the effective film coefficient is:

$$h = \frac{Q}{T_w - T_s} = \frac{2k}{s} \tag{5.18}$$

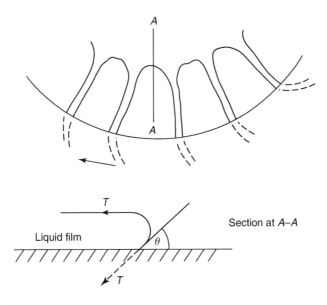

FIGURE 5.6

Schematic showing liquid film dryout on a rotating disc.

From the earlier example with water on a 0.5 m diameter disc, Equation 5.18 implies that the heat transfer film coefficient at the periphery is 43 kW/m²K, with a predicted film thickness of 28 microns. For this estimate to be realistic it is essential that the film wets the disc and does not break into rivulets. This depends upon a force balance at an incipient dryout point, as indicated in Figure 5.6. At the film stagnation point, the film momentum is potentially destroyed by the action of the component of the surface forces, parallel to the disc. Thus, for an average film velocity U_{av}, we must satisfy the following condition for rivulet maintenance:

$$T(1 + \cos\theta) > U_{av}^2 \rho s$$

where:
T is the surface tension per unit length
θ is the contact angle
ρ is the liquid density
s is the local film thickness

Coherent films are less likely as they become thinner and their velocity decreases. An inspection of Equations 5.6 and 5.7 reveals that $U_{av}^2 \rho s$ is proportional to $\omega^{2/3} M^{5/3}$. Hence, the tendency to form rivulets is less at higher disc speeds and liquid flow rates and increases with large T and small θ.

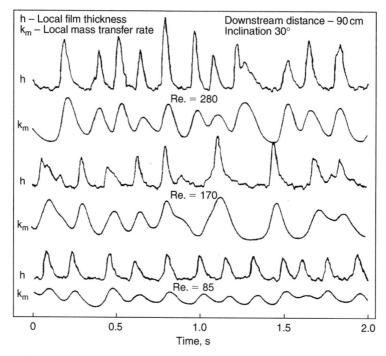

FIGURE 5.7

Simultaneous time traces of local instantaneous film thickness and transfer rate.

5.2.7 Film-flow instability

The existence of the wave structure within the film is of major practical interest, as was highlighted by some elegant experimental work conducted by Brauner and Maron (1982). They monitored the instantaneous local film thickness of a liquid flowing down a stationary inclined plane using a capacitance technique. Simultaneously, the local mass transfer coefficient was measured between the plane and the liquid, using the limiting electrolytic current method. Their plots are reproduced in Figure 5.7. It was clear that the passage of a ripple was associated with a significant enhancement of the mass transfer coefficient, as a consequence of the flow field associated with ripple propagation. An analogous phenomenon may be observed when sand particles are disturbed by wavelets in shallow seaside pools. While the phenomenon has considerable theoretical interest, its immediate practical implication is very important because it suggests that the disc heat and mass transfer performance could be enhanced still further by appropriately engineering the disc surface profile.

5.2.8 Film-flow studies

The film-flow behaviour on a disc was studied by Woods (1995), who photographed a fully wetting film of dilute ink as it travelled over a spinning glass disc. Care was taken to supply the liquid from a central axisymmetric distributor in a particularly uniform manner. After calibration, the local film thickness was inferred from the density of the photographic image at that point. Despite the care taken with liquid feed introduction, the initially smooth inner film always broke down into an array of spiral ripples, as shown in Figure 5.8. These spiral structures then broke down further until the wave pattern became utterly chaotic, provided that the disc was big enough. It is known that the film is intrinsically unstable and the phenomenon appears to be qualitatively equivalent to the breakdown of a smoke plume from a cigarette, where a stable smoke column is succeeded by a chaotic zone about 20 cm above the smoke source. The behaviour can also be observed when liquid flows over a stationary surface such as a windowpane or a dam spillway.

Woods concluded that two types of wave existed: nearly two-dimensional (2D) and three-dimensional (3D). The amplitude of the 2D spiral waves grew rapidly and therefore a theory based on the assumption of small amplitudes is not valid across the whole disc. A transition from 2D to 3D waves occurred once their amplitude reached about three to four times the local film mean thickness. Higher liquid flow rates stimulated a more rapid break-up of the wavelets. Only about 1% increase in liquid surface area was ascribed to the presence of waves. Thus, any improvement in heat/mass transfer performance generated by the waves must be due to the additional shear that they induce.

Recently, Sisoev et al. (2005) have mathematically modelled the stability and mass transfer characteristics of a film on a rotating disc. Numerical solutions of the partial differential equations which govern the hydrodynamics, and the associated mass transfer, reveal the formation of large finite-amplitude waves and imply mass transfer rates in broad agreement with experiment.

5.2.9 **Heat/mass transfer performance**

One of the attractive features of the SDR is that its high fluid dynamic intensity favours the rapid transmission of heat, mass and momentum, thereby making it an ideal vehicle for performing fast endothermic reactions which usually also benefit from an intense mixing environment. It must be noted, however, that heat transfer from the process liquid to any cooling/heating fluid behind the disc involves a second film coefficient which may severely limit the overall heat transfer rate (discussed later). Some of the more relevant recent experimental studies of spinning disc performance may now be considered.

Koerfer (1986) investigated the desorption of oxygen using a range of perforated, grooved and smooth 600 mm diameter discs at speeds up to 600 rpm. The average mass transfer film coefficient (after accounting for the contribution from the peripheral liquid spray) is shown in Figure 5.9. As can be seen, the performance of the disc with concentric grooves (10 mm wide, 2 mm deep) was significantly better than that of the smooth disc. The perforated disc had 1.5 mm holes on a triangular pitch which gave 31% open area. This resulted in even higher performance, with mass transfer coefficients up to 70×10^{26} m/s compared with 5×10^{26} m/s under equivalent conditions for the smooth disc. Some of this improvement was attributed to the extra surface area created as the film jumped across the holes.

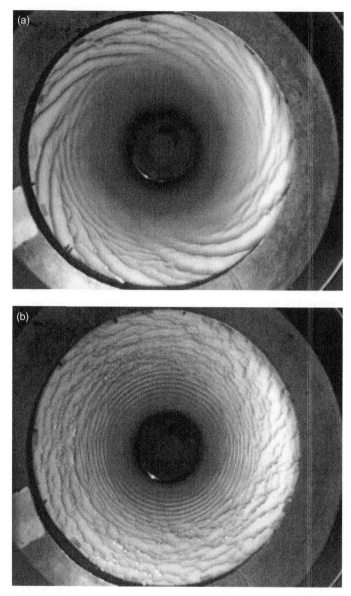

FIGURE 5.8

(a) – (f). Liquid film behaviour on a rotating disc with liquid flow rate of 19 m³/s and rotational speeds of (a) 100 rpm; (b) 200 rpm; (c) 300 rpm; (d) 400 rpm; (e) 500 rpm and (f) 600 rpm.

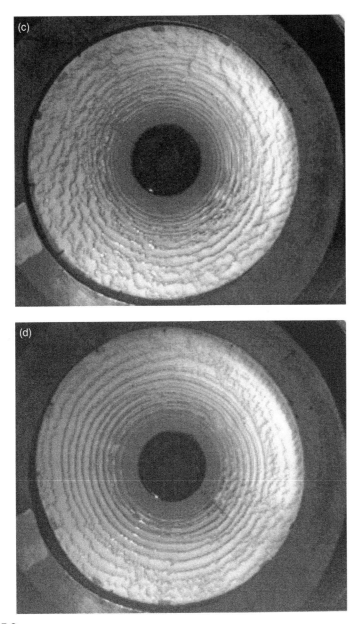

FIGURE 5.8

(Continued)

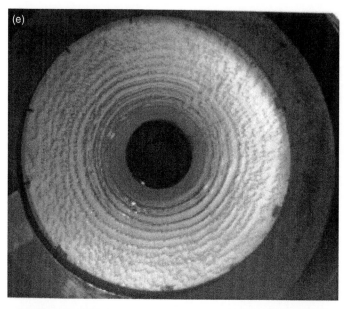

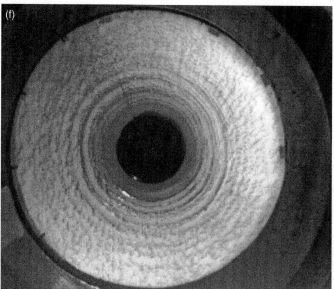

FIGURE 5.8

(Continued)

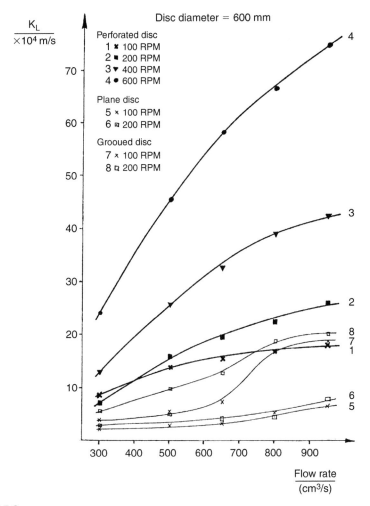

FIGURE 5.9

Mass transfer performance of a rotating disc in oxygen desorption.

Bell (1975) used the CO_2-diethanolamine system to evaluate area-averaged mass transfer performance of a range of smooth, mesh and perforated discs. Once again, the smooth disc gave the lowest mass transfer coefficient, while the perforated and mesh discs were broadly similar. For all the discs, a pronounced maximum coefficient was noted at a critical rotational speed. This phenomenon was ascribed to film breakdown in the peripheral regions of the discs as the rotational speed increased. The existence of peripheral dry areas was confirmed photographically.

Aoune and Ramshaw (1999) measured both local and average heat and mass transfer film coefficients on a smooth 0.5 m diameter disc rotating at up to 900 rpm.

For pure water at inner radial positions the heat transfer coefficient was relatively high, but it exhibited a minimum at intermediate radii before increasing to the peripheral value. Typical heat transfer coefficients were in the range 10–20 kW/m²K whereas a 40% water/60% propylene glycol mixture (with a tenfold higher viscosity) only gave 5–10 kW/m²K. Mass transfer with water was relatively insensitive to radial position, being in the range (3–9) 10^{-4} m/s. The Higbie unsteady state mass transfer model, when based on the film exposure time on the disc, grossly underpredicted the mass transfer rates. Bearing in mind the wavy nature of the flow, this suggests that much shorter exposure times are encountered in practice. The unexpectedly high inner heat transfer rates were attributed to the additional shear stress involved in bringing the feed liquid up to the disc rotational speed, which was not taken into account by the Nusselt model.

Jachuck and Ramshaw (1994) explored the influence of surface profile on the heat transfer performance of a spinning disc. Using a smooth disc as a benchmark it was shown that disc surfaces disrupted with metal powder or grooves gave a significantly improved performance, presumably due to better film mixing. The best performance at modest disc speeds was obtained with undercut grooves as shown in Figure 5.10, which were originally conceived as a means for improving the circumferential distribution of any radial rivulets. At higher disc speeds, the film radial velocity was such that liquid was projected off the disc, thereby compromising the heat transfer process.

When strongly exo- or endothermic reactions are to be performed on a spinning disc, some provision is required to supply or remove the reaction heat involved as

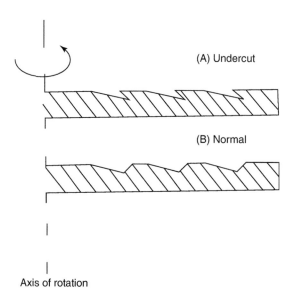

FIGURE 5.10

Types of disc grooves tested at Newcastle University.

rapidly as possible. It has been shown above that the heat transfer film coefficient associated with the process fluid is high. However, care must be taken to ensure that the thermal resistance associated with the heat transfer fluid and the conductive resistance of the disc material does not significantly restrict the overall heat transfer rate. A stainless steel disc 2 mm thick incurs a conductive resistance of around 8 kW/m²K, while an organic heat transfer fluid film coefficient may only be 5 kW/m²K. It is therefore imperative to match, or exceed, the process fluid heat transfer coefficients if possible by using thin conductive disc material and by optimising the flow conditions for the heat transfer fluid. One SDR design is shown in Figure 5.11, where it can be seen that the heat transfer fluid flows over, then under, a stationary splitter plate which is held within the rotating assembly. In order to restrict the pressure drop to reasonable levels, the inner flow gap may need to be greater than that at the periphery.

An alternative, even more effective, SDR design is shown in Figure 3.11 in Chapter 3, which is based on a spiral configuration for the heating fluid channels beneath the disc, whereby the outflow and return channels are comprised of interleaved spirals. The flow cross-section within a given spiral channel is constant and at the designers' discretion. A further important feature is that the effective disc area for heat transfer is enhanced by the inter-spiral walls which act as very efficient fins. From a mechanical point of view, the spiral fins also serve to reinforce the disc surface so that it remains undistorted by the differential pressure forces between the heating channel and the process fluid: an inherent problem with the splitter plate design.

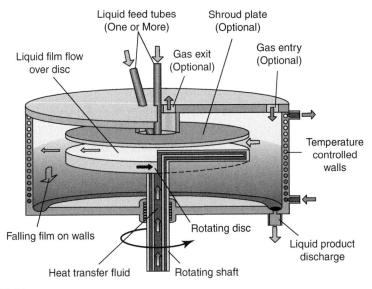

FIGURE 5.11

The spinning disc reactor with heat transfer fluids moving above and below a stationary splitter plate.

5.2.10 Spinning disc reactor applications

5.2.10.1 Strategic considerations

At the most basic level, the SDR is an extremely effective gas–liquid contacting device. This makes it ideal for performing many intensified heat or mass transfer operations and, as will be discussed later, it may be deployed as an evaporator or an aerator/desorber. However, its principal application in the process industry is likely to be as a very high performance reactor. Since the reactor is the heart of any process, the SDR can radically improve the economics and efficiency of many key processes, both in the commodity and in the fine chemical area.

In order to exert full control over the progress of a chemical reaction or physical transformation, the fluid dynamic environment must be intense enough to ensure that the mixing and heat transfer rates are faster than the intrinsic chemical kinetics. This concept is shown schematically in Figure 5.12, which illustrates the progress of a reaction represented simply as A + B→C, with the reactants A and B travelling in plug flow along a tubular reactor. When the interdiffusion of A and B is slow compared with the reaction rate, then C is produced near the original plane of A/B separation. This represents a total loss of control on two counts:

1. The A/B stoichiometric ratio varies widely across the reactor diameter. Therefore, the selectivity for the desired product C, is likely to be compromised because a more realistic reaction scheme will usually include many side reactions.
2. Most of the reaction occurs in the immediate neighbourhood of the plane of A–B separation. Thus, only a small fraction of the available reactor volume is used and an opportunity for intensification is lost.

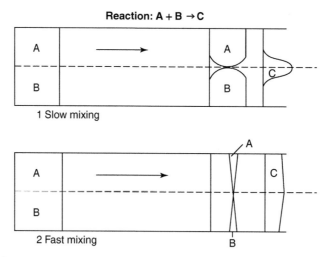

FIGURE 5.12

The influence of mixing and reaction rates on reactor behaviour.

On the other hand, when mixing is fast, the A/B ratio is uniform and control over the product spectrum can be maintained: *all* the reactor space is then used to maximum effect. Also, since the intrinsic kinetics are allowed free rein, the reactor is able to operate at the maximum intensity permitted by the specific chemical system.

While it should be self-evident that a rational reactor design demands knowledge of both the fluid dynamic environment and the detailed process kinetics, the latter are rarely available. In many instances this leads to the severe limitation of many important reactions due to an inadequate fluid dynamic intensity. Some of these are known to be fast, e.g. liquid phase nitrations, while others (incorrectly) are assumed to be slow, e.g. most polymerisations. In these circumstances, the pragmatic approach is to use a high intensity reactor for each system and then to assess the impact upon the space–time productivity. Obviously, an intrinsically slow system is resistant to further acceleration and this will rapidly become evident. One significant qualification of this contention involves the very short residence time in the SDR compared with its conventional counterparts. In certain reactions the process temperature is restricted to one that avoids product breakdown *in the time available*. Since the residence time in the SDR, when for example it is performing a polymerisation reaction, is up to 10 seconds rather than the several hours involved in a conventional stirred vessel, we must re-examine the process's temperature trajectory. A significantly higher operating temperature may well be acceptable because the undesired breakdown component may not have time to be generated. The higher temperature will reduce the liquid viscosity and accelerate the reaction. The lower viscosity may reduce the residence time still further. Therefore, the SDR can exploit a process operating envelope that is much larger than that which is accessible using conventional technology.

With regard to the processing of viscous liquids, by far the most important application of the SDR relates to the manufacture of polymers. Some examples of this are discussed in Chapter 8. However, the key processes involved are:

1. Condensation reactions
2. Radical reactions
3. Devolatilisation

The progress of a condensation reaction is controlled by equilibrium with a volatile product which, if continuously removed, drives the reaction forward. Unfortunately, as polymerisation proceeds, the liquid viscosity increases, thereby rendering the removal of the volatile component much more difficult. The batch stirred vessel which is conventionally used for polymer manufacture has a limited ability to remove a volatile component from the increasingly viscous polymer melt. On the other hand, the SDR can maintain effective mass transfer and, as will be shown later, can achieve in one pass (taking several seconds) the same increment in polymerisation as would conventionally require tens of minutes. An SDR with one or more discs on the same shaft is therefore capable of performing polycondensation extremely rapidly. The short residence time also facilitates rapid changes of product grades with minimum wastage. The rate of a polymerisation that proceeds via a series of radical reactions is controlled by the micro-mixing environment within the polymer melt. Once again, the stirred vessel is a poor means to achieve the desired intensity, whereas the SDR has an impressive capability in this respect. It is well known that ultraviolet radiation is a very effective means for radical generation and this technique has been proposed in the past for stimulating certain radical polymerisations. Unfortunately, the radiation extinction distance in a polymer melt is only a few millimetres, so a polymerisation reactor comprising a stirred vessel having a diameter of several meters is not a rational option. On the other hand, the thin polymer films that can be created and maintained on the SDR allow *all* of the polymer to receive a continuous, uniformly high radiation dose. This then maintains a very high reaction intensity for the manufacture of, for example, acrylic polymers (Dalglish et al., 1999).

Furthermore, the industrial manufacture of polymers is rarely taken to completion, and this requires unreacted monomer to be removed from the product when its viscosity is highest. Such a devolatilisation procedure is notoriously difficult because it usually involves the vacuum stripping of a stirred vessel's contents for many hours. Just as the SDR promotes the removal of the volatile component of a condensation reaction, it is also effective at dramatically accelerating the devolatilisation process. Chapter 8 also includes examples of SDRs used in fine chemicals/pharmaceuticals and precipitations/crystallisation.

5.3 Other rotating reactors

5.3.1 Rotor stator reactors: the STT reactor

In the STT system (Kreido Laboratories, 2004) two films are sheared against one another due to a significant difference in speed, as one is initially adjacent to a stationary outer cylinder (stator) and the other to the inner, rapidly rotating cylinder (rotor). Within milliseconds, the reactants are micro-mixed, speeding chemical reactions by removing mass transfer limitations between the two and allowing them to approach their inherent kinetics. An STT unit in cutaway form is illustrated in Figure 5.22.

Operating parameters such as the annular gap and rotational speed are adjusted for each application. Care must be taken to operate the STT in a regime which avoids Taylor vortices, a phenomenon which restricts complete homogeneity (see Section 5.3.2) and turbulence, which can cause significant back pressure to develop across the system. A typical specification (selected features) for Kreido's Innovator 200 series pilot reactor is given in Table 5.1.

The STT has applications in:

- Chemical synthesis: esterification. Peanut oil converted to soap in 11 s, about 300 times rate increase.
- Polymers: polyolefins. Ten times faster than in a batch reactor, at lower temperatures.

Table 5.1 Data for STT Pilot Reactor Innovator 200 Series.	
Materials for Wetted Surfaces	**SS 316L, Titanium, Hastelloy C**
Size (horizontal reactor)	61 cm W × 76 cm L × 144 cm H
Weight	136–227 kg
Stator inner diameter	64.26 mm
Rotor working length	30.43 cm
Rotor outer diameter (typical)	63.50 mm
Gap between rotor and stator	0.38 mm
Working volume	23 ml
Rotation speed	5,000 rpm maximum
Sealing (standard seal)	45–50 PSIG, 130°C, 5000 rpm

- Polyacrylates: 90% 1 conversions in less than 20 s.
- Solid synthesis: particle size control by shear rate and residence time (about 3 s).
- Bioprocessing: yeast fermentation of molasses decreased fermentation time by about 40%, by dispersing active materials to improve activity/unit volume.

Meeuwse et al. (2012) report reliable scale-up of multistage rotor stator spinning disc reactors with regard to their gas–liquid mass transfer behaviour. The configuration used was based upon multiple units rotating on a common axis. Visscher et al. (2012) have studied the liquid–liquid mass transfer phenomena in such systems, and characterised them as a function of rotational speed for a water–heptane mix, finding that the mass transfer rates can be 25 times those in equivalent packed columns, and 15 times that in micro-channel devices.

5.3.2 Taylor–Couette reactor

The Taylor–Couette reactor (TCR), sometimes referred to as a vortex flow reactor, consists of two concentric cylinders, one of which rotates, Figure 5.13.

If the rotational speed is high enough, this generates stable secondary (Taylor) vortices that have certain advantages in mixing fluids. One key advantage of this form of fluid mixing is that it opens up the possibility of achieving long residence time reactions with plug flow in a relatively compact geometry, as mixing is not dependent upon net flow velocity, as with conventional PFRs, where the mixing is dependent upon achieving turbulence. Each vortex is essentially a stirred tank in the sense that it is well-mixed throughout its volume. If operated correctly, such reactors can therefore operate as a series of well-mixed volumes, resulting in plug flow in a similar manner to CSTRs in series (Pudjiono et al., 1992). This must be tightly controlled, as outside of the narrow operating window there are a number of possible flow structures, fluid can be retained in vortex cores, and/or bypassing of vortices can occur, moving the overall RTD significantly away from plug flow (e.g. Zhu et al., 2000). Indeed, this narrow operating window would appear to be one of the TCR's greatest shortcomings.

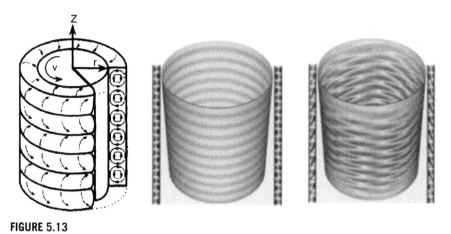

FIGURE 5.13

Taylor vortices in an annular gap (Dutta and Ray, 2004).

Advantages of TCRs include:

- Their relatively open flow structure lends itself to applications involving solids up to very high loadings, as it is not particularly prone to blockages.
- The uniformity of mixing within them compared to a stirred tank.
- High heat and mass transfer coefficients are achievable, partially due to the good radial transport engendered by the vortices.

5.3.2.1 Applications

Applications of the TCR have included single-phase liquid reactions; reactions between immiscible liquids; dispersion of solids in liquids; crystallisations; polymerisations; electrochemistry; fermentation; photochemical reactions; emulsion polymerisation; synthesis of silica particles; heterogeneous catalytic reactions and liquid–liquid extractions.

Kang et al. (2003) studied the application of the TCR to crystallisation of calcium carbonate, exploiting the high mass transfer coefficients (of carbon dioxide in this case) and, they found, enhanced nucleation. In fact, due to the mixing patterns in the reactor, the crystal nucleation and growth processes were found to be near completion in the entry region of the reactor, and the rest of the reactor only acted to reduce crystal size, presumably by attrition.

Giordano et al. (2000) report the use of such reactors for suspension of immobilised enzymes. One of the key advantages was the relatively low shear, as the particles supporting the enzymes were fragile. This low shear would also be an advantage when using whole cells and was investigated for CHO cells by Haut et al. (2003). They found that the reactor was indeed suitable for growing such cells, in two different flow regimes.

Dutta and Ray (2004) developed a photocatalytic reactor for water purification (by oxidising organic contaminants) using a titania catalyst coated onto the inner cylinder, with an ultraviolet light inside the inner cylinder activating the catalyst. The vortices ensure that the fluid has good exposure to the catalyst. This arrangement removes the need for catalyst particle separation found in previous designs.

Vedantam and Joshi (2006) give an excellent review of Taylor-Couette flow, identifying some of the shortcomings of the present state of knowledge in the art. Principally, they say, there is a lack of good experimental data on heat and mass transfer, although Dluska et al. (2001) did evaluate mass transfer coefficients for carbon dioxide/water systems experimentally, finding them to be greatly increased.

5.3.3 **Rotating packed bed reactors**

The rotating packed bed (RPB), as discussed in Chapter 6, has been principally developed for separation processes. However, successful commercial operation of an RPB for reactive stripping, by Dow Chemical Company, in the US, (Trent and Tirtowidjojo, 2001) has opened up the possibility of much wider use of the RPB for mass transfer-limited gas–liquid reactions.

The Dow process involved the reaction of chlorine with sodium hydroxide to produce HOCl and sodium chloride. Chlorine gas is absorbed into the aqueous solution of sodium hydroxide by a liquid-side mass transfer limited process. Once absorbed, the reaction is considered instantaneous. The HOCl will decompose rapidly to sodium chlorate in the presence of sodium chloride, and the aim of the RPB is to strip the HOCl into the gas phase before decomposition can commence. This allows high yield to be achieved. Previous methods used for this step, such as spray distillation, were unable to give the 80% yield required, and the rapid rate of the process means that multiple processes (chlorine absorption, reaction with sodium hydroxide and HOCl stripping) should take place simultaneously in the same item of plant. The RPB was selected, and gave a 90% yield, while allowing the stripping gas flow rate to be halved (further data is given in Chapter 9). The RPB has also been successfully used for the absorption of HOCl into water, provided that the heat of absorption can be effectively removed.

A comparison as a gas–liquid reactor was afforded by the Dow study of an unspecified reaction where the rate was known to be mass transfer limited. The rate of reaction was, specifically, controlled by the rate of reaction gas absorption into a solvent carrying the second reactant. However, when carried out in the RPB, the data output suggested that the reaction had become kinetically limited. Dow Chemicals stated that this allowed the exploration of the reaction chemistry and kinetics to provide a better understanding of the overall process. This was used to improve other mass transfer-limited reactors, and could lead to more applications for RPBs. The performance of the RPB reactor, illustrated in Figure 5.14, in comparison with an STR showed improvements in reaction rate of 3–4 orders of magnitude, with a 2–3 order increase in mass transfer.

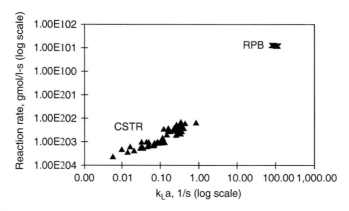

FIGURE 5.14

Reaction rate of the RPB and SDR as a function of mass transfer coefficient (Trent et al., 2001).

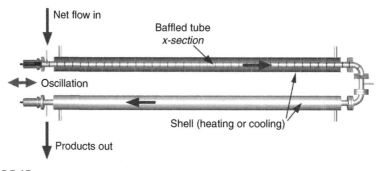

FIGURE 5.15

The oscillatory baffled reactor (OBR).

5.4 Oscillatory baffled reactors (OBRs)

The oscillatory baffled reactor (OBR), illustrated in Figure 5.15, is a form of continuous plug flow reactor, in which tubes fitted with equally spaced, low-constriction orifice plate baffles have an oscillatory motion (range 0.5 to 10 Hz) superimposed upon the net flow of the process fluid.

The most important applications of OBRs lie in running long (greater than 10 minutes) residence time processes in continuous mode. Many such processes are currently run in batch because conventional designs of continuous reactor are impractical (the reasons include cost, control and size). The OBR offers the prospect of a compact plug flow reactor with uniform, controllable mixing. Its scaleability is another distinct advantage; the mixing mechanisms do not change between

FIGURE 5.16

Typical mixing patterns in an oscillatory baffled reactor.

laboratory and industrial scale, given dynamic and geometric similarity. The combination of the baffles and the oscillatory motion creates vortical flow patterns conducive to efficient heat and mass transfer (see Figure 5.16), whilst maintaining plug flow.

On the down stroke of the piston (as in Figure 5.16) toroidal vortices are formed below the baffle. The vortices dissipate on the following upstroke of the piston, as similar vortices are formed above the baffle. This constant creation and destruction of vortices results in a well-mixed volume of fluid on either side of the baffle. This can be considered to be a small well-mixed stirred tank. If enough of these well-mixed volumes are put in series, then a plug flow residence time distribution can be achieved. However, unlike conventional tubular reactors, where a minimum Reynolds number must be maintained (to achieve turbulence to produce a flat velocity profile), the degree of mixing is independent of the net flow, allowing long residence times to be achieved in a reactor of greatly reduced length-to-diameter ratio.

The residence time distribution has been found (Stonestreet and Van der Veeken, 1999) to be a function of the net flow Reynolds number, Re_n, and the oscillatory Reynolds number, Re_o, where they are defined as:

$$Re_o = \frac{2\pi f x_0 \rho D}{\mu}$$

$$Re_n = \frac{\rho v D}{\mu}$$

Where f is the frequency of oscillation (note that this is usually a few Hz), x_0 is the centre-to-peak amplitude of the oscillation and the other variables are as in the conventional Reynolds number for flow in pipes. The ratio of the two Reynolds numbers is termed the velocity ratio, ψ, and is a measure of the degree of plug flow. Generally, OBRs are designed such that ψ is in the range 2–12 as in Figure 5.17, as this ensures good plug flow, typically equivalent to more than 10 perfectly

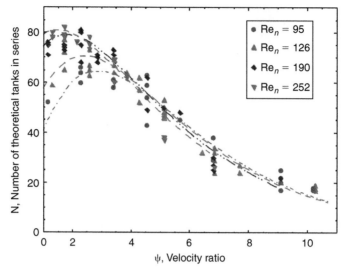

FIGURE 5.17

Residence time distributions of oscillatory baffled reactors.

stirred tanks-in-series. Due to the decoupling of mixing from net flow, plug flow is achieved in the OBR when Re_n is laminar.

An OBR can therefore be much more compact than conventional plug flow reactor designs, allowing reactor designs for longer residence times to be of practical dimensions. The main niche application is the conversion of long residence time batch processes to continuous processing, where other continuous reactor designs prove to be too costly or impractical (typically due to footprint, control or pumping duties).

This technology can demonstrate advantages in a range of unit operations, including heat and mass transfer, multiphase mixing, particle suspension, bioreactions and fermentations.

5.4.1 Gas–liquid systems

OBRs exhibit enhanced mass transfer between gases and liquids. The main mechanisms for the enhancement are increased hold-up and increased breakage of bubbles (reducing size, thereby increasing interfacial surface area and reducing rise velocity, which increases bubble residence time further). Figure 5.18 compares the mass transfer coefficient for an OBR with that of an STR for an air–water system on the basis of power density (Ni and Gao, 1996).

Mass transfer has also been shown (Ni et al., 1995) to be significantly increased in yeast cultures in the OBR. Figure 5.19 is for an air–water–yeast system.

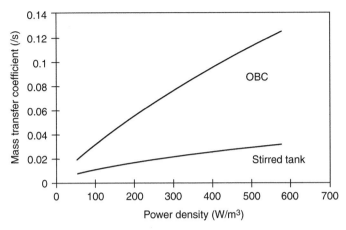

FIGURE 5.18

OBR–STR mass transfer comparison (Ni and Gao, 1996).

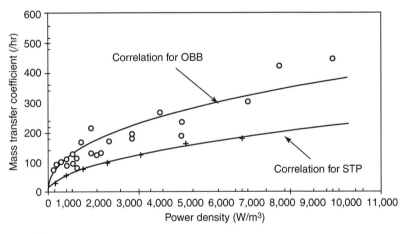

FIGURE 5.19

Enhanced K_lA due to oscillatory mixing.

5.4.2 Liquid–liquid systems

A significant advantage of oscillatory flow operation has been realised in liquid–liquid systems. The controlled and uniform mixing intensity has been shown to result in very narrow droplet size distributions in applications such as suspension polymerisation (Ni et al., 1998, 1999, 2000a; Stevens 1996). Such processes have been modelled, principally, by the population balance approach (Ni et al., 2000a,b; Stephens et al., 1997), and have been successfully correlated to experimental findings, see Figure 5.20.

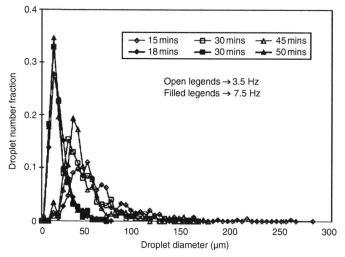

FIGURE 5.20

Typical droplet size distribution.

5.4.3 **Heat transfer**

Tube-side heat transfer coefficients can be significantly enhanced compared to flow at the same net flow Reynolds number in steady flow systems. As can be seen in Figure 5.21, the Nusselt numbers converge at higher Reynolds numbers, but operation at such high net flow Reynolds numbers would negate the design advantages of using OBRs. In the lower Re region, Nu is typically enhanced by a factor of between 10 and 30. This enhancement is due to the increased radial transport characteristic of the vertical mixing patterns. It is clear that heat transfer coefficients equivalent to those of smooth-walled tubes in well-developed turbulence can be achieved at much lower (net flow) Reynolds numbers in an oscillatory baffled system.

Mackley and Stonestreet (1995) developed a correlation between Nu and the net and oscillatory flow, Re:

$$Nu = 0.0035\,Re_n^{1/3}\,Pr^{1/3} + 0.3\left[\frac{Re_o}{(Re_n + 800)^{1.25}}\right]$$

5.4.4 **OBR design**

Harvey and Stonestreet (2002) developed a design protocol for the OBR based upon that for a shell-and-tube heat exchanger (see Figure 5.22). The main difference between this protocol and a typical shell-and-tube design is that it is based upon establishing good mixing and plug flow, which are not explicitly a part of any

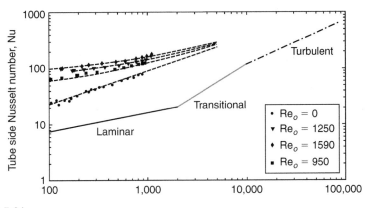

FIGURE 5.21

Comparison of Nu versus Re_n for smooth-walled tubes and oscillatory baffled reactors at varying Re_o (Mackley and Stonestreet, 1995).

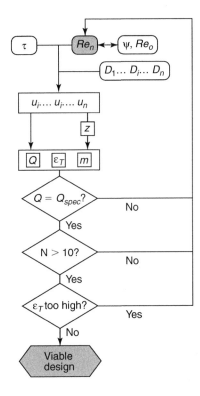

FIGURE 5.22

Mixing based design methodology for continuous oscillatory baffled reactors (Harvey and Stonestreet, 2002).

shell-and-tube protocol. These characteristics are ensured by the correct choice of the input values of Re_o, the oscillatory Reynolds number, and ψ, the velocity ratio.

Scale-up is achieved by maintaining dynamic and geometric similarity via maintaining the same values of Reynolds number. Smith (2000) demonstrated this approach successfully by measuring the degree of axial dispersion for systems at different diameters. This predictable, well-characterised scale-up is one of this reactor's great strengths, as scale-up of CSTRs/STRs is frequently difficult to predict. This is the main body of the case made for using this reactor in a number of applications, such as protein refolding (see Section 5.4.5), where scaling up of conventional technologies is difficult. Furthermore, Ni and Gao (1996), showed that mass transfer coefficients actually increased with increasing scale.

Energy dissipation was considered by Baird and Stonestreet (1995), who calculated it based on addition of power consumption terms due to oscillatory flows and to net flow with an oscillatory enhancement factor.

5.4.5 Biological applications

The mixing in OBRs is very uniform, as the radial and axial velocities are of the same magnitude (Baird et al., 2003). This has been shown to be an advantage when handling shear-sensitive materials, such as certain pharmaceutical crystals, and in flocculators (Ni et al., 2001). Low and uniform shear is one of the aspects of OBRs that should allow them to be used as bioreactors. Other advantages of the OBR for biological applications include:

1. Continuous plug flow operation for long processes, which most bioprocesses are.
2. Enhanced gas-liquid mass transfer. This is often the limiting factor in aerobic systems, and the advantages are clearly demonstrated in Section 5.4.1 for an air–water–yeast system.
3. Heat transfer enhancement (see Section 5.4.3).

Case studies in this area include:

1. Formation of pullulan, a biopolymer, by *Aureobasidium pullulans* (Ni et al., 2005). It was demonstrated (for a batch OBR) that a 50% reduction could be achieved in the time to required yield versus a conventional stirred tank.
2. Growth of *Alcaligenes eutrophus H16* for the production of the biodegradable plastic, polyhydroxybutyrate (PHB) (Mackley and Harrison, 1992), where the biomass growth matched that of a conventional laboratory scale shake flask, but is considerably more scaleable.
3. Protein refolding: it has been demonstrated that the OBR can be used for protein refolding, using a direct dilution method (Lee et al., 2001). Its performance is comparable to that of a stirred tank, but the great advantage of the OBR here is its scaleability, so the outcomes observed in the laboratory should translate to larger scale production, which is not the case with current, stirred tank technology.

4. Flocculation: oscillatory flow conditions require lower shear rates than stirred tanks to cause a given degree of flocculation. Furthermore, the narrower shear rate range in OBRs enhances control of floc size and morphology. Ni et al., (2001) demonstrated that oscillatory baffled devices could require as little as 5% of the average shear rate in an equivalent stirred tank to provide the same flocculation rate, in this case for bentonite with *Alcaligenes eutrophus*.
5. Production of an aroma compound γ-decalactone using *Y. lipolytica* cells: a meso-reactor (see Section 5.4.8) was used to demonstrate a 50% reduction in time to required conversion, attributed to enhanced liquid–liquid mixing, with the production rate shown to increase linearly within the range of conditions used, with oscillator Reynolds number (Reis et al., 2005).

5.4.6 Solids suspension

The flow patterns within the OBR can, in principle, be used to suspend solids, e.g. for use as catalysts or reactants. Suspension can be uniform or stratified as desired. Mackley et al. (1993) developed a correlation to describe the behaviour:

$$\gamma = \left[1 - \exp\left(-R\frac{V_m}{V_s}\right)\right]$$

...where V_m is the maximum oscillation velocity, V_s is the settling velocity of the particles, and R is a constant. The value γ was used to determine the uniformity of suspension in the reactor. $\gamma = 1.0$ indicated a uniform suspension, whereas $\gamma < 1.0$ indicated that a particle concentration gradient could be achieved. Note that this value could be altered dynamically by changing the oscillation velocity, V_m.

Another example of the OBRs use with solid particles is as a photochemical reactor with solids suspension, in this case the vortical flow patterns being used to suspend catalytic titania particles to convert organics in waste water. The titania needs to be activated by ultraviolet, and the reaction requires the presence of oxygen, so air is bubbled through. The gas–liquid mass transfer is enhanced by the oscillation of the fluid, as it increases hold-up time (bubble residence time) and reduces bubble size (increasing surface area and further increasing hold-up time). The flow patterns simultaneously ensure good exposure of the titania particles to the radiation from an axially located ultraviolet lamp.

5.4.7 Crystallisation

Lawton et al. (2009) report the continuous crystallisation of a model active pharmaceutical ingredient (API) by solution cooling in an OBR, finding that the OBR is advantageous for reasons of heat transfer (increased surface-area-to-volume ratio, allowing increased control and uniformity of temperature), linear scale-up and the ability to easily operate at slightly higher pressures and temperatures. They report a substantial reduction in processing time from 9 h 40 min to 12 min, which would result in a substantially smaller reactor, on a same-throughput basis.

Callahan and Ni (2012) report that baffle scraping on the inner surface of the reactor tube may be responsible for many of the differences observed between crystallisation in STRs and the OBR. Specifically, in the crystallisation considered, it appears to be a source of primary nucleation, whereas all nucleation in the STR is secondary.

Ristic and Chew (2005) report that use of oscillatory mixing rather than a stirred tank for crystallisation of paracetamol results in particles of smaller size, smoother surface and lower micro-strain (crystalline imperfections), at equivalent power density. They claim that this is due to the growing crystals being exposed to alternating directions of flow, rather than having one preferential direction, as is often the case in STRs (tangential).

To summarise, possible advantages of the use of OBRs for crystallisation include:

- Temperature control: allowing tighter control of the crystals produced.
- Predictable scale-up: allowing laboratory scale result to be translated to larger scales. This is a significant challenge in many unit operations, but in crystallisation particularly.

5.4.8 Oscillatory meso-reactors: scaling OBRs down

A relatively recent development in OBR technology is downscaling the reactor, so that it can be used for applications such as small-scale continuous production of pharmaceuticals, or as a scaleable continuous screening device. To this end, the oscillatory flow meso-reactor was developed at Cambridge (Reis et al., 2005). In its current incarnation it consists of jacketed glass tubes of the order of 5 mm in diameter and 25 cm in length. It has been demonstrated via PIV/CFD studies (Reis et al., 2005) and residence time studies (Reis et al., 2004) that it reproduces the flow patterns of OBRs at larger scales. It has also been demonstrated to effectively suspend relatively high concentrations of particles, which would allow its use for, for example, screening of solid catalysts. Indeed, the device has been demonstrated for high throughput screening of solid catalysts for biodiesel production (Zheng et al., 2007).

Phan and Harvey (2010) have developed a range of different baffle designs, which have a range of different applications, varying with the viscosity and the phases present. A particularly interesting design is the helically baffled oscillatory flow meso-reactor, which exhibits an extraordinarily wide operating window for plug flow (Phan and Harvey, 2012). Solano et al. (2012) have modelled the heat transfer characteristics of the helical oscillatory flow meso-reactor, showing that the flow is a combination of oscillation and a swirling flow and has the potential to substantially increase the heat transfer coefficient.

One well-known liquid–liquid system that was investigated, initially simply as a model system, was biodiesel production (Phan et al., 2012). It was shown that, with the tight control of conditions possible in such meso-reactors, the residence time required could be reduced to 2 min (c.f. industrial residence times of 1–2 h). The factor by which a reactor volume could be scaled would therefore be at least a factor of 120 (probably more due to other batch processing inefficiencies). Part of this was a consequence of being able to operate at higher catalyst concentrations, which were not possible without tight control of residence time, as beyond this residence time it was shown that a saponification side reaction became dominant, and converted both product (biodiesel) and reactant (vegetable oil) to soap. This effect would simply have not been observed without the tight control of residence time possible in a meso-reactor.

5.4.9 Case study

> **INTENSIFICATION OF A BATCH SAPONIFICATION PROCESS BY CONVERSION TO CONTINUOUS PROCESSING IN AN OSCILLATORY BAFFLED REACTOR (OBR)**
>
> Conversion of a batch saponification reaction to continuous processing in an OBR has been shown (Harvey et al., 2001) to result in a 100 fold reduction in reactor size, as well as greater operational control and flexibility.
>
> A 1.3 litre OBR was evaluated for an industrial batch saponification process where conversion to continuous processing in conventional tubular reactors was considered unfeasible, due to the long residence time required. The saponification reaction was the hydrolysis of a complex natural mixture of esters in an ethanol/water solvent.
>
> It was found that the OBR could achieve the batch product specification in a residence time of 12 min: one tenth that of the batch reactor (2 h). This was because the reaction was able to proceed at its inherent kinetic rate due to the improved mixing characteristics of the OBR, i.e. the reaction was no longer *mixing-controlled*.
>
> A flow-conversion model, incorporating a tanks-in-series residence time distribution and the reaction kinetics, agreed well with the experimental results, predicting that the reaction could be performed at relatively low residence times when kinetically controlled. The model led to a number of insights into the optimal operation of the OBR. The most significant prediction was that the OBR could give the desired conversion and selectivity at lower temperatures, without significant alteration to other process variables, such as residence time and molar ratio. These predictions were extensively verified by experiment. Furthermore, at these lower temperatures an improved product quality specification was achieved, as there were two competing reactions in this particular process, one producing the desired sterol, the other a similar sterol that had to be maintained below a certain level.
>
> The greatest incentive for conversion of this saponification reaction to continuous processing was safety. The batch reactor used (see Table 5.2) contained a large volume (50 m^3) of solvent, kept at a temperature above its boiling point at ambient pressure. The

---

Table 5.2 Comparison of Operating Conditions for Batch and OBR.

	Current Batch	Predicted OBR
Temperature	115°C	85°C
Pressure	1.0 barg	1.7 barg
Volume	75 m^3 (50 m^3 fill)	0.5 m^3
Residence time	2 hr	12 min

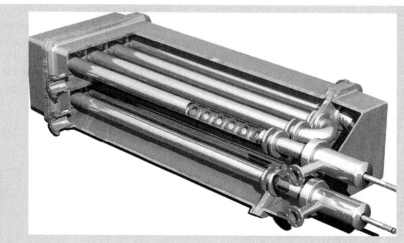

FIGURE 5.23

A conceptual design for a 500 litre industrial-scale oscillatory baffled reactor.

switch to continuous operation would greatly lower the inventories of solvent, thereby significantly improving safety. In this case the safety would be further improved, as the new operating temperature of 85°C was much closer to the ambient pressure boiling point of the solvent. The lower temperature operation would result in energy savings, although the main reduction in energy usage would primarily be due to the more efficient heat transfer. The operating conditions and size of a full scale OBR based upon the results of these trials, (assuming the same production rate and product quality) are given in Table 5.2. It should be noted that various studies have demonstrated that scale-up of OBRs is predictable (see Section 5.4.4).

The estimated size reduction results from the decrease in required residence time and the increase in occupancy: 100% for a continuous reactor, less than 10% for this particular batch reactor, as only 2 h of a 24 h cycle were taken up with the reaction, the rest consisting of cleaning, heating, cooling, emptying and idle time.

A conceptual industrial-scale unit, a 20-pass, 500 litre OBR, is shown in Figure 5.23. This would be capable of a continuous production rate of 2 te/h if a 15 min mean residence time is assumed. The design is based upon a 100 mm tube diameter fitted with standard orifice baffles (with a hole diameter of 50 mm) spaced at 150 mm. Each tube is approximately 3.5 m long. The reactor is heated by a fluid-filled outer shell, fitted with internal baffling to promote efficient fluid flow.

5.5 Micro-reactors (including HEX-reactors)

It was often assumed in the past that a high degree of turbulence equated to good mixing and this is indeed, generally true, but this is only the case for larger scale mixing phenomena, principally turbulent eddies. Diffusion can be used for mixing; but it only becomes a significant mixing phenomenon in the correct circumstances, usually when operating at very small-scale. Micro-reactors, as the name suggests, are reactors with channel sizes of the order of micro-metres, at which scale diffusion is the dominant mixing mechanism. They typically consist of a series of plates with machined or etched channels less than 1 mm in width.

The term 'micro-reactor', however, has come to refer to reactors with small channels varying in size from catalytic plate gas phase reactors involving channel dimensions of 1–2 mm, to liquid phase processors which may contain channels of 200 µm equivalent diameter. In the context of the commodity and fine chemical industry, it is unlikely that the 1 micron channels, currently being investigated at various laboratories, will be a viable process technology in view of their minute production capability, even when several thousand channels operate in parallel. This, of course, raises the issue of manifolding. The general area of micro-processor development for single-phase liquid reactions has been spearheaded by mainland Europe and the US, though the more difficult applications involving multiphase (liquid–liquid) reactions have been relatively neglected. This is partially due to the difficulties associated with ensuring uniformity of distribution of the reactants when perhaps 10,000 parallel streams are involved, with reactants having differing viscosities. Thus manifold design is absolutely critical.

The study of true micro-reactors has become a vast field, too large to be covered in this text, hence this section focuses on 'larger' micro-reactors, which extends to cover catalytic plate reactors and heat exchanger (HEX-) reactors. Since the pressure drop in laminar flow for a given mass flow rate varies as $(diameter)^{-4}$ there is a strong disincentive to use the 1 micron channels currently being considered for the latter application. Furthermore, the risk of fouling will also increase with decreasing channel width. Burns and Ramshaw (1999) described a unit where industrially competitive reaction rates can be achieved in a liquid–liquid reactor with modest bore sizes of between 127 and 254 µm.

Chemical engineering interest in this technology centres on the use of such reactors in manufacturing drugs and fine chemicals at scales of up to, for example, 500 tonnes/year, rather than for bulk chemical production, although, as illustrated later, the work of Chart Energy and Chemicals with their plate fin heat exchanger reactor has extended into the bulk application area.

The use of falling films, as in the IMM micro-reactor system shown in Figure 5.24, may seem to be a risky strategy, owing to the perceived lack of control of the film (governed by gravity). However, there are many options described elsewhere in this book for managing the flow. Loewe and Hessel (2006) described this unit's use for fluorination of toluene at 15°C, where it replaced a process that normally required operation at 270°C.

An approach adopted by Luo et al. (2007), based upon previous work in the US by Professor Adrian Bejan, uses the concept of dichotomic branching i.e. each branch divides into two sub-branches. The unit reported by the authors distributes an inlet flow (such as 300 l/h) into 256 outlet flows evenly distributed over a square cross-section of 6×6 cm². The 256 flow paths are all hydraulically equivalent. The distribution of channel diameters is optimised according to Murray's criterion, with a scaling ratio of $2^{1/3}$. The device was manufactured by stereolithography.

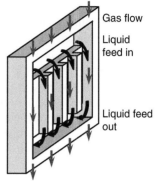

FIGURE 5.24

Steel falling film micro-reactor of IMM. Characteristics: 64 channels, each 300 μm wide, 100 μm deep and 65 mm long; surface area 20,000 m^2/m^3.

(courtesy: UCL)

Manifold optimisation is a compromise between pressure drop and porous volume. This leads to an optimal channel diameter distribution based on scaling laws, while the lengths are determined by the specification of the outlet surface. For distributors with a dichotomic structure, such as that shown, and in conditions of laminar flow, the main theoretical limitations of this treatment are that it neglects inertial effects, and thus applies only at very small Reynolds numbers. In addition to the use of this design method in manifolds for micro-reactors, Luo et al. (2007) also report interest in several other areas, including:

- Cooling radioactive fluids using compact exchangers that are later treated as waste.
- Formulation of multi-component products: continuous mixing of powders and additives, cosmetics, detergents, food.
- Controlled-size emulsions, mixing of viscous fluids.
- Generation of controlled swarms of bubbles and drops in extraction and washing processes.

5.5.1 The catalytic plate reactor (CPR)

In a catalytic plate reactor (CPR), metal plates coated with a suitable catalyst are arranged such that exothermic and endothermic reactions take place in alternate channels (Figure 5.25). These channels typically have a width of the order of millimetres and a catalyst thickness of the order of microns.

The advantages of CPR designs over conventional reactors are the high heat transfer coefficients and minimal intra-catalyst diffusion resistance. The heat transfer mechanism within a CPR is conduction through the plates separating alternate process channels and as such, is largely independent of the process gas superficial

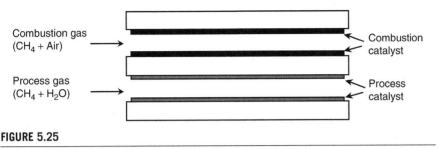

FIGURE 5.25

A pair of adjacent channels in the catalytic plate reactor.

Table 5.3 Size Reduction using Catalytic Plate Reactors for the Fischer-Tropsch Reaction.

Reactor	Reactor Volume	Yield of C5 + kg/m³/hr
Sasol	432.1	29.98
Exxon	23.8	43.84
Catalytic plate reactor	1.0	163.12

velocity. The catalyst layers within a CPR are thin, which results in minimal diffusion limitations and thus, high catalyst utilisation. These advantages result in reactors which are smaller, lighter and have a smaller associated pressure drop than conventional alternatives. The potential saving in reactor volume can be seen in Table 5.3.

5.5.1.1 Steam reforming

The feasibility of the concept of coupled endo- and exothermic reactions has been investigated *in silico* for steam reforming of methane (Zanfir and Gavriilidis, 2003). This is a suitable reaction as it is fast and highly endothermic, and can be matched to the catalytic oxidation of methane. The studies show that it should be possible to reduce the size of steam reformers by two orders of magnitude, hence this is, in principle, an excellent example of process intensification that could revolutionise this industry.

The replacement of the homogeneous combustion used in conventional reactors by catalytic combustion brings several other advantages. The reaction occurs at a lower temperature than conventional combustion, posing fewer constraints for materials of construction and producing virtually no NOx. As it is a flameless process, long radiation paths needed in conventional fired furnaces are replaced by channel dimensions of 1–2 mm, greatly reducing reactor size. The device shown in Figure 5.26 is produced by diffusion bonding a number of patterned metal shims, yielding a compact metal block that can operate at high temperatures and pressures.

FIGURE 5.26

Design of simple CPR by using stacked diffusion bonded shims (manufactured by Chart Energy and Chemicals).

5.5.1.2 Methane reforming

CPRs are also being investigated for methane reforming, using nickel-based catalysts. The technique has been shown to strongly influence both the CO:H$_2$ ratio and rate of carbon lay-down.

The conventional simple plate CPR concept suffers from two major disadvantages. The first is that it is difficult to replace the catalyst when it is exhausted; and the second is that, since the rate of heat generation decreases as the fuel is depleted, (the rate is dependant on approximately [CH$_4$]$^{0.76}$) the last section of the reactor contributes very little to the overall conversion of the fuel unless significant methane slippage is accepted.

A novel reactor design called the Hot Finger (HF) that addresses problems of catalyst deactivation was patented by Protensive. The reactor allows for some degree of axial heat integration. The unit is illustrated in Figure 5.27, detailed in Figure 5.28 and as a complete unit in Figure 5.29.

5.5.1.3 Fischer-Tropsch synthesis

A second application of the CPR is product enhancement for catalytic reactions, where the product spectrum is highly dependant upon catalyst temperature. In such an application, alternate channels contain a boiling heat transfer fluid to maintain an isothermal catalyst temperature. The hydrocarbon product spectrum produced by a Fischer-Tropsch catalyst is highly dependent upon catalyst temperature and rate of diffusion of reactants into the catalyst matrix. The reaction is highly exothermic

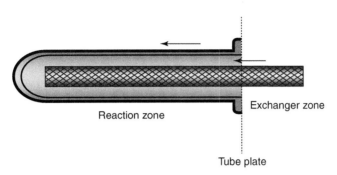

FIGURE 5.27

The Hot Finger reactor.

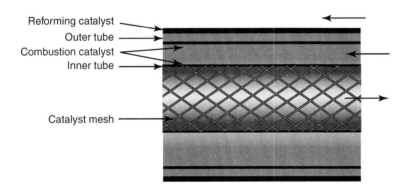

FIGURE 5.28

Details of the Hot Finger.

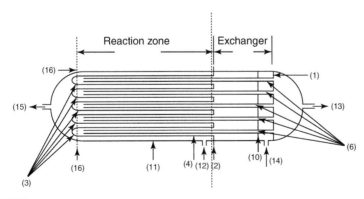

FIGURE 5.29

A complete HF reactor (UK Pat: GB 2372462). The numbers refer to descriptions found in the cited Patent.

and, if rates of heat removal from the catalyst are not sufficiently high, hot spots will form which result in degradation of the product spectrum. Studies have revealed that thin catalyst coatings attached to heat transfer surface areas within a CPR can greatly enhance the yield of desirable products per unit volume as compared to conventional fixed-bed technology. This volume saving, coupled with an overall lighter design, necessitates less ancillary equipment and the low pressure drop makes the Fischer-Tropsch CPR a potential reactor for the recovery of stranded gas reserves.

5.5.2 HEX-reactors

Heat exchanger reactors (often referred to as a HEX-reactors) are essentially compact heat exchangers used as reactors. The advantage conferred is, as in their heat transfer duties, their large heat transfer surface areas per unit volume, which when used as reactors allow them to accommodate highly exothermic reactions.

Generally HEX-reactors are advantageous for fast reactions with substantial exo- or endotherms, with sizeable by-product formation. Industrial examples include nitrations, polymerisations, hydrogenations, halogenations and aminations. Typically, such processes have by-product outputs of rates between 1–5 kg/kg of desired products (in bulk chemicals), and 5–50 kg/kg of product for fine chemicals. These processes provide the greatest opportunity for realising the benefits of HEX-reactors compared to stirred tank reactors. HEX-reactors are usually derived from existing compact heat exchanger variants, e.g. the printed circuit reactor (PCR) and Chart-kote units.

5.5.2.1 The Alfa Laval plate heat exchanger reactor

Introduced in the literature in 2007 (Haugwitz et al., 2007), the open plate reactor (OPR) is based upon the well-known Alfa Laval Plate Heat Exchanger (see Chapter 4). The OPR is specifically designed to handle highly exo- or endothermic and fast reactions. The aim is to safely produce the chemicals using highly concentrated reactant solutions. The main new features of the OPR, needed to reach this objective, are the improved heat transfer capacity and the micro-mixing conditions inside the reactor. It consists of reactor plates, inside which the reactants mix and react (see Figure 5.30), and cooling plates, inside which cold water flows. There is one cooling plate on top of each reactor plate and one below, see Figure 5.31.

In Figure 5.31 the OPR is shown from two different angles. The figure on the left illustrates the first rows of the reactor plate. Each row is divided into several cells. In the figure above, ten cells constitute one row. The primary reactant, A, flows into the reactor from the inlet on the upper left. Between the inlet and the outlet, the reactants are forced by inserts to flow in horizontal channels of changing directions. The flow inserts are specifically designed to enhance the micro-mixing and guarantee good heat transfer capacity, and are the subject of patents. The dashed vertical lines represent the cooling water channels on each side of the reactor plate. The figure on the right shows the OPR from the side, with cooling plates on each side of the reactor plate.

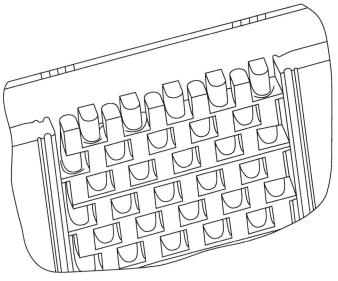

FIGURE 5.30

The internal structure of the Alfa Laval open plate reactor.

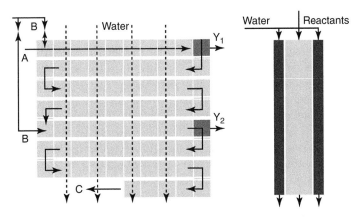

FIGURE 5.31

Left: A schematic of a few rows of a reactor plate. Reactant A is injected at top left and reactant B is injected at multiple sites along the reactor. Y1 and Y2 are internal temperature sensors used for process control and supervision. The cooling water flows from top to bottom in separate cooling plates. Right: the plate reactor seen from the side, with the reactor part in the middle and cooling plates on each side.

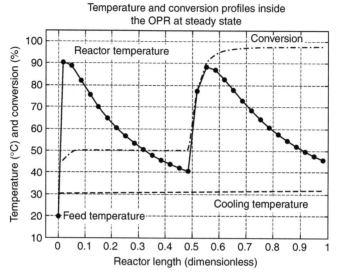

FIGURE 5.32

Simulated temperature (solid), conversion (dash–dot) and cooling temperature profile (dashed) along the OPR (n = 30) with reactant injections at two sites resulting in two temperature maxima at T = 90.4°C and 88.5°C.

The design concept for the reactor allows for great flexibility in adapting the process for new reaction schemes. The type of inserts and the number of rows in the reactor plate can be adjusted to provide the desired residence time. While the cooling plates on each side of the reactor plate have vertical flow channels, the reactor channels are horizontal, giving a cross-flow heat exchange pattern. However, the general flow direction of the reactor flow is vertical, so the heat exchange pattern can be modelled as co-current. It is also possible to have separate cooling flows with different temperatures to cool selected parts of the reactor. For some reactions it may even be beneficial to heat the last section of the OPR to further increase the conversion.

Figure 5.32 shows the performance of the OPR. Together this set of input variables leads to two temperature maxima of T = 90.4°C and 88.5°C. The conversion of reactant A reaches 50% after the first injection, due to the surplus of reactant A in the first half of the OPR. After the second injection, the conversion reaches 97.8% at the outlet of the reactor.

5.5.2.2 The HELIX reactor

The HELIX heat exchanger was developed by TNO (TNO – patent WO 92/07226). It consists of one or two helical wound tubes, as shown in Figure 5.33, usually around a straight tube. The very small radius results in significant secondary flows (Dean vortices) in the wound tubes. This leads to a great degree of radial mixing,

FIGURE 5.33

The HELIX reactor unit (tube diameter 5 mm).

thereby enhancing the heat transfer. The mixing is equally of service when it is used as a reactor.

TNO report use of this reactor for production of 1-ethyl-3-methyl-imidazolium bromide, as a model reaction, successfully demonstrating a high degree of temperature control in the reactor, in the absence of a solvent. Another model reaction evaluated, this time for crystallisation, was calcium carbonate formation. Here the particle size of the crystals was shown to be reduced and the particle size distribution narrowed substantially, principally due to the flow structures in the HELIX.

5.5.2.3 The PCR – printed circuit reactor

Heatric used the inherent plug flow characteristics of the printed circuit heat exchanger (PCHE) (see Chapter 4), to develop the In passage printed circuit reactor (IP PCR), which merges chemical reaction, fluid mixing and heat transfer into one simultaneous operation. Compared to conventional reactors, the IP PCR offers improved selectivity or productivity through a number of features, some of which are common to other catalytic plate and HEX reactors. Features worthy of highlighting in the case of the IP PCR are:

- Plug flow operation – very narrow residence time distribution.
- Close temperature control, enabling:
  - Increased reaction temperature
  - Optimised reaction temperature profile
  - Elimination of hot spots
- Reactant addition on passage-by-passage basis, with rapid mixing
- Multiple sequential reactant additions and reactions.

With regard to temperature control, the multi-fluid capability of PCRs allows the reaction zone to be bounded by several different heat exchange zones. These zones can perform distinct functions, such as preheating and quenching the reactants. If the reaction is endothermic or exothermic then the reaction zone can include a heat exchange fluid which can either add or remove heat. Fast exothermic reactions (e.g. direct oxidations, nitrations) will especially benefit from the precise temperature control available, increasing both yield and conversion.

Staged reactant addition and mixing of reactants can be achieved by etching a network of fluid distributor channels into the plates. These perform passage-by-passage mixing of process fluids. Mixing of several different fluids or staged mixing of a reactant at intermediate points can be achieved by using further sets of distributor passages, each independently fed. Heatric points out that further potential benefits of the IP PCR include:

- Reduced residence time: both reaction time and cooling (quench) or heating time.
- Suitability for design pressures in excess of 600 bar, which opens up the opportunity to operate at supercritical conditions.
- Facility to perform a sequential series of reactions in a single IP reactor, with no significant intermediate hold-up or residence time: ideal for processes involving hazardous or unstable intermediates.
- Negligible scale-up risk: more passages, not bigger passages.

5.5.2.4 The multiple adiabatic-bed PCR

The multiple adiabatic-bed PCR (MAB PCR) can operate with heterogeneous catalysis. The unit separates reaction and heat transfer, allowing each to be optimised (see Figure 5.34 and Seris et al., 2008). In a steam reformer a high degree of heat integration can be achieved using PCHE and MAB technology. The plant can be divided into four sections, as shown in Figure 5.34: the pre-reformer; the main block (including reforming, combustion and boiling); the steam drum and the low temperature water gas shift reactor (LTS). The main plant module contains nine reforming reactors, ten combustors and eleven heat exchangers in a single integrated block. The configuration of this module is the subject of a patent. The combustion occurring in this block is the only heat input to the process, providing heat not only for the endothermic reaction, but also for steam generation.

In this particular unit, the reforming reaction was carried out at an absolute pressure of approximately 2 bar, since the plant was conceived as a fuel processor for a PEM fuel cell that would operate at slightly above atmospheric pressure. The reformer catalyst was a proprietary monolith-supported, noble metal catalyst (Engelhard). The combustion side operated at atmospheric pressure and employed a proprietary palladium-based, monolith-supported catalyst (Engelhard). Heatric states that the MAB PCR offers improved selectivity or productivity on a broader basis through:

- Close temperature control, through:
  - Multiple, smaller reaction steps
  - Multiple inter-stage heat exchange
- Optional reactant addition at each step.
- Optimised catalyst bed aspect ratio.

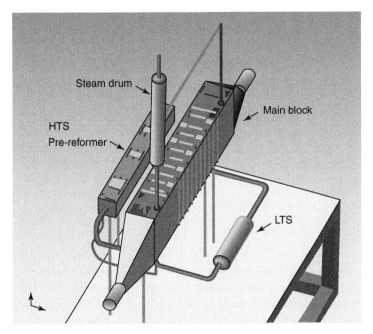

FIGURE 5.34

3D representation of the steam reforming pilot plant. The length of the main block is about 700 mm.

- Smaller catalyst particles – higher catalyst effectiveness.
- Reduced adiabatic residence time.

The MAB PCR combines two established technologies: alternating adiabatic catalytic reactor beds and heat exchangers. It is suitable for applications ranging from fuel processing to the production of fine and bulk chemicals. The PCHE cores used for heat exchange are constructed as described in Chapter 4. Staged reactant addition and mixing of reactants can be achieved by etching a network of fluid distributor channels into the plates. These perform passage-by-passage mixing of process fluids and uniformly distribute process streams into catalyst beds, in a similar way to that in the IP PCR.

5.5.2.5 The chart compact heat exchanger reactors

An alternative to coating the surface of a compact heat exchanger with a catalyst is to put small catalytically-coated pellets in the fluid channels. Rather like the larger variants, where tubes were packed with such pellets in, for example, tubular reformers, such a concept has a number of possible advantages, including easier catalyst replacement. A range of compact heat exchanger reactor units, known as ShimTec (see Figure 5.35), have been developed by Chart Energy and Chemicals.

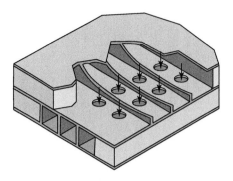

FIGURE 5.35

The injection mechanism for any ShimTec structure.

They are manufactured from thin plates/shims and stacked to form assemblies that can resist the high loads required by the diffusion bonding process. Typically they are fabricated from stainless steel, although they may also be produced as vacuum-brazed aluminium assemblies. Ranges such as ShimTec have compact multi-channel construction, resulting in high values of heat transfer surface area per unit volume, enhanced heat transfer coefficients and good mixing efficiency, in a diffusion-bonded construction that can be operated at elevated temperature and pressure operation. ShimTec units are available in a wide range of sizes and have varied performance characteristics to suit specific applications. The normal material of construction is stainless steel (316), and nickel alloy reactors have now been developed. The advantages of such designs are high temperature and pressure operation, and the flexibility of duty given by the use of multiple channels. Chart produce units with channel dimensions generally greater than 1 mm, in order to minimise the possibility of fouling.

Chart have produced three versions of the ShimTec reactor for packed catalyst applications, Types 1, 2 and 3. Type 1 offers a simple rectangular construction for low-pressure applications where the coolant or heating stream is a liquid and the process and reactant streams are gases or liquids. Type 2 is designed for use as a packed-bed catalytic reactor when all streams are gases. Type 3 (shown in Figure 5.36) is designed for high-pressure liquids.

The pins which join the hot stream separating plates to the cold stream separating plates accommodate the mechanical loads resulting from the internal pressure within the layer, provide heat conduction paths and promote turbulence for increased heat transfer. The pins are the full thickness of the shim and are arranged to be coincident in each shim layer, but the joining ligaments are off set in each shim layer. The shims are stacked to provide the desired layer height. They are closed off with solid separating plates in the case of the coolant, and perforated in the case of the reactant. The description equally applies to the pin-fin surface. This, along with Chart-Kote, is particularly good for coating with catalyst.

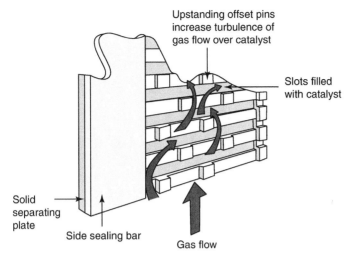

Upstanding offset pins increase turbulence of gas flow over catalyst

Slots filled with catalyst

Solid separating plate

Side sealing bar

Gas flow

FIGURE 5.36

Internal structure of the Chart-Kote variant.

The problem of providing open passageways into which the catalyst may be packed is solved by using thicker plates, these have a series of slots which, when placed one on another, form open-ended channels. If instead of solid separator plates, perforated versions are used, then fluid from an adjacent reactant stream may be injected precisely into the channels in order to control or enhance the reaction caused by the process fluid flowing through the packed catalyst.

The construction of Type 3 is similar to that of Type 1 but, due to its geometry, it is capable of working at higher pressures. A layer construction system is adopted, but the layers are typically 2–6 mm thick rather than 0.5 mm. External to the flange diameter are the four lugs that are accessible when the unit is bolted in place in the pipe/duct installation. Applications include the following:

- Reforming of methane, methanol and other hydrocarbons for the production of hydrogen for powering fuel cells.
- Providing the basis for reactor design for catalytic clean-up of hydrogen rich streams for low temperature fuel cell applications.
- Providing the basis for design of gas-shift units that can be used in conjunction with a hydrocarbon reformer to increase the hydrogen content of the reformed fuel.

The application of the catalyst in micro-reactors (in this case for steam reforming) is discussed by Kundu et al. (2007). They compare the relative merits of coating the surfaces of the reactor with a catalyst or packing the catalyst particles into the channels. They conclude that coating gives a lower pressure drop, but packing gives a better conversion at a lower temperature. Potential users will need to examine the trade-offs in making the decision whether to coat or pack.

FIGURE 5.37

The compact catalytic reactor core, showing the relatively large channels for packing with catalyst, along with the heat transfer surface. The scale can be seen with reference to Figure 5.39, the reactor measures only a few cm across.

(courtesy: Chart Energy and Chemicals)

The potential for the packing of catalyst into channels in the compact catalytic reactor is illustrated more clearly in Figure 5.37 in cutaway form. In Figure 5.38 a manifolded pipe with feeds for the process fluid(s) accommodates the reactor core.

The FinTec BAHX plate fin heat exchanger reactor variant shown in Figure 5.39 can have the catalyst packed into open channels or into the gaps between the fins. Chart, of course, has a strong background in plate-fin heat exchangers, as illustrated in Chapter 4, and this forms the basis of the reactor illustrated.

5.5.3 The corning micro-structured reactor

Corning have, over many years, developed micro-structure technologies and are able to generate features in materials with mm to micrometer channel sizes. The target is principally to develop reactors based on this expertise to replace batch

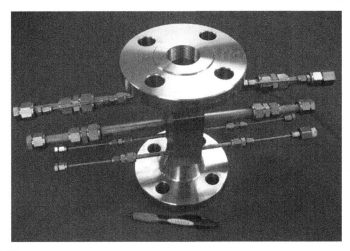

FIGURE 5.38

The Chart CCR unit located between two flanges.

(courtesy: Chart Energy and Chemicals)

FIGURE 5.39

The plate fin heat exchanger reactor from Chart is capable of bulk chemicals manufacture.

pharmaceutical plants with throughputs to 10 kte/yr upwards with micro-reactor technology. Their system was launched at ACHEMA in 2006.

The technology consists of micro-structure layers, with a mass transfer layer where the reaction takes place and a heat transfer layer for heating/cooling. Sheet size is 10×15 cm, thickness is 1–1.32 mm. Validation pressure is 20 bar and the temperature range 280–250°C. The reactors can withstand temperature excursions from 2,200°C to 1,200°C. Throughputs are typically 0.5–20l/h, with a mean of 40 tonne output/a.

Configurations include single and multiple injection reactors. In the latter case this can include a preheat section, followed by splitting of streams. This allows better temperature management along the flow path and decreases the potential for hot spots. Also it avoids local overconcentration of active species. Examples of uses include liquid–liquid polystyrene precipitation and liquid–gas reactions. Data are given by Caze (2006). For the reader interested in other micro-reactors, there are several case studies in Chapters 8–11, including the use of the Velocys unit.

Of particular interest are Corning's structured glass micro-reactors. McMullen and Jensen (2011) demonstrate the use of such Corning reactors for 500-fold scale-up from a micro-reactor, as part of a kinetic study on a model reaction (Diels-Alder reaction of isoprene and maleic anhydride in DMF).

5.5.4 Constant power reactors

Constant power reactors developed by AM Technology (see Appendix 3) employ variable geometry flow channels to regulate the process power at different stages within the reactor. This is achieved by breaking up the reactor into a series of stages, each stage having an optimised ratio of heat transfer area to channel volume for the desired reaction scheme. High surface ratios (area to volume) are used where the reaction is fast and low ratios where the reaction is slow. This approach offers significant advantages for fast and slow reactions alike. AM Technology has developed two types of constant power reactors under the collective name of 2FLO, designed to cope with both fast and slow reactions.

The arrangement delivers a constant heating or cooling demand along the length of the reactor, allowing for high thermal differences (between the heat transfer fluid and the process fluid) to be employed without creating hot or cold zones within the reactor. This delivers heat transfer performance comparable to true micro-reactors without the need for very small channels, whilst offering the benefit of a lower pressure drop and a reduced tendency to block. Alternatively, very high heating or cooling fluxes can be employed without creating hot or cold zones within the reactor. The variable channel reactor (VCR) design, Figure 5.40, allows the employment of temperature monitoring and control on each plate, increasing the flexibility of the reactor since it allows the user to alter the heating or cooling power in local zones within the reactor.

Mixing within the VCR relies on diffusion mixing in the narrow channel sections and static mixing in the wider channel sections. The VCR is ideal for fast reactions, given its very high heating/cooling capabilities, low pressure drop and reduced tendency to block. The channel width of the variable channel reactor, however, has a limited operating range. If the channel width is expanded too far, problems of back mixing or poor static mixing can be encountered.

For slower reactions a different kind of constant power reactor is used, known as the agitated cell reactor. The agitated cell reactor (ACR) shown in Figure 5.41

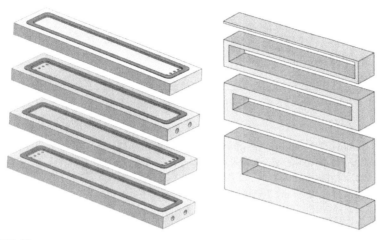

FIGURE 5.40

Variable channel reactor (VCR). The plate assembly is on the left hand side, and the figure shows the process fluid flow path.

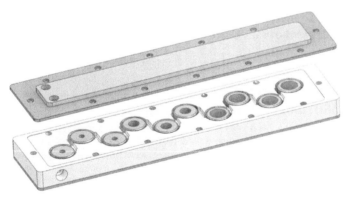

FIGURE 5.41

Agitated cell reactor (ACR).

is a form of constant power reactor where the product flows through a series of agitated cells. The concept adapts a well-rooted meso-scaled technique of the continuously stirred tank reactor to the constant power notion for greater increase in process control and stability.

Figure 5.41 shows an ACR with the heat exchanger plate removed. The ACR reactor consists of a series of cells within a block. Product flows between the cells via small channels. A cooling/heating plate seal forms the sealing face on one face

FIGURE 5.42

Flow path through a VCR followed by an ACR.

of the cell block. A cooling/heating plate can be fitted to the other face of the cell block. Alternatively, the second plate can be used for sampling and measuring temperature along the reactor.

Agitation elements are located within the reaction cells. When the cell block is mounted on a vibrating platform, the agitation elements promote mixing without the need for individual stirrer drive units. In addition, the agitators can easily be replaced with the best suited design or coated with a catalyst for the reaction of choice, with minimal downtime. Where solids are handled, gas pockets can be used to promote agitation as the cells are vibrated. By varying the size of the agitation elements, the volume of each cell can be varied according to need. This allows both the residence time per cell and the ratio of surface area to process volume within each cell to be controlled. The ACR offers a low cost solution for small CSTRs. The ability to vary the cell capacity with inserts allows more efficient separation of stages. Figure 5.42 shows a VCR followed in series by an ACR.

5.5.4.1 Case study

> ### ACETIC ANHYDRIDE HYDROLYSIS
>
> This reaction was chosen as it is a classical calibration reaction for calorimeters. It is a fast second-order exothermic reaction well-suited for studying the dynamic response of the VCR. Two feed streams (acetic anhydride and water) were simultaneously pumped into the reactor such that the mean residence time was 14.84 s. The reactor had two cooling zones, at temperatures of 100°C and 25°C. 1D integrated reaction and heat exchange models based on Nusselt correlations were developed to compare with the results. The study indicated that a conservative estimate of the heat transfer coefficient ranged from 1.52–0.36 kWm^{22}K^{21} for the given conditions. Heat release at the initial stages was satisfactorily controlled. The process composition of the end product was found to be 98.2% acetic acid, surpassing the predictions of the modelled results. Under equimolar conditions of the acetic anhydride hydrolysis, the VCR demonstrated excellent temperature control, with very high exit conversion of product. For continuous operation on a per annum basis, one reactor could produce <6.1 tonnes of acetic acid, whilst operating under safer conditions than in conventional processes, i.e. very low pressure drop throughout with product exiting at ambient temperature. The device was shown to exhibit almost identical behaviour to that of an equivalent (true) copper micro-reactor.
>
> A further AM Technology reactor case study is given in Chapter 8 (Section 8.4.10) – the Coflore reactor.

5.6 **Field-enhanced reactions/reactors**

Electric fields are, like rotational fields, techniques for active enhancement of heat and mass transfer, as discussed in Chapter 3, Section 3.4. Electric fields have been used in a number of ways in reactors, and interest in this area is increasing. Some techniques are already used in intensification; others are the subject of laboratory or pilot plant investigation. As with any enhancement technique, there are trade-offs between the benefits and the costs, and in the case of electrical enhancement methods the costs are in terms of energy (electricity) and capital (the local (e.g. microwave) power generation plant, which can be highly specialised).

5.6.1 **Induction-heated reactor**

EA Technology carried out a feasibility study of induction heating of reactor walls (Gardner, 1991). The reaction was the electrically-driven dehydrogenation of ethyl benzene to styrene. The walls of a channel packed with catalyst were inductively heated at low frequency. This reduced the temperature gradient across the wall for cases where external heating is required. The power supply for induction heating was also claimed to be much cheaper and more efficient than the equivalent high frequency systems.

5.6.2 **Sonochemical reactors**

The application of ultrasound, as well as being useful for cleaning, anti-fouling tasks and mixing, can enhance rates of chemical reactions, and can eliminate or reduce the need for catalysts. The basis of the ultrasonic reactor is to use power ultrasound to create cavitation bubbles within liquids due to rarefaction waves overcoming intermolecular forces (Mason and Cordemans, 1996). The cavitation bubbles collapse rapidly, resulting in temperatures up to 5000 K and pressures of many thousands of bar (Gogate, 2008).

One of the great advantages of sonochemistry is its ubiquity; unlike many other enhanced field techniques that require very specific fluid or catalyst properties, the only constraint for sonochemistry is that there must be a liquid present to propagate the acoustic waves. The waves, though, do not necessarily propagate very far, and this has been used as an argument for matching sonochemical applications to intensified technologies, as the length scales in intensified reactors tend to be much shorter than those in more conventional reactors.

The high temperatures produced in the cavities can produce highly reactive intermediates in the bubbles. The reactive intermediates diffuse out into the surrounding liquid and cause reactions. This is one mechanism by which ultrasound affects the course and rates of chemical reactions. It has been suggested that high-speed micro-jets created during collapse of the bubbles can cause intense convection next to solid surfaces. This is hypothesised as another mechanism by which heterogeneous diffusion-controlled reactions are accelerated. Additionally, it is believed that the micro-jets can clean passivated surfaces, cause fusion of particles or can fragment particles, thus creating additional surface area, all of which can affect the rate of catalytic chemical reactions.

Sonochemistry has often been allied to electro chemistry (sonoelectrochemistry), as many of sonochemistry's benefits are particularly suitable to electrochemical processes, e.g. degassing electrode surfaces, disrupting diffusion layer, improving transport of ions through the double layer and electrode surface cleaning (Mason and Cordemans, 1996), all of which combine to increase efficiency. These advantages again arise due to micro-jet formation near surfaces.

There are several observed effects of ultrasound on many reactions that are still not understood. Besides, there are many engineering issues (e.g. scale-up, variation of sound field in reactors, etc.) still to be resolved before ultrasound can be put to use on a commercial scale. However, reports (e.g. Gogate et al., 2004) suggest that compared to the conventional sonic horn (which may be seen in some plants as a method for keeping equipment, such as heat exchangers, free from solid foulants), novel multiple frequency reactors would be better for large-scale use in, for example, phenol destruction. The critical factor in scale-up is achieving adequate cavitational activity, which is relatively easy to obtain in the laboratory scale plant, but not necessarily at full-scale. However, there are a number of relatively new suppliers of such technology, at ever-increasing scales. The state-of-the-art in the scale-up of ultrasound devices is well-covered in Leonelli and Mason (2010). They conclude that '...there can be no doubt that the development of both microwave and ultrasound reactors have reached industrial scale'.

5.6.2.1 Biological applications of ultrasound

Although ultrasound has often been thought to destroy cells, and indeed it does, if used correctly it can actually cause increase in the productivity of biological species, even on live systems. The mechanisms of this enhancement are not necessarily well understood, but the effect is pronounced (Chisti, 2003). Note that the destruction of cells by ultrasound can also be a desirable effect: ultrasound has been used to enhance extraction of numerous chemicals from natural feeds by rupturing the cells and thereby facilitating contact between chemical and solvent (Mason and Cordemans, 1996).

The efficiency of chemical and biochemical reactions can be enhanced if extraction, for example, the transfer of chemical species from one liquid to another immiscible liquid, can be improved. Transfer can be much improved if a greater amount of surface area can be created, normally by generating very small droplets of one of the fluids. This can be achieved by using nozzles or by agitation, but electric fields can also be used to break up droplets into micron sizes. Jones and Weatherley (2003) report the hydrolytic splitting of esters to yield free fatty acids and glycerol via this technique. One of the principal requirements for effective hydrolysis is a free interface where the lipase can catalyse the reaction. Also, good mass transfer of the reaction substrate from the bulk phases is needed. In this case, a high voltage electrostatic field was used to spray the lipase into sunflower oil, which significantly increases the reaction rate. A comprehensive review of electrically enhanced mass transfer techniques is given by Weatherley (1993).

Sonochemistry is a vast area of research, so is not reviewed thoroughly in this text. For a review of sonochemistry in general, the reader is directed to Thompson and Doraiswamy's review (1999), and for its applicability to chemical processing to Mason and Cordemans review (1996).

5.6.3 **Microwave enhancement**

The use of microwaves has a number of potential advantages for reactions. They are a very efficient form of heat transfer, have been observed to accelerate reaction rate, and can heat specific parts of a reaction mixture. There are a range of mechanisms by which the application of microwaves can speed chemical reactions. The main advantage is due to extremely rapid targeted heating. One result of this is enhancement of reactions by local superheating of solvents. In heterogeneously catalysed reactions, application of microwaves can often lead to greatly increased reaction rates by intense local heating of the catalyst particles speeding the local reaction rate. There is considerable debate as to whether any mechanisms beyond simple heating have been observed, as some of the rate enhancements under microwave treatment have been difficult to explain.

The study of microwaves, particularly in organic synthesis, has grown steadily since the first publication in 1986 until the current day, where there is a recognised field of microwave chemistry, there were around 400 publications in this area in 2007 alone and over 900 in 2011 (since the first publication in 1986, there have been over 3000 publications). The reader is directed to Loupy (2002) for a comprehensive guide to microwave chemistry. The possibility of making use of the enhancements in rate, reduced temperature, etc., in an industrial setting is now on the horizon, and as such continuous reactors are beginning to be evaluated. Glasnov and Kappe (2007) state that because of the limitations in microwave penetration depth into reaction media, the key to scale-up is continuous flow systems that can operate on a smaller scale for the same production rate. They describe a range of small-scale tubular reactors. A variety of microwave reactors are now available at laboratory scale. Some designs are suitable for continuous flow, and some can be modified to run continuously. Suppliers include Anton Paar, Sineo Microwave and Milestone.

Microwave heating is being investigated as a method of supplying heat in methanol reforming. As with a range of endothermic reactions, methanol-steam reforming can be limited in rate by the ability to supply heat. Perry et al. (2002), for instance, demonstrated that productivity could be significantly increased using microwave heating by minimising the radial heat transfer effects.

As for ultrasound, the reader is directed to a recent review of both microwave and ultrasound for process intensification (Leonelli and Mason, 2010). This is particularly good for giving a snapshot of the current state-of-the-art in scaling up this technology, and a range of combined techniques.

5.6.4 **Plasma reactors**

One well-known example of using electricity to apply concentrated energy into a small area is the plasma reactor. Plasma is generated when an electric current is conducted through a gas, resulting in ionisation of the gas. The Birkeland-Eyde process for nitrogen fixation, developed in the early twentieth century, is the earliest known plasma reactor process. More recently, one of the most well-known plasma reactor uses was for the conversion of methane to acetylene, which has been updated (Fincke et al., 2002) to improve yields to 90–95% while approaching conversion efficiencies of 100%. This work has been mirrored by Kim et al. (2007) using a plasma catalytic reactor, which they call a dielectric barrier discharge (DBD) reactor. They discovered that the key to increasing the yield in this particular unit was to increase the platinum loading on the γ-Al_2O_3 catalyst. Kim and colleagues achieved a yield of slightly over 50%, although earlier workers suggested that 35% was the limit.

Plasma reactors are also of current interest for the remediation of hazardous wastes, including organic liquids such as acetone and fluorocarbons. The use of an electrothermal plasma reactor for the chemical reduction of small inorganic gases is reported by Steinbach et al. (2003). Other applications are introduced in Chapter 3.

5.6.5 **Laser-induced reactions**

One of the most common reactions induced by lasers is combustion; lasers have been studied for initiating combustion in vehicles in both petrol and natural gas-fuelled systems, and have been studied as igniters in gas turbines. Lasers are limited in the duration and level of power they can produce, unless very large facilities can be provided. The wavelengths of laser energy are not always appropriate to that required by the 'target', and much energy can be wasted. Following on from the general introduction to lasers as intensification tools in Chapter 3, Buback and Vogele (2003) reported that the high-pressure polymerisation of pure ethylene could be induced by lasers.

The development of small solid-state lasers used in consumer products may be compatible with lab-on-a-chip concepts. Zeev (2003) proposes in a patent to use optical irradiance to activate chemical reactions, with laser energy taken along optical fibres to one or more irradiators that are in contact with the reagents of chemical reactions. There are, of course, safety benefits in taking energy into sealed containers this way. The theory behind laser catalysis is explained in Vardi and Shapiro (1998).

5.7 **Reactive separations**

One of the features of intensified process plant is the opportunity offered to combine a number of unit operations in a way that is not possible, or at best difficult, with conventional process equipment. The HEX-reactor discussed above is an example, where heat removal or addition is fully integrated with reaction(s) in a single package. It should be noted that a combination of unit operations usually represents process intensification in itself, regardless of the technology used, as significant overall reductions in capital cost, equipment size, etc., can be achieved.

There are other unit operations in chemical plant, including mixing (apparent in some HEX-reactor variants), and a wide variety of separation processes, some of which are discussed in Chapter 6. In the case of the latter, the combination of reactions and separation processes may be classified (Kulprathipanji, 2002) as reactive separation processes thus:

- Reaction/distillation
- Reaction/extraction
- Reaction/absorption
- Reaction/adsorption
- Reaction/membrane
- Reaction/crystallisation

The advantages of combining unit operations include energy and capital cost reductions, and increased reaction efficiency. A selection of generic advantages and limitations is given in Table 5.4. Absorption and reaction were the first to be combined, but interest in the other combinations did not grow until about 30 years ago. A useful discussion on other multifunctional reactors, in addition to those including separation, is given by Stitt (2004).

In this section, reactive distillation, reactive extraction, reactive adsorption and membrane reactors are discussed.

Table 5.4 Advantages and Disadvantages of the Commercialisation of Reactive Separation Processes.

Advantages	Disadvantages
Enhanced reaction rates	Relatively new technology
Increased reaction conversion and selectivity	Limited applications
Reduced reaction severity	Complex modelling needs
Heat integration benefits	Increased operational complexity
Novel process configurations possible	Significant development costs
Reduced capital costs/operating costs[1]	Increased scale-up risks
Simplified separations	Extensive equipment design effort

(adapted from Kulprathipanji, 2002)
[1]*Interestingly, Stitt (2002) gives cost comparisons between a reactive distillation unit and its conventional alternative. For xylene production (150,000 tpa) the net benefit is a mere 3.8% on capital cost. This is due to the need to operate the integrated column at 30 bar, instead of atmospheric pressure. Stitt points out that this saving is nowhere near the 25–50% generally needed to justify investment in a new technology.*

5.7.1 Reactive distillation

Using distillation to remove volatile products from reactions is an effective method of increasing conversion, and consequently there are currently over 150 reactive distillation-based processes at industrial scale (range 100–3,000 kte/yr). Conventionally, the arrangement is essentially a distillation column in which the packing is either coated with a catalyst or has catalysed particles located within a supporting structure. The catalyst may also be in the same phase as the reacting species. The aim of the distillation column is to separate the products of the reaction by fractionation. Another reason for performing reaction and distillation simultaneously is to remove impurities or undesired species. Reactive entrainers may be added, to react with difficult to remove substances or to convert them to more volatile species. Advantages of reactive distillation (RD) include:

- Process simplification: the potential for capital cost reduction is clear in that RD can remove the need for separate reaction and separation steps. Prior to use of RD, for example, the methyl acetate process required two reactors and eight distillation columns. It has been succeeded by systems containing one RD column and two separating columns (Agreda et al., 1990), resulting in a significant reduction in overall capital cost.
- Improved conversion: by pushing the reaction to the product side by removing one of the product species.
- Improved selectivity: an example is the production of ethyl benzene from benzene and ethene. The ethyl benzene produced is more reactive with the ethene than the benzene, so removing it *in situ* using RD represents a simple way of preventing by-product formation (Qi and Zhang, 2004).
- Reduced catalyst requirement per unit product.
- Increased reliability, simply due to the smaller amount of equipment and particularly to no longer having equipment with moving parts.
- Reduced energy requirement.
- Removal of azeotropes.
- Multifunctionality: RD can encompass more than just these two processes.
- Reduction of by-products.
- Heat integration: operation at temperatures where heat is more useful.
- Control of highly exothermic reactions: due to operation at a species' boiling point, meaning that heat transfer coefficients are high.
- Reduced problems with hotspots and thermal runaway.

There are, however, a number of important constraints to this technique. The reaction must fit the technique, i.e. one of the products must be the most volatile component, and the reaction and catalyst must also be active at the set of conditions (temperature, pressure) dictated by the distillation. A further challenge currently is that modelling such systems is proving difficult due to the interplay of chemical and physical phenomena. Modelling is necessary as pilot-scale trials are expensive. Scale-up is a problem for RD associated with the separation of functions, as there is no technique at present for scaling up the combined unit operation of reaction and distillation.

The main industrial applications of RD are esterification, etherification and alkylations (Tuchlenski et al., 2001). Certain hydrogenations and hydrodesulphurisations are also performed in this way (Harmsen, 2007). RD is most often used for esterification reactions, the most famous example of which (and indeed the most famous example of RD in general), is the Eastman process for methyl acetate, initiated in 1980. Eastman implemented RD in a 200 kte/yr methyl acetate facility, where to this day it effectively performs five different functions. Hendershot (1999) describes how an RD process, requiring three vessels and eliminating the need for reboilers, condensers, etc., replaced a plant with 11 vessels i.e. the number of unit operations was reduced by over 70%.

Other prospective applications of RD include liquid–liquid hydrolyses; saponification; nitration; oxidation; fermentations; hydrolysis of aqueous methyl formate and dehydrogenation of cyclohexane to benzene.

5.7.2 **Reactive extraction**

Reactive extraction (RE) is the integration of reaction and solvent extraction process steps. This results in fewer process steps overall, thereby reducing capital cost. Other possible advantages that can be realised by performing extraction in this way include enhanced selectivity and efficiency. RE can be an effective way of removing a desired product from the reaction zone, thereby preventing any side reactions (Krishna, 2002).

One mode of RE is the addition of a second liquid phase in the reactor, containing a selective solvent that removes the product. Krishna (2002) quotes their use in the bromination of dialcohols in the aqueous phase, where it is necessary to prevent the second OH group from reacting with HBr to form the dibromide. By adding an HC to the reaction mixture, the HC extracts the monobromide, thus preventing it from reacting to form the dibromide.

The desire to improve the production of lactic acid, due to increasing demands for it as a monomer for the synthesis of biodegradable polymers, has led to the use of reactive extraction to improve productivity. Figure 5.43 shows a semi-batch process used for this application, which has greatly improved operation (Wasewar et al., 2003).

Surfactant-enhanced solvent extraction, using either reversed micellar solutions for liquid–liquid extraction or the use of pre-dispersed solvent extraction using colloidal liquid aprons, are interesting new directions. Haas et al. (2004) are investigating this technique for the production of biodiesel from soy flakes (although it is referred to as *in situ* transesterification), to remove the costly crushing step from the process. Currently, biodiesel is produced by producing vegetable oil by crushing and solvent extraction, then reacting the oil in the presence of a liquid alkaline catalyst with methanol (occasionally ethanol) to produce the ester (the biodiesel). Instead, in reactive extraction, the seeds would be macerated and contacted with an

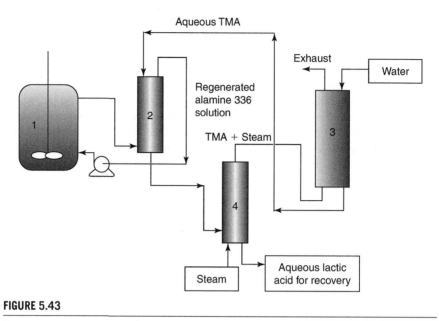

FIGURE 5.43

Reactive extraction: effect of glucose concentration on the productivity of lactic acid in batch and semi-batch modes (Wasewar et al., 2003).

alcohol (containing catalyst) directly, the alcohol taking the role of both solvent and reactant. Research at Newcastle University has focused on production of biodiesel from rapeseed (Zakaria and Harvey, 2012), as this is the dominant crop in Europe for biodiesel production, and a range of inedible crops that grow on marginal land, thereby avoiding competition between food and fuel, notably including jatropha (Kasim and Harvey, 2011).

5.7.3 Reactive adsorption

Adsorption is a basic separation technique that can be used in parallel with a reaction to, for example, increase conversion by removing a product. Reactive adsorption combines the separation role of, for example, a solid adsorbent with the reaction taking place on a different surface. The ability to remove the adsorbent or to desorb one or more of the products of the reaction can be convenient and intensive, although adsorbents do not necessarily respond rapidly. There could well be, for instance, synergy between the fluidised-bed reactors and such adsorption processes.

As discussed in Kulprathipanji (2002) there are other reactors for carrying out reactions and adsorption together (preferably continuously). The counter-current moving-bed chromatographic reactor (CMCR) is one unit, having been used for hydrogenation, with good selectivity. However, successful commercialisation is not yet reported.

5.8 **Membrane reactors**

Membrane reactors are becoming increasingly of interest. The membrane can function as a separator for recovering reaction products *in situ*, or for catalyst recovery, allowing a multifunctional role (Stankiewicz and Moulijn, 2000). As for most other reaction/separation techniques the key feature of membrane reactors is their ability to take reactions to completion by continuous product removal.

5.8.1 **Tubular membrane reactor**

One form of membrane reactor which is attracting attention is the tubular inorganic catalytic membrane (TICM) described by Centi et al. (2003). Possible applications range from refinery uses to environmental protection, and catalytic membrane reactors in general have seen a growth in uses during the last ten years. For example, the ability to selectively supply a reactant to the reaction zone and realise an optimum concentration profile to give maximum selectivity has significant benefits for reactor operation. This approach has been of benefit in direct oxidation of propane to acrolein and ethene epoxidation, but in both cases the membrane has been catalytically inert.

Putting the catalyst into the membrane pores, as in liquid-phase hydrogenation reactions, allows lower pressure operation, as better three-phase contact can be achieved. Another hydrogenation application, the reduction of nitrates or nitrites in water using a tubular membrane reactor configuration as shown in Figure 5.44, was described by Centi et al. (2003). This study showed that the membranes could be regenerated intermittently by removing them from the reactor, calcining and pre-reducing with helium/hydrogen. When reintroduced the membranes were essentially fresh.

Effluent treatment may be carried out using ceramic membrane bioreactors. However, equipment remains expensive compared to conventional activated sludge treatment methods. Although where tubular membranes are used, the performance can be improved by using tube inserts, which are effective turbulence promoters. It was found (Xu et al., 2002) that the permeate flux was increased by a factor of 2.5 without any detrimental effect on the effluent quality. Other formats for membrane bioreactors include fully stirred tanks, packed-bed columns and jet loop columns.

5.8.2 **Membrane slurry reactor**

Gas–liquid solid (GLS) and liquid–solid (LS) type slurry reactors are widely used in the chemical, fine chemical and pharmaceutical industries. Applications include hydrogenation (partial) oxidation, condensation, esterification and enzymatic conversion processes. In these slurry processes the catalysts have to be recovered externally from the reaction mixture. The external filtration step often leads to catalyst attrition and deactivation, as well as filtration and handling problems.

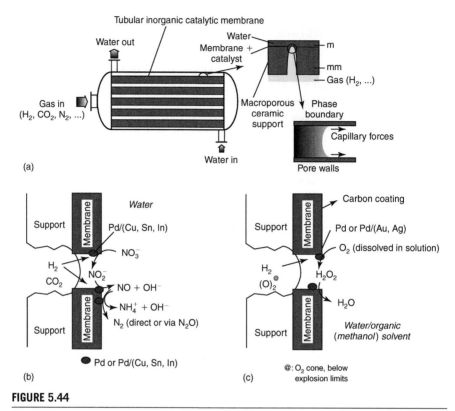

FIGURE 5.44

Configuration of a tubular inorganic catalytic membrane reactor module (Centi et al., 2003).

The additional auxiliary equipment creates a high catalyst inventory of costly catalyst in the total system. Furthermore, attrition of catalyst particles can create large fractions of fines, which can cause loss of catalyst particles and emissions to the environment. Inefficient use of the catalyst, together with wide residence time distributions, can give unwanted side product formation.

A solution to these problems is to improve the reactor, in the form of the membrane slurry reactor. In the membrane slurry reactor concept, separation and heat transfer are integrated within the chemical reactor by means of membranes and heat exchanger tubes. This enables better control of process conditions and a more efficient use of the catalyst. Thus catalyst loss can be prevented and the lifetime extended. A much more important benefit is, however, that the membrane slurry reactor can bring the use of more selective (enzymatic) catalysts and selective product recovery within reach, leading to more efficient use of raw materials, reduction of the amount of unwanted side products and emissions and optimisation of the quality of the main product. The membrane slurry reactor thus forms an important step in the transition process towards a more sustainable, green chemical industry.

The membrane slurry reactor typically consists of a housing containing hollow fibre membranes and heat exchanger tubes. By using membranes with a pore diameter of 0.5 μm or smaller it is possible to use catalyst particles from about 1 μm without entrainment. The reactants and products are removed from the reactor via the hollow fibre membranes. Retaining the catalyst particles within the reactor has the following advantages:

- No external catalyst filtration.
- No pumping of solid particles.
- Less catalyst attrition.
- No catalyst deactivation.
- Broad catalyst particle distribution, including small particles, is feasible.
- Minimal catalyst hold up.
- The required volume of the reactor will be reduced as, due to the improved filtration behaviour, it is possible to increase the catalyst loading and so decrease the required contact time, thereby reducing unwanted side reactions.

5.8.3 Biological applications of membrane reactors

Trusek-Holownia (2008) reports the use of a membrane reactor for the enzymatic hydrolysis of casein. Although such processes may never be performed at a large enough scale to justify intensification of the process by conventional economies of scale arguments, there is another driver, which is to minimise usage of the enzymes themselves as they are very expensive. In this case, use of a membrane demonstrates that the residence time required is reduced and the utilisation of substrate substantially increased.

5.9 Supercritical operation

Supercritical operation of a reactor means operating it at temperatures and pressures above the critical point of one or more of the substances present (see Figure 5.45). Supercritical fluids exhibit a combination of properties normally associated with both liquids and gases. They combine the solvent capacity of a liquid with the viscosity of a gas.

Most often the fluid is carbon dioxide, and it is used purely as a solvent. Carbon dioxide's great advantage is that its critical point is relatively easily attainable (31°C and 74 bar) and it is readily available and inexpensive. Propane is similar in this respect, as its supercritical point is 96°C and 43 bar. Water is often considered as a supercritical solvent. However, water's critical point occurs at 374°C and 200 bar, so its use is much less convenient, hence CO_2's dominance.

The key advantages of supercritical operation are:

- Ease of separation: the solvent can be separated from the reaction mixture simply by releasing the pressure to some degree.

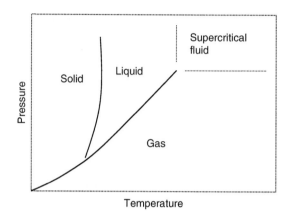

FIGURE 5.45

Equilibrium diagram showing supercritical region.

- Organic solvents, particularly halogenated solvents, which may be hazardous or have detrimental effects on the environment, are replaced.
- Enhanced and controllable solubility at supercritical conditions.
- Sequestration of CO_2.

Supercritical CO_2 has a solvating power similar to that of a light hydrocarbon, and this increases with pressure as the molar density of the supercritical CO_2 increases. Supercritical CO_2 is completely miscible with permanent gases such as hydrogen and nitrogen, leading to much higher concentrations of dissolved gases than are achievable with conventional solvents. Reactions investigated include hydrogenations, as the supercritical CO_2 overcomes the problem of the poor solubility of hydrogen in organics, which otherwise necessitates a great deal of agitation to overcome the mass transfer limitations.

5.9.1 Applications

Supercritical processing was used in 1994 in pharmaceutical production at Chalmers University. De Gussa carried out hydrogenation of fats and oils using supercritical CO_2 around the same time. Since 1995, Nottingham University in the UK, has worked with Thomas Swan and Company to develop a more generic process capable of processing tonnes/annum in the laboratory. As a consequence of this collaboration, in 2001, Thomas Swan launched the world's first continuous phase, high pressure supercritical fluid reactor for both pilot and commercial scale production. They chose catalytic hydrogenation as a prime candidate for improvement by supercritical operation, as there are a number of problems associated with conventional hydrogenation processes that can be overcome. Hydrogen is infinitely miscible with $scCO_2$, which eliminates the mass transport problems commonly associated with traditional solvent-based processing (batch or Buss loop).

Furthermore, through the appropriate choice of reaction pressure, temperature, catalyst type, residence time and stoichiometry it is possible to achieve a far higher degree of selectivity than has been observed under usual reaction conditions. One example is the preparation of 3-ethylcyclohexene by catalytic hydrogenation of the corresponding diene. Under conventional reaction conditions, the best that could be achieved was a 4:1 mixture of the desired product and the unwanted fully saturated by-product, even at only 80% conversion. Under supercritical conditions, complete conversion was achieved with 100% selectivity to 3-ethylcyclohexene, without the need for any additional solvents (Stephenson et al., 2003).

The catalytic hydrogenation of nitro-aromatic functionalities to the corresponding aniline is another synthetically desirable transformation which can yield superior results under supercritical conditions. Thomas Swan & Co has been able to demonstrate nitro group reductions with 100% conversion, with no detectable side products (including hydroxylamine derivatives), even when other hydrogen-sensitive functionalities have also been present in the molecule. Of particular significance to the fine chemical industry is the ability to reduce nitro-aromatics without the production of the corresponding hydroxylamine side-product, an ever-present problem with conventional reduction methods.

The use of supercritical CO_2 necessitates additional unit operations to allow the conditioning of the fluid so that it is at the correct conditions when the reactants are added. The reactants and solvents are mixed and then taken to the reactor. The separation is undertaken downstream of the reactor, where the conditions are no longer supercritical, much of the separation is achieved by this drop in pressure itself. As with any process change, these extra process steps must be taken into account when developing a supercritical process to ensure that the whole process is rendered more economically feasible, not just the reaction step itself.

5.10 Miscellaneous intensified reactor types

A variety of other reactors are covered in applications chapters, notably nuclear reactors (Chapter 10) and the gas turbine reactor and turbo cracker (Chapter 9). Furthermore, Table 5.5 summarises a selection of intensified reactors and their merits, limitations and potential uses. The following sections give some details of a variety of other intensified reactor designs.

5.10.1 The Torbed reactor

The Torbed is a toroidal fluidised bed (Figures 1.11, 5.46 and 5.47) used as a reactor (Groszek, 1990). As a fluidised bed there is efficient gas–solid mass and heat transfer. It has a low pressure drop, allowing process gas recirculation (Shu et al., 2000). Torbed is suitable for gas–solid reactions taking a maximum of a few minutes.

5.10.1.1 Applications

Applications of the Torbed include ore roasting (a typical unit 4 m high×3 m diameter handling 15 te/h of sulphide ore), and drying (e.g. chicken litter),

Table 5.5 Summary Table on Various Intensified Reactors.

Reactor Type	Advantages	Disadvantages	Possible Uses
Stirred tank types			
Continuous stirred tank reactor	Overcomes batch limitations; cheap; can be used in series; easy to clean; temperature control better than batch STR.	Still an STR: inherent limitations to temperature control and mixing when scaled up.	Crystallisation; bio-reactions; hazardous reactions; general pharmaceutical and fine chemical reactions.
The constant-flux reactor	Stable temperature control.	Unproven at industrial scale.	
Tubular reactors			
Plug flow reactor Fixed-bed reactors Loop reactor	Can extend residence times. Tight control of processing history.	No mixing in basic form Can get hot-spots.	Short reactions (less than a few minutes).
Oscillatory flow/baffled reactors	Continuous processing; consistent product quality; can accommodate multiple feed points; compact (up to 100 fold reduction in volume). Eight-fold reduction in reaction time claimed. Good heat transfer (for the net flow).	Suited to long reactions only (≥10min).	Continuous polymerisation; coagulation in water treatment; production of photo chemicals Biodiesel production, effluent treatment, various fine chemicals. Biological processes.
Heat exchanger reactor	Advantages below common to most types.	Small passages may block due to fouling.	See case studies.
Printed circuit reactor (PCR)	Compact; multi-stream.		
ShimTec	Good mixing; can be configured as a micro-reactor.		Highly exothermic fast reactions.
HELIX reactor	Good radial mixing.		
FlexReactor	May be readily configured to suit the needs of the reaction.		Can handle highly exothermic reactions.
Metal foam tubular reactor	Easy to configure.		Methane reforming.

Reactor type	Characteristics	Limitations/comments	Applications
Catalytic plate reactors[1]	Very compact for heterogeneous gas phase reactions.	Longevity/catalyst reactivation still under investigation.	Hydrocarbon reforming/cracking, etc.
Micro-reactors	Highly compact; easily scaled up; minimises hotspots; precise control possible; operating pressures generally not limiting.	Manifolding of multiple small channels can be difficult. Fouling/channel blockage (can be alleviated by using micro-mixers). Stabilisation of the interface between fluids.	Explosives; small-scale synthesis – alkylation, acylation, etc.; biochemical reactions, etc. Trans-esterifications. Single-phase and two-phase units available.
Reactors involving rotation			
Spinning disc reactor (SDR)	High heat and mass transfer; plug flow; intense mixing capability; short residence times; low propensity to foul; facilitates u/v irradiation.	Moving parts; most suited to 'fast' reactions.	Sulphonations; hydrogenations; chlorinations; polymerisations; crystallisation; particle size/form control.
Spinning cone reactors	As for SDR, but under certain conditions produces very small crystals.	Few examples yet.	Continuous crystallisation/precipitation.
Rotating packed bed	Excellent for mass-transfer limited processes. Uniform dispersion; high turbulence; thin films; good controllability.	Some development work still proceeding. Limited use outside odour removal/clean-up.	Crystallisation; reactive stripping.
Taylor-Couette reactors	High heat and mass transfer; good phase interaction (e.g. for emulsification); compact; others as for SDR.	Untried above laboratory scale.	Polymerisation; electrochemistry; fermentation; photochemistry.
Mop fan deduster	Large surface areas for contacting; dynamic cleaning action; compact.		Odour control reactions; flue gas desulphurisation.
Fluidised bed reactors			
Compact Torbed reactor	Efficient heat and mass transfer; can handle materials of different grades; controllable; energy-efficient; recycling facilitated; hotspots eliminated.	Limited uses so far in fine chemicals sector.	Catalyst reactivation; gasification/pyrolysis; pasteurisation (clinical waste, herbs and spices); calcination; combustion and desorption.

(Continued)

Table 5.5 (*Continued*)

Reactor Type	Advantages	Disadvantages	Possible Uses
Bio-fluid bed reactor	Features as for other types, but designed for low shear stresses.	Difficult to scale-up. A packed-bed reactor can scale-up by a factor of 50,000 for bio-reactions, but FBRs can only be scaled up by 10–100 times because of effect of size on fluidisation characteristics. Changes in stream flow-rate can affect conversion.	Cell culture.
Bubble column reactors	Overcomes fouling problems of solid packing; can give very effective gas/liquid contacting.	Poor bubble size distribution can reduce effectiveness.	Bioreactions/sludge treatment.
Vibration-enhanced unit	Ensures uniform bubble size.		Bio-reactions.
Bubble chopper	As for vibration-enhanced unit.		
Reactors using electric fields	Take thermal/acoustic energy to where it is needed.	Some variants need expensive power sources. Safety considerations.	
Induction-heated reactor	Aids start-up of some catalysts.	Scale-up uncertain.	
Ultrasonic-enhanced reactor/sonochemical reactor	Enhances cavitation.		
Microwave-heated reactor	Aids heat supply to endothermic reactions. Efficient heating method.	Hotspots can be an issue due to non-uniform heating; some materials not heated by microwaves.	Methanol-steam reforming.

Electric field-enhanced bio-reactors	Contributes to enhance extraction processes.		e.g. hydrolytic splitting of esters.

Reactors combined with other unit operations

Reactive distillation	Overcomes equilibrium limitations; cost and heat integration benefits; avoidance of azeotropes; improved selectivity; can put in more than one reactor zone (multifunctional reactor).	Not easy to find applications where the separation and the reaction conditions match; volatility constraints.	Methyl acetate from methanol and acetic acid; MTBE; ETBE (others being developed, e.g. hydrolyses, oxidative dehydrogenations).
Reactive extraction	Efficient for control of highly exothermic reactions; extraction can increase rate and yield of reaction.		Liquid–liquid hydrolyses; saponification; esterification; nitration; oxidation, etc.; fermentation reactions.
Reactive adsorption	Higher conversion; better yields.	Selection of appropriate catalysts and adsorbents; matching of process conditions; solids handling.	Hydrolysis of aqueous methyl formate; dehydrogenation of cyclohexane to benzene.

Membrane reactors

Tubular catalytic membrane	High yield; control of reactant(s) feed rate.	Can foul; can be expensive.	Oxidations; epoxidations; hydrogenations, etc.
Membrane slurry reactor	Integration of reaction, heat exchanger and catalyst filtration in one unit; self-cleaning; no catalyst deactivation; lower catalyst hold-up.		Hydrogenation; esterification.
Catalytic reactive extrusion – CALTREX	Continuous reactor function; excellent mixing and high sheared thin films produced.		Hydrogenations; polymer reactions. Others involving highly viscous fluids.

(Continued)

Table 5.5 (*Continued*)

Reactor Type	Advantages	Disadvantages	Possible Uses
Supercritical fluids	Better environmental properties than organic solvents. Can accelerate reaction kinetics. Continuous operation possible. High selectivity/conversion.	Number of full-scale plants still limited in the chemicals sector. Need CO_2 handling plant.	Hydrogenations; acylations; hydroformylations; etherifications.
Ionic liquids	Environmentally benign; avoids toxic catalysts; can be recycled; can speed up reactions.	Relatively unknown; can be difficult to separate solvate from ionic solvent.	Organic synthesis in general; green synthesis.
Miscellaneous reactor types			
Tubular reactors with catalytically-coated tube inserts	Cheap and simple.	Could foul. Pressure drop may be high.	
Enhanced reactor using heat pipes	Good heat distribution; rapid removal of heat if necessary; temperature control option.	Life of high temperature units may be limited; heat pipes are small pressure vessels.	Reforming reactions; fast exothermic bulk reactions.
Plasma reactors and laser-induced reactions	Could upgrade methane.	Scale-up difficult.	Combustion; methane upgrading.
Gas turbine reactor	Would give higher gas turbine cycle efficiency; produce chemicals from a prime mover.	Need to match very fast kinetics for the Turbo-cracker variant. Reactions matching the reheater/recuperator duties may be limited.	Production of chemicals needing exothermic reactions.
Ultra-violet (U/V) reactors	Fast reactions (used on SDR).	Limited U/V penetration.	Polymerisation; breakdown of refractory components.

[1] *The term catalytic plate reactor, like heat exchanger reactor, may be generally used for PCR-type units and a number of one-off plate reactor types.*

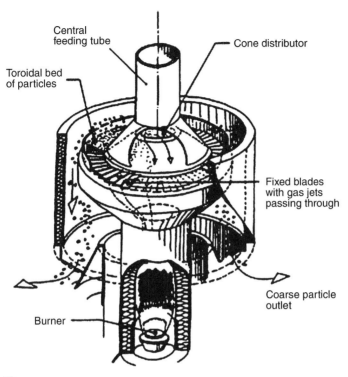

Central feeding tube

Cone distributor

Toroidal bed of particles

Fixed blades with gas jets passing through

Coarse particle outlet

Burner

FIGURE 5.46

The Torbed fluidised bed reactor.

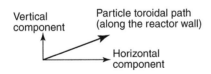

Vertical component

Particle toroidal path (along the reactor wall)

Horizontal component

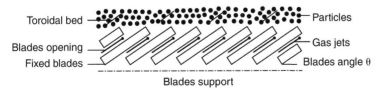

Toroidal bed

Blades opening

Fixed blades

Particles

Gas jets

Blades angle θ

Blades support

FIGURE 5.47

The fluid motion in the Torbed.

where advantages have been that the unit is considerably smaller. GMF in The Netherlands is using the Torbed instead of a prilling tower in fertiliser production. For processing 30 te/h, the original towers are 25–70 m high and 7–25 m in diameter. The Torbed system ranges in size from 6–8 m high and from 2–7 m in diameter. Other applications include catalyst reactivation, gasification/pyrolysis, pasteurisation (clinical waste, herbs and spices), calcination, combustion and desorption.

5.10.2 Catalytic reactive extruders

Reactive extruders allow the reactive processing of materials that are highly viscous, without the need for large quantities of solvent (as would be the case if done in an STR). The process of extrusion, whereby materials are blended by high shear mixing and subsequently forced through a discharge arrangement, has been developed to a high degree of sophistication. Modern extruder designs facilitate accurate control of the temperature and pressure conditions within a system, and with viscous materials, such as polymers, the pressures developed make it is possible to utilise the extruder as a continuous chemical reactor.

Two such reactions, of considerable commercial significance and potential subjects for study, are the hydrogenation of vegetable oils and the removal of unsaturated bonds in ABS polymer. Catalyst systems that are immobilised on the working surfaces of the extruder have been developed to prevent a difficult, expensive catalyst recovery step. By minimising catalyst loss in this way it is possible to employ highly active catalysts based on precious metals whose use is presently precluded on economic grounds. The factors which militate against the use of fixed heterogeneous catalysts in a conventional reactor, i.e. low effective surface area, long diffusion paths and the formation of boundary layers, will be of little significance in an extruder. This is because the reactants are continuously presented to the catalyst in the form of an intimately mixed, highly sheared thin film.

5.10.3 Heat pipe reactors

Reactions do not necessarily proceed uniformly, and uneven rates of reaction can lead to poor product quality and increased by-products. Some intensified reactor types overcome the lack of uniformity of performance by careful design, but larger reactor vessels are inherently non-isothermal. This can be overcome by attempting to move heat from those most exothermic parts of the reaction to areas where the reaction is less fast. The heat pipe (see Chapter 11) is a passive heat transfer device that can be used to implement this in a reactor. Sometimes called a super thermal conductor, the heat pipe uses an evaporation–condensation cycle to transfer heat over a distance with minimal temperature drop. The fluid in the pipe is selected to suit the reactor operating temperature; liquid metals may be used for high temperature reactions (at 600° C to 1,000° C plus), for instance.

One of the first references to heat pipes in chemical reactors (they had been used in nuclear reactors), was in a methanation plant (Biery, 1977). Heat pipes were

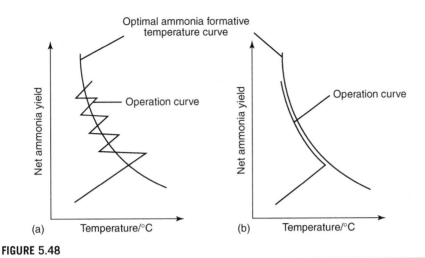

FIGURE 5.48

Temperature profiles in an ammonia conversion process: (a) original (b) heat pipe modified.

inserted next to rows of cylindrical ceramic catalyst columns, so that they took heat from hotspots on the catalyst. The vertical heat pipes then transferred the heat to a steam generator, located in the top of the reactor vessel. The steam was recovered for process use.

Zhang and Zhuang (2003) proposed heat pipes to improve the performance of ammonia converters for fertiliser production, suggesting that a 6% increase in yield (from 13 to 19%) could be achieved. The heat pipe allows much more even temperature distribution and reaction rates in the reactor, and allows one to approach the ideal temperature curve shown in Figure 5.48 (b).

Alternatively, heat pipes may be used to take heat out of highly exothermic reactions. In most cases, the heat pipes are close to, or in contact with, separate catalysts. In some examples, the heat pipes could be coated with a catalyst, as for finned tubes (see Chapter 3).

The use of heat pipes in chemical reactors is discussed in detail by Reay and Harvey (2012).

5.11 Summary

The reactor is the key unit operation in most chemical processes, and it has been the target of most of the attention with regard to PI. In this chapter, a selection of the most common intensified reactors are introduced and applications discussed (see also the applications chapters for more detailed case studies), and the table over the preceding six pages summarises the state-of-the-art in intensified reactor design.

References

Agreda, V.H., Partin, L.R., Heise, W.H., 1990. High-purity methyl acetate via reactive distillation. Chem. Eng. Prog. 86, 40–46.

Aoune, A., Ramshaw, C., 1999. Process intensification: heat and mass transfer characteristics of liquid films on rotating discs. Int. J. Heat Mass Trans. 42, 2543.

Baird, M.H.I., Stonestreet, P., 1995. Energy-dissipation in oscillatory flow within a baffled tube. Chem. Eng. Res. Des. 3 (A5), 503–511.

Baird, M.H.I., Harvey, A.P., Mackley, M.R., Ni, X., Rama Rao, N.V., Stonestreet, P., 2003. Mixing through oscillations and pulsations: a guide to achieving process enhancements in the chemicals and process industries. Chem. Eng. Res. Des. 81, 373–383.

Bell, C., 1975. The Hydrodynamics and Heat Transfer Characteristics of Liquid Films on a Rotating Disc, vol. 1. Newcastle University, Newcastle upon Tyne, UK, (PhD Dissertation).

Biery, J.C., 1977. Methanation: With High Thermodynamic Efficiency Energy Recovery. Los Alamos Laboratory Report LA-6656-MS. Los Alamos Laboratory, USA, (January).

Brauner, N., Maron, B.M., 1982. Characteristics of inclined thin films, waviness and the associated mass transfer. Int. J. Heat Mass Transf. 25, 99–110.

Buback, M., Vogele, H.-P., 2003. Laser-induced high-pressure polymerisation of pure ethylene. Die Makromolekulare Chemie. Rapid Commun. 6 (7), 481–484.

Burns, J.R., Ramshaw, C., 1999. Development of a microreactor for chemical production. Trans IChemE 77, 206–211. (Part A).

Callahan, C.J., Ni, X., 2012. Probing into nucleation mechanisms of cooling crystallisation of Sodium Chlorate in a stirred tank crystallizer and an oscillatory baffled crystallizer cryst. Growth Des. 12 (5), 2525–2532.

Carslaw, H.S., Jaeger, J.C., 1959. Conduction of Heat in Solids, second ed. Oxford University Press, Oxford, p. 101.

Caze, P., 2006. Engineered reactors for chemicals industrial production. Presentation at Fourteenth PIN Meeting, Grangemouth, Scotland. see <www.pinetwork.org>.

Centi, G., Dittmeyer, R., Perathoner, S., Reif, M., 2003. Tubular inorganic catalytic membrane reactors: advantages and performance in multiphase hydrogenation reactions. Catal. Today 79–80, 139–149.

Chisti, Y., 2003. Sonobioreactors: using ultrasound for enhanced microbial productivity. Trends Biotechnol. 21 (2), 89–93.

Dalglish, J., et al., 1999. Photo-initiated polymerisation using a spinning disc reactor. Third International Conference on Process Intensification for the Chemical Industry. BHR Group Conf. Series, Pub No 38.

Dluska, E., Wronski, S., Hubacz, R., 2001. Mass transfer in gas-liquid Couette-Taylor flow reactor. Chem. Eng. Sci. 56, 1131–1136.

Dutta, P.K., Ray, A.K., 2004. Experimental investigation of Taylor vortex photocatalytic reactor for water purification. Chem. Eng. Sci. 59, 5249–5259.

Fincke, J.R., Anderson, R.P., Hyde, T., Detering, B.A., Wright, R., Bewley, R.L., et al., 2002. Plasma thermal conversion of methane to acetylene. Plasma Chem. Plasma Process. 22 (1) (March).

Gardner, D., 1991. Electricity: a new power source for catalytic reactors. Chem. Br., 1120–1122. (December).

Giordano, R.L.C., Giordano, R.C., Cooney, C.L., 2000. Performance of a continuous Taylor-Couette-Poiseuille vortex flow enzymic reactor with suspended particles. Process Biochem. 35 (10), 1093–1101.

Glasnov, T.N., Kappe, C.O., 2007. Microwave-assisted synthesis under continuous-flow conditions. Macromol. Rapid Commun. 28, 395–410.

Gogate, P.R., 2008. Treatment of wastewater streams containing phenolic compounds using hybrid techniques based on cavitation: a review of the current status and the way forward. Ultrason. Sonochem. 15, 1–15.

Gogate, P.R., Mujumdar, S., Thampi, J., Wilhelm, A.-M., Pandit, A.B., 2004. Destruction of phenol using sonochemical reactors: scale up aspects and comparison of novel configuration with conventional reactors. Sep. Purif. Technol. 34, 25–34.

Groszek, M.A., 1990. The Torbed process: a novel concept in heat and mass transfer. Proceedings of the International Deep Mining Conference: Innovations in Metallurgical Plant, Johannesburg, SAIMM, pp. 191–195.

Haas, M.J., Scott, K.M., Marmer, W.N., Foglia, T.A., 2004. In situ alkaline transesterification: an effective method for the production of fatty acid esters from vegetable oils. J. Am. Oil Chem. Soc. 81 (1), 83–89.

Harmsen, G.J., 2007. Reactive distillation: the front-runner of industrial process intensification. A full review of commercial applications, research, scale-up, design and operation. Chem. Eng. Process. 46, 774–780.

Harvey, A.P., Stonestreet, P., 2002. A mixing-based design methodology for continuous oscillatory baffled reactors. Chem. Eng. Res. Des. 80, 31–44.

Harvey, A.P., Mackley, M.R., Stonestreet, P., 2001. Operation and Optimisation of an Oscillatory baffled reactor. Ind. Eng. Chem. Res. 40, 5371–5377.

Haugwitz, S., Hagander, P., Noren, T., 2007. Modelling and control of a novel heat exchanger reactor, the open plate reactor. Control Eng. Pract. 15, 779–792.

Haut, B., Ben Amor, H., Coulon, L., Jacquet, A., Halloin, V., 2003. Hydrodynamics and mass transfer in a Couette-Taylor bioreactor for the culture of animal cells. Chem. Eng. Sci. 58 (3–6), 777–784.

Hendershot, D.C.,1999. Designing safety into a chemical process. Proceedings of the Fifth Asia Pacific Responsible Case Conference and Chemical Safety Workshop, Shanghai, 7–10 November.

Jachuck, R.J.J., Ramshaw, C., 1994. Process intensification: heat transfer transfer characteristics of tailored rotating surfaces. Heat Recovery Syst. CHP 14, 475.

Jones, E., Weatherley, L.R., 2003. Modelling the kinetics of lipase-catalysed oil hydrolysis in an electrically enhanced liquid-liquid system. J. Chem. Technol. Biotechnol. 78, 194–198.

Kaneko, H., et al., 1993. Water electrolysis under microgravity condition by parabolic flight. Electrochim. Acta 38 (5), 729.

Kang, S.H., Goo, S., Lee, S.G., Jung, W.M., Kim, M.C., Kim, W.S., et al., 2003. Effect of Taylor vortices on calcium carbonate crystallisation by gas-liquid reaction. J. Cryst. Growth 254 (1–2), 196–205.

Kasim, F.H., Harvey, A.P., 2011. Influence of various parameters on reactive extraction of *Jatropha curcas* L. for biodiesel production. Chem. Eng. J. 171 (3), 1373–1378.

Kim, S.-S., Lee, H., Choi, J.-W., Na, B.-K., Song, H.-K., 2007. Methane conversion to higher hydrocarbons in a dielectric-barrier discharge reactor with $Pt/\gamma\text{-}Al_2O_3$ catalyst. Catal. Commun. 8, 1438–1442.

Koerfer, M., 1986. Hydrodynamics and mass transfer of thin liquid films flowing on rotating perforated discs. Departmental Report, Chem. Eng. Dept., Newcastle University, Newcastle upon Tyne, UK.

Kreido Laboratories, 2004. *Our STT Technology*. <http://www.kreido.com/stt.htm>.

Krishna, R., 2002. Reactive separations: more ways to skin a cat. Chem. Eng. Sci. 57, 1491–1504.

Kulprathipanji, S. (Ed.), 2002. Reactive Separation Processes Taylor & Francis, New York.

Kundu, A., Ahn, J.E., Park, S.-S., Shul, Y.G., Han, H.S., 2007. Process intensification by micro-channel reactor for steam reforming of methanol. Chem. Eng. J. doi: 10.1016/j/cej.2007.02.026.

Lawton, S., Steele, G., Shering, P., Zhao, L., Laird, I., Ni, X., 2009. Continuous crystallisation of pharmaceuticals using a continuous oscillatory baffled crystalliser. Organic Proc. Res. Dev. 13, 1357–1363.

Lee, C.T., Mackley, M.R., Stonestreet, P., Middelberg, A.P.J., 2001. Protein refolding in an oscillatory baffled reactor. Biotechnol. Lett. 23, 1899–1901.

Leonelli, C., Mason, T.J., 2010. Microwave and ultrasonic processing: now a realistic option for industry. Chemical Eng. Process. 885–900.

Levenspiel, O., 2007. Chemical Reaction Engineering. Wiley.

Loewe, H., Hessel, V., 2006. New approaches to process intensification: microstructured reactors changing processing routes. Proceedings of the Second Process Intensification and Innovation Conference, Christchurch, New Zealand, 24–29 September.

Loupy, A., 2002. Microwaves in Organic Synthesis. Wiley-VCH.

Luo, L., Tondeur, D., Le Gall, H., Corbel, S., 2007. Constructal approach and multi-scale components. Appl. Therm. Eng. 27, 1708–1714.

Mackley, M.R., Harrison, S.T.L., 1992. A pulsatile flow bioreactor. Chem. Eng. Sci. 47, 490–493.

Mackley, M.R., Stonestreet, P., 1995. Heat transfer and associated energy dissipation for oscillatory flow in baffled tubes. Chem. Eng. Sci. 50, 2211–2224.

Mackley, M.R., Smith, K.B., Wise, N.P., 1993. The mixing and separation of particle suspensions using oscillatory flow in baffled tubes. Chem. Eng. Res. Des. 71 (A6), 649–656.

Mason, T.J., Cordemans, E., 1996. Ultrasonic intensification of chemical processing and related operations: a review. Chem. Eng. Res. Des. 74 (A5), 511–516. (July).

Matsushima, H., et al., 2003. Water electrolysis under microgravity: Part I experimental technique. Electrochim. Acta 48, 4119.

McMullen, J.P., Jensen, K.F., 2011. Rapid determination of reaction kinetics with an automated microfluidic system. Organic Proc. Res. Dev. 15 (2), 398–407.

Meeuwse, M., van der Schaaf, J., Schouten, J.C., 2012. Multistage rotor-stator spinning disc reactor. 58 (1), 247–255.

Ni, X., Gao, S., 1996. Scale up correlation for mass transfer coefficients in pulsed baffled reactors. Chem. Eng. J 63, 157–166.

Ni, X., Gao, S., Pritchard, D.W., 1995. A study of mass transfer in yeast in a pulsed baffled bioreactor. J.Biotechnol. Bioeng 45, 165–175.

Ni, X., Zhang, Y., Mustafa, I., 1998. An investigation of droplet size and size distribution in methylmethacrylate suspensions in a batch oscillatory baffled reactor. Chem. Eng. Sci. 53, 2903–2919.

Ni, X., Zhang, Y., Mustafa, I., 1999. Correction of polymer particle size with droplet size in suspension polymerisation of methylmethacrylate in a batch oscillatory baffled reactor. Chem. Eng. Sci. 54, 841–850.

Ni, X., Bennett, D.C., Symes, K.C., Grey, B.D., 2000a. Inverse phase suspension polymerisation of acrylamide in a batch oscillatory baffled reactor. J. Appl. Polym. Sci. 76, 1669–1676.

Ni, X., Cosgrove, J.A., Arnott, A.D., Greated, C.A., Cumming, R.H., 2000b. On the measurement of strain rate in an oscillatory baffled column using particle image velocimetry. Chem. Eng. Sci. 55, 3195–3208.

Ni, X., Cosgrove, J.A., Cumming, R.H., Greated, C.A., Murray, K.R., Norman, P., 2001. Experimental study of flocculation of bentonite and alcaligenes eutrophus in a batch oscillatory baffled flocculator. Trans. IchemE. 79 (Part A).

Ni, X., Gaidhani, H.K., McNeil, B., 2005. Fermentation of pullulan using an oscillatory baffled fermenter. Chem. Eng. Res. Des. 83 (A6), 640–645.

Nusselt, W., 1916. Die oberflachen kondensation des wasserdampfes. Z. ver. deut. ing. 60 (p. 541, and p. 569).

Perry, W.L., Datye, A.K., Prinja, A.K., Brown, L.F., Katz, J.D., 2002. Microwave heating of endothermic catalytic reactions: reforming of methanol. AIChE J. 48 (4), 820–831.

Phan, A.N., Harvey, A., 2010. Development and evaluation of novel designs of continuous mesoscale oscillatory baffled reactors. Chem. Eng. J. 159 (1–3), 212–219.

Phan, A.N., Harvey, A.P., 2012. Characterisation of mesoscale oscillatory helical baffled reactor–experimental approach. Chem. Eng. J. 180, 229–236.

Phan, A.N., Harvey, A.P., Eze, V., 2012. Rapid production of biodiesel in mesoscale oscillatory baffled reactors. Chem. Eng. Technol. 35 (7), 1214–1220.

Pudjiono, P.I., Tavare, N.S., Garside, J., Nigam, K.D.P., 1992. Residence time distribution from a continuous couette-flow device. Chem. Eng. J. Biochem. Eng. J. 48 (2), 101–110.

Qi, Z., Zhang, R., 2004. Alkylation of benzene with ethylene in a packed reactive distillation column. Ind. Eng. Chem. Res. 43, 4105–4111.

Reay, D.A., Harvey, A.P., 2012. The role of heat pipes in intensified unit operations. Appl. Therm. Eng. doi: 10.1016/j.applthermaleng.2012.04.002 (2012).

Reis, N., Vicente, A.A., Teixeira, J.A., Mackley, M.R., 2004. Residence times and mixing of a novel continuous oscillatory flow screening reactor. Chem. Eng. Sci. 59 (22–23), 4967–4974.

Reis, N., Harvey, A.P., Mackley, M.R., Teixeira, J.A., Vicente, A.A., 2005. Fluid mechanics and design aspects of a novel oscillatory flow screening meso-reactor. Chem. Eng. Res. Des. 83, 357–371.

Ristic, R.I., Chew, C.M., 2005. Crystallisation by oscillatory and conventional mixing at constant power density. AIChE J. 51 (5), 1576–1579.

Seris, E.L.C., Abramowitz, G., Johnston, A.M., Haynes, B.S., 2008. Scaleable, microstructured plant for steam reforming of methane. Chem. Eng. J. 135 (Suppl. 1), S9–S16.

Shu, J., Lakshmanan, V.I., Dodson, C.E., 2000. Hydrodynamic study of a toroidal fluidised bed reactor. Chem. Eng. Process. 39, 499–506.

Sisoev, G.M., Omar, K., Matar, O.K., Lawrence, C.J., 2005. Gas absorption into a wavy film flowing over a spinning disc. Chem. Eng. Sci. 60 (7), 2051–2060.

Smith, K.B., 2000. The Scale-Up of Oscillatory Flow Mixing. University of Cambridge, UK, (PhD Thesis).

Solano, J.P., Herrero, R., Espin, S., Phan, A.N., Harvey, A.P., 2012. Numerical study of the flow pattern and heat transfer enhancement in oscillatory baffled reactors with helical coil inserts. Chem. Eng. Res. Des. 90 (6), 732–742.

Stankiewicz, A., Moulijn, J.A., 2000. Process intensification: transforming chemical engineering. Chem. Eng. Prog., 22–34. (January).

Steinbach, P.B., Manahan, S.E., Larsen, D.W., 2003. The chemical reduction of small inorganic gases in an electrothermal plasma reactor. Microchem. J. 75, 223–231.

Stephenson, P. et al., 2003. Patent WO 03099743. Asymmetric hydrogenation under supercritical conditions. Publication date 4 December.

Stevens, G.G., 1996. Suspension Polymerisation in Oscillatory Flow. University of Cambridge, UK, (PhD Thesis).

Stitt, E.H., 2002. Alternative multiphase reactors for fine chemicals. A world beyond stirred tanks? Chem. Eng. J. 90, 47–60.

Stonestreet, P., Van der Veeken, P.M.J., 1999. The effects of oscillatory flow and bulk flow components on residence time distribution in baffled tube reactors. Chem. Eng. Res. Des. 77 (A8), 671–684.

Thompson, L.H., Doraiswamy, L.K., 1999. Sonochemistry: science and engineering. Ind. Eng. Chem. Res. 38, 1215–1249.

Trent, D., Tirtowidjojo, D., 2001. Commercial operation of a rotating packed bed (RPB) and other applications of RPB technology. Proceedings of the Fourth International Conference on Process Intensification in the Chemical Industry, BHR Group, pp. 11–19.

Trusek-Holownia, A., 2008. Production of protein hydrolysates in an enzymatic membrane reactor. Biochem. Eng. J. 39 (2), 221–229.

Tuchlenski, A., Beckmann, A., Reusch, D., Dussel, R., Weidlich, U., Janowsky, R., 2001. Reactive distillation–industrial applications, process design and scale-up. Chem. Eng. Sci. 56, 387–394.

Vardi, A., Shapiro, M., 1998. Laser catalysis with pulses. A Phys. Rev. 58 (2), 1352–1360. (August).

Vedantam, S., Joshi, J.B., 2006. Annular centrifugal contactors: a review. Chem. Eng. Res. Des. 84 (A7), 522–542.

Visscher, F., van der Schaaf, J., de Croon, M.H.J.M., Schouten, J.C., 2012. Liquid–liquid mass transfer in a rotor-stator spinning disc reactor. Chem. Eng. J. 185–186, 267–273.

Wasewar, K.L., Pangarkar, V.G., Heesink, A.B.M., Versteeg, G.F., 2003. Intensification of enzymatic conversion of glucose to lactic acid by reactive extraction. Chem. Eng. Sci. 58, 3385–3393.

Weatherley, L.R., 1993. Electrically enhanced mass transfer. Heat Recovery Syst. CHP 13 (6), 515–537.

Woods, W.P., 1995. The Hydrodynamics of Thin Films Flowing Over a Rotating Disc. Newcastle University, Newcastle upon Tyne, UK, (PhD Dissertation).

Xu, N., Xing, W., Xu, N., Shi, J., 2002. Application of turbulence promoters in ceramic membrane bioreactor used for municipal wastewater reclamation. J. Membr. Sci. 210, 307–313.

Zakaria, R., Harvey, A.P., 2012. Direct production of biodiesel from rapeseed by reactive extraction/*in situ* transesterification. Fuel Process. Technol. 102, 53–60.

Zanfir, M., Gavriilidis, A., 2003. Catalytic combustion assisted methane steam reforming in a catalytic plate reactor. Chem. Eng. Sci. 58, 3947–3960.

Zeev, K.E., 2003. Method and apparatus for optically activating chemical reactions. Russian Patent RU2210022, published 8 October.

Zhang, H., Zhuang, J., 2003. Research, development and industrial application of heat pipe technology in China. Appl. Therm. Eng. 23 (9), 1067–1083.

Zheng, M., Skelton, R.L., Mackley, M.R., 2007. Biodiesel reaction screening using oscillatory flow meso-reactors. Process Saf. Environ. Prot. 85 (B5), 365–371.

Zhu, X., Campero, R.J., Vigil, R.D., 2000. Axial mass transport in liquid/liquid Taylor-Couette-Poiseuille flow. Chem. Eng. Sci. 55, 5079–5087.

Intensification of Separation Processes

6

OBJECTIVES IN THIS CHAPTER

The principal objectives in this chapter are to discuss the several ways in which separation processes, ranging from distillation and evaporation though drying and crystallisation, and, in this second edition, electrolysis, can be intensified. There are process-specific methods such as liquid–liquid extraction, absorption, and adsorption that must also be considered. The range of technologies is substantial, including high gravity fields (the HiGee distillation unit) and microwaves for drying. Techniques familiar to chemical engineers, including divided wall distillation columns (some say 'dividing wall') and new packing, are also covered. The aim is also to introduce the concepts so that they can be appreciated in the later applications chapters.

The chapter concentrates on 'active' intensification methods, such as rotation and electric fields, as these are the approaches that can produce radical changes in process plant design and performance, as with the spinning disc reactor described in Chapter 5.

6.1 Introduction

After reactions, separations are the most important unit operations within the chemical industry. Separations are also essential components of many other process industry sectors, in particular food and drink; paper and board; and textiles – where the removal of liquids in drying and the concentration of liquids (e.g. in soft drinks manufacture) are highly energy-intensive.

Distillation is the most significant separation process within the process industries. It is a major energy user in the chemicals sector where, together with drying and evaporation, it consumed the equivalent of a million tonnes of oil in the UK in 2004. Distillation columns are used in many operations, the main ones being:

- Crude oil distillation.
- Distillation of petroleum fractions.
- LPG separation and gas processing.
- General hydrocarbon separation.
- Aromatic separation.
- Water/organics separation.
- Water/inorganics separation.

A number of approaches can be adopted to reduce energy use in the columns, two major examples of which can involve process intensification. The first of these, improvements to the distillation column population to make the units more efficient rather than wholesale replacement by intensified alternative separation methods, is the most likely route in the short- to medium-term, covered briefly below. In the longer term, alternative plant or separation techniques are needed. Potentially, the most important of these is HiGee – and in Chapter 8, where a number of power generation concepts are discussed, the use of HiGee (and other intensified techniques) for carbon capture is introduced.

There are a variety of other 'conventional' separation methods used for evaporation and drying processes and these can also benefit from the application of PI. Topics such as dewatering using centrifuges and the application of electrical enhancement methods to speed up drying are also discussed below.

6.2 Distillation

6.2.1 Distillation – dividing wall columns

Dividing (or divided – the term varies) wall columns (DWCs) are a convenient way of putting two distillation columns into one shell, the 'wall' separating the two functions. Their use is applicable where three or more components require separating. Normally, this needs two or more columns, with each column having its own reboiler and condenser – which is energy-inefficient, capital intensive and takes up a lot of space. Kaibel (2007) points out that there are two main types of dividing wall columns. The simpler type represents a column which has the dividing wall assembled either at the upper or lower end of the column (visible in Figure 6.1 in the

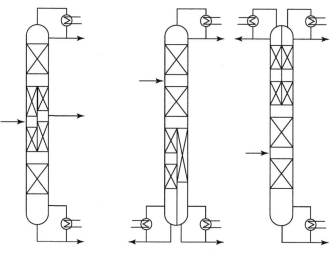

FIGURE 6.1

Configurations of dividing wall distillation columns (Kaibel, 2007).

second and third columns). The first applications on a production scale took place only in 2004, although the configuration was proposed in the 1930s. The more common type of dividing wall column is shown on the left-hand side. The dividing wall is placed in the middle section of the column, above and below the feed and the side draw.

By overcoming thermodynamic inefficiencies inherent in direct sequencing of columns and allowing some direct contact heat transfer, energy use is reduced and capital spending on heat exchangers and columns is minimised. Energy savings are typically 30% – perhaps higher (up to 40%) for close boiling point fluids. Capital costs may be reduced by, typically, 25%. The structure of a dividing wall inside a column is shown in Figure 6.2, and dividing wall columns have also been proposed for reactive distillation, a topic discussed in Chapter 5, in addition to azeotropic and extractive distillations.

This latter type of distillation was the subject of a recent analysis by AkzoNobel Research (Suszwalak and Kiss, 2012) in the context of bioethanol dehydration. The study showed that extractive DWC was technically feasible and could yield highly pure bioethanol (>99.8%wt). Additionally, an energy saving in excess of 10% was predicted, with the added benefits (that would be reflected in capital cost savings)

FIGURE 6.2

A view inside a dividing wall column (Kaibel, 2007).

of elimination of a reboiler and the use of only one column shell. It was recommended that the use of extractive DWCs would be particularly interesting in new bioethanol plant construction.

An interesting perspective on dividing wall columns, which nicely affords us an update on the application of these efficiency units, is given by researchers, again at AkzoNobel Research, and at the University Politehnica in Bucharest (Kiss and Bildea, 2011). The team points out that only a few companies, led by Montz and BASF, use this type of column, in spite of the fact that it can separate three components in a single tower. The main thrust of the work of Kiss and Bildea, 2011 is directed at overcoming a barrier to what they perceive as limiting the uptake of the technology – insight into the operation and control of DWCs. Again, conservatism takes over and companies select large column solutions.

The study concludes on a highly positive note, giving guidelines for selecting DWC control strategies. Several strategies were examined, including a number of multi-loop proportional integral derivative (PID) control strategies. These were effective in handling persistent disturbances quite quickly, but model predictive control (MPC) was superior.

6.2.2 Compact heat exchangers inside the column

The concept of using heat exchangers for both heat and mass transfer in multi-stage rectification has been around for over 50 years, but more recently companies such as Costain Oil, Gas and Process Ltd have used highly compact heat exchangers, particularly plate-fin units, for a variety of cryogenic duties including air separation and ethylene recovery. The reflux heat exchanger is functionally equivalent, stated Finn (1994), to the multi-stage rectification section of a distillation column. The advantages over conventional distillation are several, but two main benefits arise:

- Small temperature differences exist between the condensing feed stream and the streams providing refrigeration. (Large temperature-driving forces lead to inefficiencies in conventional columns.)
- The CHE has a large number of partial condensation stages, so temperature and composition differences between vapour and liquid are small and separation takes place close to equilibrium conditions.

Some may argue the degree of intensification achieved using this technique, but the results are impressive; Finn quoted a greater than 99% recovery of the required olefins and much reduced downstream energy needs for fractionation. It is now appropriate to examine in more detail the use of high gravity fields to aid separations – a technique that has been briefly introduced in earlier chapters – exemplified by HiGee.

The use of plate-fins comes into the work carried out at ECN and Delft University of Technology in The Netherlands on heat-integrated distillation columns (HIDiCs). Bruinsma et al. (2012) compared an HIDiC with plate-fin internals with one utilising a plate-packing configuration using structured packing supplied by Sulzer ChemTech.

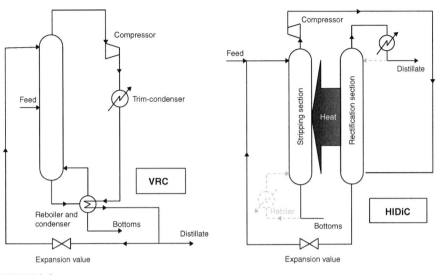

FIGURE 6.3

Heat integrated distillation column (shown next to one incorporating mechanical vapour recompression – a heat pump) (Bruinsma et al., 2012).

Figure 6.3 shows the HIDiC (and also an MVR column, incorporating a mechanical vapour recompression circuit). MVR is also used in the HIDiC, where vapour from the stripping section is compressed upstream of the rectification section – and it is suggested that this can be 30–50% more efficient in terms of energy use than the basic MVR system.

The unit tested revealed that the plate-packing HIDiC was substantially better than the one incorporating plate-fin structures. This was reflected in better heat transfer, separation and pressure drop per stage. The superiority was attributed to poor liquid distribution in the plate-fin unit, leading to dry spots and an increase in the height equivalent to a theoretical plate (HETP). It would be interesting to explore the performance of an optimised plate-fin unit, too.

6.2.3 Cyclic distillation systems

An alternative route to intensified distillation has been introduced recently by a group in The Netherlands and Ukraine (Maleta et al., 2011). It is claimed that operating in this manner, the cyclic distillation process introduces two different modes of operation within a distillation column. The first, as described by the authors, occurs when the vapour is flowing upwards through the column, but the liquid remains stationary on the plates within the column. The second part of the cycle is called a liquid flow period, during which vapour flow ceases and reflux and feed

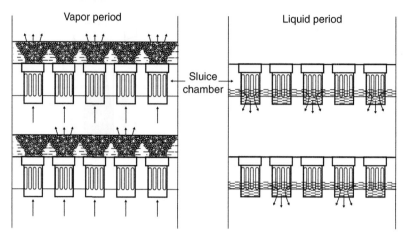

FIGURE 6.4

Cyclic distillation – schematic showing the vapour period, followed by the liquid period (Maleta et al., 2011).

liquid are supplied. In this phase the liquid holdup drops down to the tray below. The schematic of these two modes of operation is illustrated in Figure 6.4.

In a case study reported in the paper Maleta et al., (2011) modelled an industrial-scale cyclic distillation column in the food sector (at the Lipnitsky Alcohol Plant, Ukraine). The distillation column strips water from ethanol to produce 30,000 l/day of food grade ethanol and analyses suggested that the cyclic column operation, using a column with 15 trays, would have a separation efficiency 200–300% greater than a conventional bubble cap tray column with 56 trays. An exercise on a separate column that concentrates the impurities was also highly positive in its conclusions.

6.2.4 HiGee

As its name implies, the HiGee rotating packed-bed (Mallinson and Ramshaw, 1979; Ramshaw, 1993) comprises a rotating torus of packing which has a relatively high specific surface area compared to conventional column packing (Figure 6.5). Liquid is injected onto the inner cylindrical packing surface from a stationary set of nozzles and, after percolating through the packing, it leaves the rotor and enters the machine case at the prevailing peripheral speed – typically 40–60 m/s. It then leaves by a suitable drainage point which must be designed so as to prevent liquid accumulation capable of interfering with the rotor operation. Vapour to be contacted enters the machine casing and is forced to flow radially inwards and leave via the rotor centre. As shown in Figure 6.5, a mechanical seal must be installed to block any vapour bypass flow around the rotor. The seal may be a lubricated face seal or

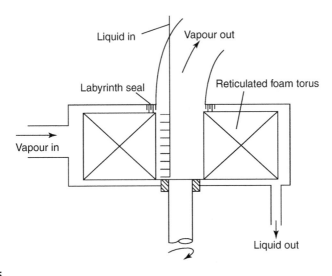

Liquid in

Vapour out

Labyrinth seal

Reticulated foam torus

Vapour in

Liquid out

FIGURE 6.5

The original HiGee rotating liquid contactor.

the labyrinth type, the latter being preferred in view of the fairly high relative velocity at the seal contact point.

The packing employed must be able to withstand the body forces generated within the rotor, in order to maintain balanced conditions while having a specific area in the range 500–5,000 m²/m³ and a high voidage (>85%). Contenders for this duty are:

- Reticulated metal foam.
- Composite layers of gauze or expanded metal.
- Wound layers of fibrous material.

In general, the packing characteristics will be uniform throughout the rotor, though some grading may ultimately be attractive in order to manipulate the local flooding conditions. The strategic basis for the use of high acceleration fields for the intensification of multi-phase operation is outlined in Chapters 8, 9 and 10 where application data are given. The specific application involving rotating packed-beds is best explained in the context of the Sherwood Flooding Chart (Sherwood et al., 1938) which is shown in Figure 6.6. The chart was originally derived in order to explain fluid behaviour in packed absorption and distillation columns in which liquid trickled downwards against an up-flowing gas or vapour. As the flow increases, so does the interfacial interaction and it becomes more difficult for liquid to move downwards through the packing. In the limit, liquid is prevented from any movement and it simply accumulates within the column – a condition known as flooding. At this point the column becomes ineffective as a mass transfer

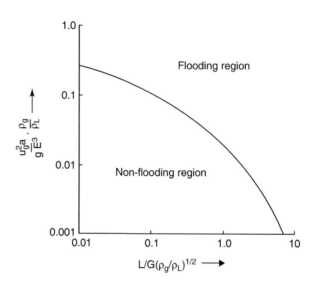

FIGURE 6.6

The flooding correlation for gas–liquid flow in packed beds.

device, because most of the interfacial area is lost. The Sherwood Chart for an inviscid liquid plots L/G $(\rho_g/\rho_l)^{\frac{1}{2}}$ as abscissa against the ordinate, where:

L = liquid mass flux
G = gas mass flux
ρ_g = gas density
ρ_l = liquid density
a = packing area per unit volume
g = applied acceleration
E = packing voidage (capital E in the figure)
u_g = gas velocity

The experimental results for random and stacked packing under terrestrial acceleration are shown in Figure 6.7, which distinguishes a lower non-flood zone from the upper flooding zone. It should be noted that for a packed column operating under terrestrial acceleration, both the applied acceleration and the total cross sectional area for fluid flow remain constant. Therefore, once the flooding condition is reached, flooding occurs *throughout* the column and a sudden substantial increase in pressure drop is experienced.

The abscissa X of the Sherwood plot for a given process system is fixed by the process requirements which dictate the liquid and gas fluxes, L and G. Hence, for example, with a distillation column operating at total reflux L = G and, since ρ_l/ρ_g, 1000, we have X ~ 0.03 and Y (flood) ~ 0.2 for random packing. For a given system, whether under terrestrial or enhanced acceleration, this implies that X is constant.

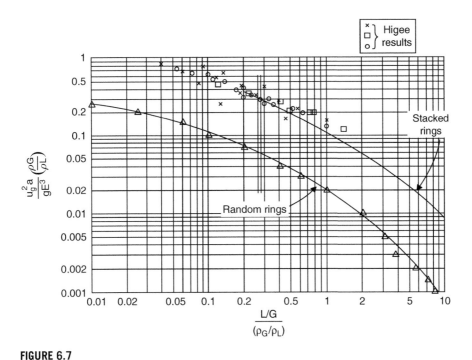

FIGURE 6.7

Experimental results showing HiGee flooding correlation.

The centrifugal acceleration generated within a rotating bed is $g = r\omega^2$ and it therefore increases towards the periphery. Noting that E^3 for a high voidage packing will vary only over a limited range (say 0.5–0.75), a thousandfold increase in g implies that at the flooding limit either u_g may be increased by a factor ~30 or the packing-specific surface area can increase by a factor ~1,000. The first case allows both the gas and liquid flow to increase about thirtyfold, since L/G is fixed, whereas the second case suggests that packing having a very high surface area can be used. This intuitively should lead to greatly improved mass transfer and reduced theoretical stage heights. In practice, both a and u_g should be increased so as to create a roughly equi-axed toroidal packing.

In view of the largely radial nature of the gas and liquid flows within the packing, the mass fluxes in the inner rotor zones are higher than those at the periphery. This, together with the radial variation of the centrifugal acceleration noted above, means that the tendency for the packing to flood is highest at the inner radius and least at the periphery. As a consequence, the flooding behaviour and the overall pressure drop characteristics of the packed rotor are very different from those of a stationary column, for which the flooding limit is reached simultaneously throughout the packing. The design of a HiGee contactor should therefore be based on an appropriate fraction of the flows at the inner flooding condition. In the absence of an abrupt increase in the overall pressure drop, an experimental determination of the flooding condition for the rotor is quite difficult. However, it has been observed that an increased liquid agitation is detectable at the inner packing radius, coupled with a characteristic hissing noise as the flooding condition is approached. The results shown in Figure 6.7 were obtained on this basis and broadly correspond to those achieved with stacked packing in a column exposed to terrestrial acceleration. Fluid fluxes at flooding are of the order of tenfold higher than those in a column, with transfer unit heights of around 1.5 cm for gas film-limited systems and about 4 cm for liquid film-limited systems.

The fluid flow behaviour within the rotor has been studied in some detail by Burns and Ramshaw (1996) who initially performed photographic investigations followed by hold-up and velocity measurements. Using diluted emulsion paint and black polyurethane sponge packing, it was shown that the liquid injected on the inner packing surface attained the local rotational speed after about 1 cm penetration. Beyond this, at the lower rotor speeds, much of the liquid flow occurred in the form of discrete rivulets rather than being distributed uniformly over the packing surface. As rotor speed increased, liquid tended to flow as 'flying droplets' across the packing voids in contrast to the previous rivulet/film flow. This probably accounts for the somewhat surprising observation that the best packing from a mass transfer point of view was not necessarily that having the greatest specific surface area. Coarse packing (e.g. with a surface area ~1,000–2,000 m^2/m^3) was generally better, presumably because 'flying droplets' were encouraged with their additional surface area. It will be recognised that coarse packing has the additional advantage of lower pressure drop.

The early development of the HiGee concept by ICI was based on a prototype unit for the ethanol/propanol distillation system. This used two rotors of Retimet (a Ni/Cr reticulated foam) having the dimensions:

ID	300 mm
OD	800 mm
Axial thickness	300 mm

After laborious improvements to the liquid distribution system in order to establish uniformity, over 20 theoretical distillation stages in the 25 cm depth of packing was achieved. The presence of any flow shadows within the packing, such as could be caused by tie bolts, caused significant performance impairment. Clearly maldistribution is a major issue both for conventional packed columns and rotating packed beds.

This can be appreciated when we consider a column with maldistribution to be equivalent to two ideal columns operating in parallel, each with L/G on either side of the overall mean value (Figure 6.8). More severe maldistribution corresponds to a bigger difference in the slope of the operating lines, i.e. L/G. For close-boiling mixtures, the number of theoretical stages becomes particularly sensitive to a given change in L/G as a 'pinch' situation is created and approached. Thus, for a given mass transfer capacity, i.e. a given number of installed theoretical stages, the separation capacity of one or both columns is progressively compromised. In view of

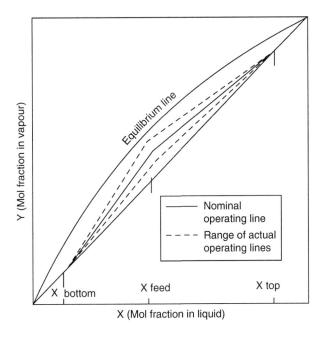

FIGURE 6.8

The impact of maldistribution on packing performance.

this, only about 20 stages can be prudently expected within one rotor unless some redistribution technique can be applied. As a consequence, the separation of close boiling mixtures such as propane/propylene requiring ~200 theoretical stages cannot be performed in one rotor, however large. A feasible alternative is to install multiple packed-beds on one shaft and devise a suitable system for conducting liquid from the periphery of one to the interior of its successor along the shaft – and vice versa for the vapour flow.

After their photographic studies, Burns et al., (2000) went on to measure liquid hold-up and velocity within the rotating packing (typical packing is shown in Figure 6.9) using a conductivity technique pioneered by Basic and Dudukovic (1995). Two pairs of circular gauze electrodes were inserted into a non-conducting packing. The liquid volume fraction was inferred from the resistance of the electrode pairs, while the average liquid velocity between the pairs was measured by timing the transit of a pulse of conducting tracer liquid. It was shown that liquid velocities within the packing were in the order of 1 m/s giving a residence time, 1 second. Liquid hold-up decreased roughly in inverse proportion to the packing radius so that peripheral values, 5% were achieved even when the inner rotor face was flooded.

In addition to the care which has to be taken to maintain excellent liquid distribution, HiGee has a further operational characteristic which concerns the liquid and gas behaviour in the eye of the rotor. The flows within the packing correspond

FIGURE 6.9

The packing typically used in the HiGee machines.

to a *forced* vortex in which the gas and liquid streams rotate at virtually the same angular velocity as the packing. Relative to the packing, the flows are nearly radial because the lateral Coriolis acceleration ($2v_r\omega$ where v_r is the radial velocity) is small compared to the radial centrifugal acceleration. However, at the centre there is little frictional interaction to compel the fluids to conform to the packing angular velocity, so a free vortex is generated, leading to relatively high tangential gas velocities because angular momentum (tangential velocity × local radius) is maintained as the radius decreases. These can disrupt the liquid flow leaving the distributors and, in the limit, remove the liquid before it reaches the packing surface. The liquid distributor must be designed to overcome this problem.

There are two illustrations of the original HiGee machine in Chapter 1.

6.2.4.1 Principal HiGee operating features

The radial thickness of the rotor is directly related to the separation duty and the corresponding number of stages required, noting the stage heights given above. The hydraulic capacity depends upon the flooding rates at the inner radius and the inner cylindrical surface area. Typically, operation at about 50% of the flooding rates can be prudently adopted. Generally, the pressure drop incurred/theoretical stage will be equivalent to that generated by a trayed column rather than the lower value of its packed equivalent. Hence, the HiGee characteristics of most interest are:

- Small footprint/low height.
- Residence time ~1 second.
- Very low process inventory.

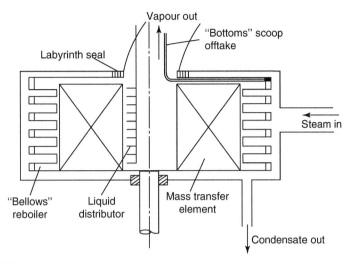

FIGURE 6.10

HiGee with a peripheral reboiler.

Performance is not influenced by changes in orientation, e.g. when barge or ship-mounted. Applications which are particularly suited to HiGee are therefore where space, weight or hazardous inventory cause significant concern. This could include many offshore or sea-going installations. HiGee may be difficult to justify when relatively cheap column construction materials such as mild steel or polymers will suffice. However, if stainless steel or other more expensive corrosion resist-ant material is needed, then HiGee becomes attractive. Finally, the pressure drops within a rotor will largely rule it out for high vacuum duties in its present state of development. It is more appropriate for operation at or above one bar. Since it has proved impractical to provide feed/off-takes at a mid-radial position, distillation duties require two separate rotors, preferably on one shaft, with an intermediate liq-uid supply.

In its present configuration the HiGee rotating packed-bed concept only applies the elevated acceleration to the mass transfer element of a distillation operation. However, it is worth noting that the associated reboiler and condenser functions both involve multiple phases. They could therefore benefit from inclusion within the rotor, leading to a peripheral reboiler and a central condenser. This would lead to a further development of the PI strategy by the provision of multi-functional units, see Figure 6.10.

Among the valuable work done in the Far East since the first edition, the research at the Research Centre for High Gravity Engineering and Technology, located within Beijing University of Chemical Technology stands out, (see Appendix 4). Research by Wang et al. (2007) led to what was claimed to be a new HiGee – a rotating zigzag bed (RZB). The RZB is claimed to function without

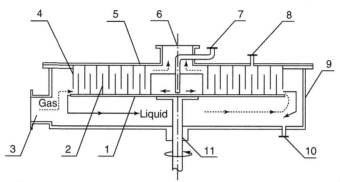

1. Rotational disc. 2. Rotational baffle. 3. Gas inlet. 4. Stationary baffle.
5. Stationary disc. 6. Gas outlet. 7. Liquid inlet. 8. Intermediate feed.
9. Rotor casing. 10. Liquid outlet. 11. Rotating shaft.

FIGURE 6.11

The rotating zigzag bed HiGee variant of Wang et al.

liquid distributors and can accomplish intermediate feeds in continuous distillation processes. The rotor (Figure 6.11) couples a rotating disc with a stationary disc each of which has vertical baffles, as shown in the figure. The gas and liquid flows are as in a conventional RPB, and if intermediate fluid injection is needed it can be done easily at locations on the static disc (5 in the figure). Further work was then directed at continuous distillation applications, but it was realised that the RZB unit did not use the best packing, in terms of high gas–liquid interfacial area, and the unit exhibited a lower mass transfer rate than the conventional RPB.

It should be noted that the RZB is very similar to the centrifugal absorber unit described by Chambers and Wall in 1954, as shown in Figure 12.32 in Coulson and Richardson (1978). The reader who has studied Chapter 1 will also see an internal geometry similarity with the Moss still in Figure 1.9.

While integrated HiGee rotating packed beds (RPBs) were designed by Ramshaw and others many years ago, more recently in Beijing. Luo et al., (2012) have taken this concept a step further and studied a two-stage counter-current RPB (TSCC-RPB) for continuous distillation.

Illustrated in Figure 6.12, the new unit has a rotating disc (4 in the figure) and a stationary disc (3 in the figure) in each rotor to enhance contact between the liquid and the packing. This also led to increased contact time, raising the mass transfer rate. In this unit, stainless steel wire mesh is used as packing material. After obtaining fundamental hydrodynamic data, an acetone–water system was used to determine the separation efficiency. The paper compares the two-stage RPB with other concepts and the data for efficiency are given in Table 6.1.

The efficiency was defined as NPM $= N_T/n(r_o-r_i)$

where N_T is the number of theoretical plates and n is the number of stages. r_i and r_o are inner and outer radii of the rotor.

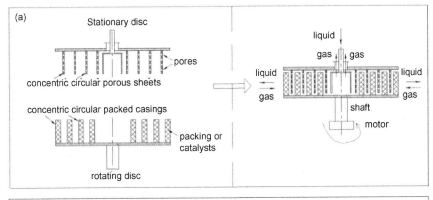

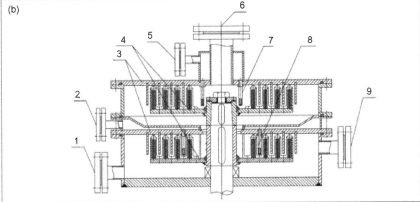

FIGURE 6.12

Schematic diagrams of (a) rotor, and (b) two-stage counter-current rotating packed bed.
(1) is the liquid outlet, (2) liquid inlet 2 – intermediate feed, (3) stationary disc, (4) rotating
disc, (5) liquid inlet 1 (reflux feed), (6) gas outlet, (7,8) liquid distributors and (9) gas inlet
(Luo et al., 2012).

Table 6.1. Comparison of TSCC-RPB with other Distillation Equipment

Equipment	Rotational Speed (rpm)	Distillation System	Efficiency	Reference
Packed column	0	Acetone–water	4	Li (1998)
2 RPBs	1,500–3,000	Ethanol–isopropanol	50–57 (estimated)	Short (1983)
2 RPBs	0–1830	Alcohol–water	20–136	Li et al., (2008)
RZB	600–1,400	Ethanol–water	20–38	Wang et al., (2008)
TSC-RPB	600–1,400	Acetone–water	24–58	–

(Adapted from Luo et al., 2012).

The team suggested that although the pair of RPBs could give a superior performance, their heat losses might be greater than the TSCC-RPB, as the flow paths in the latter would be shorter.

6.2.4.2 Phase inversion

The HiGee concept was originally based on a continuous gas phase in which liquid was dispersed over the packing. As noted above, this currently results in liquid being discharged into the machine case at the rotor tip speed. While the corresponding loss of kinetic energy can perhaps be accepted for the modest flows encountered in most distillation and absorption duties, it can become an issue when very large flows of water need to be deaerated for oilfield reinjection. In this case the flows are ~500 tonnes/hour and a technique for energy recovery may be attractive. This can be largely achieved by inverting the phases so that gas is dispersed as bubbles within a continuous liquid immersed in the packing (Peel et al., 1998). A schematic arrangement is shown in Figure 6.13, where the rotor is divided into two compartments, one containing the packing and the other allowing liquid to return to the inner radius for discharge. The arrangement operates essentially as a centrifugal lute which permits liquid to leave with only about 10% of the kinetic energy

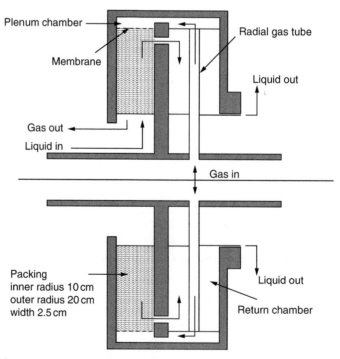

FIGURE 6.13

0.5 m machine design.

(Reproduced from Peel, J. et al., (1998). Trans I Chem E, Vol. 76, Part A).

corresponding to the tip speed. It goes without saying that the return chamber must be fitted with radial baffles to suppress the free-vortex which would otherwise occur. A significant pressure gradient is created within the rotor and the gas must be supplied at a sufficient pressure to overcome this. However, for absorption/stripping duties which involve sparingly soluble gases, the gas mass flow is much smaller than that of the liquid, so the energy implications are expected to be acceptable. Indeed, in many cases the stripping gas, e.g. methane or nitrogen, may be available at a sufficient pressure.

Peel et al., (1998) and Ramshaw and Thornton (1967) also describe a study carried out to investigate bubble impact with simulated packing. It was shown that there was a critical bubble diameter below which bubble breakdown did not occur. The critical value was a function of density difference, collision velocity and the applied acceleration. It was predicted, and experimentally demonstrated, that higher acceleration environments could reduce the critical size and thereby lead to a much smaller size distribution. Provided that the bubbles could reach a substantial fraction of their terminal velocity before impact with packing elements, this implies that high volumetric mass transfer rates should be achievable. The reticulated packing cells must be larger than the bubble diameter in order to realise this condition. This design of rotor (1 m diameter, 0.015 m axial thickness) has been operated as a pilot plant unit at Newcastle University in the UK, where it reduced the oxygen concentration in 15 tonnes/hour simulated seawater from 9 ppm to 0.5 ppm.

6.2.4.3 Modelling of rotating packed beds

It is always helpful when someone attempts to model newly developed thermal or thermodynamic concepts so that sub-routines can be added to commercial software packages. One of the authors was heavily involved in a project on micro/compact heat exchangers some years ago with this aim in mind.

It is therefore encouraging to see that the RPB distillation column has come under some scrutiny, by workers in Brazil, with a view to simulate the equipment in the commercial software Aspen Plus V7, (Prada et al., 2012).

The results, using the Fortran sub-routine that took into account the mass transfer coefficients of the two phases, confirmed that the HiGee unit gave substantial reductions in column height and volume compared to a conventional static column. The distillation selected as a test case was ethanol–water and the correlations used for the volumetric mass transfer coefficients in the RPB were by Reddy et al., (2006) and Rajan (2008), based upon experimental measurements on RPBs. The height equivalent of a theoretical plate in the HiGee case was predicted to be 0.0055, while for the conventional column it was 0.3246.

6.3 Centrifuges

It is not intended to dwell upon the conventional centrifuge in this book. However, the disc-bowl type is worthy of mention, as is the ultra-centrifuge used in the nuclear reprocessing field (see also Chapter 10). All centrifuges are active intensification techniques, and those familiar with centrifuge technology may see parallels between this and HiGee (and spinning disc reactors) that may allow some lateral thinking in the development of yet more innovative unit operations, or combinations of the same.

6.3.1 Conventional types

The usual duty of a centrifuge is to raise the settling velocity of bubbles or particles in order to intensify a phase separation such as slurry concentration or dewatering. In general, process centrifuges operate at moderate accelerations in the range of 1,000–10,000 g. The units illustrated in Figure 6.14 rotate at around 2,000 rpm and are used for molasses processing in the sugar industry (Day, 2004).

The intensification from rotation using commercial centrifuges for evaporation, allows one to reduce the surface to 20% of that of a falling film evaporator, and a reduction in residence time to 12% of that needed in a falling film unit. They can also deal with highly viscous fluids, to over 20,000 cP, and have a small retained volume. Some have facilities for cleaning in place (CIP).

The key to the quality of the material being processed, important in food and other sectors, is a combination of low evaporation temperature and a short residence time. The rotation helps to ensure constant wetting rates and film thickness, avoiding burn-on. Low and well-defined vapour velocities are also important to avoid product loss. Concentration ratios of 25:1 can be achieved in a single pass,

FIGURE 6.14

These conical continuous basket centrifuges operate at around 2,000 'g'.

and there is a very low thermal impact on the product. The penalty is energy – because the unit is a single-effect machine and thus needs a high temperature difference. Energy use can be reduced by using a conventional pre-evaporator as an economiser – this reduces energy use by 66%. The reader interested in following recent developments in centrifuges is advised to read the overview by Harald Anlauf of the University of Karlsruhe (Anlauf, 2007).

6.3.2 **The gas centrifuge**

At very high centrifugal accelerations (10^5–10^6g) gas mixtures tend to stratify, with the higher molecular weights accumulating at the periphery. Thus, the gas or ultra centrifuge is used for isotope separation, allowing the enrichment of uranium in the manufacture of nuclear fuel. The alternative techniques, such as diffusion, are much less efficient. Even so, the separative power of the centrifuges is very modest and depends principally upon their peripheral speed and axial length. In order to accomplish a typical U_{235}/U_{238} separation duty a large cascade of centrifuges is required, which may comprise many thousands of interconnected units. Full details of gas centrifuges used in the nuclear reprocessing industry are given in Chapter 10 (Section 10.2).

6.4 **Membranes**

Membranes separate the components of a gas or liquid mixture on the basis of their different permeabilities through the membrane. Only a pressure or concentration driving force (as opposed to heat energy) is required to make the separation. The energy needs in some membrane processes approach the thermodynamic minimum, making them highly energy-efficient[1]. Examples of membrane concepts are shown in Figure 6.15.

In extraction, for example, membranes are beneficial where a high degree of solute removal is necessary, and where the solvent is the low boiling point component in the distillation recovery step. The membrane behaves like a high efficiency packing with an extremely large surface area for mass transfer.

The actual basis for separation can be the differential solubility in the membrane material or the membrane 'hole' size. When separating gases, hollow membrane fibres are popular, as the small diameter of the fibres allows them to withstand the substantial driving pressures involved. Separation of the components in liquid systems usually relies on some form of spirally wound membrane sheet and spacer. Once again, substantial differential pressures are involved across the membrane surface and this conventionally requires the use of some form of spacer to prevent closure of the channels. Not only do these spacers (e.g. coarse woven fabric) enhance the total pressure drop to be overcome, but they can also contribute to polarisation with its consequent performance loss.

Work at the University of Dortmund (Buchaly et al., 2007), in Germany, has introduced membrane separation units downstream of a reactive distillation process, as illustrated in Figure 6.16. The authors studied the heterogeneously catalysed

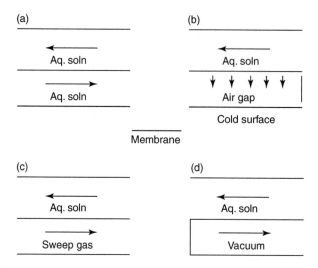

FIGURE 6.15

Common configurations used for membrane distillation: (a) DCMD, direct contact membrane distillation; (b) AGMD, air-gap membrane distillation; (c) SGMD, sweep gas membrane distillation; and (d) VMD, vacuum membrane distillation (Caputo et al., 2007).

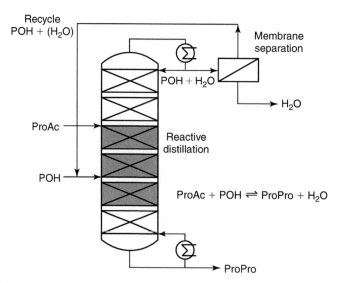

FIGURE 6.16

The membrane separation unit used, in this case, downstream of a reactive distillation column.

n-propyl propionate synthesis from 1-propanol and propionic acid. Experiments were carried out on a pilot scale membrane unit to examine how effective the membrane was for the dewatering of the binary 1-propanol-water mixture (see Section 6.5 on drying for other membrane dewatering data).

Although not applied to generic applications of membrane separations, the use of ultrasound has been shown, in one case, to enhance the operation of membranes. The particular case reported by George et al., (2008) suggested that improvements of up to 62% could be achieved in pervaporation mass transfer, although the impact on gas diffusion separation efficiency was not measurable.

In the energy field, currently of topical interest is the work in Russia reported in 2003 (Belyaev et al., 2003) on successfully combining membrane air separation with coal gasification to allow poor quality coals to be combusted without excess emissions.

As discussed in Chapter 3, the use of combined enhancement/intensification methods can contribute substantially to even greater improvements in performance – an opportunity that is by no means fully exploited in either research or practical application. Membranes can also be combined with chemical reactions, as discussed in Chapter 5, and this has now extended to nano-sized channels in ceramic membrane structures (Teplyakov et al., 2006).

Air separation has been the subject of studies using membranes at ECN in The Netherlands (Vente et al., 2006). Dynamic filtration is an area where membranes have been proposed. Jaffrin (2012) examines ways of increasing membrane performance by using rotation and/or vibration. It is claimed in this review that dynamic filtration increases the permeate flux and also positively affects the membrane selectivity (compared to cross-flow filtration). These benefits are at the expense of higher cost and complexity, although the earlier limits on disc size (affecting the available membrane area) have been overcome using ceramic discs. 'Areas of over $120\,m^2$ are now feasible' Jaffrin says, 'Applications include fractionation of milk proteins, removing micro-algae from seawater and the concentration of calcium carbonate suspensions'.

The classification of membranes in the context of carbon capture as process intensification units is discussed in Chapter 8.

For a brief overview of continuous separations, with an emphasis on intensification, a study by TNO in The Netherlands reviews a number of options, including HiGee and a system based upon micro-channels (Roelands and Ngene, 2011). The micro-channel concept is taken further in the context of a micro-evaporator, by TNO and Delft University of Technology (Mihailovic et al., 2011).

6.5 Drying

Dryers are integral parts of a large number of industrial processes and are used in sectors ranging from pharmaceuticals to textiles and ceramics. We commonly associate drying with solids, but gases frequently require drying and dehumidification in industry and in buildings, drying is growing in importance.

After distillation and evaporation, drying is the next major process energy user and spans a wider variety of sectors than distillation. Drying of solids normally involves removal of water, but in general any liquid removed from a solid is classified as drying. The separation is carried out most frequently by either mechanical means, e.g. pressing or thermally – the application of heat via a hot air stream, for example. Gas drying can be carried out by a number of thermodynamic cycles, such as vapour compression and adsorption. Gas dehydration is important in natural gas processing, for example, and membranes may be used. Where a proportion of the liquid present is removed, the term 'dewatering' may be applied. It is commonly associated with sludge and cake materials.

The majority of solids dryers use heat to remove moisture from the solid objects, which may range from micro-particles to large blocks of material, such as plasterboard. The general aim of dryer improvements is to intensify the process – to dry the product in as short a time as possible, using the minimum amount of energy, without degrading the product in any way. Because passing hot air over the surface of a solid is not particularly effective as a moisture removal method (unless the particles are small), techniques which excite the moisture molecules directly within the solid, rather than relying upon the solid thermal conductivity, are seen as intensive procedures. Electric fields can be used here, as discussed below. The reader may also consider processes such a freeze-drying and fluidised-bed drying (see the Torbed, Chapter 5) which can, depending upon the nature of the product, both speed up moisture removal and reduce the size of drying plant. Where energy efficiency is important, heat pump-assisted dryers may also be effective – there are a number of variants of these, some of which are discussed later in this chapter.

Crystallisation may be related to drying, in that moisture removal from a solution can lead to a solid being formed. The processes used, however, are different although they may be classified, like drying, as separations. Crystallisation has not received the attention given to other separation methods, in the context of either energy saving or PI.

6.5.1 Electric drying and dewatering methods

In this section, electrical drying methods are briefly discussed – examples of applications are discussed in Chapter 10. As an example, the industrial microwave dryer exhibits similar benefits to the domestic microwave oven – the energy is taken directly to the inside of the object(s) being heated, rather than having to be conducted from the surface of the particle or body. Microwave drying can be combined with conventional (hot air) drying to finish off the process, and a combination of steam and microwaves has been used where cooking is needed, for example with chickens. Radio frequency (r.f.) energy can also be used for drying, as can resistance heating. A further discussion of electrical enhancement is given in Chapter 3.

6.5.1.1 Microwaves

Some of the best examples of microwave drying applications originate in the food industry (see Chapter 10). Pasta drying, as an example, is a very slow process using conventional means, taking 10–20 hours at 40°C in large hot air recirculating ovens. In the US, a 1,500 kg/h pasta oven using 60 kW of microwave power, together with hot air, dries the pasta in 15 minutes. As well as increased production rates, the dryer is smaller and the product is bacteriologically more acceptable.

The use of microwaves as, part of a hybrid drying system can bring additional savings, as described in a study of cast ceramic ware (Anon, 1992). Here, cast ceramic ware used both microwave energy and vacuum to aid the drying process. Interestingly, reductions in drying processing times approaching those achieved for reactions with spinning disc reactors (SDRs) (see Chapter 5) were achieved, ware drying times being reduced from hours to minutes. Energy use was 26% of the conventional system.

Because one is using a relatively expensive form of energy in delivering microwave or r.f. (radio frequency) heating, they are often used solely as the final – more difficult – stage of moisture removal. In one case the drying of onions required moisture levels to reduce from 80% to 10%. It was most cost-effective to use the microwave dryer for the final drying, from 20% to 10% moisture, resulting in a 30% reduction in energy costs of the final drying stage and lowering the bacteria count by 90%.

6.5.1.2 Radio frequency fields

Capenhurst.tech supplemented mechanical dewatering (e.g. filter pressing) of slurries by using an r.f. field and a vacuum, which of course allows easier removal of water vapour. The system should reduce processing times, improve product quality and reduce the need for ancillary equipment. Applications are envisaged in the food, chemicals, pharmaceutical and agricultural sectors. R.f. assisted dewatering could, say Capenhurst.tech, have uses in effluent treatment.

6.5.2 Membranes for dehydration

Several processes need to be carried out before natural gas can be safely released into transmission pipelines. The removal of water is one of these and an improved system based upon membranes, already successfully used for CO_2 removal, was tested some years ago by the US Gas Technology Institute (GTI). The system was expected to result in a 50–70% reduction in size and weight of the dehydration unit.

Shell Global Solutions is another company which has studied membranes for conditioning of natural gas (Rijkens, 2000; Rijkens and Sponselee, 2001). Among the membrane types which can be used for gas separation/drying are hollow fibre units. Membranes have been used by Shell Oil, who applied a Kvaerner-designed membrane contactor successfully for dehydration. Reductions in absorber weight of up to 70% were claimed.

It is interesting to note that the increase in contacting area per unit volume using a membrane (see the table below), can be compared with the similar results achieved with compact heat exchangers (see Chapter 4). The gas/liquid tower may be regarded as analogous to the shell and tube heat exchanger, while the membrane surface area density is similar to that of the more compact exchangers, such as the

PCHE or Marbond units. (Some readers may like to ponder on the reasons for this similarity – are their fluid dynamic limitations in both items of plant, or manufacturing challenges?)

Contactor type	Size (m²/m³)
Gas/liquid tower	100–250
Membrane	500–1,500

Shell Global cite the following features of membranes for gas drying/dehydration:

- They can remove water down to very low concentrations – e.g. 1%.
- The selectivity of the membranes is very high – for example, only 0.5% of the methane stream might permeate the membrane with the water.
- A separate sweep is required to collect the water and take it away from the membrane. This is done by vacuum or another pick-up gas, such as nitrogen.

6.6 Precipitation and crystallisation
6.6.1 The environment for particle formation

The operation of crystallisers and precipitators is critically dependent upon the supersaturation environment prevailing within the crystal magma, because this influences both the nucleation of new particles and the growth of those that already exist. When solute diffusion controls the crystal growth process, the growth rate is proportional to the prevailing supersaturation. However, an additional reaction resistance often arises which involves the activation energy associated with the incorporation of solute molecules into the advancing growth layer. In this case, the growth has a higher order dependence on supersaturation, typically 1.5–2.5 (Mullins, 1972). On the other hand, the nucleation process, whether primary or secondary, has a 2–9 order dependence on supersaturation. Hence, high supersaturations favour nucleation rather than growth, leading to small crystals. It will be recognised that the excellent mixing and heat transfer capabilities of SDRs make them attractive for producing small or nano-sized particles which are currently of intense industrial interest.

Supersaturation can be generated in several ways. Perhaps the simplest technique is merely to cool a solution saturated at a higher temperature. Alternatively, supersaturation can be created by removing the solvent or adding an antisolvent for systems where the solubility is only a weak function of temperature. Finally, supersaturation can be created by reaction – either between two liquids or a liquid and a gas. In all these cases, the intense environment created within an initially crystal-free liquid film moving in plug flow over the disc can generate very high supersaturation and consequently, small and fairly uniform crystals.

6.6.2 **The spinning cone**

This characteristic of the SDR may be attractive in several industries (e.g. pharmaceuticals and coatings), where the product quality is intimately related to the fineness of the crystals and the tightness of the size distribution. The concept has been tested in a spinning cone precipitator (Hetherington et al., 2001), which shares most of the characteristics of a spinning disc (see Figure 6.17) except that the centrifugal acceleration vector is not aligned to the cone surface. Barium sulphate was generated by mixing equimolar solutions of $BaCl_2$ and Na_2SO_4 in a central reservoir. A thin liquid/slurry film flowed to the cone rim, from where it was collected and subsequently analysed in a Malvern Mastersizer. Equivalent batch experiments were performed, for the purpose of comparison, in a 50 ml agitated beaker. At a supersaturation of 500, defined as:

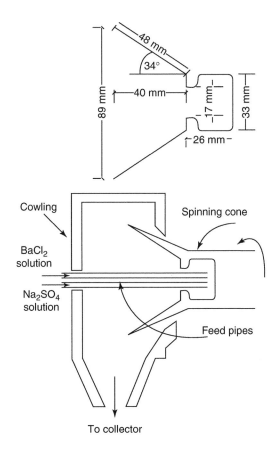

FIGURE 6.17

Spinning cone reactor layout.

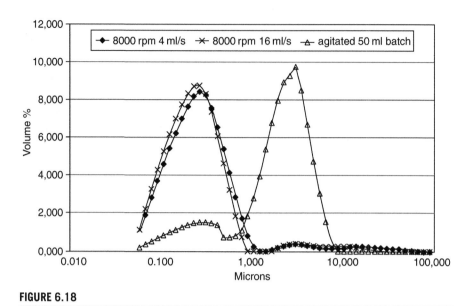

FIGURE 6.18

BaSO$_4$ size distribution for a supersaturation of 5,000.

Molar concentration of Na$_2$SO$_4$ or BaCl$_2$/Molar solubility of BaSO$_4$, the cone produced crystals at 6,000 rpm that had a Sauter mean diameter of 3.2 microns, compared to 6.85 microns from the batch runs. However, at a supersaturation of 5,000, the batch yielded a Sauter mean diameter of 0.75 microns, compared with 0.18–0.32 microns from the disc of the cone. The particle size distributions reproduced in Figure 6.18 highlight the cone behaviour at 8,000 rpm more starkly, with a decided shift to 0.1–1 microns, compared with 1–10 microns in the batch.

Thus it can be seen that, for many systems, spinning precipitators hold out the prospect of generating the crystal size distributions that have considerable industrial interest. This view is further reinforced by some earlier work by SmithKline Beecham (Oxley et al., 2000), mentioned above and discussed as a case study in Chapter 8.

6.6.3 Electric fields to aid crystallisation of thin films

Thin films of silicon are used for active matrix liquid crystal displays. The TFT (thin film transistor) screens which save us carrying heavy VDUs between computer rooms are in major use for such systems, and cost reduction in manufacture is important to ensure good market penetration.

Research by Rao and Sun (2004) suggested that the process known as metal-induced lateral crystallisation (MILC) a good way of obtaining pure thin films but which takes 10–20 hours–can be speeded up using electric fields. These overcome the slow thermal diffusion of NiSi$_2$, which governs the speed of crystallisation, allowing the time to be reduced. Another benefit of the use of this method is the way the applied field can encourage directional crystal growth. This aids electrical mobility in the device. In the case here, microwave annealing was used to produce crystal films in 50 minutes. More recently, Hong and Ro (2007) have encouraged even more rapid crystal formation by electrically heating the thin film.

6.7 Mop fan/deduster

As discussed in Chapter 6, the rotating packed-bed is an intense device for providing counter-current flow within a packing having a relatively high specific surface area. From a thermodynamic point of view, counter-current flow is beneficial when operating a distillation or absorption process because local concentration differences and the associated entropy gain may be minimised. Therefore, when high efficiency is critical, the counter-current approach is adopted. However, there are some phase disengagement and mass transfer duties where efficiency is not of paramount importance and for which co-current flow of the phases can be tolerated. This avoids the flow restrictions imposed by flooding considerations and leads to a simpler device. It was against this background that Byrd (1986) invented the mop fan, initially for lime dust capture at ICI's quarry near Buxton (UK).

6.7.1 Description of the equipment

The device is based on a conventional fan, but has a novel impeller comprising a radial fibre mop which captures the particles as it rotates. This is shown in Figure 6.19(b) where a comparison is made with a conventional fan (a). The mop itself is shown in Figure 6.19(c). The centre of the mop is irrigated with an appropriate liquid, so that when a particle collides with a fibre, it is trapped within the draining liquid film and flushed to the machine casing. The knock-out capability of the device depends on the rotor speed and the particle and fibre diameter, as shown below. However, at normal fan speeds around 50% of the 2 micron particles are removed.

As might be expected, the pumping efficiency of the mop is only about 60% of that for a correctly bladed fan, but this performance handicap is compensated by its dual capability as a fan/disengagement machine. It also has a significant, though modest, mass transfer capacity which corresponds to about 3 transfer units (for a 30 cm diameter unit). Despite this restriction, the device is well suited to simple mass transfer duties such as flue gas treatment, e.g. SO_2/CO_2 removal. It is currently available in fibreglass/polypropylene and in metal from Begg Cousland Ltd in the United Kingdom (see Appendix 3).

6.7.2 Capture mechanism/efficiency

The entrapment of dust or mist particles by the mop fibres occurs by an inertial mechanism, since the relative velocity between the incident gas and the mop fibres

(a) High capacity centrifugal fan with conventional rigid impeller

(b) High capacity centrifugal fan with novel flexible fibre impeller

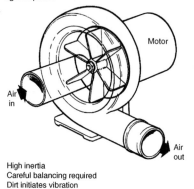

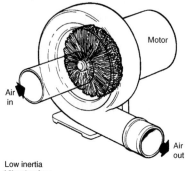

High inertia
Careful balancing required
Dirt initiates vibration
Dangerous failure mode
High cost

Low inertia
Vibration free
Self-cleaning
High safety factor
Low cost
Special scrubbing facility

(c)

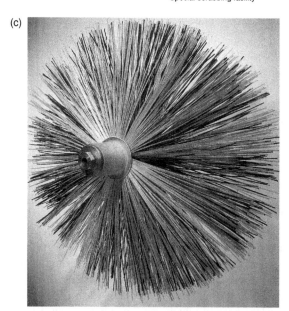

FIGURE 6.19

(a) A conventional fan; (b) the mop fan showing its internal structure; (c) the actual mop.

is a substantial fraction of the tip speed of the mop. The particle capture process can be appreciated from Figure 6.17, which sketches the air flow around a fibre and shows an upstream particle as it begins to negotiate its way around the fibre, which is shown in circular cross-section. If the mass within an imaginary particle surface corresponded to the fluid density, then the particle would follow the streamline as shown and no collision would occur. However, in general, particles suspended in

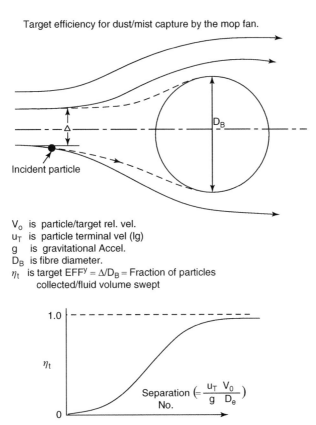

Target efficiency for dust/mist capture by the mop fan.

V_0 is particle/target rel. vel.
u_T is particle terminal vel (lg)
g is gravitational Accel.
D_B is fibre diameter.
η_t is target $EFF^y = \Delta/D_B =$ Fraction of particles
        collected/fluid volume swept

FIGURE 6.20

The mop fan particle capture process.

an air stream will be *denser* than the fluid and they will therefore be substantially greater than the fluid they displace. Despite this, the forces exerted on the particle by the fluid will remain the same, so that the accelerations imposed on the particles are reduced, i.e. the particles tend to move in straighter trajectories, as shown by the dotted line in Figure 6.20. Larger particles experience lower acceleration, and a greater proportion collide with the fibre. A target or capture efficiency can be defined as the ratio between the 'collision corridor' (within which all incident particles collide) and the fibre diameter. The actual target efficiency for a range of situations can be estimated from the charts in Figure 6.21 which have been derived for various target shapes and particle terminal velocities.

An interesting corollary to the above ideas arises when the particle is *less* dense than the fluid, e.g. bubbles in a liquid. In this case, the particle mass is *less* than the equivalent fluid volume, so the accelerations imposed are greater than those experienced by the fluid. This results in the bubbles deviating from the fluid streamlines, in the opposite sense to the denser particles. As a consequence, the zone

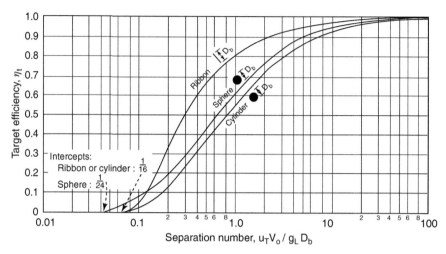

FIGURE 6.21

Capture efficiencies for various target shapes.

immediately upstream of an obstacle is depleted of bubbles, whereas dust tends to accumulate on the upstream face. This phenomenon can be exploited to design agitators for stirred vessels so that they can maintain their pumping efficiency under heavily gassed conditions.

6.7.3 Applications

As an example of the use of the mop deduster, we can look at odour control and dust removal in terephthalic acid production, this being a constituent of the production of polyester fibres. Silos or hoppers are filled by pneumatic or mechanical conveying. Raw material feed hoppers, in the preparation section, empty into the mixing process, and finished product silos are discharged to bagging stations. The filling and emptying of these silos generates vent gas which has to be treated before being released to atmosphere.

Such vent gas treatment involves the removal of large amounts of dust and, due to the presence of acetic and benzolic acid, there is also an odour problem. With typically $10\,g/m^3$ of dust, a fabric filter may be installed as a primary stage. Alternatively, or in series, a wet cyclone or scrubber may be installed. It is not unusual for the effluent water to be fed to the mixer in the preparation section, to recover product.

The mop fan/deduster extraction unit[2] made in 304 L stainless steel has proved to be ideal for this application. The wetted fibre brush, rotating at high speed, creates an extremely dynamic gas contact, which removes solids, absorbs or dehumidifies gas and acts as its own air mover – all in a single piece of plant. In the vent gas steam, it scrubs the odour (100% odour removal using a 5% NaOH solution spray) and removes any residual dust. The unit is located in the silo roof to minimise ducting. Interestingly, there is no extra pressure drop – in fact it has been shown that, with pneumatically fed hoppers, the air movement induced by the mop assists the vacuum effect in the silo.

A second application involving gas clean-up is in biomass gasification product gas. Research in England, at the University of Nottingham and in three German laboratories (Zhang et al., 2011) has successfully demonstrated that the mop fan, in conjunction with electrostatic precipitation (ESP) can successfully clean-up the product gas. The principal contaminants are fine particles; tars and alkali metals; chlorine compounds; nitrogen compounds; and sulphur-based compounds. Approaches, depending upon the contaminant, have included catalysis and sorption. Alkali contaminants can be stripped and condense onto small particles that are then removed by cyclone.

Examining the performance of the system for a number of tar species (fluorene, phenanthrene, pyrene and naphthalene), concentrations were up to $120 \, mg/Nm^3$ before mop fan quenching and ESP, reducing to $40 \, mg/Nm^3$ for naphthalene, the other three being zero, downstream of the mop fan/ESP combination. It is suggested that the combination has the potential to address all the contaminants arising from the gasification.

6.8 Electrolysis
6.8.1 Introduction

It has been pointed out in Chapter 3 that multi-phase operations benefit significantly when they are designed to operate in an elevated acceleration environment. This applies to an electrochemical duty in particular, where low cell voltage, high current efficiency and high current density are highly desirable, though in practice they are usually mutually exclusive. Electrolysis with a liquid electrolyte is a pre-eminent example of a multi-phase operation, since one or both of the electrodes discharges a gas which interferes with the passage of current from one to the other. This interference has three components. The first centres on the occlusion of the electrodes or membrane by adherent bubbles. In this respect, the phenomenon is directly analogous to nucleate/film boiling where there is a maximum heat flux which depends on the insulating effect of the gas film or bubble layer on the heat transfer surface. Any attempt to increase the heat flux simply increases the blanketing effect of the bubble layer up to the point where burn-out of the heating surface occurs.

The second effect arises simply from the presence of a substantial volume of non-conducting bubble volume within the electrolyte. This depresses the effective electrolyte conductivity according to the following relationship given by De La Rue and Tobias (1959):

$$\rho_g = \rho_o \left(1 - f_g\right)^{-3/2}$$

Where ρ_g and ρ_o are the gassed and ungassed electrolyte conductivities and f_g is the gas volume fraction.

The third effect may be regarded as beneficial since it depends on the agitation caused by the bubbles when they detach from the electrodes or membrane.

The mass transfer coefficient is enhanced because the energy released by the rising bubbles is applied exactly where it will have most effect i.e. the electrode boundary layer. This enhanced mass transfer usually means that the current efficiency is improved as the desired reaction path is followed rather than a parasitic alternative.

These phenomena together limit the extent to which the electrode spacing can be reduced in order to lower cell resistance and voltage, as any attempt to do so will quickly result in a gas-logged cell (if only terrestrial acceleration is applied). This idea provides the key incentive for exploring alternative levels of acceleration to cell operation.

6.8.2 **The effect of microgravity**

Recently, some Japanese workers have demonstrated two techniques for achieving microgravity operation. Unfortunately, the duration of the tests was severely restricted by practical considerations, as will be apparent later. The first study was by Matsushima et al., (2003) who performed water electrolysis for 8 s in a 490 m drop shaft in a series of experiments designed to explore operation in deep space. They presented photographs showing the electrodes (7 mm × 5 mm) of an

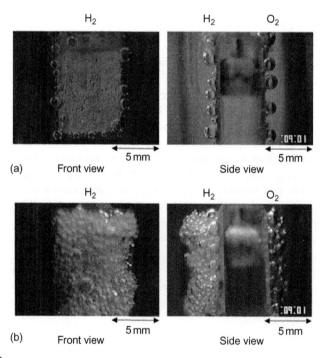

FIGURE 6.22

Comparison of gas bubble evolution in alkaline electrolyte under (a) terrestrial gravity; and (b) microgravity conditions, at 8 s after starting water electrolysis (−0.8 V vs. RHE, 25 wt.% KOH).

alkaline water electrolysis cell under both terrestrial and micro g operation These are shown in Figure 6.22. It is obvious that as time progressed the electrodes became more congested with bubbles, but as the operating time was limited to only 8 s, the full impact of the micro g operation could not be demonstrated. It is also worth pointing out that with more realistic electrode sizes, bubble blinding could be expected to occur even more rapidly. It will be intuitively recognised that as the non-conducting bubbles accumulate between the electrodes, the effective electrolyte conductivity will be reduced. This concept is encapsulated quantitatively in the equation shown above.

The alternative way of creating microgravity conditions is by following a parabolic trajectory during an aeroplane flight. Kaneko et al., (1993) managed to maintain microgravity for 20–25 s using this technique. They performed water electrolysis with an acid electrolyte and found, unsurprisingly, that after a short induction period, the current density was sharply reduced. They also found that the size of the bubbles adhering to the electrodes was much greater under microgravity conditions.

The above experimental results convincingly demonstrate that microgravity conditions are disastrous for the effective operation of gas-generating electrochemical cells. They also imply that cells which are designed to exploit a high gravity environment could have a dramatically improved performance.

6.8.3 The effect of high gravity

As far as can be ascertained, the idea of exploiting centrifugal acceleration fields for gas disengagement was first proposed for water electrolysis by Thompson (1929) in a very brief patent. This document presented a simple sketch but gave no performance details. Much later, a patent by Hoover (1964) described a rotating chlor-alkali cell. However, again no performance data were revealed. The concept was further extended by Ramshaw (1993) who also operated a chlor-alkali disc cell with a Nafion membrane. The rotating unit comprised a 20° segment of a 0.6 m diameter spinning disc, is shown in Figure 6.23. The opposing planar electrodes were coated with an appropriate electro-catalyst. Bearing in mind that the standard ICI FM 21 cell voltage was 3.17 V when operating at $3.0 \, kA/m^2$, the rotating cell required 2.7 V when running at 200 g (Figure 6.23), or 900 rpm (Figure 6.23). This represented a significant energy saving and was an encouraging endorsement of the HiGee strategy. At $5.0 \, kA/m^2$ the voltage saving was even greater. There appeared to be little point in exceeding this level of acceleration for this particular cell configuration as the voltage was not reduced much when higher rotational speeds were explored. It was noted that the energy required to rotate a full-scale cell system at the modest speeds envisaged would be small compared with the electrochemical energy saved, particularly with a bipolar design.

Cheng et al., (2002a,b) investigated chlor-alkali electrolysis in a conventional laboratory centrifuge using acidified brine at temperatures up to 353 K. The relationship between cell voltage and acceleration was evaluated using electrodes which comprised RuO_2, TiO_2 mesh and Pt/Ti mesh. Once again, very significant voltage reductions were generated when accelerations up to 190 g were applied.

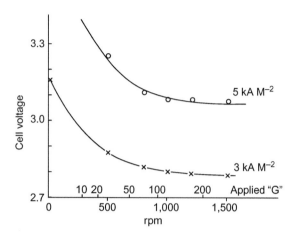

FIGURE 6.23

The voltage characteristics of a rotating chlorine cell (Ramshaw, 2003).

For example, a reduction of 0.6 V was observed at a current density of 6.0 kA/m^2. In addition, the electrode mass transfer coefficient increased from 1.28 to 3.89 m/s. This feature is expected to increase the current efficiency of chlorine production.

Cheng et al., (2002a,b) also investigated water electrolysis with brine in the laboratory centrifuge. Once again it was shown that centrifugal fields provided a powerful technique for improving the electrolytic process, especially when nickel foam electrodes were used with concentrated potassium hydroxide solutions at 353 K.

The above studies demonstrate that there is a significant body of evidence which confirms that the application of enhanced acceleration to an electrolytic process can provide major improvements to its energy efficiency. However, the experimental set-up in the above studies was not representative of industrial units. In the centrifuge, the electrodes were immersed in rotating thimbles which contained a limited volume of electrolyte. Steady operation could not be maintained for long periods as there was no consideration of many practical issues such as how to replenish the electrolyte in the cell or how to supply an array of rotating electrodes with large currents. The work reported below addresses some of the practical issues of designing an industrial-scale rotating electrolyser.

6.8.4 Current supply

Considerable thought was given to the question of current supply to the rotating cell as the envisaged level at industrial scale is expected to be at least 1,000 A for 0.5 m diameter cells in a bipolar array. Several connection arrangements were considered and carefully appraised. These included slip ring assemblies, molten metal baths and mercury-based encapsulated rotary connectors. The last two approaches received detailed attention.

A slip ring assembly was initially thought to be the obvious contender for the proposed duty. Unfortunately, with a maximum carbon brush current density of only $1 A/cm^2$, the size, cost and frictional loss of a suitable slip ring configuration was unacceptable. Attention was then focussed on low-melting point solder type of alloys.

For the molten metal bath trials Cerrobend[TM] was identified as the alloy of choice as its melting point was only about 343 K and it had a relatively low content of toxic cadmium. It appeared to be capable of coping with high currents while operating with low mechanical drag and low voltage drop. The experimental arrangement consisted of a disc fitted to the shaft which dipped into a narrow bath of molten alloy. The alloy was heated to a target running temperature by immersed cartridge heaters and the system was shrouded by a cover plate to retain any alloy spray and cadmium fumes. Early results were encouraging with voltage losses less than 0.1 V at currents of 100 A. Unfortunately, after several hours of operation at temperatures around 353 K, the alloy melt developed a mousse-like consistency which ultimately caused the bath contents to overflow. Despite exhaustive enquiries and investigations, no means could be found to avoid this problem so the molten alloy experiments were also abandoned.

A commercially available encapsulated mercury-based rotary connector supplied by Nova Ltd was then evaluated. The unit tested was capable of transmitting up to 250 A at rotational speeds up to 1,200 RPM while incurring very little mechanical drag. It retained the mercury in a hermetically sealed cylinder and did not therefore present a safety problem. The dimensions of the connector were 31 mm diameter, 28 mm long. The units were installed at each end of the shaft and operated very effectively and reliably. Tests showed that the voltage drop across the connector at a current of 200 A and a rotating speed of 800 RPM was only 0.06 V. The heat generated in the connector was low so the temperature rise during the tests was very limited. More importantly, a 100 mm diameter version of the connector should be easily capable of dealing with the currents ~1,000 A which will be needed in an industrial scale version of the bipolar array.

6.8.5 Rotary electrolysis cell design

The cell is shown in Figures 6.24 and 6.25 and described in detail in Lao et al., (2011). It is symmetrical about its mid-plane which corresponds to the porous polypropylene membrane used to separate the anode and cathode compartments so that the hydrogen and oxygen products did not mix, while allowing transmission of the electrolyte between the anode and the cathode compartments. The gas bubbles rose inwards to the hollow machine shaft where they were allowed to mix before leaving via the end of the shaft. It must be emphasised that this arrangement was only a temporary expedient for the sake of simplicity. In this laboratory unit, with the modest gas production involved, the hazard represented by the flammable gas mixture was unlikely to be significant. This approach is clearly unacceptable for a full-scale machine which would require the shaft to be 'compartmentalised' to ensure separate paths for the hydrogen and oxygen.

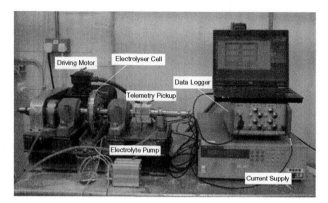

FIGURE 6.24

The rotary electrolyser test rig.

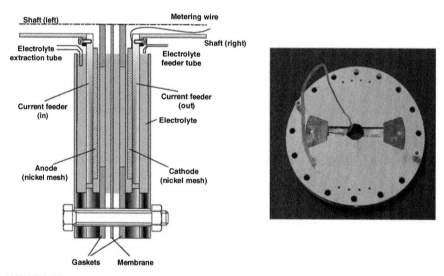

FIGURE 6.25

Schematic of the rotary electrolyser cell (left) and arrangement of the actual electrode (right).

If the full disc area were to be electrochemically activated then the current demand for this cell diameter would be up to 600 A. This level was deemed to be too high for the rotary connectors ultimately employed in this laboratory study. In view of this, most of the cell-projected area was rendered inactive by incorporating a non-conducting blank sheet which left two 45° sectors active with an effective area of 22 cm², as shown in Figure 6.25. The current was supplied to and removed from the electrodes by stainless steel sheets which were in contact with the electrode mesh. An adjustable gap was provided to make gas disengagement more or less difficult. This was arranged by incorporating polypropylene spacers of varying thickness between the electrodes and the membrane. At the cell periphery, the electrolyte solution flowed through a set of holes from the feeder channel to the removal channel via the anode and the cathode compartments. When operating, the feeder channel contained an annular pool of electrolyte which was continuously fed from a stationary tube near the machine shaft. With the assistance of a peristaltic pump, the electrolyte was supplied at flow rates of around 2×10^{-4} kg/s. The electrolyte, having passed through the cell, entered the removal channel which contained another annular pool of electrolyte with its inner radius dictated by the position of a stationary pitot abstractor tube. It was expected that with continuous electrolyte replenishment and efficient gas disengagement, the operating temperature and concentration could be held steady.

6.8.5.1 Instrumentation for the rotary cell

The key data required were the anode/cathode voltage difference (excluding any losses in the current feeder system), and the cell electrolyte operating temperature. Although it was recognised that this information could be abstracted using a multi-channel slip ring assembly, it was also known that such a system could corrupt the data by a combination of frictional heating and the Peltier effect. In view of this, it was decided to use a digital telemetry system provided by Astech Ltd, which was capable of measuring voltage to +/−2.5 mV and temperature to +/−1.0 K.

6.8.5.2 Static electrolysis cell

In order to provide a basis for comparison with conventional technology, a set of tests was also performed with a static cell under similar conditions as those used for the rotary cell. A sketch of the static rig is shown in Figure 6.26. The electrode dimensions were 5×5 cm, though this could be reduced if necessary by using a blocking non-conducting film. The membrane material was that used for the rotary cell and again the electrode gap could be adjusted by using spacers of different thickness.

Although the configuration and electrolyte circulation arrangement of the static cell was rather different to that of the rotary cell it was considered to be a useful reference for the investigation into gas disengagement and the effect of a centrifugal field.

6.8.6 The static cell tests

6.8.6.1 Effect of electrode structure

A major influence of electrode structure on cell voltage was observed with the static cell. Figure 6.27 shows the cell voltage data obtained for two different designs, namely nickel mesh on the stainless steel feeder plate, and then with the feeder plate only. The conditions for the experiment were: electrolyte, 7.7 M KOH solution,

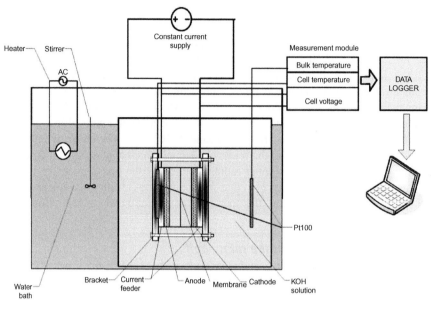

FIGURE 6.26

Schematic of the static water electrolysis rig.

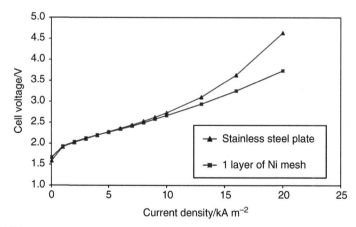

FIGURE 6.27

Cell voltage against current density for the static cell for different electrodes. The conditions for the experiment were as follows: electrolyte, 7.7 M KOH solution; bulk temperature, 348 K; inter-electrode space, 2.0 mm. Square solid dots represent the experiment data for electrodes made of nickel mesh upon stainless steel current feeding plate, and triangular solid dots for electrodes made of stainless steel current feeding plate only.

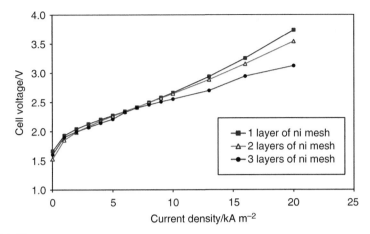

FIGURE 6.28

Cell voltage versus current density for the static cell with electrodes comprising different number of nickel mesh layers. The conditions for the experiment were as follows: electrolyte, 7.7 M KOH solution; bulk temperature, 348 K; inter-electrode space, 2.0 mm. Square solid dots represent the experiment data for electrodes made of one layer of nickel mesh upon stainless steel plate; triangles for electrodes of two layers nickel mesh upon stainless steel plate; and round solid dots for electrodes of three layers nickel mesh upon stainless steel plate.

electrode spacing 2 mm. As can be seen from Figure 6.27, at a solution bulk temperature of 348 K, with a current density of less than 10 kA/m^2 there were no significant voltage differences between the two electrode structures. However, notable differences were apparent when the current density was higher than 10 kA/m^2. This behaviour was attributed to (a) the different electrode material, and (b) the extra surface area conferred by the nickel mesh. These factors obviously became more critical as the rate of bubble generation increased. It was presumed that the effect would be observed at lower current densities if smaller electrode spacing had been used.

Further tests were performed to check the effect of additional electrode area. Figure 6.28 shows data obtained with electrodes comprised of different numbers of nickel mesh layers. Under the conditions used previously, the cell voltage tended to reduce with increased numbers of mesh layers at any given current density, though the effect was most marked at high current density. This result shows that high electrode areas are conducive to improved cell efficiency. However, with only gravitational acceleration available, this approach could be self-defeating due to the increased gas logging encountered with what effectively becomes a porous electrode.

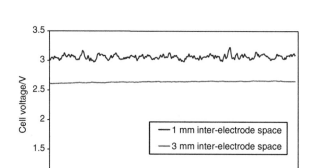

FIGURE 6.29

Cell voltage traces for the static cell with different interelectrode spaces at a current density of $10 \, kA \, m^{-2}$. Heavy solid line represents the trace for the electrodes with inter-electrode space of 1.0 mm and thin line for the electrodes with the space of 3.0 mm.

6.8.6.2 Effect of electrode spacing

Figure 6.29 shows the cell voltage/time traces obtained from the static cell under the same conditions as before, with electrode gaps of 3 mm and 1 mm. Not only was the cell voltage less with the 3 mm gap, but also showed much less variability. Presumably this was due to easier inter-electrode gas bubble disengagement and a lower gas void fraction. Once again this highlights the design conflict when trying to achieve high cell efficiency by reducing the electrode spacing because close spacing becomes self-defeating.

6.8.7 The rotary cell experiments

6.8.7.1 The effect of centrifugal acceleration

Figure 6.30 shows the dependence of the cell voltage on centrifugal acceleration obtained with the rotary cell. The electrodes were comprised of three layers of nickel mesh supported on a stainless steel current feeder plate with an electrode gap of 1 mm. The current density varied from 2.3 to 22.5 kA/m². As can be seen, the benefits of increased acceleration in this case were mainly confined to the lower acceleration ranges, beyond which there were diminishing returns for a given electrode configuration. This behaviour was confirmed by the data shown in Figure 6.31(b). An insight into cell operation at lower accelerations is shown in Figure 6.31(a). Here the cell voltage is plotted against time. It is apparent that at higher current density and lower acceleration the time trace is much more irregular – presumably reflecting poor gas disengagement. At lower gas generation rates (i.e. current densities) or higher accelerations the voltage trace is far less erratic.

The benefit of increased electrode area is implied by Figure 6.32 where operation with 1 and 3 layers of nickel mesh are contrasted. Clearly, electrodes with a

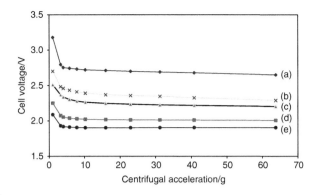

FIGURE 6.30

Cell voltage against centrifugal acceleration (in g) for the rotary cell at different current densities, with electrodes made of nickel mesh upon stainless steel plate. Different curves correspond to data from different current densities of (a) $22.5\,kA\,m^{-2}$; (b) $13.5\,kA\,m^{-2}$; (c) $9.0\,kA\,m^{-2}$; (d) $4.5\,kA\,m^{-2}$; and (e) $2.3\,kA\,m^{-2}$. $7.7\,M$ KOH solution was used; solution bulk temperature, $344\,K$.

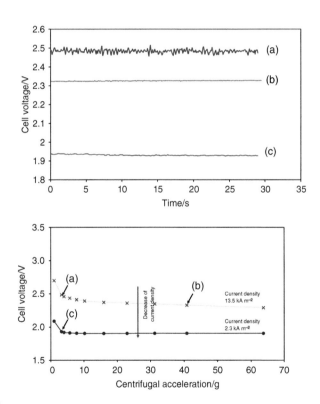

FIGURE 6.31

Cell voltage traces for the rotary cell at different current densities and rotation accelerations (top) and their positions on the curves of cell voltage versus centrifugal acceleration (bottom). Electrode material, nickel mesh upon stainless steel plate; electrolyte, $7.7\,M$ KOH solution; bulk temperature, $344\,K$. (a) current density of $13.5\,kA\,m^{-2}$ and acceleration of $3.2\,g$; (b) $13.5\,kA\,m^{-2}$ and acceleration of $41\,g$; and (c) $2.3\,kA\,m^{-2}$ and acceleration of $3.2\,g$.

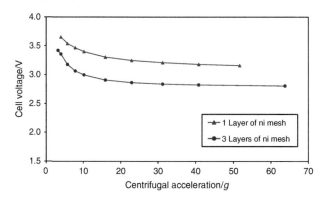

FIGURE 6.32

Cell voltage against centrifugal acceleration for the rotary cell with different electrode structures. Line with solid triangles represents data from electrodes made of one layer of nickel mesh upon a stainless steel plate; line with round dots represents electrodes of three layers of nickel mesh upon a stainless steel plate.

large area density are able to reduce cell voltage, provided all the area is fully functional. It would be expected that if many mesh layers are installed, only the first few would be electrochemically active by analogy with as metal plating baths which tend to lay their deposit on asperities rather than hollows.

In summary, the current density tested with the experimental rotary electrolyser was up to 22.5kA/m^2, based on the projected electrode area. At a centrifugal acceleration of 41 g, the electrolyser can be operated at a current density of 13.5kA/m^2 without causing bubble congestion in the 1.0 mm inter-electrode space, as the gas disengagement is facilitated by the enhanced acceleration field. This value is much higher than the 4.5kA/m^2 which is typical for a fully optimised industrial unit. The thickness of a rotary cell with electrodes comprising 3 layers of nickel mesh was about 5.0 mm. Hence, for a bipolar unit with 10 cells in series, the total active length is only 50 mm. It will be recognised that with further development, very high power densities can be achieved while maintaining competitive cell energy efficiencies.

It was noted that the rotary unit achieved most of its voltage reductions in these experiments at relatively low accelerations, typically in the range of 15–30 g. This value was much less than that reported by Ramshaw (1993) who showed that, for a differently designed rotary brine electrolyser, accelerations of 100–200 g were needed to approach its asymptotic voltage. It is believed that this behaviour is a direct consequence of the difficulty in removing the generated gas from the system which then requires the application of higher accelerations to overcome it. The cell features which contribute to this difficulty are:

a) High current density.
b) High electrode area density.
c) Substantial cell radial depth or height.

An indication of impending difficulty in affecting gas removal was the voltage time-trace which became very irregular in the presence of severe cell congestion.

Recognising the above, it is not entirely surprising that Ramshaw's rotary chlorine cell (which had a radial depth of ~7 cm compared with the ~3 cm depth of the rotary water electrolysis cell) required much higher levels of acceleration to reach its voltage asymptote.

Although the nickel electrodes used were uncatalysed, operated at one bar pressure and at about 340 K, the rotary cell voltages observed were comparable with those achieved in fully optimised, pressurised industrial units. This suggests that ultra-compact, fully developed, bipolar rotary cells could save large quantities of electric power.

6.9 Summary

This chapter deals with an important area of PI – that of separations, with the addition of electrolysis in this edition. There are many types of equipment used, including a range of active methods, and in common with other unit operations, many can fulfil dual (or multiple) roles, such as separation plus reactions. An appreciation of this fact is a major step in becoming fully aware of the potential of PI, in radically changing process plant.

References

Anlauf, H., 2007. Recent developments in centrifuge technology. Sep. Purif. Technol. 58, 242–246.

Anon,1992. Microwave/vacuum drying of cast ceramic ware. Energy Efficiency Office Best Practice Programme, New Practice Final Profile 40, September.

Basic, A., Dudukovic, M.P., 1995. Liquid holdup in rotating packed beds: examination of the film flow assumption. AIChE 41 (2), 301–316.

Belyaev, A.A., et al., 2003. Membrane air separation for intensification of coal gasification process. Fuel. Process. Technol. 80, 119–141.

Bruinsma, O.S.L., Krikken, T., Cot, J., Saric, M., Tromp, S.A., Olujic, Z., et al., 2012. The structured heat integrated distillation column. Chem. Eng. Res. Des. 90, 458–470.

Buchaly, C., Kreis, P., Gorak, A., 2007. Hybrid separation processes: combination of reactive distillation with membrane separation. Chem. Eng. Process. 26, 790–799.

Burns, J.R., Ramshaw, C., 1996. Process intensification: visual study of liquid maldistribution in rotating packed beds. Chem. Eng. Sci. 51 (8), 1347.

Burns, J.R., Jamil, J.N., Ramshaw, C., 2000. Process intensification: operating characteristics of rotating packed beds: determination of liquid hold-up for a high-voidage structured packing. Chem. Eng. Sci. 55, 2401.

Byrd, G.,1986. European Patent No. 176287 (A).

Caputo, G., Felici, C., Tarquini, P., Giaconia, A., Sau, S., 2007. Membrane distillation of HI/H2O and H2SO4/H2O mixtures for the sulphur-iodine thermochemical process. Int. J. Hydrogen Energy 32, 4736–4743.

Cheng, H., Scott, K., Ramshaw, C., 2002a. Chlorine evolution in a centrifugal field. J. Appl. Electrochem. 32, 831–838.

Cheng, H., Scott, K., Ramshaw, C., 2002b. Intensification of water electrolysis in a centrifugal field. J. Electrochem. Soc. 149 (11), D172–D177.

Coulson, J.M., Richardson, J.F., 1978., third ed. Chemical Engineering, 2 Pergamon, Oxford.

Day, N., 2004. Why centrifuges play an important role in the production of sugar. Filtr. Sep. 41 (8), 28–30. October.

De La Rue, R.E, Tobias, C.W, 1959. On the conductivity of dispersions. J. Electrochem. Soc. 106, 827.

Finn, A.J., 1994. Enhancing gas processing with reflux heat exchangers. Chem. Eng. May.

George, B.J., Pereira, N., Al Massum, M., Kolev, A.D., Ashokkumar, M., 2008. Sensitivity enhancement in membrane separation flow injection analysis by ultrasound. Ultrason. Sonochem. 15, 151–156.

Hetherington, P. et al., 2001. Process intensification: continuous production of barium sulphate using a spinning cone precipitator. Fourth International Conference on Process Intensification for the Chemical Industry. Bruges, Belgium. BHR Group Conf. Series, September 10th.

Hong, W.-E., Ro, J.-S., 2007. Millisecond crystallization of amorphous silicon films by Joule-heating induced crystallization using a conductive layer. Thin Solid Films 515, 5357–5361.

T. Hoover, 1964.US Patent 3119759.

Jaffrin, M.Y., 2012. Dynamic filtration with rotating discs, and rotating and vibrating membranes: an update. Curr. Opin. Chem. Eng. 1, 171–177.

Kaibel, B., 2007. Distillation – dividing wall columns *Encyclopaedia* of Separation Science. Elsevier, Oxford, (1–9).

Kaneko, H., et al., 1993. Electrochim. Acta 38, 729.

Kiss, A.A., Bildea, C.S., 2011. A control perspective on process intensification in dividing-wall columns. Chem. Eng. Process. 50, 281–292.

Lao, L., Ramshaw, C., Yeung, H., 2011. Process intensification: water electrolysis in a centrifugal acceleration field. J. Appl. Electrochem. 41 (6), 645. et seq.

Li, Q.S., 1998. A research and application of the new type high efficiency acetone distillation column. J. Beijing Univ. Chem. Technol. 25, 2–5. (In Chinese).

Li, X.P., Liu, Y.Z., Li, Z.Q., Wang, X.L., 2008. Continuous distillation experiment with rotating packed bed. Chinese J. Chem. Eng. 16, 656–662.

Luo, Y., Chu, G.-W., Zou, H.-K., Xiang, Y., Shao, L., Chen, J.-F., 2012. Characteristics of a two-stage counter-current rotating packed bed for continuous distillation. Chem. Eng. Process. 52, 55–62.

Maleta, V.N., Kiss, A.A., Taran, V.M., Maleta, B.V., 2011. Understanding process intensification in cyclic distillation systems. Chem. Eng. Process. 50, 655–664.

Mallinson, R. Ramshaw. C. ,1979. European Pat No 2568B.

Matsushima, H., Nishida, T., Konishi, Y., Fukunaka, Y., Ito, Y., Kuribayashi, K., 2003. Water electrolysis under microgravity - pat 1. Experimental technique. Electrochim. Acta 48, 4119–4125.

Mihailovic, M., Rops, C.M., Hao, J., Mele, L., Creemer, J.F., Sarro, P.M., 2011. MEMS silicon-based micro-evaporator. J. Micromech. Microeng. 21 doi: 10.1088/0960-1317/21/7/075007.

Mullins, J.W., 1972. Crystallisation, second ed. Butterworths, London.

Oxley, P., et al., 2000. Evaluation of spinning disc reactor technology for the manufacture of pharmaceuticals. IEC Res. 39 (7), 2175.

Peel, J., Howarth, C.R., Ramshaw, C., 1998. Process intensification: HiGee sea water deaeration. Trans. I. Chem. E. 76 (Part A), 585.

Prada, R.J., Martinez, E.L., Maciel, M.R.W., 2012. Computational study of a rotating packed bed distillation column In: Bogle, I.D.L. Fairweather, M. (Eds.), Proceedings of Twenty Second European Symposium on Computer Aided Process Engineering, June Elsevier, London,Oxford, pp. 17–20.

Rajan, S.K., 2008. Limiting gas liquid flows and mass transfer in a novel rotating packed bed Department of Chemical Engineering. Indian Institute of Technology, Kanpur, India, (M. Tech. Thesis).

Ramshaw, C., 1993. The opportunities for exploiting centrifugal fields. Heat Recovery Syst. CHP 13 (6), 493.

Ramshaw, C., Thornton, J.D., 1967. Droplet breakdown in a packed column Part I: the concept of critical droplet size. I. Chem. E. Symp. Ser. 26, 73.

Rao, R., Sun, G.C., 2004. Microwave annealing enhances Al-induced lateral crystallization of amorphous silicon thin films. J. Cryst. Growth. 273 (1–2), 68–73. (December 17).

Reddy, K.J., Amit, G., Rao, D.P., 2006. Process intensification in a HIGEE with split packing. Ind. Eng. Chem. Res. 47, 8840–8846.

Rijkens, H.C.,2000. Membrane developments for natural gas conditioning. Proceedings of the Symposium: Process Intensification: a challenge for the Process Industry, Rotterdam, October, 10.

Rijkens, H., Sponselee, J., 2001. Membrane developments for natural gas dehydration. NPT Procestechnologie 2, 51–53. (March–April).

Roelands, C.P.M. and Ngene, I.,2011. Continuous intensified separations – the next logical step in PI. Proceedings of EPIC 2011, IChemE Symposium Series No. 157, IChemE, Rugby.

Sherwood, T.K., Shipley, G.H., Holloway, F.A.L., 1938. Flooding velocities in packed columns. Ind. Eng. Chem. 30, 765.

Short, H., 1983. New mass transfer find is a matter of gravity. Chem. Eng. 90, 23–29.

Suszwalak, D.J-P., Kiss, A.A., 2012. Enhanced bioethanol dehydration in extractive dividing-wall columns In: Bogle, I.D.L. Fairweather, M. (Eds.), Proceedings of Twenty Second European Symposium on Computer Aided Process Engineering, June Elsevier, London,Oxford, pp. 17–20.

Teplyakov, V., et al., 2006. Intensification of gas phase catalytic processes in nano-channels of ceramic catalytic membranes. Desalination 199, 161–163.

E., Thompson,1929.US Patent 1701346

Vente, J.F., Haije, W.G., IJpelaan, R., Rusting, F.T., 2006. On the full scale module design of an air separation unit using mixed ionic electronic conducting membranes. J. Membrane. Sci. 278, 22–71.

Wang, G.Q., Xu, Z.C., Yu, Y.L., Ji, J.B., 2008. Performance of a rotating zigzag bed – a new HiGee. Chem. Eng. Process. 47, 2131–2139.

Wolny, A. and Kaniuk, R. ,1995. Intensification of the evaporation process by electric field. Proceedings of the First International Conference on Science, Engineering and Technology of Intensive Processing, University of Nottingham, 18–20 September, pp. 79–82, University of Nottingham.

Zhang, W., Liu, H., Hai, I.U., Neubauer, Y., Schroeder, P., Oldenburg, H., et al., 2011. Int. J. Low Carbon Technol. 7, 69–74.

Intensified Mixing

OBJECTIVES IN THIS CHAPTER

There are several types of mixer, some giving greater intensification than others. As with separations, it is often the active enhancement methods that can produce the greatest intensification of mixing, but passive methods currently dominate. Both types are described here and recent developments directed at intensify mixing are also introduced.

7.1 Introduction

Mixing is one of the less exciting areas of process engineering, but its importance in a range of sectors cannot be underestimated. Without effective mixing, many food products would not be possible and chemical processes would, at best, consume much additional time and energy. Some years ago it was observed by John Middleton, then BHR Solutions mixing consultant (see Appendix 4), that attention to the mixing process typically yielded increases in plant productivity of 10–20%, in some cases reaching 40%.

The stirred tank, of course, uses the stirrer for mixing, but the performance, even with modern designs of paddles, does little to ensure highly uniform reactions – the main point of stirring (or mixing) the fluid(s) within the pot. The most common types of mixer are in-line units and rotor stator mixers. Other mixer types considered below include variants based upon ejectors, fluidics, types using venturi aeration and ideas based upon spinning discs.

Mixing is frequently linked to reactions, and in earlier chapters, as well as in the discussion of PI applications, the reactor (or heat exchanger reactor) is designed, ideally, to have good mixing characteristics. The static tube insert introduced in Chapter 3 is an example where mixing can be used to intensify heat transfer and thus reduce the size of a heat exchanger for a given duty. Good mixing is necessary for uniform and efficient reactions. Often mixing and reaction take place in the same location or the mixer may be located upstream of the reactor. The terminology can be confusing, with some mixers actually being mixer reactors, and some being heat transfer enhancement devices. Mixers are essential components of some tubular reactors (see Section 2.10). The PDX unit – discussed in Chapter 10 (see Section 8.3.8) on the food industry – is commonly described as a reactor, but can be simply a mixer, albeit an intense one!

7.2 Inline mixers

In-line mixers are used for continuous mixing in a fluid stream – tending to operate with much smaller continuous inventories than batch or semi-batch stirred tanks. The uniformly distributed turbulent flow in in-line mixing units helps to ensure that the bubbles or drops generated within them tend to have a controllable size distribution within a narrow range. Most mixers are located in pipe work or tubes.

7.2.1 Static mixers

There are many types of static mixer available commercially, all with similar operating principles. Most units have a number of mixing elements, the combination of element configurations depending upon the mixer type. These are inserted into the pipework at the point where mixing is needed. Their main advantage is the lack of moving parts, keeping operating and capital costs relatively low. Cleaning is, of course, required where product or feedstock changes may be involved.

Static mixers are used to achieve good homogeneity between two or more streams, which can help in achieving good conditions upstream of, for example, a catalyst where reactions between the two are taking place. The intensification process using these mixers allows the components of the stream to be highly mixed within a distance of a few pipe diameters, with no external energy input apart from the modest additional pressure drop.

Andrew Green of BHR Group (Green et al., 2001), a leading UK laboratory on mixing and reactor technologies, compared the static mixer, where 100s of W/kg in mixing energy could be delivered, with a stirred tank, where 1–2 W/kg was more typical. The up to 100-fold higher mass transfer rates in static mixers (compared to stirred tanks) is characterised by uniform energy dissipation, plug flow and good radial mixing.

Static mixers are perhaps the simplest and most versatile of process intensification equipment, with application in reactions where at least one phase is a liquid. They are tube inserts that use the pumping energy/pressure drop to induce mixing and can be roughly divided into three categories:

1. Turbulent flow mixers that rely on the vortices shed from tabs positioned on the walls of the device. They promote mixing in an axial direction and so approximate well, to plug-flow devices.
2. Laminar flow mixers that physically redistribute, stretch and fold the fluid.
3. Those used for both regimes.

Such mixers are marketed as heat transfer enhancement devices by several companies, including Cal-Gavin, whose product is called HiTran (see Chapter 3 and Appendix 3). The in-line static mixer has also proved to be effective in gas–liquid contacting (a phenomenon explained fully in Chapter 6). Al Taweel et al. (2005) used a specific design of static mixer to generate narrow-sized liquid–liquid dispersions and succeeded in generating contact areas of the order of 2,200 m^2/m^3 in the

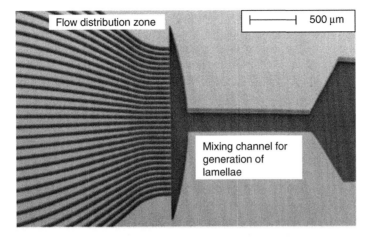

FIGURE 7.1

Micro-mixers in glass and stainless steel (Hessel et al., 2005).

region close to the woven wire-screen mixer elements. The performance of the system with regard to industrially relevant streams such as those containing surfactants had yet to be characterised.

Hessel and colleagues at IMM Mainz (2005) have investigated mixers at the micro-scale. At the top of Figure 7.1 is a slit type micro-mixer made in glass, with flow rate capabilities in the range 10–1,000 ml/h. The lower units are stainless steel mixers for pilot and production scale uses, with capabilities an order of magnitude greater than the micro unit illustrated.

At the lower end of the mixing scale we are within the micro-fluidics regime as discussed in Chapters 3 and 10. The application of active enhancement methods to increase or improve mixing at these scales can, of course, be fruitful. These may include ultrasound and electrokinetic forces. As with any flows at the small scale, particularly if multiple phases are involved, the possible adverse effect of thermophoresis and electrophoresis may work against mixing.

7.2.1.1 *Mixing in the context of micro-fluidics*

It is interesting to note that mixing is moving into the micro-fluidics regime. Naturally enough, such static mixers are classed as passive, as opposed to active, and the problem with mixing at such a small-scale is the absence of turbulence. Thus the mixing relies upon diffusion, which can of course be slow if the diffusivity of the fluid being mixed is low. Recent research in Germany and The Netherlands on passive chaotic micro-mixers (Sarkar et al., 2011), selecting specifically the staggered herringbone micro-mixer (SHM) as a case to examine – a type that has proved to be successful at this scale. In this type of mixer, 'herringbones' are placed inside micro-channels. In a typical channel size examined by the research team – 96 microns × 192 microns × 1536 microns in length – the lattice Boltzmann method was used to describe the fluid flow, which is likened to chaotic advection. This is the enhancement process brought about by the herringbone structures. It was found that the mixing length decreased by over 30% using the herringbone structure to enhance the mixing above that of simple diffusion.

7.2.1.2 *Example of a mixer heat exchanger*

In many conventional types of heat exchangers, the cooling or heating of laminar-flowing, viscous fluids can be difficult. This is due to the formation of boundary layers which can sharply inhibit thermal exchange. For applications of the above kind, there has been a notable increase in the use of static mixer heat exchangers which provide continuous radial mixing, and prevent, or at least strongly reduce, the build-up of boundary layers. A further extremely positive feature is the narrow residence time spectrum, which can be a decisive factor in the processing of temperature-sensitive viscous media, e.g. polymers.

KOCH mixer heat exchangers are available either in monotube or multitube versions. These static mixers increase the heat transfer coefficient by a factor of 6 to 8 compared with empty pipe configurations. They are used not only for rapid heating up applications, (for example, the heating of polymer solutions during manufacture of low density polyethylene and similar types of plastics) but also for the efficient cooling of viscous media.

The KOCH mixer reactor type SMR is a special type of mixer heat exchanger. Its mixing elements are formed of a series of hollow tubes through which the heat transfer medium flows. The arrangement of the tubes in the apparatus is such that they induce strong radial mixing. The SMR is used mainly either as a polymerisation reactor or a cooler. In the SMR, the fluid being treated under plug flow conditions with a narrow residence time spectrum is continuously subjected to mixing

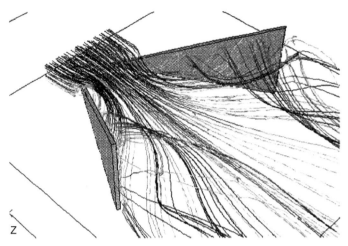

FIGURE 7.2

Streamlines downstream of a vortex generator.

across the entire cross-section of the flow, and at the same time heat is either being introduced or removed. Besides purely cooling tasks, the SMR reactor is employed everywhere where a reaction has to be carried out under defined time/temperature conditions.

As with much of the intensification work associated with mixing, the end use is a chemical reactor. Work in France on the intensification of heat transfer and mixing using heat exchangers incorporating vortex generators (Ferrouillat et al., 2006) revealed that such devices – which could also be used as intensified reactors – were very efficient in terms of heat transfer enhancement and macro-mixing, but micro-mixing was less effective. (Vortex generators are static enhancement devices and are used in plate-fin heat exchangers, amongst others, to improve mixing and heat transfer.) Figure 7.2 shows the streamlines in the wake of two vortex generators. Re is 4600.

7.2.2 Ejectors

Ejectors, for gas–liquid contacting, consist of four main sections. A rotating distributor takes the pumped liquid and orientates and stabilises its flow before it passes through a nozzle that provides a high velocity jet of fluid to create suction in the gas chamber and entrain gas into the ejector. In the following mixing tube the liquid jet attaches itself to the tube wall resulting in a rapid dissipation of kinetic energy. This creates an intensive mixing zone or shock where the high turbulence produces a fine dispersion of bubbles with a large interfacial area for mass transfer. High-energy dissipation associated with this fast process gives excellent mixing, which is carried out for a period downstream in the diffuser.

Static mixers and gas–liquid ejectors are plug-flow devices and this can improve selectivity for consecutive reaction schemes. Due to their very high-energy dissipation rates they can increase the mass-transfer coefficient by an order of magnitude compared to stirred tanks. The large levels of turbulent energy dissipation produced in these high intensity mixers acts to reduce the dispersion size, dramatically increasing the interfacial area – typical gas bubble sizes range from 0.5 to 2.0 mm, compared to 1.0 to 5.0 mm in stirred tanks and bubble columns. These factors combined give such devices the very major benefit of enhanced mass transfer rates of typically 10–100 times those of an STR. A significant benefit of high-intensity in-line mixers is that they have no moving parts, therefore sealing, high pressure and hazardous materials are less of a problem.

Wu et al., in Australia (2007), used a form of ejector nozzle developed at CSIRO to intensify the mixing of oxygen into slurry, and this has been successfully applied in a leaching tank at a gold refinery.

A form of injector rather than ejector is the Pursuit Dynamics PDX mixer/reactor. This is described in Chapter 10 (Section 10.3.7) and is successfully used in areas such as mashing systems in breweries. Based on PDX reactor technology, the PDX mashing system uses direct steam injection to heat and mix the mash. According to the company *The efficient and effective transfer of heat eliminates formation of burn-on in the mash tun and mixes the mash to a uniform homogeneity unrivalled by current conventional mash agitation technology. The PDX mashing system converts standard boiler steam to food-grade via the steam conditioning unit and the production of the beer by direct steam injection conforms to the German Purity Law (Deutsches Reinheitsgebot).*

7.2.3 Rotor stator mixers

A rotor stator mixer has, as its name implies, a rotor component that moves relative to a fixed stator. The gap of a few mm between the two allows very high shear rates to be obtained, implying high energy dissipation. The rotor stator mixer is able to perform reactions in very short times, sometimes less than 1 second. Its high energy dissipation rates make it suitable for disintegration, homogenisation, solubilisation, emulsification, blending and dispersion duties. Units can be applied in either batch processes for recirculation or continuous reaction processes where short residence times are needed. They can help reduce mixing times for liquids up to 1,000,000 cP. In liquid–liquid applications they can reduce droplet sizes below 1 micrometer – significantly increasing mass transfer rates.

Recent research has been directed at studying the micro-mixing characteristics of these mixers, (Jasinska et al., 2012), where tests were carried out on the Silverson 150/250 rotor stator mixer, a double-screen type. Encouraging results were obtained using CFD modelling, giving qualitative agreement with experimental data. This allows one to make a proper choice of the concentrations of the reactant and flows. The work was in collaboration with Unilever Research and Development.

The Taylor-Couette reactor (see Chapter 5) is also a rotor stator mixer, but is discussed separately. Units such as the Marbond HEX-reactor demonstrate mixing plus reactions plus heat transfer in one unit, and these are also discussed in Chapter 5.

7.3 Mixing on a spinning disc

Recent experiments on rotating discs have shown that the apparatus can be used to produce crystals of a specific size where mixing is a critical stage in the operation. The intensity of mixing plays a fundamental role in determining the crystal size distribution of the final product. In order to obtain very small crystals in a narrow size range, a homogeneous nucleation is required, that is, the occurrence of a very high nucleation rate in the absence of crystals. Further data on spinning discs as crystallisers are given in Chapter 6 and 8. For measurements of the behaviour of thin liquid films on a rotating disc, see Ghiasy et al. (2012), which deals primarily with the heat transfer aspects. Although these, of course, are critical to mixing in that the viscosity of the fluid on the disc is a function of temperature.

7.4 Induction-heated mixer

The mixer used, for example, in a stirred tank, can be given a boost in performance by generating heat in the blades of the mixer itself. Conventionally, in a stirred vessel, the heat input is via a jacket, with conduction of heat through the vessel wall and some forced convection thereafter in the 'mixed' product. Using electrical induction heating via a coil, the paddles themselves can be heated, allowing local reductions in viscosity.

7.5 Summary

The chapter on mixers is comparatively short – not because mixers are unimportant, but because mixing is such an integral part of other intensive processes (heat exchange, reactions and crystallisation/precipitation for example that the topic is addressed continuously through other equipment and application chapters. The point is well made by Wu et al. (2007), who stress in their review that PI needs to be implemented via increased mixing, as well as heat and mass transfer – hence the success in this area of the spinning disc and the oscillatory baffle reactors.

References

Al Taweel, A.M., Yan, J., Azizi, F., Odebra, D., Gomaa, H.G., 2005. Using in-line static mixers to intensify gas–liquid mass transfer processes. Chem. Eng. Sci. 60, 6378–6390.

de Weerd, K., 2001. Intensification of gas-liquid reactions. NPT Procestechnologie (2), 38–41. March–April.

Ferrouillat, S., Tochon, P., Garnier, C., Peeerhossaini, H., 2006. Intensification of heat transfer and mixing in multifunctional heat exchangers by artificially generated streamwise vorticity. Appl. Therm. Eng. 26, 1820–1829.

Ghiasy, D., Boodhoo, K.V.K., Tham, M.T., 2012. Thermographic analysis of thin liquid films on a rotating disc: approach and challenges. Appl. Therm. Eng. 44, 39–49.

Green, A., Johnson, B., Westall, S., Bunegar, M. Symonds, K., 2001. Combined chemical reactor/heat exchangers: validation and application in industrial processes. Proceedings of the Fourth International Conference on Process Intensification in the Chemical Industry, 215–225, BHR Group, Cranfield.

Hessel, V., Lowe, H., Schoenfeld, F., 2005. Micromixers: a review on passive and active mixing principles. Chem. Eng. Sci. 60 (8–9), 2479–2501.

Jasinska, M., Baldyga, J., Cooke, M., Kowalski, A., 2012. Application of test reactions to study micromixing in the rotor-stator mixer. Appl. Therm. Eng. <http://dx.doi.org/10.1016/j.applthermaleng.2012.06.036> (in press).

Sarkar, A., Narvaez, A., Hating, J., 2011. Numerical optimisation of passive chaotic micromixers. Proceedings of the Third Micro and Nano Flows Conference, Thessaloniki, Greece, 22–24 August, 2011.

Wu, J., Graham, L.J., Noui-Mehidi, N., 2007. Intensification of mixing. J. Chem. Eng. Japan 40 (11), 890–895.

Application Areas – Petrochemicals and Fine Chemicals

8

OBJECTIVES IN THIS CHAPTER

In this and the three subsequent chapters, applications of the technologies discussed in earlier chapters of this book are described. Some remain research projects, but in a number of cases industrial uses are detailed. In the majority of instances it will be seen that process intensification can give impressive improvements in one or more process characteristics, such as unit operation size, output quality, environmental impact, etc.

In this chapter, we concentrate on the chemical process industries, ranging from petrochemicals and bulk chemicals to fine chemicals and pharmaceutical products.

However, a major new section relates to the use of process intensification in carbon capture, from power stations (predominantly) but equally relevant to a variety of process plants, as well as offshore activities (see Chapter 9).

8.1 Introduction

The chemical industry covers a very wide range of processes, incorporating several types of unit operations, most of which can benefit from PI technology. The diversity of products is equally wide-ranging, and the sector is increasingly involved in the use of biofeedstock and the remediation of effluent and other pollutants using PI methods.

The chapter is divided into a number of sections, dealing with petrochemical refineries, bulk chemicals (those generally made in large quantities and sometimes associated with refineries that supply feedstock, but not necessarily in large plants of the types historically seen), fine chemicals and pharmaceuticals (which are made in smaller quantities either using batch or continuous processes) and bioprocessing (the increasing use of biological methods to process chemicals).

Potential energy savings using PI have already been discussed in some detail in Chapter 2, but recent data on the energy use and CO_2 emissions created by the chemicals sector throw some light on the production processes that are the worst performers and on the importance of very good selectivity in processes, in cutting losses. This is one benefit that PI confers on some processes, and should not be neglected as an argument for investing in PI unit operations. The data collected by Neelis et al. (2007), working in The Netherlands, was based upon an analysis of the

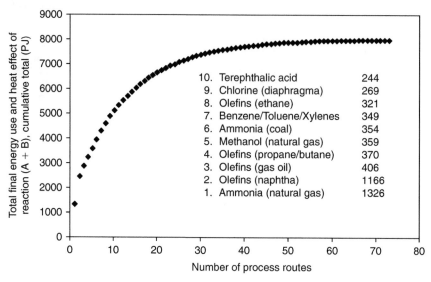

FIGURE 8.1

Distribution of the sum of total final energy use (A) and heat effect of reaction (B) worldwide over the 73 processes included in the Dutch study. The numbers listed on the right-hand side are the sums of A plus B for these processes.

73 principal production processes for the 52 most important bulk chemicals, which between them cover at least 70% of the energy use within the whole sector. Data were collected for Western Europe and the world, as well as for their home country.

Figure 8.1 shows the top ten processes in terms of energy use plus heat of reaction – led by ammonia and naphtha. These are worldwide totals and represent 65% of the total energy use in bulk chemicals. The authors also present more details on the nature of the energy. For example, in steam cracking for olefins and in chlorine production, part of the energy is used to drive the endothermic cracking reactions and is thus converted to chemical energy. Elsewhere, apart from some dehydrogenations, conversions tend to be exothermic. An important aspect of some of the important (energy-wise) exothermic processes, including acrylonitrile and ethylene oxide, is that they have poor selectivity, and the Dutch group estimated that, worldwide, the energy loss from poor selectivity was about 50% or 500 PJ/annum. Overcoming the associated over-oxidation of feedstock and particularly the output of less desirable by-products (that may need more processing) are areas that PI might help to tackle.

8.2 Refineries

In the oil refining sector, many of the generic pressures to cut costs and improve margins are the same or even more challenging than in the upstream processes.

A further feature of the refiner's interest lies in the move towards green fuels, such as city diesel and reformulated gasoline. The production of these could place a major demand on capital, energy and refinery real estate. There is a need to take a more aggressive approach, beyond simply following the crowd. PI techniques offer opportunities through reduction of process pipework and civil works, as discussed in Chapter 2 (see Table 2.1), and through heat integration.

The petrochemicals sector is perhaps ultimately the area of greatest potential for PI, but one in which the large investment in conventional process plant over the years can inhibit any enthusiasm to adopt new process technology. Here, the demands of physical processes are coupled most closely with the chance to manage and exploit reaction kinetics in different ways. The generic benefits of PI might well be harnessed to increase the effectiveness of large-tonnage commodity chemical production, but additional specific opportunities exist to differentiate through enhanced product properties, and the possibilities of distributed processing. The upstream part of petrochemicals has perhaps the 'dirtiest' fluids in the sector, and fouling is often a concern inhibiting PI.

8.2.1 Catalytic plate reactor opportunities

Looking from a broad viewpoint, a common theme of most successful examples of PI relates to the need to overcome any process restrictions imposed by inadequate heat or mass transfer rates. This gives rise to the need for short conduction and/or diffusion path lengths. However, having devised techniques for enhancing critical transfer processes, we are in a strong position to minimise the thermodynamic irreversibilities associated with temperature, pressure and concentration differentials. These represent parasitic features of any process system which tend to compromise the energy efficiency of the process and they can be significantly reduced by a deliberate PI strategy. For example, there are several important processes in which large furnaces are used to provide the endotherm for either hydrocarbon pyrolysis or reforming. Being large, the furnaces are operated at about 1 bar, with heat being recovered from the flue gas by a cascade of heat exchangers. Combustion enthalpy is generated at the adiabatic flame temperature and then degraded to a temperature of around 900°C in the process gases.

A more rational approach would be to utilise the temperature potential more effectively by incorporating the furnace into a gas turbine cycle, perhaps using a catalytic plate reactor or the equivalent. This could be achieved with *pressurised* combustion, in which the furnace acts as the combustor of a conventional gas turbine cycle. The design of the catalytic plate reactor (see Chapter 5) lends itself well to operation at elevated pressure in view of its millimetre-sized gas channels. Thus, power may be extracted from the high-grade heat before it is used at a lower temperature to drive the process.

Approximately 30% of the CO_2 generated in a typical oil refinery is due to the removal of coke, which is deposited on the catalyst particles within the fluid-bed catalytic cracker. Coke is removed by continuously removing contaminated particles and burning off the carbon in a second fluidised bed. Reactivated (and red-hot) particles are contacted with liquid crude to give a vapour/particle mixture which is returned to the cracker. Since coke deposition is highly dependent upon the surface temperature, typically doubling for each 10°C rise, hydrocarbon vaporisation by contact with a red-hot surface is not an ideal approach. Once again, it may be worth considering the catalytic plate reactor in which the catalyst is immobilised and where minimal surface temperature excesses are both feasible and acceptable.

8.2.2 **More speculative opportunities**

Two emerging technologies in the area of PI might impact all stages of the hydrocarbon value chain. Firstly, at the small-scale, the concepts of lab-on-a-chip and plant-on-a-chip open up many new possibilities for online real-time analysis and screening, and distributed processing, possibly even down hole, as discussed in the next chapter on offshore PI. Secondly, at the larger scale, the concept of combining gas turbine, heat exchanger and reactor technologies offers some very interesting possibilities in gas processing and conversion, as well as in power generation, as discussed in some detail later in this chapter.

8.3 **Bulk chemicals**

The phrase 'bulk chemicals and process intensification' may seem to be a contradiction in terms. However, the examples given below show that practical opportunities exist for applying PI in this sector. Even more radical thinking, such as moving away from central to local production, may lead to vastly increased application of PI technology across the whole chemicals sector. It is also important to show that the chemical industry, with the help of PI, can grow with reduced risk, as briefly explained below.

One of the recurrent features of the commodity chemical industry is its instability or cyclic nature. This is, in large part, due to the tendency for many firms to make investment in new plant simultaneously, to follow a rising market, only to see prices fall when overcapacity results. The conventional view of a world-class plant is that it must be big in order to capitalise on the 0.6 power rule, which states that capital costs only increase at (production rate)$^{0.6}$, as discussed (with an example) in Chapter 2. Intensified plants, however, with their lower capital costs, should allow smaller scale plants to compete economically with their standard counterparts. Thus, in following a rising market, the capital investment can be made in smaller increments and be much less risky. This should facilitate a more orderly market development without recurrent bouts of feast and famine.

Table 8.1 Comparison of Conventional and Rotating Packed Bed Plant in Chinese Tail Gas Cleaning Plant

	Foaming Plate Tower Technology	**RPB Technology**
Investment	1	0.40
Footprint	1	0.33
Power consumption	1	0.42
NH_3 consumption	1	<1
Quality of product	More sulphate	Less sulphate
Entrainment separator	Yes	No need

8.3.1 Stripping and gas clean-up

As shown in the examples below, rotating intensified plant (e.g. HiGee), are used in bulk chemicals plant. The benefits of HiGee, in particular, are substantial, as illustrated in Table 8.1.

8.3.1.1 Chinese tail gas cleaning plant–acid stripping using an RPB

The first example concerns a plant in China, where sulphur dioxide needs to be removed. The data in Table 8.1 suggest that the process is considerably superior to the current operation. A rotating packed-bed (RPB) machine capable of handling $3000\,m^3/h$ of gas during an absorption process was installed and operated in 1999, at the Zibo Sulphuric acid plant (Shandong province, P. R. China), parallel to the existing tower system for tail gas cleaning of SO_2.

Figure 8.2 is the flow diagram of the existing tower system and RPB machine system in parallel. Tail gas containing 5,000 to 12,000 ppm SO_2 (adjustable, according to liquid product desired) is first absorbed by a rich ammonium sulphite-bisulphite solution (containing total sulphite of 700 g/l) in a foaming plate tower. The 750 g/l sulphite-bisulphite solution exiting from this first tower is a reducing agent product sold to a pharmaceutical plant. Tail gas exiting from the first tower containing 30–60% of the original SO_2 is further absorbed by lean sulphite-bisulphite solution (containing total sulphite of 400 g/l) in a second foaming plate tower and exits to the stack through an entrainment separator. The residual SO_2 in the stack gas is 500–1,000 ppm.

The global chemical reaction of the process is as follows:

$$SO_2 + H_2O + 2NH_3 = (NH_4)_2SO_3$$

$$(NH_4)_2SO_3 + SO_2 + H_2O = 2NH_4HSO_3$$

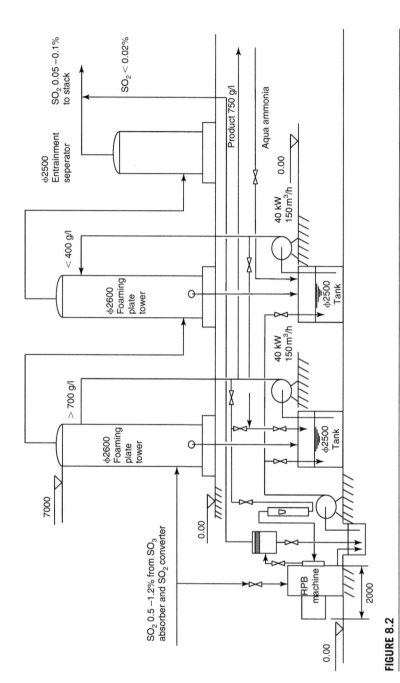

FIGURE 8.2

Flow diagram of existing absorption tower and rotating packed-bed replacement.

FIGURE 8.3

Photograph of the RPB machine at Zibo.

The RPB machine, shown in Figure 8.3, is mounted in parallel with the tower system and can be substituted for either the first or the second tower for industrial testing. The results of the RPB technology compared to existing tower technology were as follows:

1. The degree of equilibrium of absorption[1] is close to 100% for different concentrations of absorbing solutions and different SO_2 contents in the inlet gas, while the results of existing technology (tower system) are about 80% or less.
2. The SO_2 content in the exit gas is less than 200 ppm in operations within the scope of this practice (this can be lower, approaching zero, providing the absorbent's equilibrium SO_2 pressure is zero). The SO_2 in the effluent gas of the existing technology is of the order of 500 to 1,000 ppm.

Based on this practice, for a 300 t/d sulphuric acid plant tail gas (40,000 m³/h), with 750 g/l ammonium sulphite-bisulphite solution as a product, three towers in series would be needed in order to meet the same level of SO_2 in the effluent gas as two RPBs in series (see also Table 8.1).

[1]Degree of equilibrium of absorption = (actual absorption efficiency)/(absorption efficiency when in equilibrium with the inlet absorbent) × 100%.

8.3.1.2 Acid stripping using HiGee at Dow Chemicals

Three HiGee units were installed by Dow Chemicals in the US for stripping hypochlorous acid, used as the reactive chemical in a process. The rotating packed-bed used for stripping was selected after other alternatives, such as a spray distillation tower, were rejected for capital cost reasons. The packing selected was a woven wire screen, with the gas flowing counter-current to the liquid. Various packing surface areas were tested, with $2,000–3,000\,m^2/m^3$ area densities, but these exhibited no performance differences so the lower value was used.

The pilot plant showed >90% yield, compared to 80% on a conventional spray distillation unit, with stripping gas-use halved. The economics of scale-up looked favourable, but risk issues such as 'will it work?' and 'is it reliable?' needed addressing. The Chinese experiences of such plant described above provided some support for scale-up. The same packing was used as in the pilot plant, with a variable speed motor drive. With regard to reliability, speeds were <400 rpm and the partner was a centrifuge manufacturer with a proven drive train.

The production machine yield was 93–96%, this being regarded as highly satisfactory, and after six months operation the unit has proved to be very reliable. It is easy to start up and shutdown, and to operate. Pressure drop is <50% of that expected. Height of a transfer unit is 8 cm (4 cm was expected), but this is the same as the pilot plant. Three units were installed, two being adequate for production. There have been no control problems. It took four years from conception to having the process online. (See also section 5.3.3 in Chapter 5 for data on rotating packed beds in the context of reactors.)

8.3.1.3 Removal of NO_x in industrial tail gas

More recent work in China (Li et al., 2010) examined the use of RPBs to absorb NO_x arising from nitric acid towers and nitration reactors. The tail gas can be absorbed using water in conventional columns – the wet method – but the absorption rate is low and the resulting nitric acid too dilute for ready reuse. Oxidising agents, pointed out by the authors, can be introduced to speed up the absorption. These might include chlorine and oxygen and ozone. Ozone converts NO_x to absorbable species.

Formed during the process, HNO_2 is unstable in the liquid phase and gives off NO. If ozone is introduced, the HNO_2 is oxidised to HNO_3, increasing NO_x absorption. It is this reaction that Li and his co-workers point out is fast and mass-transfer limited (the absorption rate of the ozone). This directed the team towards the RPB.

The source of the NO_x was a nitration reaction at the Yinguang Chemical Company. The RPB used had a diameter of 300 mm (packing i.d. 100 mm, o.d. 290 mm) and packing height 200 mm. The packing comprised steel wire mesh.

The experiments covered a range of parameters, including ozone concentration, RPB speed (0–1,500 rpm), inlet NO_x concentration (200–240 g/m^3) and the g-forces developed. It was found that removal rates of NO_x in excess of 90% were achieved using the RPB – the best with a packed column being around 85%. Although no size comparisons were made, the RPB would be substantially smaller than the static column in practice.

8.3.1.4 Use of the 'Mop Fan' on an ammonium nitrate granulator dryer

The Begg Cousland Becoflex gas scrubber, or mop fan, is described in Chapter 6. Application of this device, invented at ICI, is now widespread, and it is used in prilling towers and ammonium nitrate dryers – see below:

Drying duty:
Combined total airflow: 64,000 m³/h from the granulator dryer section. Comprising a flow of 17,000–25,000 m³/h per each of 2 scrubber units and a flow of 7,000–15,000 m³/h per each of 2 smaller scrubber units. All 4 scrubbers are operating in parallel and discharge into a collector duct and effluent to a common discharge vessel. All materials are of 304L stainless steel.

Duty and performance:

- Operating temperature: 52°C.
- Inlet ammonium nitrate dust load: 15,000–20,000 mg/Nm³.
- Inlet ammonia load: <50 mg/Nm³.
- Scrubbing solution: 25–50% ammonium nitrate solution + nitric acid.
- Spray per scrubber max. 160 L/min at 1 bar.
- Efficiency guarantees met:
  - <10 mg/Nm³ ammonium nitrate particles.
  - <5 mg/Nm³ ammonia Pressure loss 100 mm.

One of the largest uses of prilling towers is in the production of fertilizer. In a typical process, concentrated ammonium nitrate (NH_4NO_3) solution is delivered from the evaporator to the top of the prilling tower (which rival large distillation columns in height) and is sprayed downwards from the top of the tower. The inlet air rising in the prilling tower cools the falling, sprayed solution, which then forms into solid prills. The prills fall to the bottom of the tower for collection. At the top of the tower, around the sprayer heads, the ammonium nitrate decomposes into NH_3 and HNO_3, which recombines quickly into NH_4NO_3 mist and is entrained by the air going on up to the roof fans and thus out to atmosphere.

Mist loads, according to the Begg Cousland Case Study, can vary greatly according to the design and age of the tower, from 150 to even 2000 mg/m³ with a small particle size, e.g. 1 micron.

In this instance, the first stage of mist removal used an irrigated 'Becoil' Demister/'Becone' Coalescer in 304L stainless steel or special alloy. A similar system was adopted for a urea spray in a prilling tower (Web1, 2012).

8.3.2 Intensified methane reforming

The methane reforming reaction is one of the most important in the petrochemical industry as it is the prelude to many subsequent syntheses. It is highly endothermic and may be written as:

$$CH_4 + H_2O \rightleftharpoons CO + 3H_2$$

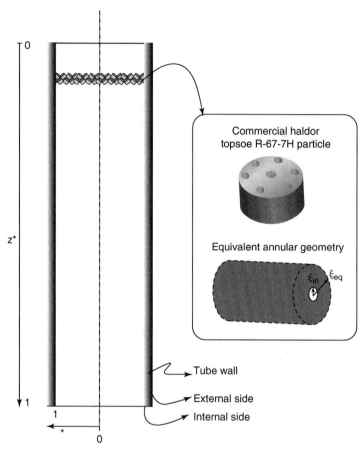

FIGURE 8.4

Layout of a conventional reformer furnace – the tubes are typically about 10 cm diameter.

In order to provide the process endotherm, the reaction is conventionally performed over catalyst pellets contained in tubes within a radiant furnace, as shown in Figure 8.4 (Pedernera et al., 2003). The heat radiated from methane flames between rows of tubes must penetrate the tube wall and ultimately reach the reaction sites within the catalyst pellets. Tube spacings are dictated by the (poor) emissivity of the methane flame and are approximately 2 m. Thus, a typical reformer furnace capable of producing 1,000 tonnes/day of synthesis gas has a volume of several 1,000 m^3. In this case, the furnace size is controlled by considerations of heat transfer intensity, rather than those of the intrinsic catalyst kinetics.

An alternative to the foregoing design is the catalytic plate reactor in which thin layers of combustion and process catalysts are deposited on opposite sides of a heat exchanger surface (see Chapter 1, Figure 1.12(b)). The combustion heat is generated catalytically on the surface and can be conducted readily to the endothermic process reaction (see Chapter 5 for more details). Extensive work at Newcastle University has shown that both process and combustion reaction rates equivalent to heat fluxes of over $10 \, kW/m^2$ can be easily achieved. On this basis, a catalytic plate reformer capable of producing 100 tonnes/day of syngas would have a matrix volume of around $12 \, m^3$. Since the combustion pressure is no longer restricted to 1 bar, the combustion system may be incorporated into a gas turbine arrangement in order to recover power. The relatively low combustion temperature ($=850°C$) means that only trivial quantities of nitrogen oxides are generated, thereby benefiting the environment.[2] The concept was jointly developed by Newcastle University and BG in order to provide hydrogen for a 200 kW fuel cell.

8.3.3 The hydrocarbon chain

At the first meeting of the UK Process Intensification Network (PIN), Brian Oswald of BP gave an overview of some of the potential opportunities for PI at various stages in the hydrocarbon business chain, from reservoir to marketplace. It highlighted some specific instances where the attributes of PI might be harnessed, but stressed that these ideas were, at the time, still in their early stages of development (see Chapter 9 for references).

Upstream, in hydrocarbon production and transportation, the emphasis is mainly on physical processes – separation, fluid flow, heat transfer. PI offers safety and environmental benefits through reduced inventories and higher energy efficiency. In offshore production, the move into ever deeper waters has renewed the emphasis on floating systems, where space, weight and sensitivity to motion are major constraints (see Chapter 9).

The ability to manufacture chemicals at the site where they may be needed as feedstock or for other applications, is one of the major benefits brought to the marketplace by process intensification. The lab-on-a-chip and 'desktop' chemicals plant are terms familiar to those involved with intensified unit operations, but a new term applied to process plant came to the fore in 2005 – the pocket-sized plant. In this case, the researchers were referring specifically to nitric acid plants (Perez-Ramirez and Vigeland, 2005).

The specific unit developed in Norway was able to simultaneously carry out the separation of oxygen from air and ammonia oxidation to NO. This reactive separation process was carried out using mixed conducting membranes, based upon lanthanum ferrite-based perovskite materials. Shown in Figure 8.5, the separation of oxygen from air is carried out at the feed side, while the NO is produced on the permeate side of the membrane by reacting ammonia with the separated oxygen. NO selectivity is 98% and no harmful N_2O is produced. The temperature range of operation is 1,000–1,333 K, presenting possible challenges.

---

[2] The very high combustion gas temperatures in conventional plant lead to high NO_x emissions.

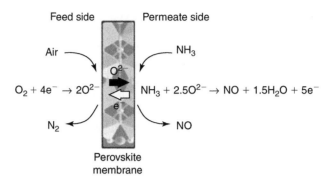

Feed side   Permeate side

Air ——   —— NH_3

O^{2-}

$O_2 + 4e^- \rightarrow 2O^{2-}$ $NH_3 + 2.5O^{2-} \rightarrow NO + 1.5H_2O + 5e^-$

e^-

N_2 NO

Perovskite
membrane

FIGURE 8.5

Reactive separation using mixed conducting membranes (Perez-Ramirez et al., 2005).

In practice, the designers suggest that hollow fibre membranes might be used in the pocket plant, or monolithic membranes where the surface area density is 500–4,000 m^2/m^3. Such compact reactors already exist for other uses (see Chapter 5).

8.3.4 Reactive distillations for methyl and ethyl acetate

Although not at the intensified scale of the nitric acid plant described above, some regard reactive distillation as the most widely applied process intensification technique today. A review by Harmsen (2007) cites 146 units in operation in 2006, based upon the technology of CDTECH. The technology is discussed in Chapter 5.

The integration of reaction and distillation within a single unit operation is not associated with such major reductions in unit operation size, as might be possible with high gravity distillation, for example. Harmsen quotes an Eastman reactive distillation column with a height of more than 80 m and 4 m diameter. This has a capacity of over 200,000 tonnes of methyl acetate per annum. Other applications of reactive distillation include selective hydrogenation, desulphurisation of petrol and aircraft fuel and the alkylation of benzene with propylene.

In Spain, Calvar and colleagues (2007) have used a packed-bed reactive distillation unit (see Figure 8.6) for the esterification of acetic acid with ethanol – giving ethyl acetate. As with conventional pure distillation columns, the variables that were examined included feed composition and reflux ratio. The benefit of combining the two unit operations was attributed to the relative ease with which products are removed from the reaction mixture. In this particular application there were limits to the purity of the ethyl acetate obtained, depending on the azeotrope composition.

8.3.5 Formaldehyde from methanol using micro-reactors

The partial oxidation of methanol to produce formaldehyde is a common industrial reaction that has attracted the attention of those working in PI, from both the use of

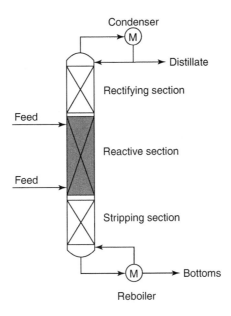

FIGURE 8.6

Schematic diagram of a reactive distillation column.

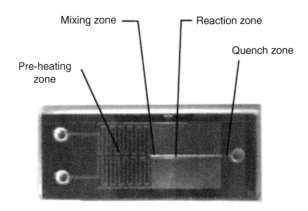

FIGURE 8.7

The silicon–glass micro-reactor with silver catalyst. Size is 25 mm wide × 63 mm long.

micro-reactors for kinetic studies (Cao and Gavriilidis, 2005) and safer operation (Patience and Cenni, 2007).

The work at University College London by Cao and Gavriilidis was based on the use of ion-etched micro-reactors made in silicon (see Figure 8.7). One of the principal reasons for the experiments was the lack of data on the kinetics of the catalytic selective oxidation of methanol to formaldehyde over a silver catalyst at relevant industrial reactor temperatures, which of course inhibits the design of reactors at the industrial-scale able to match the kinetics. The reactor used had a channel of 600 microns in width and the depth is determined by the etching time.

High conversion, of the order of 75%, and a selectivity of 85% were obtained with residence times of only 3 minutes. Conversion decreased with increasing methanol concentration. It was concluded that the use of micro-reactors for studying highly exothermic reactions was beneficial and that they were also valuable in kinetics studies.

8.3.6 Hydrogen peroxide production – the Degussa PI route

Degussa, the German chemical company (see also Section 2.4.4.1) manufactures typically 600,000 tonnes annually of hydrogen peroxide (H_2O_2). The conventional process for this chemical, which is used as a bleaching agent and also as a feedstock for chemicals in the detergent sector, involves the making of H_2O_2 from hydrogen and oxygen via hydroquinone, an intermediate. The new process, called direct synthesis of hydrogen peroxide (DSHP), reacts the hydrogen and oxygen directly after they have been fed into liquid methanol using a nano-catalyst. This single-step process, which is the PI aspect of the development, also necessitated developments to overcome the perceived explosive hazard of directly mixing H_2 and O_2. A modular reactor was developed for this purpose that allows good control of heat transfer, as well as materials transport, keeping operation outside the explosive region. The process moved from a pilot plant in 2005 to a moderate scale-up in 2006 and planned commercialisation in 2007.

8.3.7 Olefin hydroformylation – use of a HEX-reactor

It is only relatively recently that data have started appearing on comprehensive studies of the use of intensified unit operations in industrially relevant reactions – particularly for bulk production. One such study involved a comparison between the stirred batch reactor process and a heat exchanger reactor (HEX-reactor). This was reported by a consortium involving Cardiff University, Givaudan, Johnson Matthey, and Chart Energy and Chemicals (who supplied the HEX-reactor), and is presented in Enache et al. (2007). The hydroformylation reactions examined are used for the production of detergents, soap and surfactants – totalling millions of tonnes per annum.

The particular incentive for this research was the desirability of removing the solvent in which the catalyst is dissolved – toluene – for reasons of environmental protection. If the gas–liquid reaction could be carried out under solvent-free conditions, and additionally, the gas–liquid mass transfer could be enhanced, it was believed that benefits could result in several areas. The particular study reported involves the use of a homogeneous catalyst dissolved in the liquid phase. The PI arose from several sources – in the words of the researchers:

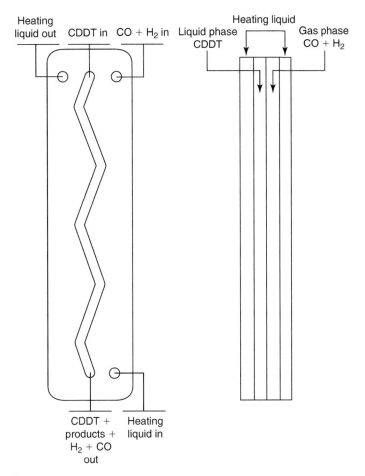

FIGURE 8.8

The HEX-reactor scheme, using the Chart Energy and Chemicals unit.

Omission of a solvent minimises the volume of the reacting liquid, the thin channel design will enable enhanced mass transfer rates and thus increased rate of reaction while the high heat transfer surface area and heat transfer coefficients are required to accommodate the increased heat transfer duty.

Results from the stirred batch reactor and the HEX-reactor were compared in terms of reaction rate and selectivity. The HEX-reactor schematic is shown in Figure 8.8.

The HEX-reactor had 2 mm × 2 mm square channel sections, and the gas phase was injected through a perforated plate running the length of the channels. Interconnected heat transfer channels in the form of a matrix were embedded in the rector body (see Chapter 5 for a description of Chart Energy and Chemicals' products). The liquid flow rate (the liquid being cyclododecatriene [CDDC] containing dissolved catalyst) was 15 ml/min, gas flow 440 N ml/min and pressure of the order of 30 bar. Reactor volume was 3 ml and the gas:liquid ratio was 1:1. The batch unit, on the other hand, was charged with 70 ml of the liquid and 1.06 g of catalyst and batch reaction time was several hours at about 80°C.

Full data are given in the referenced paper, but the work led to kinetic models of the data from both reactors, allowing a comparison of reaction rates (in terms of turnover frequency) and selectivity. The HEX-reactor was found to be free of mass transfer limitations that had affected the performance of the batch unit and the reaction rate of the HEX-reactor calculated on the basis of this turnover frequency was 10–15 times higher than that of the stirred tank. The selectivity to the monoaldehyde product was also better in the HEX-reactor, being several percent higher at the greater conversion rates of the CDDC.

The retention of solvents for this process need not be too damaging to the environment. Jin et al. (2006) in the US have used CO_2-expanded liquids (CXLs) as solvents in this process. CXLs are less damaging to the environment and overcome some of the mass transfer limitations of conventional solvents (e.g. toluene), resulting in enhanced reaction rates. The team found that the use of CO_2 enhanced the yield of aldehydes by 25%.

8.3.8 Polymerisation – the use of spinning disc reactors

The production process for polymers and the benefits that spinning disc reactor (SDR) technology can bring to it have been introduced in Chapter 5. In the following sections, the use of SDRs for the manufacture of a specific polymer, polystyrene, is discussed, followed by data on polyesterification.

8.3.8.1 Manufacture of polystyrene

The manufacture of polystyrene from various grades of prepolymer has been performed (Boodhoo et al., 1995, 1997, 2000a; Vicevic et al., 2007) on a 36 cm brass SDR using the arrangement shown in Figure 8.9. A series of concentric grooves were machined in the disc surface in order to improve liquid mixing within the film. The reaction operates via free radicals, which were initiated in this case using benzoyl peroxide. In the first instance, a series of batch runs was performed in a conventional laboratory-scale stirred vessel in order to produce a calibration curve (Figure 8.10) of conversion versus time. This vessel was then used to produce about 200 ml of prepolymer at a range of conversions that was supplied to the inner spinning disc surface over a period of about 30 seconds. The SDR was heated from below by a stationary radiant ring, and the polymer produced was collected in a cooled annular trough surrounding the disc. The styrene was diluted with about 16% w/w toluene in order to reduce the viscosity.

Figure 8.10 also shows the increment in polymer conversion in one pass over the disc as a function of the initial conversion in the preliminary batch. It can be seen that the equivalent batch time that can be ascribed to one pass on the disc

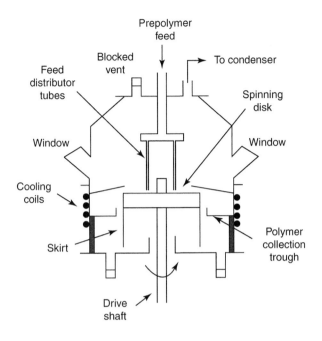

FIGURE 8.9

Schematic of an SDR styrene polymeriser.

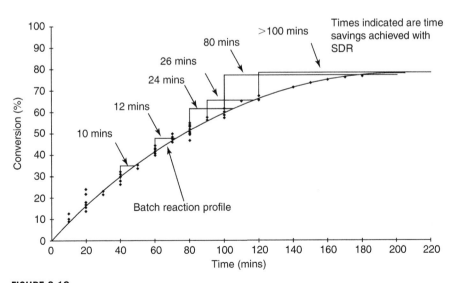

FIGURE 8.10

Free-radical polymerisation of styrene and the time savings using an SDR.

increases (up to 58 minutes) as the initial conversion increases to 63%. This implies that the benefits of the SDR become more marked as the polymer viscosity increases. It is envisaged that the process can be scaled-up either by using a larger disc or by mounting several discs on one shaft. The latter approach (i.e. several discs in series) does, however, involve the problem of transferring polymer from the peripheral collection trough to the centre of the next disc. An alternative may be to operate discs on one shaft in parallel. For the experiments just described, the feed rate was roughly 5–10 ml/s, the equivalent to an output of up to 250 tonnes/year on a continuous basis, although at this early stage this should not be considered the ultimate limitation.

The fundamental reasons for the high performance of the SDR are still a matter for debate. The significance of micro-mixing and the consequent improved probability of radical interaction have already been mentioned. However, another factor is expected to be the divergent character of the flow on the disc. This may be expected to align the polymer molecules and thereby encourage the juxtaposition of the reactive groups.

The work recently reported (Vicevic et al., 2006) will, it is believed, form the basis of *Future Polymer Processes*. Carried out in conjunction with Sheffield University, in the UK, the examination of the kinetics of the process has allowed the continuous polymerization rate on the SDR to be improved to such an extent that a 10% change in conversion could be achieved on a disc after only one pass across it – a residence time of just 2 seconds. The exciting prospect offered by such processing speeds is, in the words of the authors, that it will provide an opportunity for flexible, mobile plants encouraging distributed manufacture.

8.3.8.2 Polycondensation

The reaction between maleic anhydride and ethylene glycol has been studied as an example of polycondensation on an SDR (Boodhoo and Jachuck, 2000b). Since the reaction proceeds on an equilibrium basis, in order to drive it to completion the water produced must be eliminated from the increasingly viscous polymer melt. The grooved brass 36 cm disc described earlier for the polystyrene experiments was used at a temperature of 200°C and a disc speed of 1,000 rpm. As before, the experimental procedure involved the establishment of a benchmark batch calibration against which the subsequent disc runs could be compared. A typical acid number plot versus batch time is presented in Figure 8.11. As the acid number decreases, the conversion to polymer increases. The water of reaction was removed from the polymer film by maintaining a large nitrogen purge to the vapour space. This technique, rather than the application of a vacuum, was the preferred method for reducing the water vapour partial pressure above the disc surface.

It can be seen that the increment in polymerisation following one pass in the SDR corresponds with many minutes of reaction in the small batch reactor used as a reference. This is particularly encouraging because the mass transfer intensity in the laboratory stirred reactor is likely to be much greater than its industrial-scale equivalent, and it therefore provides a demanding benchmark for the spinning disc performance.

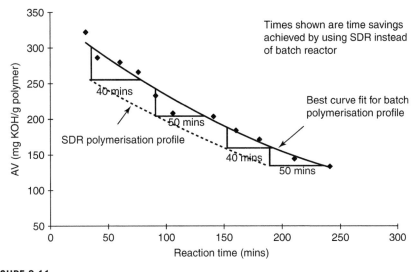

FIGURE 8.11

Time savings in the SDR for polyesterification.

8.3.9 Akzo Nobel Chemicals – reactive distillation

Reactive distillation is portrayed by some as a major area of importance in process intensification – the topic is introduced in Chapter 5. Akzo Nobel in The Netherlands has been studying how the combined process could aid a situation where market forces led to the fact that a byproduct from a plant had become commercially more attractive than the primary product.

In order to satisfy the change of emphasis of product, a conventional design based on a reactive distillation column was employed, followed by a normal column. A fully integrated design that combines reaction and distillation into one reactive dividing wall column (see Chapter 6) was selected to avoid the production of two useless byproducts. Compared to the base case using two distillation columns, this study showed that a 35% saving in capital cost could result, with a 15% energy cost saving.

8.3.10 The gas turbine reactor – a challenge for bulk chemical manufacture

8.3.10.1 Background

As has been mentioned earlier in this book, the gas turbine, particularly in its closed cycle form, is an excellent example of process intensification. The energy input is via an exothermic reaction, heat recuperation which, when practised, can be via compact heat exchangers. Energy conversion involves high speed rotation. Energy densities and power-to-weight ratios are very high, as suggested earlier in this chapter, particularly for aero-derivative units, and the performance of such systems is largely insensitive to changes in orientation. It could well function as a chemical reactor (as well as providing heat and power).

The principal components making up the closed cycle are a compressor, used to compress the gas working fluid; a heater, which raises the temperature of the compressed gas prior to its entering the third major component; a turbine. Expansion through the turbine provides energy to drive the compressor and external users (e.g. a generator). After leaving the turbine, the lower pressure/temperature gases are cooled further in a heat exchanger before being recompressed.[3]

One of the limiting features of a gas turbine, and one which designers strive to overcome in the search for improved efficiency, is the ability of turbine blades to withstand the high inlet gas temperatures needed to improve cycle performance (although concerns about NO_x formation may also limit rises in temperature in the future). Moving blades and guide vanes employ advanced cooling techniques, frequently based on bleed air flowing through complex internal passages, for maintaining materials at acceptable temperatures.

Interestingly, an effective cooling method would be an endothermic reaction inside or on the outer surface of the blades. Unpublished work in the mid-1990s by scientists working in Israel has shown that this could become an acceptable technique – tests with methane over a nickel catalyst suggested that blades could be cooled by up to 300 K more than using conventional techniques. In this case the experiments were carried out with the catalyst on the outside of the blade, with methane injected into the boundary layer. The idea to use the endotherm to generate a useful product, however, was not considered, and for practical reasons it would be most convenient to have such a reaction inside the blade.

Another interesting synergy occurs in compact heat exchanger (CHE) technology – the technology of the new generation of compact heat exchangers, such as the Heatric PCHE and Chart MarBond heat exchanger and their respective heat exchanger reactor derivatives, does, however, offer another exciting possibility – the heat exchanger reactor as compressor intercoolers or exhaust gas recuperators (see also Chapters 4 and 5).

Intercooling between compressor stages improves efficiency, and an endothermic reaction on one side of a CHE, using an appropriate catalyst, could give improved compressor air cooling – in each stage of compression, if necessary. Similarly, recuperation could be enhanced if exothermic reactions could be initiated within the CHE recuperator. Of course, one would need to analyse the benefits of simple recuperation against those of perhaps reducing the recoverable heat energy, but producing a value-added chemical product instead. Figure 8.12 shows the various components in the gas turbine that might be converted from heat exchangers to heat exchanger reactors.

---

[3] A basic open-cycle gas turbine, as used for aircraft propulsion, has a combustion chamber as the heater, and gases from the turbine are exhausted to the atmosphere without recycling. The compressor is used to raise the pressure of the combustion air.

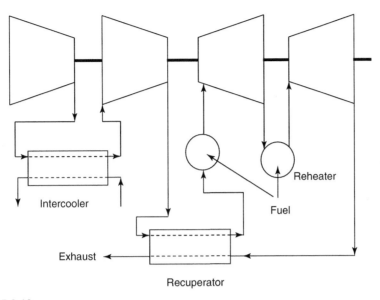

FIGURE 8.12

The recuperated, reheated and intercooled gas turbine.

The basic concept of a gas turbine reactor arose out of an investigation into catalytic combustion carried out on behalf of the UK Energy Efficiency Office and seven industrial co-sponsors (Reay and Ramshaw, 1992). Here, work on catalytic combustion as a way of reducing emissions in gas turbines was reviewed, as was the heat exchanger reactor, subsequently supported within the UK Energy Efficiency Best Practice Programme with BHR Group (a leading PI laboratory) and others. The idea then gelled to bring catalytic reactions and heat transfer together wherever possible in a gas turbine, so that it could benefit in terms of cycle efficiency, while retaining heat production/recovery (and/or the production of refrigeration), but adding the benefit of chemicals production – combined heat, power and chemicals.

By adding reactions, as indicated below, the gas turbine could become an even more versatile unit and it could benefit from associated efficiency improvements. As a result, a feasibility study was partially funded by Future Energy Solutions, acting for the UK Energy programme (Anon, 2002).

8.3.10.2 Factors affecting feasibility

There are a number of important aspects of gas turbine and related technologies which will have a strong bearing on the technical feasibility of carrying out reactions in a gas turbine (and/or its heat exchangers). These include blade cooling and general turbine/compressor blade technology, including materials, compact reactor technology and the types of reactions which can be effectively performed in compact reactors. Of course, the influence (positive or negative) that such a role will have for the conventional performance of the gas turbine, in particular, for thermodynamic efficiency and power output, etc., should not be neglected. The competing concepts reviewed as part of the study included the following, some of which are briefly discussed below:

- Blade cooling.
- Intercooling.
- Gas turbines linked to reactors.
- Peripheral reactors and less conventional propulsion systems.
- Advanced gas turbine cycles.

8.3.10.3 Recuperative steam cooled gas turbine

Steam injection into gas turbines is commonly employed. Westinghouse Electric (Anon, 1997a) proposed to use the coolant, which is both delivered to and extracted from the blades in the form of steam, to aid combustion and NO_x reduction by using it as injected steam. It also improves power output. The steam is generated in a waste heat boiler before being fed to the stationary blades, thereafter being injected into the combustion chamber. *This suggested that it may be worth considering a reforming reaction within the static blades to cool them, the reaction leading to a fuel 'supplement' which could boost the performance of the gas turbine.*

8.3.10.4 Blade cooling with endothermic fuel

The patent filed in 1991 by the United States Air Force (USAF) (Anon, 1991) comes close to the concept which allowed the subject of this study to germinate. The USAF proposed a hollow gas turbine rotor blade with a catalyst coated on the inner surface. The cooling would be effected by vaporisation and decomposition of an endothermic fuel introduced into the hollow blade.[4] The concept is illustrated in Figure 8.13. The endothermic fuel is defined as one which is liquid and which decomposes in the presence of a catalyst into two or more gaseous compounds different chemically from the liquid. At least one of the gaseous compounds is, of course, combustible in this case. The liquid fuels covered in the patent include methylcyclohexane, methanol, n-heptane or JP7 (kerosene fuel). An additional feature of using a liquid fuel as the feedstock is that latent heat of vaporisation contributes to cooling, in addition to the endothermic reaction.

The liquid fuel is sprayed onto a catalyst layer coating the inside surface of the blade. The reaction and vaporisation take heat from the surface of the blade. Gases exit at the blade tip or can leave via a trailing edge slot. They are then 'directed to the engine to be burned as fuel'. *One concept considered in the UK study was the burning of fuel in stages throughout the turbine – an idea with which this, or a derivative, might be compatible.*

---

[4]Those of you who have read the section on heat exchanger reactors in Chapter 5 will realise that the hollow blade may be regarded as being equivalent to the heat exchanger reactor in this case.

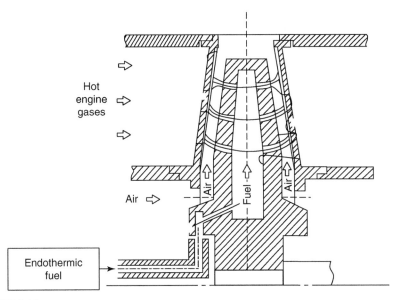

FIGURE 8.13

Endothermic reaction in a blade to effect cooling.

8.3.10.5 Gas turbine plus fuel cell

US patent 5693201 (Anon, 1997b) describes a turbine system with a fuel cell located between the compressor and turbine. Data obtained to date suggest that the compressor is used to raise the pressure of one medium which is fed to the fuel cell, the second medium being fed directly into the cell. The electrochemical reaction, as well as producing electricity, gives a sufficiently high exhaust gas temperature to allow the reject gases to be expanded through a turbine, driving the compressor. The concept is variously described as an electrochemical combustor replacement (ECCR) or as a fuel cell for combustor replacement (FCCR).

An earlier patent from the US Department of Energy describes a different approach to linking fuel cells and gas turbine cycles (Anon, 1993). An indirectly heated gas turbine (GT) cycle is followed by a fuel cell cycle, the heated air from the turbine being used to directly heat the fuel cell cathode. The hot cathode recycled gases provide a substantial part of the heat required for indirect heating of the compressed air used in the GT. A separate combustor provides the balance of the heat needs. Hot gases from the fuel cell reduce the GT fuel-needs and also the NO_x emissions. Residual heat from the fuel cell may be used in a steam cycle or for absorption cooling.

Precision Combustion Inc., in the US, has examined coupling a polymer electrolyte membrane (PEM) fuel cell to an autothermal reforming reactor (ATR), with a gas turbine being used to generate additional power (Lyubovsky and Walsh, 2006). The aim, to produce pure hydrogen from hydrocarbon fuels, is aided by the use of membrane-based hydrogen separation downstream of the ATR. The high pressure stream from the membrane is combusted before being expanded through a turbine and the system can be coupled to a PEM fuel cell to give electricity generation efficiency greater than 40%. From the PI point of view, the catalytic partial oxidation reactor of compact dimensions operating at around 8 atmospheres pressure with a power density greater than 15 MW/litre that had been demonstrated by the company suggested that even greater compactness could be achieved at design operating pressures almost double this value.

8.3.10.6 Reforming and reheat combusting in a gas turbine

In this system (Anon, 1992) a mixture of steam and a combustible effluent such as methane is reformed or partially oxidised to produce a hydrogen-rich fuel. This fuel is then used for primary combustion and also for reheat upstream of the final turbine stage. The reheat is effected by injection of the hydrogen through cooling orifices of the first turbine, upstream of the final turbine stage, and/or from the trailing edges of the stationary vanes or rotating blades. This allows auto-ignition. The patent also proposes that the hydrogen-rich fuel can be used for blade cooling before re-injection.

In what is called a reheat enhanced gas turbine, the California Energy Commission proposed a decade ago (Anon, 1997c) to reform or partially oxidise a mixture of steam and a combustible hydrocarbon (e.g. methane) to produce a hydrogen-rich fuel used for the primary combustor and a reheat combustor upstream of the turbine final stage. Auto-ignition of hydrogen-rich gas into the gas flow path between the stages is carried out.

8.3.10.7 Methane plate reformer coupled with a gas turbine

One of the concepts briefly examined during studies carried out in the last fifteen years (Reay and Ramshaw, 1992; 1998) concerned linking the reforming of methane in a catalytic plate reactor to a gas turbine combustion chamber. As the reaction is endothermic, a second set of channels is used to carry out the combustion of stoichiometric methane-air on a film of combustion catalyst deposited on the reactor wall. The importance to gas turbines is twofold – high pressure combustion gases can be expanded in the turbine to give more power than that needed to compress the methane-air mixture and the plate reactor can take place at a lower temperature, allowing a substantial reduction in NO_x.

The idea seems to have been copied in the US, with the California Energy Commission supporting work at Maryland and UC-Berkeley Universities, with a small grant, on a catalytic combustion reactor incorporating steam reforming. The secondary stream of H_2 produced would be used, it was suggested, for downstream flame stabilisation. Solar Turbines and Caterpillar Technology Centre were involved in taking this forward.

8.3.10.8 Partial oxidation gas turbine cycle (POGT)

This cycle involves an initial partial combustion reaction on the fuel gas at higher pressure. Both catalytic and non-catalytic partial combustion techniques can convert natural gas to a syngas mixture at reaction temperatures similar to current GT combustors, but at higher pressures – up to 100 bar. The hot syngas is then expanded down to normal gas turbine inlet pressure and fed to the combustor where

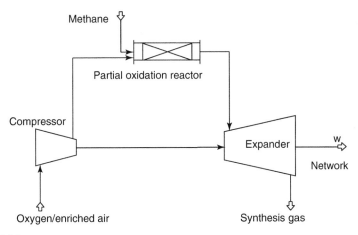

FIGURE 8.14

The POGT principle (Cornelissen et al., 2006).

the combustion is completed in the usual way. The cycle requires a novel high pressure expander. A POGT was patented by Jarix Company in Belgium (Anon, 1989). The POGT reactor is shown in Figure 8.14.

In a study by ECN in The Netherlands (Korobitsyn et al., 1998), it is indicated that the POGT has lower NO_x levels and also a reduced work output requirement from the compressor (as less excess air is needed). Later work in The Netherlands (Cornelissen et al., 2006) examined the application of the POGT to methanol synthesis and dimethyl ether, with an evaluation of the net power output included. It was concluded that a 12% gain in thermal efficiency and a cost decrease of 7% was achieved compared to competing solutions. However, the need to increase the capability of turbines in terms of turbine inlet temperature was highlighted as a prerequisite for further development.

8.3.10.9 Selected concepts

Several of the above systems are complex and require large integration efforts in order to be successful. The emphasis on the changing of heat exchangers to HEX-reactors in the UK study was seen as a first step towards using gas turbines to produce useful chemicals, and this approach is believed to be largely compatible with current gas turbine technology, as discussed below.

- Intercooling and reheating improve specific output rather than efficiency, although efficiency improvements would occur over a narrow range of cycle conditions.
- Recuperation improves efficiency, but there are difficulties in achieving the necessary heat transfer conditions, particularly for modern designs that have low exhaust gas temperatures.
- Intercooling or reheating can be used to overcome the recuperation heat transfer problem. Practical cycles would normally involve a combination of techniques which would always include recuperation.

- The current commercial preference is to use turbine exhaust heat in a steam bottoming cycle rather than for recuperation. The turbine plus steam bottoming cycle is known as a combined cycle.

Of the above features, the combination of heat input to the cycle with exothermic reactions in reheaters and recuperators was seen as producing substantial cycle efficiency increases and useful chemicals. This would involve a variety of heat exchanger reactors and similarly configured compact/intensified reactors – the effect of a reheater on efficiency, for example, is given in Table 8.2. The compact reformer/combustion unit, the intensified reactor, was attractive for emissions reduction and efficiency improvement.

The study concluded the following with regard to the gas turbine reactor in which the heat exchangers were converted into HEX-reactors.

Intercooling

- Low temperature endothermic reactions (less than 400 K) are required to avoid loss in efficiency.
- Reaction temperature is less critical with respect to specific output.
- The system works best with high compression ratios.
- It is the least attractive of the three techniques which have been considered because of the difficulties of finding suitable chemical reactions.

Table 8.2 Reheater Impact on Gas Turbine Efficiencies

Turbine inlet temperature: 1000°K
Pressure ratio: 10:1
Simple cycle efficiency: 29.9%

HP Turbine Outlet Pressure (Bar)	HP Turbine Outlet Temperature (K)	LP Turbine Inlet Temperature (K)	Cycle efficiency (%)
8	944	994	35.6
6	878	928	35.7
6	878	978	40.2
4	793	843	35.3
4	793	893	39.0
4	793	943	42.7
4	793	993	46.4
2	668	718	33.4
2	668	768	35.5
2	668	818	37.5
2	668	868	39.5
2	668	918	41.6
2	668	968	43.6

Recuperation

- It is independent of exhaust temperature (compared with conventional recuperation).
- Very high efficiency gains are possible.
- The process could be designed for a wide range of exothermic reaction temperatures.
- It is worthy of further consideration.

Reheating

- High efficiency gains are possible.
- The process could be designed for a wide range of exothermic reaction temperatures.
- The quality of exhaust heat increased, which improves the potential for conventional recuperation or the use of bottoming cycles.
- It is worthy of further consideration.

While it was difficult to identify specific gas turbines which might benefit from the technologies at this early stage, some observations can be made concerning the size of gas turbine reactors (based upon electrical output). These are provisionally assessed as:

- For petrochemicals and bulk organics – multi-MW.
- For fine chemicals – 100 kW plus.
- For offshore processing (e.g. gas–liquids, gas–other chemicals, gas–wire[5]) – 100 kW–10 MW.

As indicated, the selection of an endothermic reaction for the intercooler duty proved difficult. Obvious candidates operate at around 200°C, which is too high a temperature. It is possible that some bioreactions might be used, but this is considered a high risk exercise, and while compact bioreactors are under investigation, they are currently incompatible with conventional chemicals production infrastructures.

Blade cooling is necessary to improve cycle efficiency, higher turbine inlet temperatures being the goal of much turbine R&D. The importance to chemicals production is less clear cut. Endothermic reactions at high temperatures are common, and could be used for blade cooling.

8.3.10.10 The 'Turbo-cracker' – an ICI concept

Studies were carried out by ICI many years ago into using a gas turbine as a hydrocarbon cracker for ethylene production. The aim was to use the rapid heating and cooling available in a gas turbine to carry out a chemical reaction. The heat for an endothermic reaction is provided by adiabatic compression and rapid expansion in the turbine gives the rapid cooling necessary to maintain high

---

[5]Gas–wire is a term used in the sector for generating electricity from gas feedstock.

selectivity. The concept was revisited three years ago, and it was concluded that for a 600,000 tonne/annum cracker, ethylene costs could be significantly reduced.

The latest work on the Turbo-cracker was reported in November 2007 (Wang et al., 2007); this is an ongoing project at Cranfield University in the UK. Simulations carried out in 2007 showed that the principal advantage was that one Turbo-cracker could replace four to five conventional thermal cracking furnaces; it is also much smaller in size. Challenges to be addressed are the uncertain kinetics of thermal cracking of propane and naptha under higher pressures and the fact that compressor power demands are greater than that generated by the turbine.

8.3.10.11 Cogeneration of ethylene and electricity

The Energy Research Centre of The Netherlands (ECN) examined the feasibility of a number of routes to ethylene production, including the use of gas turbine reactors and heat exchanger reactors (Hugill et al., 2005). The aim of the study at ECN was to see if the economics of the production of ethylene from natural gas by the oxidative coupling of methane (OCM) could be improved, as at the time it was considered economically unfeasible. The study was based on the work of Swanenberg (1998), who looked at two ways of cogenerating ethylene and electricity. The route selected for analysis by Hugill and co-workers is shown in Figure 8.15.

The OCM reactor is a fluidised-bed unit (FBR), in this variant cooled using high pressure steam. The evaluation was based upon a plant using 211 kt/year of pure methane and 2408 kt/y of air (used for oxygen production) and producing 106 kt/y of ethylene and 44 MW of electricity. This combined plant was compared with the separate production of ethylene by conventional cracking and a gas-fired power generation plant. Realistic pricing of feedstock, electricity and plant were used to obtain an economic comparison between the two schemes. It was concluded that the cracker would be more profitable than a cogeneration plant, but the latter could become more attractive if the price of oxygen dropped and that of electricity rose.

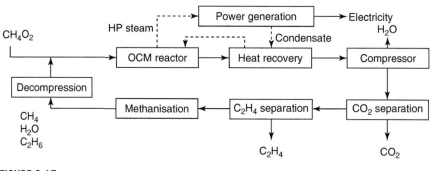

FIGURE 8.15

The cogeneration scheme proposed by Swanenberg.

The research team then looked at options for improving the cogeneration process, examining in particular what might be done to the OCM reactor. From the point of view of PI, the use of the fluidised-bed concept as a reactor should give good performance, but Hugill pointed out that a multi-stage packed-bed with inter-stage oxygen injection could lead to better selectivity. Bubble column reactors were also discussed.

Familiar to readers of earlier chapters is the HEX-reactor, proposed as another alternative to the fluidised-bed reactor. Potential advantages include a closer approach to isothermal plug flow, better catalyst utilisation, avoidance of catalyst attrition and possibly better selectivity. The unit would also be an order of magnitude more compact. One drawback might be the catalyst life in a HEX-reactor, as the FBR has partial catalyst replacement. However, since this report was produced, it is probable that the life of HEX-reactors could be sufficiently attractive to warrant reconsideration. Another variant suggested was the gas turbine reactor – differing from the main type discussed above in that the OCM reactor would replace the conventional combustion chamber so that electricity is generated from the heat of reaction. The problem perceived was that a pressure of at least 10 bar was needed in the reactor, which might give low selectivity. A further option is an oxygen-permeable membrane reactor; the ECN group believed that this could be a viable alternative to the FBR-based OCM. ECN also looked at the equipment needed for separations downstream of the reactor, but this is not addressed here.

Combining electricity generation and chemicals production, whether it involves PI or not, only makes sense if the commodity prices are appropriate. This was brought out by Cooper (2010) who examined the integrated gasification combined cycle (IGCC) in the context of its production of power and chemicals, using the syngas from the gasifier as feedstock for the latter. He concluded that (using current technology) the combined production of methanol and electricity that he examined only made sense if the electricity prices were low and methanol prices high – a situation that did not exist at the time. Nevertheless, the report did suggest that other chemicals could be more economically viable, and IGCC cycles are a useful way of dealing with the vast US coal reserves in an environmentally friendly way.

8.3.11 Other bulk chemical applications in the literature

Many other potential uses of process intensification have been proposed, some of which have been examined in some detail. They include the following:

- Methyl acetate from acetic acid and methanol.
- Hydrogen peroxide distillation.
- Liquid phase hydrogenation.
- Multi-purpose polymerisation reactor.
- Tubular reactor for polymerisation process for production of polyaniline.

A wide-ranging assessment of potential applications of micro-structured reactors by Loewe and Hessel (2006) indicates other areas where PI could be applied, including the following, some of which are categorised as fine, rather than bulk, chemicals:

- Fluorination of aromatic compounds.
- Carboxylations.
- Bromination of thiophene (giving a solvent-free process route).
- Continuous manufacture of nitroglycerin (as a medicine).
- Polymerisations (applied by DSM Fine Chemicals).

8.4 Fine chemicals and pharmaceuticals

As discussed in Chapter 2, Section 2.4.4, the pharmaceutical and fine chemicals industries are under increasing pressure to respond more quickly to market opportunities and changes. For example, once a bioactive molecule has been discovered and patented, there are roughly 20 exclusive years within which heavy R&D costs can be recovered and profits can be earned, but approval by authorities is needed at each scale-up stage.

In order to realise the vision of a single unit for R&D and production, it is imperative that the entrenched batch/stirred vessel industrial culture is overcome. This should be done by convincing the research chemists that the new PI technology is sound, allowing the 'desktop' process to be adopted *when new molecules are discovered*, rather than being imposed after the process has been developed and based upon laboratory-scale stirred vessel.

Taking as an example pharmaceuticals, the key factors are product cost, product quality and process reliability. Regulatory risks associated with safety, the environment and good manufacturing practices also have to be considered. Quick transfer of R&D to the manufacturing process is very important. There are obvious advantages in PI here. PI can deliver improved and consistent product quality, lower manufacturing investment, reliable and simplified technology transfer, high safety and minimisation of the environmental impact. The critical features are:

- Laboratory-scale to full-scale in one step!
- No inter-batch cleaning.
- Longer effective patent life.
- Just-in-time manufacture.

8.4.1 Penicillin extraction

Penicillin extraction is an example of the application of process enhancement via electrostatic fields. Solvent extraction as a means of recovery and purification of penicillin-G and a number of other antibiotics is well-established for these products. Use of liquid–liquid extraction involves operation at close to ambient temperature and equipment-intensive processes such as precipitation and crystallisation can be kept to a minimum. The high selectivity offered by chemically suitable solvents has furthered the exploitation of extraction as a recovery route. In the case of penicillin-G (pen-G), extraction into a number of solvents is well known and these include n-butyl acetate, amyl acetate, chloroform, and methyl iso-butyl ketone.

Substantial enhancements of the extraction rate of pure pen-G from buffer solution and of crude pen-G from untreated mycelial culture broths into dichloromethane at pH 4 are achievable at laboratory scale in continuous spray extraction columns. Untreated broths of moderately high viscosities (up to 6 cP) may be successfully processed by this technique. The electrostatic forces, in this case applied at the spray nozzle, exceed intermolecular forces holding large droplets together. The droplets then disintegrate into an electrostatic spray of micro-droplets, which allows a substantial increase in mass transfer. This greatly speeds up – by up to an order of magnitude – the extraction by the solvent.

Figure 8.16 shows the relationship between the mass transfer of product (K_{da}) and applied voltage for penicillin extraction in a 25 mm diameter × 0.65 m height spray column. The importance of electrostatic spraying in the improvement of overall mass transfer rate is clearly demonstrated (Weatherley, 1993).

Much of the activity reported above took place several years ago, and recent papers (see for example Kirschneck, 2007) still highlight the reasons why the sector is reluctant to introduce innovative technologies – the author, who is with the company Microinnova Engineering in Austria (see Appendix 4) cites the regulation of the pharmaceuticals sector as a primary handicap. The author reports on laboratory studies that have shown a yield improvement potential of about 15% and argues that, based on the dominance of the chemicals and solvents in the production costs (typically 80% of the total), the payback period on the investment in PI plant can be less than two years – although many in industry will have come across numerous

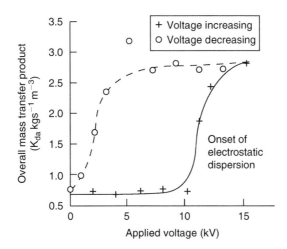

FIGURE 8.16

Electrically-enhanced extraction of penicillin-G into methylene chloride.

cases where a payback period of the order of two years is considered unacceptable. Microinnova has developed liquid–liquid applications with throughputs of 3 tonne/h (operating from 2005) and scale-up has successfully been carried out for liquid–liquid and liquid–gas reactions.

8.4.2 AstraZeneca work on continuous reactors

The pharmaceuticals company AstraZeneca has recently reported on research that has taken place over a number of years in addressing a switch from batch or semi-batch processing to continuous operation. Wernersson (2007) stated that continuous stirred tank reactors (CSTRs), tubular reactors and micro-reactors had been evaluated. Substantial testing has also been done using the Alfa Laval plate reactor (see Chapter 5). Critical variables examined as a function of reactor type have included heat transfer capacity, mixing capacity and flow characteristics.

Applications of Alfa Laval plate-type continuous reactors to pharmaceutical synthesis were for a liquid–liquid reaction where an alcohol is oxidised to an aldehyde using sodium hypochlorite – resulting in an increased yield compared to the batch operation – and a ring closure reaction. Here the reaction was carried out at a temperature above the solvent's normal boiling point, and again a substantially higher yield was obtained. When reactions were undertaken in micro-reactors, very high heat transfer capability was observed, but these would be used in different applications to the Alfa Laval reactor.

8.4.3 Micro-reactor for barium sulphate production

Among the interesting PI developments at Newcastle University in the UK is a project to use micro-reactors to produce barium sulphate particles. These are used in papermaking, paints and pigment production, and are ideally of less than 20 microns in size. The work on micro-reactors follows earlier trials using spinning disc and spinning cone reactors. The team (McCarthy et al., 2006) concluded that the aqueous precipitation of barium sulphate gave mean crystal sizes as low as 0.2 microns diameter in micro-reactor channels of 0.5 mm internal diameter. Conversions in excess of 60% could be achieved under certain supersaturation conditions. It was concluded that the potential of narrow channel based micro-reactors for providing a continuous source of ultra-fine particles for many applications was considerable. Scale-up to many thousands of channels operating in parallel could give large throughputs of value to industry. Another barium compound required in particulate form can be made using SDRs, as described below, and further data on crystallisation of compounds such as barium sulphate are given in Chapter 6.

8.4.4 Spinning disc reactor for barium carbonate production

Barium carbonate is used as a raw material for synthesising other ceramics used in electronics and magnetic materials. The opportunities for its use could be increased

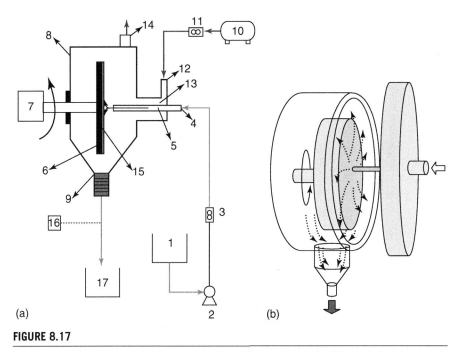

FIGURE 8.17

(a) A schematic diagram of a high gravity system for producing $BaCO_3$ particles via a carbonation route; (b) the SDR.

if an economical production route for the manufacture of fine $BaCO_3$ crystals could be developed. Tai et al. (2006) demonstrated the production of barium carbonate using a spinning disc reactor (SDR).

The equipment consisted of a 12 cm diameter SDR enclosed in a cylindrical acrylic chamber (see Figure 8.17). CO_2 was fed with the barium hydroxide slurry into the chamber, which formed a barium carbonate slurry, the latter being directed at the centre of the disc (via a distributor) that rotated at speeds of 600–2100 rpm. On the disc the dissolved barium carbonate reacted with the absorbed CO_2 to produce barium carbonate particles.

The influence of rotation speed and feed rate on the segregation index is shown in Figure 8.18. The segregation index, which is low for good mixing (zero for perfect mixing) and high (unity) for poor mixing, shows that near-perfect mixing occurs throughout the range of disc speeds for low feed rates, and improves with speed for 900 ml/min flow rates. The low feed rate also was compatible with the smallest particle sizes.

In Chapter 6, reference is made to the spinning cone for making barium sulphate in crystalline form, and details are given concerning the process parameters used and measurements made during the experiments.

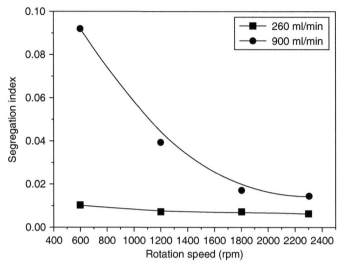

FIGURE 8.18

A comparison of the segregation index for two feed rates of the slurry under different rotation speeds.

8.4.5 Spinning disc reactor for producing a drug intermediate

The intense heat and mass transfer environment that can be established within the liquid film flowing over the disc allows high selectivities and conversions to be achieved when fast liquid-phase reactions are performed. Very encouraging results were achieved in an industrial study of a phase-transfer-catalysed (p-t-c) Darzen's reaction to produce a drug intermediate, as described below.

Almost a decade ago SmithKline Beecham (SKB) – now Glaxo SmithKline – set up a development group to look at new processes and carry out trials on pilot plant. As a leading manufacturer of pharmaceuticals, challenges such as shorter market exclusivity time for drugs and an increasing flow of competing preparations suggest that an obvious activity within R&D to meet these challenges would be PI. In a rapidly moving sector – 15–40% of current sales value arises from new drugs approved in the last five years and R&D represents up to 20% of the value of sales – investment decisions by the company to maintain its position in what is a worldwide market then worth £200 billion needed to be correct! In pharmaceuticals, the following considerations are of prime importance:

- Product cost, quality and process reliability.
- Regulatory risks (safety/environment/good manufacturing practices).
- Manufacturing investment.
- Rapid technology transfer for R&D to production.

Within the sector there is a mix of batch and continuous processes, the former offering flexibility (possibly overcome in continuous mode using the spinning disc

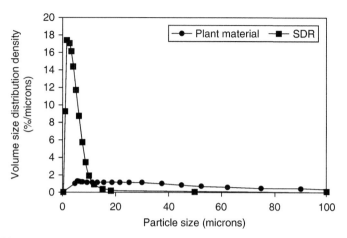

FIGURE 8.19

Recrystallisation of an API – comparison of size distributions.

reactor (SDR)), and low-volume production. Where continuous processes can be improved using PI, and where batch operations can become continuous by using PI technology, the following benefits were seen for the company:

- Improved and consistent product quality.
- Lower manufacturing investment.
- Reliable and simplified technology transfer – from R&D to production.
- Safety and minimisation of the environmental impact.

The spinning disc reactor (see also Chapter 5) was used by the company to produce an active pharmaceutical ingredient (API) that was required in the form of crystals with a specific size distribution (Oxley et al., 2000). The basis of the work was a 15 cm diameter spinning disc, on which crystallisation was induced by adding an antisolvent. The improvements observed for the SDR output, in terms of crystal size distribution, are shown in Figure 8.19, where a comparison is made with the normal plant material. Overall, the PI solution compared to 'conventional' production methods gave the following quantified benefits:

- 99.97% reduction in reaction time
- 99.21% reduction in plant volume
- 93.33% reduction in impurity level
- 6.83% reduction in reaction temperature

It was found during the trials that the stainless steel disc was subject to crystal scaling after a few runs. However, a thin layer of PTFE suppressed this without significantly impairing the disc's heat transfer performance. (For the reader interested in the specific needs in the pharmaceuticals sector, papers such as that by Hall and Stoker (2003) provide useful guidance on plant design.)

8.4.6 **SDR in the fragrance industry**

A more recent study has involved a 20 cm diameter SDR with a catalytically-activated surface to perform the rearrangement of α-pinene oxide to campholenic aldehyde (Vicevic et al., 2001), which is an important intermediate used in the fragrance industry. The comparative performance of the batch reactor and the SDR is shown in Table 8.3. For equivalent conversion and selectivity, the unoptimised SDR gave a much higher throughput than the equivalent batch reactor and avoided the need to separate catalyst slurry from the product. Figure 8.20 shows the variation of selectivity as a function of disc speed and feed flow rate; Figure 8.21 gives the conversion levels achieved. While conversion falls from 100% at the higher flows and speeds, presumably due to the reduced liquid residence time on the disc, the selectivity increases. Thus it might be expected that a larger disc (or a sequence of

Table 8.3 Comparison of the Best SDR Runs with Batch Results for Conversion of α-Pinene Oxide to Campholenic Aldehyde

	Batch Process	**SDR (Continuous)**
Process time (s)	300	1
Processed feed (kg/h)	1.2	209
Conversion (%)	100	100
Selectivity (%)	64	62
NOTE:	Catalyst separated from the product mixture	No loss of catalyst

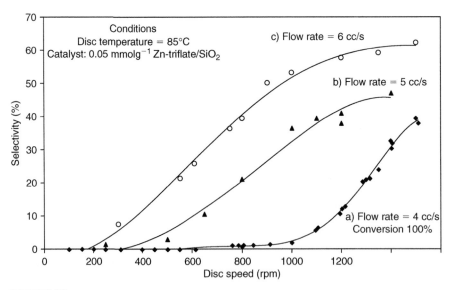

FIGURE 8.20

Selectivity towards campholenic aldehyde at 85°C, at various feed rates.

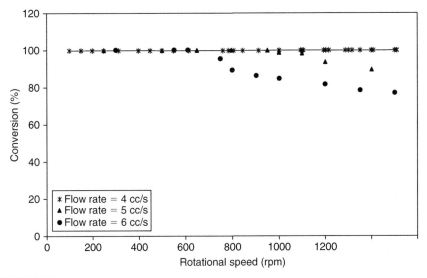

FIGURE 8.21

Conversion of α-pinene oxide at 85°C, and various disc speeds.

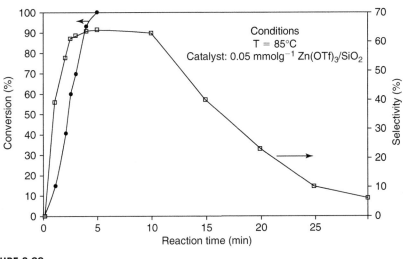

FIGURE 8.22

Batch reaction: conversion and selectivity towards campholenic aldehyde.

small discs) could combine high conversion and high selectivity. The batch reactor performance is summarised in Figure 8.22, where it can be seen that 100% conversion requires five minutes (equals one second on the disc) and a maximum selectivity of 65% is reached.

Table 8.4 Comparison of the Reactions Using Conventional and Microwave Heating (Wharton, 2011)

Reaction	Time (Conventional)	Time (Microwave)	Temperature (°C)	Yield (%)
Suzuki coupling	2 h		120	92
		2 min	160	99
		1 min	155	98
Dihydroprimidine	8 h		120	35
		4 min	145	71
Ionic liquid	4 h		120	95
		1 min	200	95

8.4.7 A continuous flow microwave reactor for production

As will be obvious from other chapters, the use of microwave energy can make a substantial contribution to speeding up processes such as drying and reactions and improving product quality. C-Tech Innovation in the UK reported recently (Wharton, 2011) on their development of a large-scale continuous-flow microwave reactor that can process up to 1 t/day. Broadly, the C-Tech unit uses a travelling wave microwave applicator that enables the designer to accommodate high powers and relatively large bore pipes for carrying the material being processed – and the designers point out that (as with other PI plant) it allows one to go from the laboratory to production without long-winded development steps.

Three reactions were initially investigated:

1. A Suzuki reaction.
2. A dihydropyrimidine cyclisation.
3. An ionic liquid synthesis.

The impact of using the microwave heating is illustrated in Table 8.4.

The impressive reaction time reductions using the microwave unit led to substantial energy savings. Data were used by the author that Croda had supplied, on the basis of their average batch reaction energy consumption, and a comparison suggested that for the Suzuki and ionic liquid cases, the microwave process used 89% less energy.

8.4.8 Ultrasound and the intensification of micro-encapsulation

Pharmaceutical products are being increasingly 'micro-encapsulated', the formulations being enclosed in very small particles formed by coatings of a 'wall' material around the active ingredient. Dalmoro et al. (2012) in their most interesting review, state that this is done for a variety of reasons – the one with which most of us recognise is to minimise the taste of the drug, or possibly to ease its passage into the bodily distribution system.

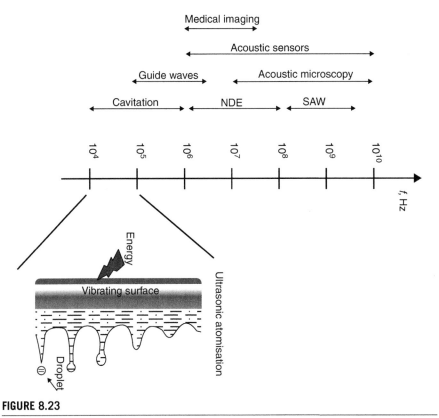

FIGURE 8.23

Ultrasonic frequency ranges, type of application and ultrasonic atomisation mechanism (Dalmoro et al., 2012).

There are several methods used to micro-encapsulate active ingredients, ranging from solvent evaporation to spray drying and fluidised beds. Some of these sound as if they fall within the PI range of methodologies already, but the researchers highlight ultrasonic atomisers as the key to their micro-encapsulation process. As identified in Chapter 3, ultrasound can operate at a frequency within a very large bandwidth – from hundreds of Hz to many MHz. As shown in Figure 8.23 the process can occur over a range of frequencies. The authors cite several research programmes covering a variety of frequencies. Bovine serum albumen (BSA) microspheres were made using an ultrasound atomiser working at 100 kHz, giving droplets of 20 microns diameter at a rate of 0.1–1.0 l/h. By combining spray drying and atomisation at a similar frequency, with a similar product, the drying chamber size could be much reduced and much higher yields (>80%) were achieved compared to spray drying alone. As with other PI techniques, such as SDRs for crystallisation, the reproducibility of the particle size was an added benefit – in this instance the mean diameter averaged 18.25 microns, with a deviation of plus or minus 1.05 microns.

Other products produced with the aid of ultrasonic micro-encapsulation include anti-inflammatory drugs and 'gentle' processing of protein formulations. For the reader interested in a more detailed analysis of extraction using ultrasound intensification – here for extracting glycyrrhizic acid from licorice root, see Charpe and Rathod (2012).

8.4.9 Powder coating technology – Akzo Nobel powder coatings Ltd

Akzo Nobel, working in the North East of England with Newcastle University (Dissanayake et al., 2012), has been investigating how PI could be used to improve the manufacturing route to powder coating (or powder paint). Powder paint is used widely as an alternative to solvent-based paints, which of course create environmental and health concerns. Thus by using powder coating, VOC emissions are eliminated.

The current production process employs a number of unit operations in order to carry out dry mixing, melt extrusion, flake granulation and fine grinding, to name a few. As with many such processes, they suffer from essentially random particle size distributions and shape, which affects the uniformity of the powder coating film. There are other drawbacks to the process that also affect the finished quality.

The intensification process that has resulted in better pigment dispersion and a much more even distribution of particles is based upon non-isothermal, flow-induced phase inversion – FIPI. Phase inversion involves transition between, for example, an emulsion of water-in-oil and an oil-in-water one. Flow-induced phase inversion is being investigated to produce nanoparticles for drugs, for example. Akay, the co-author working at Newcastle University, found that non-isothermal FIPI differed from the isothermal FIPI, and the former is characterised as, in the authors' words: "*...a powder coating composition in (the) molten state is phase inverted (crumbled) by a thermomechanically induced melt fracture to produce particles by subjecting the melt to a well-defined flow field within small processing volumes.*" It was found that rapid cooling could allow the formation of much smaller particles, as well.

The results were sufficiently attractive in terms of particle size span and flowability to suggest that the process, subject to further development, could have commercial scale potential.

8.4.10 Chiral amines – scaling up in the Coflore flow reactor

One of the few recent new reactor types that has entered the market in recent years is the Coflore agitated tube reactor (ATR). This is a dynamically mixed reactor which relies on loose agitator elements (see Section 5.5.4 in Chapter 5) and mechanical shaking of the reactor body to ensure effective mixing. In overcoming some of the drawbacks of conventional rotating mixers in tubes, Coflore reactors have demonstrated low pressure drop, high mixing efficiency with long reaction times and the ability to handle slurries and gas/liquid mixtures.

A full description of the biochemistry involved in the reaction is given by Gasparini et al. (2012), but the basis is selective oxidation biocatalytically of the D-amino acid flavin adenine dinucleotide (FAD). The outcome is L-amino acid.

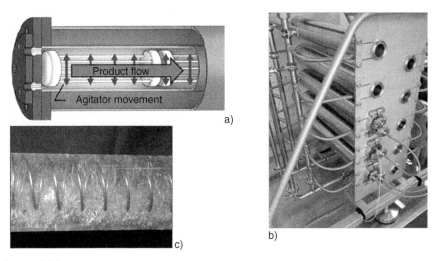

FIGURE 8.24

The Coflore ATR. (a) Section of a tube with agitator; (b) assembly; (c) gas/liquid mixing (Gasparini et al., 2012).

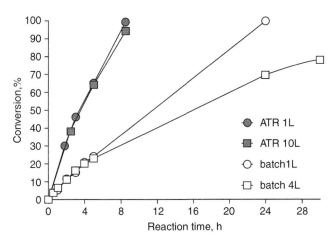

FIGURE 8.25

Effects of scale up in continuous and batch operation. Enzyme loading = 21 g/l, batch stirrer speed = 400 rpm, ATR agitation frequency = 2 Hz, oxygen flow = 0.25 l/min/l_{vessel} for batches and ATR 1l. For 10l ATR, oxygen flow = 0.075 l/min.

The Coflore reactor is illustrated in Figure 8.24 where the tube containing the agitators is shown. These give strong radial mixing when the tubes are subjected to lateral shaking – gas/liquid mixing being shown in Figure 8.24(c). With regard to the outcome Figure 8.25 shows the effect of scale up in both continuous and batch modes. Throughputs for the ATR are up to 10l and it can be seen that conversions are good and time substantially reduced. For the 10l throughput, the oxygen consumption is much reduced, too. The commercial implications of this are reduced capital cost and operating costs, and reduced catalyst consumption.

8.4.11 Plant-wide PI in pharmaceuticals

As discussed elsewhere in this book, the concept of plant-wide application of process intensification, rather than 'tinkering' with a single (or a few) unit operation(s) is the ideal approach in order to maximise the benefits. This has been suggested by Patel et al. (2011) working at Imperial College London and AM Technology (who feature above with their novel reactor types).

The authors suggest that PI and process systems engineering are becoming aligned, and they have analysed the route to manufacture of Ibuprofen compound as one where this alignment will permit consideration of PI in the complete primary manufacturing stage, which also involves crystallisation – not just in the reactor. They hint at the extension of this to other novel pharma production lines – an exciting scenario for chemists and chemical engineers.

The reader interested in PI using pervaporation will find the Conference paper by Clavey (2011) of interest. He reports on work at DSM Nutritional Products on an energy-efficient route to carrying out the acetalisation reaction, without leaving large amounts of water to be removed. A 49% increase in conversion rate from a carbonyl compound was achieved, too.

8.5 Bioprocessing or processing of bioderived feedstock
8.5.1 Transesterification of vegetable oils

An interesting review of cavitational reactors (see Chapter 5) by Gogate (2007) highlights trans-esterification to produce biodiesel from vegetable oils. The study reported by Gogate involves using acoustic and hydrodynamic cavitation methods to the alky esters of fatty acids that are used as the basis of biodiesel fuels.

The reactor used for the sonochemical work was a conventional ultrasonic bath that many readers will recognise as a piece of equipment for cleaning components. The hydrodynamic cavitation reactor was a vessel with an orifice plate in the main feed line to generate cavitation. While fully acceptable yields of 97–99% were achieved with the sonochemical reactor (and also, except for the case of peanut oil feedstock, with the hydrodynamic cavitation reactor), the most interesting data related to the energy used. These are compared in Table 8.5. The energy efficiency is defined as the yield in kg of product per kJ of energy used in the reactor.

8.5.2 Bioethanol to ethylene in a micro-reactor

One of the stimulants to the use of bio-based feedstock in chemical manufacture is the increasing price of oil. The attractiveness of using renewable resources as the

Table 8.5 Comparison of Energy Efficiency for Biodiesel Production

Technique	Time (min.)	Yield (%)	Energy Efficiency (kg/kJ)
Acoustic cavitation	10	99	8.6×0^{-5}
Hydrodynamic cavitation	15	98	3.37×10^{-3}
Conventional stirred vessel (electric drive to stirrer)	180	98	2.27×10^{-5}
Conventional with a heater to sustain reflux	15	98	7.69×10^{-6}

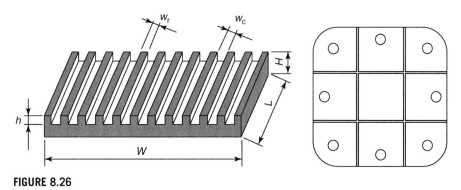

FIGURE 8.26

Micro-channel reactor configuration. W, width of total channels 44.5 mm; w_c, width of single channel 1.0 mm; w_f, width of fin 0.5 mm; h, depth of channel 1.25 mm; H, thickness of chip 2 mm; L, length of channel 30 mm.

basis of the production of essential chemicals should also not be underrated. Chen et al. (2007), in China, have successfully used micro-channel plate reactors to produce ethylene via the catalytic dehydration of ethanol over $TiO_2/\gamma\text{-}Al_2O_3$ catalysts.

The cracking furnaces commonly used in industry for ethylene production are rather large and give relatively low yields of ethylene at high reaction temperatures. The resulting high energy use and low utilisation of plant capacity, point out Chen and his colleagues, leads to increasing interest in the catalytic dehydration route, which so far has been largely confined in its application to Brazil and India. The micro-reactor geometry used in the Chinese work is illustrated in Figure 8.26.

8.5.3 Base chemicals produced from biomass

For the reader interested in the latest ideas concerning how biomass can be used as the basis for production of bulk chemicals, it is strongly recommended that he/ she consults the work carried out by a group of four universities in the Netherlands, and the University of York in the UK. In an extensive paper (Sanders et al., 2012), the eight authors present a number of case studies, including the production of hydrogen from waste biomass streams and the production of base chemicals from amino acids. A novel combination of microwaves and supercritical extraction is also proposed for green and sustainable processing.

8.6 Intensified carbon capture
8.6.1 Introduction

An area of chemical engineering technology that has perhaps grown most rapidly within the past few years is that of carbon capture. The main aim of carbon capture technology is to reduce the CO_2 emissions from all plants that burn fossil fuels – steelworks, offshore facilities, chemical plants and, perhaps the largest challenge, power generation plant on shore, in particular, but not limited to, coal-fired power stations. Once the effluent is captured, it is then sequestered or stored, hence the acronym CCS that is most commonly use to define the whole process from capture to, in some cases, locking the CO_2 away in holes below the sea bed (such as depleted gas or oil fields).

8.6.2 Carbon capture methods

There are a number of approaches to carbon dioxide capture (or minimisation of its production in the first place). Several methods, if not all, have been of interest to process intensification engineers, not least because of the massive capital and running costs (CAPEX and OPEX) associated with such plant, particularly on large industrial processes and power generation plant.

Some post-combustion carbon capture technologies are listed in Figure 8.27, and Table 8.6 gives some relative merits of the technologies.

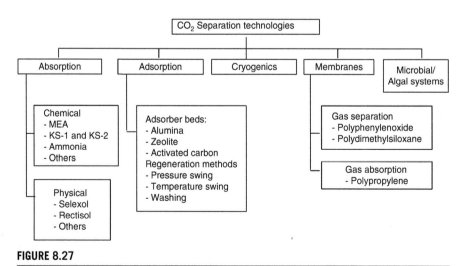

FIGURE 8.27

Post-combustion carbon dioxide separation technologies (Fraser, 2011).

Table 8.6 Advantages and Disadvantages of Some Post-Combustion Capture Methods (Fraser, 2011)

Technology Option	System Requirements	Advantages	Disadvantages
Chemical absorption	Absorber and distillation sections. Chemical solvent (MEA, DEA, HPC, KOH, dNaOH)	– Suitable for dilute CO_2 streams i.e. power plant flue gas – Operates at ambient pressures – Somewhat commercially viable and proven efficiencies	– High absorption energy demands – Solvent degradation is a common issue yet to be fully resolved
Physical absorption	Absorber and separation sections. Physical solvent(Selexol)	– Less energy intensive – Solvents are less susceptible to degradation	– Requires high operating pressure – Works better with gas streams having high CO_2 content
Adsorption	Adsorption bed	– Very high CO_2 removal is possible	– Requires very high operating pressures – Costly
Membranes	Membrane filter	– Efficient – High surface/volume ratio	– Requires very high operating pressures – Very costly

8.6.2.1 Pre-combustion capture

Pre-combustion carbon capture involves the removal of carbon dioxide before any combustion takes place. This process is generally made up of three stages:

1. Conversion of the fuel to hydrogen and carbon monoxide.
2. Reaction of the carbon monoxide with water in a 'shift conversion' to produce carbon dioxide.
3. Separation of the carbon dioxide from hydrogen to produce an emission-free combustible fuel, typically syngas from coal. The removed CO_2 is generally liquefied and sent to a storage facility (Rutherford and Simons, 2010).

8.6.2.2 Post-combustion capture

Post-combustion carbon capture involves sequestering carbon dioxide from flue gas following combustion. This involves transforming a low pressure/concentration stream of CO_2 into a more concentrated process stream via methods such as absorption. This steam is then usually compressed to allow for transportation or further processing. A simple CO_2 treatment process sees the flue gas treated with a gas scrubber, where the absorbent within the column takes up the CO_2 leaving a cleaned gas stream. The absorbent and CO_2 then pass to a regeneration unit which heats the mixture and releases a gas stream of high purity CO_2. Typical absorbents for this process are amines or carbonates. More data and flow sheets are given in Section 8.6.3.

The main advantage of post-combustion technology is that it can be incorporated to existing power plants, allowing a reduction in infrastructure and less of an impact on the environment. It has attracted the most attention in the context of PI.

8.6.2.3 Oxy-fuel combustion

The third method of carbon capture known as oxy-fuel combustion capture involves the combustion of a fuel in pure oxygen, rather than air which is predominantly nitrogen. This itself is a form of process intensification as the removal of the non-combustible nitrogen within the air allows for a smaller volume of gas for the same volume of fuel, resulting in a reduced inventory of equipment. The removal of nitrogen also allows for a reduced energy requirement to bring the gas up to the correct combustion temperature.

The oxy-fuel combustion process gives more complete combustion, which results in the flue gas being more pure in carbon dioxide. Conventional separation of oxygen and nitrogen is often very energy intensive and therefore defeats part of the purpose of the intensification process. However a method known as chemical looping can be used, discussed later.

A selection of the most common post-combustion capture methods is given in Figure 8.27. It is obvious from the processes involved, for example absorption, adsorption and membranes, that PI could have much to offer carbon capture. A useful comparison of three principal techniques, absorption, membranes and chemical looping is given by Rezvani et al. (2009).

8.6.3 Intensification of post-combustion carbon capture

It is fair to say that the economic removal of carbon dioxide from power plant flue gas is probably one of the most urgent problems facing the energy industry at the present time. The gas and liquid flows are colossal for a typical 500 MWe unit, being about 600 kg^3/s flue gas and 4 t/s liquid. In addition, the power needed to drive the capture process using conventional technology is likely to be a substantial fraction of that being generated. These two features impose severe demands on any intensified approach to this problem. For less onerous gas cleaning duties the conventional approach is to use an appropriate liquid to absorb the CO_2 and then to steam-strip the strong liquor to recover virtually pure CO_2 for use elsewhere or for sequestration in underground reservoirs. The absorbent usually chosen is Monoethanolamine (MEA) in a 30% water solution which reacts reversibly with the CO_2. Absorption occurs in a low temperature (~40°C) mild steel column and the stripping operation is performed in a separate column/reboiler at >100°C. A typical flow sheet is shown in Figure 8.28. The MEA solution concentration is determined by the maximum liquid viscosity and corrosivity which can be tolerated by the mild steel absorption and stripping towers as both of these properties increase significantly with concentration. An intermediate heat exchanger is incorporated between

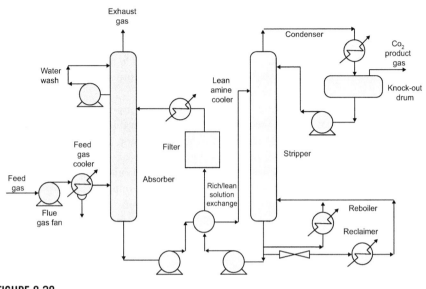

FIGURE 8.28

Post-combustion carbon capture using MEA – typical plant layout.

the absorption and the stripping towers to recoup heat from the hot liquid returning to the stripper.

8.6.3.1 The absorber

It will be recognised that the absorption, stripping and boiling/condensation functions which are the key elements of the capture process, are all multi-phase operations which are highly susceptible to an enhanced acceleration environment. Since the rotor needed for given duty is much smaller than a standard mild steel column, relatively corrosion-resistant stainless steel can be considered as an economic alternative construction material. Any problem associated with liquid viscosity can be overcome by the higher shear forces which can be generated in the rotor. This immediately raises the prospect of operating the process in one or more packed rotors, with the gas and liquid streams in co-current or counter-current flow. A key advantage of the former approach is that the rotor will propel the flue gases up the stack, rather than impose a parasitic pressure drop as in a conventional absorption column. This avoids the need for a large and powerful gas blower. Whether this can be done depends entirely on the kinetics and equilibrium characteristics of the absorption process.

Thus, for example, if the reaction kinetics are so fast so as to be considered to be instantaneous, then the gas and liquid composition on either side of the interface can be considered to be in equilibrium. If, in addition, the CO_2 solubility in the liquid phase is high, then the absorption process will be solely controlled by the fluid dynamics in the gas film and independent of those in the liquid phase. In this case the gas phase CO_2 concentration at the interface is close to zero. This situation is likely to prevail when the amine concentration and the reaction temperature are high. As pointed out earlier practical access to the use of high amine solution concentrations, with their associated viscosity and corrosion problems, is facilitated by the elevated acceleration fields generated within rotors.

The difficulty of the mass transfer duty is measured by the number of 'transfer units' (NTU) involved. For the situation above when a flue gas concentration of around 15% is to be reduced to about 1.5%, the expression for the NTU involved reduces to $\ln(10)=2.3$. Unpublished work by Ramshaw at ICI using the ammonia–water system, another gas film controlled absorption process, showed that the height of a transfer unit was approximately 1.5 cm. This implies that a rotor packing depth of <5 cm would suffice for our present purpose. Strong experimental support for this interpretation of the absorption process is provided by the work of Chambers and Wall (1954) who tested a rotating vaned disc which was interleaved with a stationary equivalent, to absorb CO_2 from a gas stream, using neat MEA. The liquid was fed to the centre of the disc and moved outwards in a series of sprays from the tip of the vanes. Further support is provided by Jassim et al. (2007) who explored the behaviour of a packed 40 cm diameter rotor using MEA solution concentrations ranging from 30% w/w to 100%. They observed a dramatic fall in the HTU with MEA concentrations above 70% w/w This was taken to imply that the use of higher amine concentrations did indeed accelerate the reaction kinetics to the degree needed to satisfy the above hypothesis.

Preliminary calculations for the operation of a rotary packed absorber – a rotating packed bed (RPB) – quickly reveal the importance of having a low pressure drop within the packing, primarily to reduce the power loss. Even with a packing depth conservatively estimated as 10 cm, the packing volume fraction must be only ~1%. At the same time the packing must be mechanically robust enough to withstand the stresses imposed by the rotation, without significant deformation. A packing comprised of knitted wire or a very open woven structure could well be an appropriate solution to this problem. Previous work by Burns and Ramshaw (1996) and Burns et al. (2000) with a relatively high density packing (>5% v/v), showed that there were two principal flow patterns depending on the flowrate and the centrifugal acceleration. At high flow and low acceleration the liquid moved over the packing as a film whereas at low flow and high acceleration, droplets tended to detach and jump from one packing element to the next – hence forth described as the 'flying drop mode'. In the first case the area available for mass transfer corresponds roughly to that of the packing, and in the second case the area is governed by the size of the drops detaching from the wire packing elements – a situation analogous to dew on a spider's web.

An inspection of the extensive solubility data provided by Jou et al. (1995) shows that for the CO_2 loading range which is likely to be of interest – say 0.2–0.4 moles CO_2/mole MEA, the equilibrium CO_2 partial pressure above a 30% MEA solution at 40°C is small compared with that in the gas phase. Since the CO_2

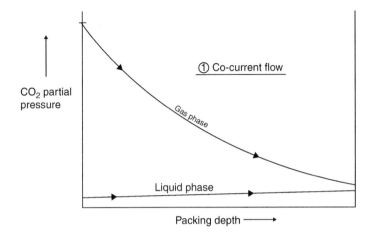

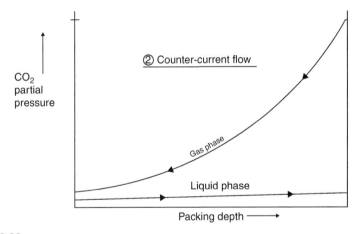

FIGURE 8.29

Concentration profiles for co/counter current flow.

partial pressure above the liquid phase is small compared with that in the gas phase, the profile of concentration difference driving the mass transfer is similar for both co- and counter-current contact – see Figure 8.29. This implies that either co- or counter-current gas/liquid flow is equally effective for performing the absorption duty. In view of this and the comments above co-current flow is chosen, with the liquid and gas flowing radially outwards and then discharging into a volute casing to recover as much of the gas pressure head as possible.

A further advantage of using concentrated amine solution is that it allows much smaller liquid flows to be used because the CO_2 carrying capacity of each cubic metre of concentrated solution is so much greater than that of the conventional 30% solution usually employed. This has important ramifications when sizing the rotors. Preliminary calculations suggest that the absorption duty for the flue gas from a 500 MWe gas fired unit could be performed by two 4 m diameter rotors approximately 4 m in length. This compares with a conventional process needing two 20 m towers of 9 m diameter.

8.6.3.2 The reboiler/condenser

The function of the reboiler/condenser is to steam-strip CO_2 from the strong MEA solution arriving from the absorber. The resultant steam/CO_2 mixture is then condensed to give pure CO_2 and water condensate. The process must be driven by an outside source of steam in order to maintain a water balance and supply the heat needed to reverse the absorption reaction and provide the latent heat of vaporisation. Since three 2-phase processes are involved (boiling/condensation/stripping), they can in principle be operated more intensely in a rotating unit. Clearly, the reboiler needs to be mounted outboard of the stripper packing so that the steam can travel inwards to perform the stripping function, as shown schematically in Figure 8.30. It is proposed to locate an annular plate reboiler near the rotor periphery so that it can partially vaporise the stripped solution percolating through from the stripper packing.

Rather than running full the reboiler is designed to operate on a film basis so that both the boiling and condensation functions are performed filmwise on opposite sides of the heat exchanger plates. In this respect the operation is broadly similar to the heat and mass transfer to thin liquid films on spinning discs which is known to be very effective (Aoune and Ramshaw, 1999). This design allows full advantage to be obtained from the enhanced acceleration field at the rotor periphery. The stripped solution, now with a depleted CO_2 content, leaves the rotor via an overflow weir, the radial location of which determines the liquid level within the rotor. Since the stripper operates at a much higher temperature than the absorber (typically ~100°C rather than ~40°C), its performance is even less likely to be impeded by the reaction kinetics than that of the absorber.

8.6.3.3 Intermediate heat exchanger

An intermediate heat exchanger is needed between the stripper and the reboiler to partially recover heat from the hot stripped solution returning to the absorber, thereby maximising the thermal efficiency of the overall process. Since no phase change occurs in this intermediate exchanger its function can probably be best performed intensively by a printed circuit type of unit, such as that manufactured by Heatric Ltd, provided that the liquid streams remain free of potentially clogging particles. This design of exchanger depends for its effectiveness on the high area density and small channel dimensions which can be achieved, though the relatively high viscosity associated with strong MEA solutions must be carefully considered. As already noted this high solution strength is associated with much lower liquid

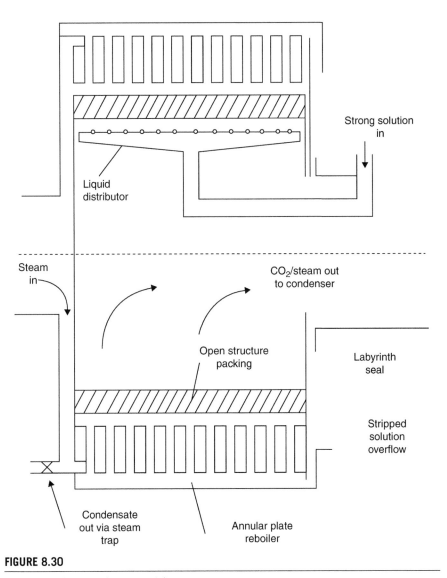

FIGURE 8.30

The reboiler/stripper (not to scale).

flows which substantially reduce the duty which has to be fulfilled by the heat exchanger, thereby allowing its size to be further reduced.

8.6.3.4 The way ahead

The above ideas represent a credible approach to the treatment of the large volumes of flue gas which are generated by gas and coal-fired power plants. Each of the operations discussed has been proven at the pilot-scale, as shown in the references already provided. The intensified version of the process promises to have a major impact on the capital cost and energy consumption of the plant due to:

- Much smaller equipment.
- Elimination of a large gas blower.
- Greatly reduced liquid flows.

Notable areas where further exploratory work is desirable include:

- Determination of the gas pressure drop in the very open-structure packing which is proposed.
- Determination of the 'flying droplet' size which is likely to be generated within the packing and the influence of the higher liquid viscosity associated with strong MEA solutions.
- Confirmation that the proposed design of reboiler is feasible.

8.6.3.5 Other current and recent research on intensified carbon capture using RPBs

Substantial work on RPBs has been carried out on a small-scale by various companies in several countries. Work in Taiwan on CO_2 absorption from indoor air using alkanolamine as the absorbent in an RPB is reported (Cheng and Tan, 2011). The concentration of carbon dioxide was reduced from 1000 to 100 ppm consistently in a bed rotating at 1200 rpm and a gas flow rate of 50 l/min. The height of the bed was 20 mm, and the inner and outer diameter 76 and 160 mm, respectively. Other work using alkanolamines in Taiwan is reported by Cheng and Tan (2009), while different bed configurations are examined by Lin and Jian (2007) and Lin and Chen (2011).

In China, as discussed earlier in this book, the Beijing University of Chemical Technology has a Research Centre concentrating upon HiGee-type technologies (Zhao et al., 2010). As well as discussing applications such as that detailed in Section 8.3.1.1, work on CO_2 absorption is also mentioned. This is covered in much more detail by Yi et al. (2009) at the same laboratory, where a model for the absorption was developed. It was suggested that the model could also be applied to use with absorbents other than MEA.

A review of the patent literature is often a valuable way of obtaining gems of information on new technologies, and PI is no exception. Statoil ASA, a most progressive company based in Norway, has patented a rotating desorber for removing CO_2 from an absorption fluid. The unit incorporates a reboiler, located between the inner core and the outer circumference (Anon, 2011a). Illustrated in one form in Figure 8.31, this shows a cross-section view about a vertical axis of rotation. At the lower level is the stripper unit (12 – a rotating packed bed) where CO_2 is desorbed from the rich absorbent (e.g. MEA), distribution being via nozzles (5). The reboiler is section 14, on the periphery of the RPB. Heat input to the reboiler is via steam line 4, leaving as condensate in 6, 116 is the condenser. A further patent (Anon, 2011b) covers a combination of spray absorber and rotating desorber, specifically

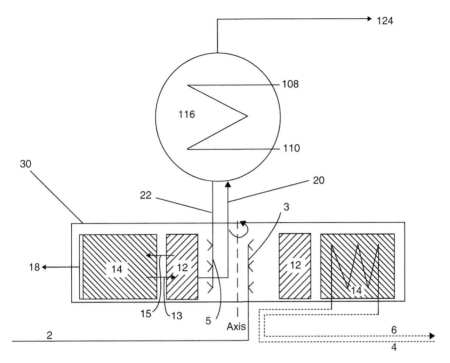

FIGURE 8.31

Statoil ASA's rotating desorber wheel.

directed at higher concentrations of amine absorbents that are normally used, in this case in concentrations of 61–100%.

Although not directly related to RPBs, the work of the Institute of Reaction Engineering at TU Dortmund (Professor David Agar) is worthy of mention in the context of the desorber energy demand. The group (Zhang et al., 2012) has been examining the use of specific solvents in the category of thermomorphic biphasic types (TBS) which contain lipophilic ('sucking') amines as active components. These perform as extracting agents, enhancing desorption at temperatures below the solvent boiling point. Zhang et al. suggest that because of the potentially low regeneration temperatures (about 90°C), waste heat could be used for desorption. A most interesting aspect of this work was the attempt to improve the carbon dioxide release by using agitation and ultrasound[6] (see Chapter 3). These, together with nucleation, enhanced desorption by various amounts. Ultrasound is discussed briefly below in the context of mineral carbonation.

---

[6]This is not the only reference to ultrasound in the context of carbon capture – as discussed in the context of mineral carbonation for carbon capture in Section 8.6.4.4, it can be used to enhance convective mass transfer and thus carbon dioxide sequestration.

8.6.4 Intensification of carbon capture using other techniques

As hinted at in Section 8.6.2, the several methods for carbon capture are all subject to intensification studies, and full data would warrant a book in its own right – it suffices here to give a few examples and to direct the reader to appropriate sources of further information.

8.6.4.1 Membranes

Keeping with post-combustion capture, the use of membranes is receiving substantial attention. Again the reason for their application is to reduce the size of the absorption unit. In a paper detailing omissions in research in this area, carried out by two groups in Norway and France (Favre and Svendsen, 2012), the authors suggest that little data are available to quantify the PI benefits using membranes, in spite of much concentrated research over several years, revealing a wide range of estimates for the intensification factor (that relates directly to size reduction compared to a packed column) of 1.5–8. These are mainly associated with MEA/CO_2 systems and membranes made of PTFE, PP or PVDF. The authors suggest that a formal comparison with a baseline absorption column figure for the volumetric CO_2 absorption capacity of $1 \, mol/m^3$ should be made for all the membranes used, and they highlight new membranes based upon high temperature polymers or microporous inorganic membranes. They suggest that technology transfer from the active gas absorption and membrane contactor fields should benefit the CC community.

One company out with the power generation sector that has been examining the use of membranes for post-combustion CC is Neste Oil Oyj in Finland. A most comprehensive dissertation covering work at Lappeenranta University of Technology (Topcu, 2010), under Professor Ilkka Turunen examined the potential for PI in the Finnish petrochemical industry. A membrane-based absorption process was examined in order to capture the CO_2 from the flue gases of a fluid catalytic cracking (FCC) unit. The flue gas flow rate was 180 t/h, of which 14% by volume was CO_2. MEA was used to absorb the gas and a PVDF polymer membrane of length 20 cm was selected. Simulation suggested that up to 97% of the CO_2 could be removed using this approach.

8.6.4.2 Pressure Swing Adsorption (PSA)

Thakur et al. (2011) have reported on intensification of carbon capture using a duplex PSA system, limited at this stage to modelling. PSA involves a number of steps – pressurisation, adsorption, blowdown and purging. The authors concede that the equipment for PSA is bulky, and intensification is directed at reducing the size of the plant. This can be done using the duplex system, where mass transfer takes place over the whole bed. The authors have managed to substantially reduce inactive zones using their modified cycle, claiming that this could lead to one or two orders of magnitude reduction in bed size and lower energy needs.

As an aside it would be interesting to see whether this approach could benefit other applications where adsorption beds tend to be large and slow in their reaction to gas inputs (see for example Chapter 11).

8.6.4.3 Chemical looping combustion and chemical looping reforming

Chemical looping combustion (CLC) is an alternative to direct fuel combustion, one that is suggested will give significant cost reduction per kg CO_2 captured, mainly due to the fact that the stream of flue gases leaving the fuel reactor (see Figure 8.32(a) where the basic principle is illustrated), is nearly pure CO_2 after the water is removed. From a concept first proposed in the 1980s, chemical looping has developed to pilot scale demonstration units operating at the 10 kW range for 100s of hours at Chalmers University, with a 120 kW unit designed and built at the Technical University, Vienna. As described by Ocone (2010), standard CLC systems work by circulating the oxygen carrier metal between two fluidised bed reactors; typically one being a circulating fluid bed, the second a bubbling bed reactor. Transfer of solid oxides between the two is via two 'loop seal' valves separating the two beds from air and fuel gas feeds. Modelling (see Ocone, 2010) and experimental analysis show clearly that heat transfer from oxide particles to gas, and recovery from the solid as it passed between two thermally different reactors is critical to the overall efficiency of the CLC system.

A small rotating bed reactor for chemical looping combustion was designed, constructed and tested at SINTEF in Norway (Hakonsen and Blom, 2011) as a way

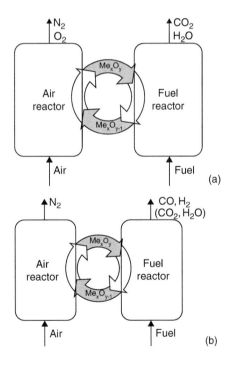

FIGURE 8.32

Schematic of chemical looping combustion (a) and reforming (b) (Moldenhauer et al., 2012).

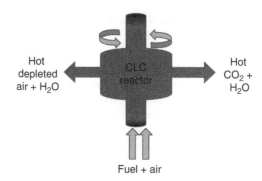

Hot depleted air + H$_2$O

CLC reactor

Hot CO$_2$ + H$_2$O

Fuel + air

FIGURE 8.33

The SINTEF rotating bed CLC reactor (Hakonsen and Blom, 2011).

of intensifying the chemical looping combustion process. Illustrated schematically in Figure 8.33 this uses a CuO/Al$_2$O$_3$ oxygen carrier and methane as fuel. Process parameters such as bed rotating frequency, gas flows and reactor temperature were varied to find optimal performance of the prototype reactor. The paper indicates that 90% CH$_4$ conversion and >90% CO$_2$ capture efficiency based on converted methane have been obtained. Stable operation was accomplished over several hours, and it was also found that stable operation can be regained after intentionally running into unstable conditions. Potential CO$_2$ purity obtained is in the range of 30–65% – mostly due to air slippage from the air sector – which seems to be the major drawback of the prototype reactor design. Considering the prototype nature of the first version of the rotating reactor setup, SINTEF believed that significant improvements could be made to further avoid gas mixing in future modified and up-scaled reactor versions.

It is worth the interested reader examining some of the other ideas for rotation (and for enhancement of static bed performance) that have attracted interest with solid carriers, such as the 'rotating regenerator' concept of Toshiba. Again, without overstressing the technology transfer aspects, the switching between static beds was perfected many years ago by UK companies involved in regenerative gas burners, and large rotating regenerators as used in power plant have been examined for CC duties by companies such as Hitachi, who of course, have the experience of constructing large rotating plant of such a type.

Returning to chemical looping, Moldenhauer et al. (2012) also examined the closely related chemical-looping reforming (CLR). They schematically illustrate this as shown in Figure 8.32(b), and explain that the principal difference between CLC and CLR is that the latter produces synthesis gas (syngas) – used by some as a fuel or feedstock gas, in spite of it having an energy density about half that of natural gas. In CLR there is insufficient air added to the air reactor to fully convert the fuel, making instead hydrogen and CO (main constituents). In the work at

Chalmers University (Sweden), the fuel used was kerosene, demonstrating that a liquid fuel could be fed directly into a CLR. So, if intensification can be applied to the CLC process, it may be feasible in CLRs too – a claim made for chemical looping dry reforming (CLDR) by researchers at the US's University of Pittsburgh (Bhavsar et al., 2012). In their case, the oxidiser is fed with CO_2 and the fuel reactor uses natural gas. Unlike the earlier CLR the output contains no hydrogen, giving CO and water.

8.6.4.4 *Mineral carbonation*

Mineral carbonation has for many years been seen as a cost-effective and technically viable route to carbon (or CO_2) sequestration. The process, a reaction between carbon dioxide and alkaline solids, can be carried out using a wide variety of solid materials. Calcium oxide is perhaps the most common, together with other oxides such as MgO, natural materials and alkaline waste materials – for example fly ash, cement kiln waste and steelworks slag. For a full, if slightly dated, literature review of the topic, see Huijgen and Comans (2003). Some of the minerals formed by mineral carbonation can have downstream uses, as well.

The work by Santos et al. (2011), reported at a European Process Intensification Conference, investigated whether ultrasound could enhance the kinetics of carbonation. A modest increase, compared to that achieved by mechanical agitation, resulted in the rate of carbonation increasing from 0.59–0.65 g-$CaCO_3$/min. Further combining ultrasound with magnesium ions supported the optimum formation of pure aragonite crystals (most commonly, calcite is formed during carbonation). An extended version of the presentation, recently published (Santos et al., 2012), examines energy efficiency and scale-up considerations.

8.7 **Further reading**

The area of chemicals production, spanning bulk, fine and bio chemicals, is massive, and has attracted a great deal of attention from those working in process intensification. Much of the research is at a small-scale (in terms of chemical output) and it would take a separate volume to detail all the work carried out during the past few years. However, some recent works are listed here in terms of the chemical being produced or the process being intensified, with references included in the list at the end of this chapter.

> *Esterification reactions*. Reactive stripping in monolith reactors is described by Schildhauer et al. in Delft (2005). Esterification of rosin and glycerol has been enhanced by Spanish researchers (Ladero et al., 2007).
> *Conversion of biomass-based syngas to alcohols*. Micro-channel reactors for converting the gas from biomass into methanol and higher alcohols are described by Hu et al. in the US, (2007).
> *Micro-reactors in applications in drug discovery etc.* Collaboration between TU Eindhoven and the Vita Nova Institute in Brazil has resulted, as one outcome, an

excellent review paper on the use of micro-reactors in the pharmaceutical sector (Baraldi and Hessel, 2012). Within the context of flow processing, as opposed to the normal batch approach in the sector, the authors argue that the time-to-market can be positively influenced by the ease with which the process can be scaled up. Numerous examples are given, and the transfer of the technology to a Brazilian Institute is highlighted and benefits to the community there discussed.

Microwave processes – adsorption selectivity. Legras et al. (2012) report upon microwave irradiation for improving adsorption selectivity in co-adsorption processes. These might include carbon capture or the removal of volatile organic compounds (VOCs). A new process combining several adsorbent beds is proposed, and the benefits of microwave-assisted desorption discussed. A useful review of the intensification of desorption processes (also relevant to adsorption heat pumps – see Chapter 11) is given by Cherbanski and Molga (2009). They recommend the use of fluidised bed adsorbers instead of fixed bed ones, as being more compatible with the microwave systems.

Ozone injection to intensify NO_x absorption. A study in Poland (Skalska et al., 2012) examined a number of abatement methods for the range of NO_x gases, concentrating upon emissions from the chemical industry. The process was successfully intensified by adding ozone – the emissions arising from a nitric acid plant. Up to 30% intensification of the absorption was achieved, the ozone being added upstream of the absorption into sodium hydroxide. A further benefit claimed was that no waste product was formed.

Production of lower alcohols. German researchers discuss options for intensifying the production of sec-butyl alcohol (Pfeuffer et al., 2007).

Pharmaceutical opportunity for the spinning disc reactor. Research at Newcastle University (Vicevic et al., 2008) in the UK indicates that the use of the SDR for the catalytic isomerisation of α-pinene oxide to campholenic aldehyde can be successful and can also minimise unwanted side reactions – a route, say the authors, to greener catalytic reactions – this follows on from the application reported in Section 8.4.6.

In a chapter in which we introduce carbon capture, it would be remiss of us not to make reference to the use of renewable energies such as solar thermal energy – biomass is mentioned earlier – as a source for driving reactions. the International Energy Agency (IEA) Solar Heating and Cooling Programme has a Task 49/Task IV within it that is called 'Solar process heat for production and advanced applications'. As we are completing this chapter, a workshop is being held in Graz, Austria on 'Solar process Heat and Process Intensification - Applications in the Food Industry'. The activity is at an early stage, but if we go back in time a few years to 2001, we can find work in Israel and Switzerland, (the Weizmann Institute of Science, and ETH Zürich respectively) on solar-driven chemical reactors. While 'intense' may not have been the word used to describe these reactors, as examples of PI they were rather good demonstrations (Steinfeld and Epstein, 2001).

8.8 Summary

The applications of PI in the chemical process sector encompass bulk and fine chemicals and a range of unit operations. Dominant are high gravity systems (HiGee and spinning disc reactors) and micro- or HEX-reactors, with a growth in interest (and useful application) of microwaves. In many cases, the examples are one-offs, but the replication is bound to increase in future years as environmental and production demands grow and new intensified reactor types, some of which are highlighted above, come into operation. The impact of environmental pressures is particularly evident in the new section on carbon capture, where PI technology will inevitably have a role to play in several of the important methods for power plant/process clean-up.

References

Anon, 1991. US Patent 5125793. Turbine blade cooling with endothermic fuel. Assigned to the USA as represented by the Secretary of the Air Force, filed 8 July.

Anon, 1992. US Patent 5313790. Endothermic fluid based thermal management system. Assigned to Allied-Signals Inc., Morris County, NJ, filed 17 December.

Anon, 1993. US Patent 5449568. Indirect-fired gas turbine bottomed with fuel cell. Assigned to US Department of Energy, filed 28 October.

Anon, 1997a. US Patent 5640840. Recuperative steam cooled gas turbine method and apparatus. Assigned to Westinghouse Electric Corporation, Pittsburgh, PA, 24 June.

Anon, 1997b. US Patent 5695319. Gas turbine. Assigned to Hitachi, Japan, issued 9 December.

Anon, 1997c. US Patent 5590518. Hydrogen-rich fuel, closed-loop cooled, and reheat enhanced gas turbine power plant. Assigned to California Energy Commission, Sacramento, CA, issued 7 January.

Anon, 1989. European patent registration 08901102. Turbine a Gaz Isotherme à Oxidation Partielle. Assigned to J. Ribesse, Jarix Company, 11 October.

Anon, 2002. Combined heat, power and chemicals: the feasibility of gas turbine reactors. Report by David Reay and Associates in association with the partners. Agreement E/CA/00340/00/00/4576, ETSU. Final report June.

Anon, 2011a. Patent WO 2011/005118 A1, rotating desorber wheel. Assigned to Statoil ASA, Filed 11 July 2010, Published 13 January 2011.

Anon, 2011b. Patent WO 2011/005117 A1. Compact Absorption-Desorption Process and Apparatus using Concentrated Solution. Assigned to Statoil ASA, Filed 9 July 2010, Published 13 January 2011.

Aoune, A., Ramshaw, C., 1999. Process intensification: heat and mass transfer characteristics of liquid films on rotating discs. Int. J. Heat Mass Tran. 42, 2543.

Baraldi, P.T., Hessel, V., 2012. Micro reactor and flow chemistry for industrial applications in drug discovery and development. Review Paper. Green Process Synth. 1, 149–167.

Bhavsar, S., Najera, M., Veser, G., 2012. Chemical looping dry reforming as novel, intensified process for CO_2 activation. Chem. Eng. Technol. 35 (7), 1281–1290.

Boodhoo, K.V.K., et al., 1995. Process intensification: spinning disc polymeriser for the manufacture of polystyrene. First International Conference on Process Intensification in the Chemical Industry. BHR Group Conf. Series. Pub. No. 18.

Boodhoo, K.V.K., Jachuck, R.J.J., Ramshaw, C.A., 1997. Spinning disc reactor for the intensification of styrene polymerisation. Second International Conference on Process Intensification in Practice. BHR Group Conf. Series. Pub. No. 28.

Boodhoo, K.V.K., Jachuck, R.J.J., 2000a. The spinning disc reactor for condensation polymerisation. Green Chem. J. 2, 235–244.

Boodhoo, K.V.K., Jachuck, R.J.J., 2000b. Process intensification: spinning disc reactor for styrene polymerisation. Appl. Therm. Eng. 20 (12), 1127–1146.

Burns, J.R., Ramshaw, C., 1996. Process intensification: visual study of liquid maldistribution in rotating packed beds. Chem. Eng. Sci. 51 (8), 1347.

Burns, J.R., Jamil, G.N., Ramshaw, C., 2000. Process intensification: operating characteristics of rotating packed beds – determination of hold-up for a high voidage structured packing. Chem. Eng. Sci. 55, 2401.

Cao, E., Gavriilidis, A., 2005. Oxidative dehydrogenation of methanol in a microstructured reactor. Catal. Today 110, 154–163.

Chambers, H.H., Wall, R.G., 1954. Some factors affecting the design of centrifugal gas absorbers. T I Chem. Eng-Lond 32 (S96).

Charpe, T.W., Rathod, V.K., 2012. Extraction of glycyrrhizic acid from licorice root using ultrasound: process intensification studies. Chem. Eng. Process. 54, 37–41.

Chen, G., Li, S., Jiao, F., Yuan, Q., 2007. Catalytic dehydration of bioethanol to ethylene over $TiO_2/\gamma-Al_2O_3$ catalysts in microchannel reactors. Catal. Today 125, 111–119.

Cheng, H.-H., Tan, C.-S., 2009. Carbon dioxide capture by blended alkanolamines in rotating packed bed. Energy Procedia 1, 925–932.

Cheng, H.-H., Tan, C.-S., 2011. Removal of CO_2 from indoor air by alkanolamine in a rotating packed bed. Sep. Purif. Technol. 82, 156–166.

Cherbanski, R., Molga, E., 2009. Intensification of desorption processes by use of microwaves – an overview of possible applications and industrial perspectives. Chem. Eng. Process. 48, 48–58.

Clavey, T., 2011. Process intensification by using pervaporation. Proceedings of European Process Intensification Conference, EPIC 2011, Manchester, UK. IChemE Symposium Series No. 157, pp. 45–47.

Cooper, H.W., 2010. Producing electricity and chemicals simultaneously. Chem. Eng. Prog., 24–32. (February).

Cornelissen, R., Tober, E., Kok, J., van de Meer, T., 2006. Generation of synthesis gas by partial oxidation of natural gas in a gas turbine. Energy 31, 3199–3207.

Dalmoro, A., Barba, A.A., Lamberti, G., d'Amore, M., 2012. Intensifying the microencapsulation process: ultrasonic atomisation as an innovative approach. Eur. J. Pharm. Biopharm. 80, 471–477.

Dissanayake, B., Morgan, A., Akay, G., 2012. Process intensification in particle technology: characteristics of powder coatings produced by nonisothermal flow-induced phase inversion. AIChE J. 58 (4), 1060–1068.

Enache, D.I., Thiam, W., Dumas, D., Ellwood, S., Hutchings, G.J., Taylor, S.H., et al., 2007. Intensification of the solvent-free catalytic hydroformylation of cyclododecatriene: comparison of a stirred batch reactor and a heat-exchanger reactor. Catal. Today 128, 18–25.

Favre, E., Svendsen, H.F., 2012. Perspective: membrane contactors for intensified post-combustion carbon dioxide capture by gas–liquid absorption processes. J. Memb. Sci., 1–7. 407–408.

Fraser, D., 2011. Process Intensification Assignment - Opportunities for Process Intensification in the area of Carbon Capture. Heriot-Watt University, Edinburgh.

Gasparini, G., Archer, I., Jones, E., Ashe, R., 2012. Scaling up biocatalysis reactions in flow reactors. Organic Process Research and Development, Special Issue: Continuous Processes 2012. dx.doi.org/10.1021/op2003612

Gogate, P.R., 2007. Cavitational reactors for process intensification of chemical processing applications: a critical review. Chem. Eng. Process. doi: 10.1016/j.cep.2007.09.014

Hakonsen, S.F., Blom, R., 2011. Chemical looping combustion in a rotating bed reactor – finding optimal process conditions for prototype reactor. Environ. Sci. Technol. 45 (22), 9619–9626.

Hall, S., Stoker, A., 2003. APIs: new plants for new products. Pharm. Eng. May–June.

Harmsen, G.J., 2007. Reactive distillation: The front-runner of industrial process intensification. A full review of commercial applications, research, scale-up, design and operation. Chem. Eng. Process. 46, 774–780.

Hu, J., Wang, Y., Cao, C., Elliott, D.C., Stevens, D.J., White, J.F., 2007. Conversion of bio-mass-derived syngas to alcohols and C_2 oxygenates using supported Rh catalysts in a microchannel reactor. Catal. Today 120, 90–95.

Hugill, J.A., Tillemans, F.W.A., Dijkstra, J.W., Spoelstra, S., 2005. Feasibility study on the cogeneration of ethylene and electricity through oxidative coupling of methane. Appl. Therm. Eng. 25, 1259–1571.

Huijgen, W.J.J., Comans, R.N.J., 2003. Carbon dioxide sequestration by mineral carbonation. ECN Report ECN-C–03-016, February 2003, ECN, The Netherlands.

Jassim, M., Rochelle, G., Eimer, D., Ramshaw, C., 2007. Carbon dioxide absorption and desorption in aqueous monoethanolamine solutions in a rotating packed bed. Ind.Eng. Chem.Res. 46, 2823.

Jin, H., Subramaniam, B. Fang, J., 2006. Process intensification by employing CO_2-expanded liquids (CXLs) as alternative solvents: a case study of homogeneous catalytic hydroformylation of higher olefins. Proceedings of the Second Conference on Process Intensification and Innovation, Christchrch, New Zealand, September 24–29.

Jou, F.Y., Mather, A.E., Otto, F.D., 1995. The solubility of CO_2 in a 30 mass percent monoethanolamine solution. Can. J. Chem. Eng. 73 (1), 140.

Kirschneck, D., 2007. Micro-reactors on production scale: a technical and economical status report for pharmaceutical applications. Proceedings of the European Congress of Chemical Engineering (ECCE-6), Copenhagen, 16–20 September.

Korobitsyn, M.A., et al., 1998. Analysis of a gas turbine with partial oxidation. Proceedings of the Forty third ASME International Gas Turbine and Aeroengine Congress, Sweden, June. Paper ASME 98-GT-33.

Ladero, M., de Gracia, M., Alvarero, P., Trujillo, F., Garcia-Ochoa, F., 2007. Process intensification in the esterification of rosin and glycerol. Proceedings of the European Congress of Chemical Engineering (ECCE-6), Copenhagen, 16–20 September.

Legras, B., Polaert, I., Thomas, M., Estel, L., 2012. About using microwave irradiation in competitive adsorption processes. Appl. Therm. Eng. doi: 10.1016/j.applthermaleng.2012.03.034

Li, Y., Liu, Z., Zhang, L., Su, Q., Jin, G., 2010. Absorption of NO_x into nitric acid solution in rotating packed bed. *Separation Science and Engineering*. Chinese J. Chem. Eng. 18 (2), *244–248.*

Lin, C.-C., Jian, G.-S., 2007. Characteristics of a rotating packed bed equipped with blade packings. Sep. Purif. Technol. 54, 51–60.

Lin, C.-C., Chen, Y.-W., 2011. Performance of a cross-flow rotating packed bed in removing carbon dioxide from gaseous streams by chemical absorption. Int. J. Greenhouse Gas Control doi: 10.1016/j.ijggc.2011.02.002

Loewe, H. Hessel, V., 2006. New approaches to process intensification – microstructured reactors changing processing routes. Proceedings of the Second Process Intensification and Innovation Conference, Christchurch, New Zealand, September 24–29.

Lyubovsky, M., Walsh, D., 2006. Reforming system for cogeneration of hydrogen and mechanical work. J. Power Sources 157, 430–437.

McCarthy, E.D., Dunk, W.A. Boodhoo, K.V.K., 2006. Tailoring particle properties through precipitation in a narrow channel reactor. Proceedings of the Second Process Intensification and Innovation Conference, Christchurch, New Zealand, September 24–29.

Moldenhauer, P., Ryden, M., Mattisson, T., Lyngfelt, A., 2012. Chemical looping combustion and chemical looping reforming of kerosene in a circulating fluidised-bed 300 W laboratory reactor. Int. J. Greenhouse Gas Control 9, 1–9.

Neelis, M., Patel, M., Bach, P., Blok, K., 2007. Analysis of energy use and carbon losses in the chemical industry. Appl. Energy 84, 853–862.

Ocone, R., 2010. Modelling the hydrodynamics of chemical looping reactors. Proceedings of First International Conference on Chemical Looping, IFP-Lyon, France, 17–19 March.

Oxley, P., et al., 2000. Evaluation of spinning disc reactor technology for the manufacture of pharmaceuticals. IEC Res 39 (7), 2175–2182.

Patel, M.P., Shah, N., Ashe, R., 2011. Plant-wide optimisation and control of a multi-scale pharmaceutical process. Computer Aided Chem. Eng. 28, 714–717.

Patience, G.S., Cenni, R., 2007. Formaldehyde process intensification through gas heat capacity. Chem. Eng. Sci. 62, 5609–5612.

Pedernera, M.N., Pina, J., Borio, D.O., Bucala, V., 2003. Use of a heterogeneous two-dimensional model to improve the primary steam reformer performance. Chem. Eng. J. 94, 29–40.

Perez-Ramirez, J., Vigeland, B., 2005. Lanthanum ferrite membranes in ammonia oxidation. Opportunities for 'pocket-sized' nitric acid plants. Catal. Today 105, 436–442.

Pfeuffer, B., Petre, D., Kunz, U., Hoffmann, U., Turek, T. Holl, D., 2007. Production of sec-butyl alcohol by olefin hydration – a candidate for process intensification? Proceedings of the European Congress of Chemical Engineering (ECCE-6), Copenhagen, 16–20 September.

Reay, D.A. Ramshaw, C., 1992. An investigation into catalytic combustion. Contractors' Report, Submitted to ETSU, Harwell, August.

Reay, D.A., Ramshaw, C., 1998. Gas turbine reactor study. Contract Report to BP Exploration April.

Rezvani, S., Huang, Y., McIlveen-Wright, D., Hewitt, N., Deb Mondol, J., 2009. Comparative assessment of coal-fired IGCC systems with CO_2 capture using physical absorption, membrane reactors and chemical looping. Fuel 88, 2463–2472.

Rutherford, A. Simons, S., 2010. Scottish Carbon Capture & Storage. Retrieved October 15, 2011, from Scottish Centre for Carbon Storage: <http://www.geos.ed.ac.uk/sccs/capture/precombustion.htm>

Sanders, J.P.M., Clark, J.H., Harmsen, G.J., Heeres, H.J., Heijnen, J.J., Kersten, S.R.A., et al., 2012. Process intensification in the future production of base chemicals from biomass. Chem. Eng. Process. 51, 117–136.

Santos, R., Ceulemans, P., Francois, D. Van Gerven, T., 2011. Ultrasound-enhanced mineral carbonation. Proceedings of European Process Intensification Conference, EPIC 2011, Manchester, UK. IChemE Symposium Series No. 157, pp. 108–116.

Santos, R., Francois, D., Mertens, G., Elsen, I., Van Gerven, T., 2012. Ultrasound-enhanced mineral carbonation. Appl. Therm. Eng. doi: 10.1016/j.applthermaleng.2012.03.035.

Schildhauer, T.J., Kapteijn, F., Moulijn, J.A., 2005. Reactive stripping in pilot scale monolith reactors–application to esterification. Chem. Eng. Process. 44, 695–699.

Skalska, K., Miller, J.S., Ledakowicz, S., 2012. Intensification of NO_x absorption process by means of ozone injection into exhaust gas stream. Chem. Eng. Process. <http://dx.doi.org/10.1016/j.cep.2012.06.007>

Steinfeld, A., Epstein, M., 2001. Light years ahead. Chem. Br. 37 (5), 30–32.

Tai, C.Y., Tai, C.T., Liu, H.S., 2006. Synthesis of submicron barium carbonate using a high-gravity technique. Chem. Eng. Sci. 61, 7479–7486.

Thakur, R.S., Kaistha, N., Rao, D.P., 2011. Process intensification in duplex pressure swing adsorption. Comput. Chem. Eng. 35, 973–983.

Topcu, E., 2010. Potentials of process intensification in the Finnish petrochemical industry. Master's Thesis, Faculty of Technology, Lappeenranta University of Technology, Finland.

Vicevic, M., et al., 2001. Process intensification for green chemistry: rearrangement of alpha-pinene oxide using a catalysed spinning disc reactor. Fourth International Conference on Process Intensification for the Chemical Industry. Bruges, Belgium BHR Group Conf. Series, 10 September.

Vicevic, M., Novakovic, K., Boodhoo, K.V.K. Morris, A.J., 2006. Kinetics of styrene free radical polymerisation in the spinning disc reactor. Proceedings of the Second Conference on Process Intensification and Innovation, Christchurch, New Zealand, September 24–29.

Vicevic, M., Boodhoo, K.V.K., Scott, K., 2007. Catalytic isomerisation of a-pinene oxide to campholenic aldehyde using silica-supported zinc triflate catalysts. II. Performance of immobilised catalysts in a continuous spinning disc reactor. Chem. Eng. J. 133, 43–57.

Vicevic, M., et al., 2008. Kinetics of styrene free radical polymerisation in the spinning disc reactor. Chem. Eng. J. 135, 78–82.

Wang, M., Ramshaw, C. Yeung, H., 2007. Feasibility study of a turbo-cracker. Presentation at 15th PIN Meeting, Cranfield University, 21 November. See <www.pinetwork.org>

Weatherley, L.R., 1993. Electrically enhanced mass transfer. Heat Recovery Syst. CHP 13 (6), 515–537.

Web1, 2012. Begg Coulsland Markets web site.<http://www.beggcousland.com/markets_applications2.cfm?ApplicationID=17> (accessed 22.08.12.).

Wernersson, M., 2007. Small scale continuous reactors in process development. Proceedings of the European Congress of Chemical Engineering (ECCE-6), Copenhagen, 16–20 September.

Wharton, Y., 2011. Microwave chemistry – out of the lab and into production. Proceedings of the European Conference on Process Intensification, Manchester, 2011. IChemE Symposium Series No. 157, pp. 117–121.

Yi, F., Zou, H.-K., Chu, G.-W., Shao, L., Chen, J.-F., 2009. Modeling and experimental studies on absorption of carbon dioxide by Benfield solution in rotating packed bed. Chem. Eng. J. 145, 377–384.

Zhang, J., Qiao, Y., Agar, D.W., 2012. Intensification of low temperature thermomorphic biphasic amine solvent regeneration for CO_2 capture. Chem. Eng. Res. Des. 90, 743–749.

Zhao, H., Shao, L., Chen, J.-F., 2010. High-gravity process intensification technology and application. Chem. Eng. J. 156, 588–593.

Application Areas – Offshore Processing

9

OBJECTIVES IN THIS CHAPTER

The demands in terms of plant safety, reliability, efficiency and most importantly, size and weight, grow as one moves from onshore locations to the offshore sector. In this chapter, the different approaches by companies to offshore development – including downhole and whether processes should be carried out on ships, platforms or even back onshore – will help readers to appreciate the examples given later of intensified unit operations used in such situations.

9.1 Introduction

Offshore processing covers a variety of technologies used in several totally different locations, the common feature being that they are not based on land, but on or in the sea, or below the seabed. The processing may be carried out on an oil or gas rig – a platform raised above the sea, standing on legs or anchored in position. PI activities centred round printed circuit heat exchangers (PCHEs) started, as far as the commercial market is concerned, on such platforms. Processing may increasingly be carried out on a ship – an FPSO (floating production, storage and offloading vessel) – subsea or downhole (downhole being in the pipework below the seabed or in the reservoir region). Some companies – see below – feel that the ease of processing onshore suggests that this would be the preferred location, rather than away from land. In some locations, particularly in the US, onshore oil wells are common, and an example of deaeration that is given below is applied to onshore secondary oil recovery – in fact, it is in the area of separations of gases and liquids that the offshore processing sector has adopted active PI technologies. Compact heat exchanger technology has been extended by several companies, most notably, Heatric and Velocys, to heat exchanger reactor concepts that can be used offshore (examples are given later).

A use of PI plant that started in the 1980s, with some gusto, is the application of HiGee technology offshore for seawater deaeration/deoxygenation – needed before it is returned to oil wells – and other duties. In China, the activity in offshore HiGee has grown considerably, while a US company, GasTran Systems, is using the high gravity technology in its rotating packed-bed unit. Others, such as Professor Waldie at Heriot-Watt University (UK), Twister BV in The Netherlands and US researchers at Michigan State University (all discussed below) use static means, such as cyclones for imparting the gravity forces on fluids.

An area that has not received a great deal of attention, but one which would warrant implementation of PI, is desalination. An early example of this was Hickman's centrifugal vapour compression evaporator, discussed also in Chapter 1. As an offshore or near-shore process, intensified desalination plant would find a ready market in many parts of the world – perhaps a challenge for readers.

It may be argued that the ultimate result of the intensification of offshore plant would be to make it disappear – either by doing away with it and using multiphase pumping to carry the fluids ashore, or, as indicated earlier, to carry out the processes on the seabed or subsea. Multiphase metering permits produced fluids to be metered locally, so they could then be combined with flows from other wells to share pipelines and processing infrastructure.

9.2 Some offshore scenarios

9.2.1 A view from BP a decade ago

Brian Oswald, before he retired from what was then BP Amoco, followed up the quote of Terry Lazenby, BP chief engineer, at a UK Heat Exchanger Action Group meeting in 1998 (cited earlier) by giving some profound insights into how process intensification might impact upon the whole petrochemicals route, from exploration to product delivery (Oswald, 1999). Looking at oil and gas primary processing, Oswald summarised the situation afforded by PI in three sentences:

- Do it differently – intensify, integrate and revisit the basics.
- Do it elsewhere – downhole, subsea/remote, in the pipework, and downstream.
- Do it less – challenge conventional specifications.

With regard to downhole processing, one of the situations discussed in this chapter, Oswald's concepts are summarised in Figure 9.1. PI might offer a number of benefits here, including the unlocking of additional reserves, at present largely determined, if conventional methods are used, by the price of a barrel of oil, and the production of higher-value products at the wellhead. If one could bring only the wanted materials to the surface, there would be a reduced impact on the environment and demand for surface facilities. Also, as the conditions downhole are different to those above the sea or on land, particularly in terms of higher temperatures and pressures, these 'free' energy sources might be commandeered to drive the intensified processes.

9.2.2 More recent observations – those of ConocoPhillips

As with all new technologies and attempts to change the mindset within industries, realism can have a sobering effect on enthusiasts. A more recent meeting of the UK Process Intensification Network, held at Heriot-Watt University, Edinburgh, was largely dedicated to the offshore sector (Anon, 2004). The opening talk at the meeting was by Mike Swidzinski of ConocoPhillips. At the time, he was technology

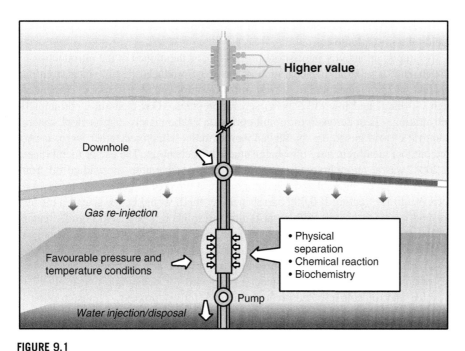

FIGURE 9.1

Activities envisaged in downhole processing (Oswald, 1999).

team leader for North Sea business with the company, and was able to give an off-shore industry perspective of current practices and challenges, also highlighting how these might be related to PI. His observations are highly relevant to any discussion on PI offshore and are paraphrased below.

He began by giving an overview of the developments in offshore extraction, commencing in the 1970s, when working at sea bed depths of 100–150 m, where temperatures were about 4°C, to the 1990–2000 period, where depths were several times greater and local temperatures close to freezing point (these were not down hole temperatures). In the last four years to 2004, the trend had been towards sea-bed gas gathering at 1,000–1,500 m depth.

In describing the various processes undertaken from the wellstream fluid source to onshore processes (gas flows being typically 160 million standard cubic feet/day (mscf/d)), it was pointed out that these necessitated a pressure reduction from 750+ barg to 70 barg, which could be considered as an inefficiency (of course, some may see this pressure reduction as an opportunity for energy recovery). Other changes necessary in the transport process were the addition of chemicals, for wax and corrosion control, and fluid measurement (wax of course may inhibit intensification because of perceived fouling, if not controlled). A paper at the same meeting by a researcher at Heriot-Watt University suggested that the oscillatory baffle reactor

(OBR, see Chapter 5) could be used to induce flow disturbances in a pipeline that could inhibit wax build-up.

An important activity in offshore processing, as mentioned in the introduction to this chapter, is separation. Separation of gas from produced liquid, water from oil, residual oil from water, and dissolved solids from oil, are all necessary. There were high (40 barg) and low (13.5 barg) separation stages. Gas clean-up – principally dehydration – is undertaken to inhibit corrosion of the mainly carbon steel system. Removal of acid gases, e.g. by alkyanolamine units (which are rather large) is carried out, but membranes are now being studied for clean-up. The use of membranes, in 2004, was at an embryonic stage offshore. The separation of residual oil from water is driven by internal and regulatory control directed at getting the oil level to zero. A unit for treating 100,000 barrels per day (bpd) would be the size of a modest university lecture theatre. With regard to taking solids or scale from liquid streams, corrosion needs to be prevented, as this can lead to bacteria incubation. Chemical treatment is necessary, including the addition of wax and corrosion inhibitors.

In gas processing, dehydration and compression are necessary (albeit inefficient) process stages. Glycol contactors are used for dehydration. The glycol is recycled and, of course, needs regenerating. A further process – transportation – involves recompressing the process gases, pumping liquid hydrocarbons and the disposal of waste products/chemical treatment.

All of the above involve heavy, large equipment and all have inherent inefficiencies. One challenge is to reduce weight within certain constraints. In particular, there must be no reduction in equipment availability, reliability and integrity. Risks to personnel and the wider plants must not be increased. Capital and operating costs should not be higher, and from the health and safety point of view, it would be ideal if one could engineer out the need for people to be located offshore.

ConocoPhillips were aware of several initiatives to simplify and intensify these processes. The PCHE was compared with a TEMA shell and tube unit as an example of PI. However, processing on the seabed (the particular project being called VASPS) was one approach. Here, a dummy well would be drilled, and a separator element put in it. The energy in the well would be used to create a vortex and separate the gas from the liquid. This has been done offshore in Brazil, and is seen as 'on the horizon'. Progress is hampered by the need for a reliable electro-submersible pump and a nearby host (within 15 km) to receive the liquid and to supply the power. Another seabed concept was H-sep, which was downhole gravity separation. Aker-Kvaerner has studied this: it involves taking fluids from the production zone, separating them into hydrocarbons and a water–gas mixture and pumping these to the surface. This was an improvement over current practice, but there is still the need for a nearby host facility to supply the power and receive the production output.

In responding to the question, can we do better?, interestingly, and perhaps controversially, as far as environmentalists are concerned, Mike Swidzinski said Can we move processing to shore?". Before addressing this, he asked "Why do we process?". The threats to smooth production flow are several – hydrates, corrosion, plugging, scale and complex multiphase regimes. If one can stabilise all the components into, for example, stable slurry that is easily transportable, the concept of cold-flow might be realised. There is Norwegian interest in this concept, which involves flow over long enough distances to ensure that most of the time the fluid(s) flowing, is spent at seabed temperature. This would get rid of most of the above problems, and it would eliminate the need for pipeline insulation – one could use bare carbon steel pipes. The well pressure energy could be employed – enough for several tens of kilometres transport.

Regarding the timescale for implementation, it was believed that the developments would be seen between 2007 and 2010, thereafter removing the need for surface or floating facilities. In summary, three points were made:

- Conventional oil and gas processing philosophy does not encourage more PI, etc.
- Subsea processing can be an opportunity for PI, but the need for a nearby host is a problem.
- Controllable slurry transport over long distances may be a way forward – leading possibly to onshore processing – this would dispense with offshore superstructures.

The reader may wish to make his or her own conclusions concerning the impact these three statements has had, or will have, on process intensification offshore. Examples later in the chapter may suggest that point three is some way off, and PI would have an important role to play in the interim.

9.2.3 **One 2007 scenario**

Looking at a Scottish call for research proposals some years ago in the area of innovative subsea technologies, areas of general interest are: enhanced subsea recovery, ultra-deepwater activities, subsea processing and equipment operation in high temperature and/or high pressure environments. PI could feature strongly in a number of areas as follows:

Enhanced subsea recovery:
    Subsea desalination.
    Downhole compression/separation.
Subsea processing:
    Power generation systems.
    Improved processing systems for subsea use.
    Subsea water injection and treatment.

The reason for the strong interest in enhancing subsea recovery is that recovery rates for wells are frequently only 70% of the equivalent from platform-drilled wells. Cost-effective ways for increasing these recovery rates are needed. In the case of subsea processing, the need is to transfer much of what is done on platforms onto the seabed. In addition to cost reduction, it would reduce the number of personnel stationed in dangerous environments.

9.3 Offshore on platforms or subsea

9.3.1 Setting the scene

In offshore production, the move into ever-deeper waters has renewed the emphasis on floating systems, where space, weight and sensitivity to motion are major constraints. If equipment is to be moved onto the seabed, or even into the wells themselves, PI offers some key enabling technologies. Perhaps the need to find step-out solutions will be enough to overcome the resistance to unproven technologies which has dogged PI in the offshore area in the past.

The cost benefits achieved by intensifying plant are substantial, ultimately reflected in the weight and size of the platform supporting the production and processing modules. Some fifteen years ago, this was appreciated in the case of the PCHE, where an example used for compressor after-cooling was said to offer a five-fold reduction in size and weight. The space and weight saving also led to a cost saving of £750,000, compared to a conventional upgrade (Anon, 1991). The principal areas of intensification offshore relate to:

- Compact separations.
- Compact heat transfer plant.
- Multiphase metering and pumping.
- Compact subsea and satellite installations.

For separations, hydrocyclones have become an accepted technology for the treatment of oily water, where they reduce inventories and space and weight needs by an order of magnitude. From the safety viewpoint, they are very simple and much less liable to leak than older style flotation units. Rotary hydrocyclones can give even higher g-forces for effective separation, although maintenance of seals and the high-speed plant may be regarded as a drawback. HiGee and micro-distillation plant are also options – see later.

Downhole pumps, compact heat exchangers, rotating and in-line separators are already in use. In particular, the drive to improve physical separations and reduce size and cost has lead to the use of some intensified equipment. There is use of rotating separation equipment in the UK offshore sectors, but again overseas, notably China as already mentioned, leads the way both offshore and onshore.

As further use is made of natural gas, be it by liquefaction and transport, gas-to-liquid conversion, or gas-wire conversion into electric power, the use of floating production platforms will necessitate compact plant. An example of a new approach offshore is the conversion of gaseous-liquid products to avoid flaring, for instance using a catalytic plate reactor. Offshore processing may be the ultimate destination of novel technologies mentioned in the petrochemicals overview above. PI is therefore increasingly a target for the sector processing natural gas supplies, and soon possibly including methane hydrates, of which there is a 200 years supply at the current rates of gas consumption.

This drive to improve separations and reduce size could conceivably extend to offshore reactions, to convert something worthless offshore to something valuable onshore. Novel topside, downhole and pipeline techniques for both separation and reaction are possible. One may be downhole heavy crude oil processing, as discussed below.

9.3.2 Downhole heavy crude oil processing

Heavy crude is often regarded as an anathema to process intensification engineers, because it fouls things up and, for example, would not be passed through a compact heat exchanger. However, it has been examined as an obvious candidate for downhole processing by major oil companies. Examples of this were reported in the mid-1990s by Texaco (Weissman, 1997; Weissman and Kessler, 1996), the intention being to catalytically upgrade the oil so that, as an example, impurities might be left in the ground.

Texaco's idea was to upgrade the crude either in a reservoir near the well hole or to do it in the well-bore. The discussions in the papers examine the catalysis involved in reacting oil and co-reactants such as hydrogen or water over a heterogeneous catalyst in these locations. An interesting suggestion was that naturally-occurring minerals in the rock or metals in the oil might act as catalysts – easing the process somewhat. The authors studied the economics of downhole treatment of crude oil and found that, if high sulphur oils could be processed there and at least 50% of the sulphur could be removed, the operation might be cost-effective. Experiments were carried out and it was concluded that the concept was economically and technically feasible, the economics depending upon the quality of the crude in the reservoir.

9.3.3 Compact heat exchangers offshore (and onshore)

There have been a number of attempts to develop highly compact heat exchangers for offshore (and onshore oil and gas field) processing. Heatric pioneered the activity with the printed circuit heat exchanger (PCHE) (see also Chapter 4), although the attempt by Rolls Royce to market a titanium equivalent failed. There may be increasing competition from Velocys (see below) but the PCHE has become well-established in several platform duties and is currently dominant in the marketplace. An early PCHE application offshore (although not the first – that was for Esso in 1989) was on the Jude platform operated by the then Phillips Petroleum UK (Knott, 1993). Six PCHEs were used, five on cooling duties and one as a fuel heater in the gas treatment process – the value of the contract was £1.7million.

Apart from the technical and thermodynamic benefits of the PCHE, the great attraction of this unit for offshore activities is the small size, or 'footprint' compared to the routinely-used shell and tube heat exchangers. The reduction in structural and support requirements probably far outweighs, in cost-benefit terms, the closer approach temperatures, energy savings and other features of CHEs. This can be worth £millions on a large platform in deep water. The ability to use less material than the competition, when exotic materials are required, is also a strong feature of the PCHEs attractiveness in such arduous conditions.

While onshore fields are less common in Europe than offshore facilities, one well-documented example of the former is a titanium PCHE unit used to service one of the few onshore oil and gas fields in use in the UK – the BP Amoco Wytch Farm Gathering Centre in Dorset. It was found that high ambient temperatures in the summer limited the plant throughput to 77,000 bopd, about 10% less than production in winter. The limitation was due to the performance of an air cooler upstream of the gas dehydration molecular sieve unit. The implication of this was that, in summer, the gas temperatures, and hence the water content, were higher and could not be handled. The option selected to overcome the limitation was to install a trim cooling stage downstream of the existing air cooler which would operate during the summer. Filtered and deoxygenated sea water was available for the cooling duty and the warmed sea water could be reused in desalters.

The PCHE was chosen because space constraints were critical (see Figure 9.2) and the above-ground location favoured the 1 tonne PCHE rather than the 20 tonne equivalent shell and tube exchanger. Prior to installation, a series of trials on the prevention of biofouling in narrow channel heat exchangers carried out at another BP Amoco terminal showed that control by chemical injection was acceptable in the titanium PCHE (Anon, 1998; Reay, 1999).

Another area pioneered by Heatric in offshore processing using PCHEs is the use of coolers and dewpoint control in gas processing. A typical production stream

FIGURE 9.2

The PCHE on the gas gathering plant at Wytch Farm, UK.

from the wellhead contains a mixture of oil, condensate, gas and water. Initial separation to try to isolate the useful output, in this case the gas, produces a hot gas stream that is likely to be saturated with both hydrocarbon liquid and water. It is necessary to drop both the hydrocarbon and water dewpoints below the lowest possible export pipeline temperature, otherwise two-phase flow may occur, and hydrates, one of the undesirables (as is waxing) in offshore processing, may form.

In the Heatric process, the warm wet pressurised gas from the inlet separator is pre-cooled in the PCHE and then throttled in a Joule-Thomson (JT) valve to a lower pressure. The drop in pressure produces a cooling effect and both hydrocarbon liquids and water condense out of the gas. The two-phase stream passes to a separator where the liquids are removed. The cold dry gas from the separator is returned to the exchanger to chill the incoming warm, wet gas (Figure 9.3). Refrigeration is sometimes introduced to reduce the pressure drop used by the JT in chilling the gas and thereby retain more available pressure for downstream utilisation. Hydrate formation may also be suppressed by glycol injection, upstream of the heat exchanger.

9.3.4 Extending the PCHE concept to reactors

As discussed in Chapters 4 and 5, the PCHE has been extended to function as a reactor – the PCR or printed circuit reactor, and variants can be used for processes that are relevant to offshore applications (see also Section 9.7). One application that Heatric has targeted is a gas-to-methanol process applied to improved oil recovery – a necessary step already highlighted in Section 9.2.3. The motivation for looking at reactors for use in association with enhanced oil recovery is based upon

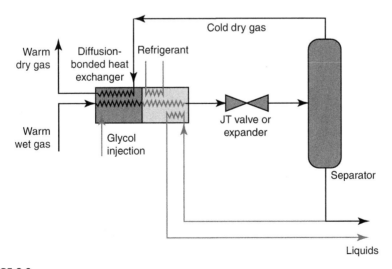

FIGURE 9.3

The Heatric multi-stream heat exchanger upstream of the JT valve.

the perception that the amount of CO_2 in the oil being extracted will increase due to the use of wells as deposits for this greenhouse gas – coming within the term 'CO$_2$ sequestration'. As well as the presence of the gas in associated produced gas, Heatric points out (Banister and Rumbold, 2005) that some North American operators already use CO_2 well fracturing as a means to aid improved oil recovery.

Heatric proposed a compact gas-to-methanol unit in which steam reforms the high CO_2 content natural gas and converts it to methanol, with the hydrogen-rich tail gas being burnt to heat the reformer. Heatric stated that with fields such as the North Sea Sleipner West gas field having 9% CO_2 in the reservoir, 76% of the cost of a North Sea CO_2 capture and oil recovery project could be attributed to onshore CO_2 capture and offshore transport. If the gas-to-liquid compact reactor process could be employed in an effectively CO_2-tolerant way, the gas export pipelines from the wells could be used for CO_2 transport to the wells, helping to make this route to CO_2 sequestration more cost-effective. Other Heatric ideas for offshore uses include offshore production of methanol for hydrate suppression.

9.3.5 HiGee for enhanced oil recovery – surfactant synthesis

The High Gravity Engineering and Technology Unit at Beijing University of Chemical Technology has recently extended their activities to intensification associated with enhanced oil recovery (EOR), although in a different way to Heatric (Zhang et al., 2010). There the group has been working on a rotating packed bed (RPB) in order to synthesis petroleum sulphonate (PS) that is a surfactant used effectively in chemical flooding – a technique to reduce the surface tension between the oil and the water in the well, allowing them to be more readily separated. The PS is added to the water being injected into the well to enhance recovery.

Falling film and stirred tank reactors are conventionally used to make PS, although the reaction is fast (and exothermic). Mass transfer, micro-mixing and rapid heat removal are features of successful reactor operation – the feeds being sulphur trioxide and, in this case, a petroleum fraction from crude oil from the Shengli field.

The HiGee experimental set-up is illustrated in Figure 9.4.

The outcome of these tests was highly positive. Data from an STR experiment are compared with results from the RPB rig in Table 9.1. The principal conclusions were that: firstly, the solvent plays an important role in the HiGee reaction, and this confirmed a feature of the RPB – that uniform reactant distribution is necessary for fast exothermic reactions. Secondly, as side reactions can be enhanced, too, care should be taken to determine optimum reaction time and other conditions. Thirdly, above certain values of rotation speed and circulation time, the benefits were less obvious. And finally, strong intensification was observed, resulting in higher product quality and process efficiency compared to the use of STRs.

Further data on HiGee – via the GasTran system for deaeration of water – are given in Section 9.3.6.2. This is also relevant to EOR – see also Harbold and Park, 2008.

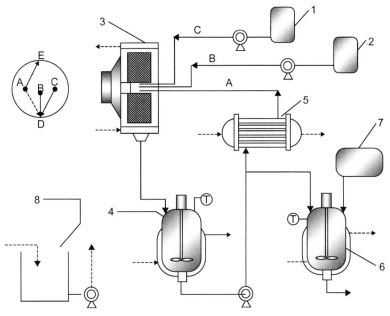

FIGURE 9.4

The experimental rig to test HiGee sulphonation: (1) feedstock storage tank; (2) sulphonating agent storage tank; (3) RPB; (4) circulating tank; (5) heat exchanger; (6) neutralisation stirred tank; (7) ammonia storage tank; (8) circulating cooling system (Zhang et al., 2010).

Table 9.1 Comparison of RPB and STR Performance (Zhang et al., 2010)

Reactor Type	Active Matter (wt.%)	Unsulphonated Oil (wt.%)	Volatiles/ Inorganic Salts (wt.%)	Operation Type	Average Residence Time	Energy Consumption Per Unit o/p
RPB	45.3	23.5	25/6.2	Continuous	15 min	0.8
STR	30.2	39.8	25/5	Semi-cont.	6 h	1.0

Note: The interfacial tension in the case of the RPB was 4.5×10^{-3} and for the STR 6×10^{-3}.

9.3.6 Deoxygenation using high gravity fields

In oxygen extraction, contacting plays a necessary and important role in the process. The most common techniques for contacting are packed, plate or bubble columns, or agitated vessels. These can be large, particularly when large throughputs are involved. The use of enhanced acceleration, with g-forces over 1,000 g, using either rotating packing or tangential injection of liquid into a static vessel enables significant intensification to be obtained. Examples of these two techniques are discussed below.

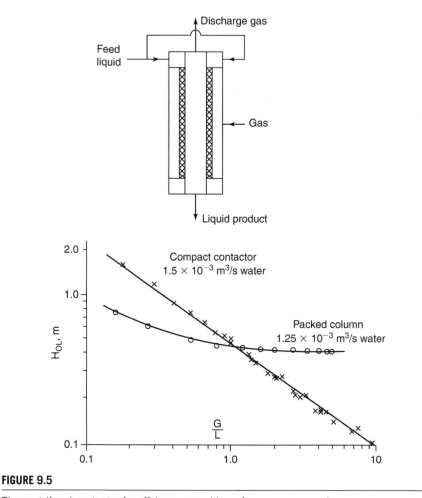

FIGURE 9.5

The centrifugal contactor for offshore use, with performance comparison.

9.3.6.1 Centrifugal oxygen removal using liquid injection for 1,000 g production

In the injection system (Waldie, 1995), liquid is fed tangentially to form an annular spinning layer on the inside of a permeable tube. Gas passes inwards through the permeable wall and bubbles through the liquid. Centrifugal forces encourage the inward motion of the lower density gas and the formation of a central gas core. With appropriate design, the liquid leaving the base of the unit is free of gas, and the gas leaves through a central tube at the other end (see Figure 9.5).

Processing intensities for the deoxygenation of water can be over two orders of magnitude higher than for packed or bubble columns, particularly for high values of gas/liquid volumetric flow rates. The unit is designed for offshore applications, where space is at a premium. In terms of performance factors, the volumetric mass transfer coefficient, $K_l a$ increases as $G^{0.69}$, where K_l is the overall liquid mass transfer coefficient (kg mol/m^{3}s), a is the specific surface area (m^2/m^3), and G is the total gas flow rate (m^3/s). Comparison with other plant can be made for different values of G/L, where L is the liquid flow rate (m^3/s). The economics of the system become very attractive when used offshore, due to reduced footprint. The footprint can be further reduced by miniaturising the hydrocyclone and putting it downhole.

9.3.6.2 HiGee for deoxygenation

Deoxygenation or deaeration of seawater using HiGee technology started at Newcastle University in the UK, some decades ago, and later involved Glitsch, (the Dallas-based US company now called Koch-Glitsch) in trials of rotating packing for a number of HiGee applications; and Conoco with the Orkney Water Test Facility, where a 750 tonne/day deaerator was proposed. These included, as well as oxygen removal, putting ammonia into water, CO_2 absorption, H_2S absorption into amines and methanol/ethanol distillation, as well as volatile organic compound (VOC) stripping (Fowler, 1986; Fowler and Khan, 1987). Latterly, GasTran Systems has taken up the mantle in the US.

Recently, the use of HiGee for seawater deoxygenation has been extended into practice in China (Chen et al., 2003), with commercialisation beginning in 1998, based upon work at Beijing University of Chemical Technology (see also Chapter 1).

Initial work involved a 50 t/h HiGee unit using natural gas at 0.07 MPa to deoxygenate water from the Yellow River, which had a 6–12 ppm content of oxygen. The oxygen content was reduced to less than 0.05 ppm. The natural gas was then reinjected back into the gas network. Operating parameters examined included gas/liquid ratio (G/L), (the most significant), liquid flow rate and rotational speed. It was found that when G/L was around unity (L = 50 m^3/h), the oxygen content in the outlet water dropped as rotational speed increased, but for high G/L (>2) the speed had minimal effect. An optimum speed of 960 rpm was found. Gas pressure drop and HiGee power consumption were also investigated. As G/L approached 2, with a gas flow rate of 90 m^3/h, the power used was 20 kW at 900 rpm. Under this condition the oxygen in the outlet gas was 1,500 ml/m^3 (0.15%).

Scale-up led to two HiGee units for deoxygenation of well water in the Shengli oil field of the China Petrochemical Corporation. Water throughput/unit was 250 t/h. The rotor inner diameter was 600 mm and the outer diameter 900 mm. Power consumption was typically 120 kW and the outlet O_2 content was, <5 parts/billion, compared to an inlet level of 8–10 ppm. Data on the relative benefits of this unit compared to conventional technology are given in Table 2.2 in Chapter 2. The units are shown in Figure 9.6 and, with the platform shown earlier in Chapter 2.

The vacuum tower system that the HiGee unit replaced had two critical problems:

1. High structural loading, especially wind loading on the platform due to weight and height, which increased the investment in platform structure.
2. The need for a reducing agent to further lower the residual oxygen (this led to increased operating cost).

FIGURE 9.6

6,000 t/d seawater deaerator rotating packed-bed unit on platform.

The US company, GasTran Systems, has been associated with a rotating packed-bed unit for deaeration of water flood for secondary oil recovery that can be used on onshore oil wells. The motivation is corrosion prevention and hence cost reduction. It has been estimated that the annual cost of corrosion in the US oil and gas industry is of the order of $1.4 billion, with a sum of $320 million going to capital expenditures for corrosion control (data supplied by CC Technologies, see Appendix 5). The primary cause of corrosion in the downhole tubing, surface production and processing piping, is the presence of gases and chemicals in the water flood and produced water. Dissolved oxygen, carbon dioxide and hydrogen sulphide, both individually and in combination, do most of the corrosion damage. Chemical corrosion inhibitors can be expensive, with a typical oil field spending $7 million or more per year. Chemical dosing pump breakdowns and insufficient feed monitoring are common in the industry, leading to either overfeed or underfeed of chemicals over long periods. More importantly, the byproducts of some chemical additives produce sulphate compounds, exacerbating H_2S production in anaerobic conditions and ultimately degrading the value of the oil. Metal passivation, the alternative sometimes employed, necessitates the use of expensive pipe and/or tube alloys.

GasTran Systems uses the rotating packed-bed concept for deaeration, thus significantly reducing the costs and improving the economics of onshore water flooding. By performing the majority of work in removing oxygen, CO_2 and H_2S, the GasTran Systems equipment can effectively protect against corrosion in water flood applications. The RPBs can be easily adapted to strip dissolved gases with either methane, nitrogen, or under vacuum. Features of the GasTran Systems unit relevant to the oil well deaeration application are:

- Reducing dissolved oxygen and CO_2 to less than 2 parts per billion (ppb).
- Excellent H_2S removal – 99.5% with a single pass.

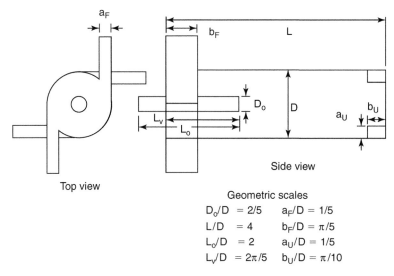

Geometric scales

D_o/D	$= 2/5$	$a_F/D = 1/5$
L/D	$= 4$	$b_F/D = \pi/5$
L_o/D	$= 2$	$a_U/D = 1/5$
L_v/D	$= 2\pi/5$	$b_U/D = \pi/10$

FIGURE 9.7

Dimensions of a 5 mm diameter hydrocyclone (volume = 0.393 ml).

- Highly scalable operation with turn-down ratio of 95%.
- Reaches steady state performance in seconds.
- Low operating pressure requirements.
- Stainless steel construction with alternatives available.
- Lightweight and small footprint.

9.3.6.3 Miniature hydrocyclones

Some years after Professor Waldie's work, the idea of using hydrocyclones to effect liquid–liquid separation downhole was examined. The development was motivated by the need to deal effectively with produced water – that arising from oil and gas wells and 'contaminated' with these products. If the water can be extracted from the oil or gas and reinjected into the sea, cost savings could be substantial. Similarly, if the system could handle well outputs where the water content is very high (water:oil ratios of up to 50:1 in mature wells) the exploitation of such mature wells could be encouraged.

In order to implement this, in areas where the diameter of equipment would be limited by the hole size, miniaturisation was required. Petty and Parks (2004) at Michigan State University in the US stated that large diameter production wells could accommodate hydrocylones of 20–76 mm diameter, but these are too large for small diameter production wells. Also, the turbulent nature (Re around 10,000) of the flow in these can make the water–oil mixture difficult to separate. The work of Petty and Parks looked at the modelling of hydrocylones of diameters of less than 10 mm (see Figure 9.7). These would operate, it is claimed, in parallel, to handle the high flow rates of large wells with high water to oil outputs. The miniature cyclones would, when stacked, have a capacity of greater than 10 m^3/h, but each one would have a capacity very much less than 0.2 m^3/h and a very low pressure drop (<1 bar). To handle large flows, many hundreds would be needed.

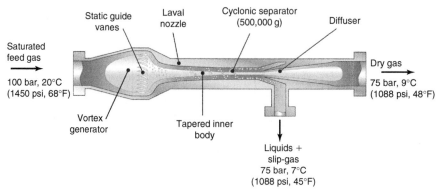

FIGURE 9.8

Cross section through the Twister.

The simulation suggested that the capacity of 413 small hydrocyclones was equivalent to that of a single 76 mm diameter hydrocyclone – about 124 litres/min. The total volume of the 413 units would be only 162 cm³, while a large single unit would occupy 7,600 cm³.

9.3.6.4 Supersonic gas conditioning – the Twister

The Twister, produced in The Netherlands by Twister BV (Betting and Epsom, 2007; Van Eck and Epsom, 2006), also incorporates a cyclone stage for gas/liquid separation, downstream of an expander and upstream of a recompression unit, packaged as shown in Figure 9.8.

Saturated feed gas enters a vortex generator to induce swirl before it is cooled adiabatically, condensing out the moisture, in a Laval nozzle. The cooled gas/liquid (or fine droplets) mixture is subjected to increased swirl in the nozzle which generates g-forces of about 500,000 g, whereupon the liquid is bled off with slip-gas before the bulk of the now-dry gas is diffused to dissipate the kinetic energy and increase the static pressure at the exit. The separation efficiency is typically 95%.

The first commercial Twister installation was implemented by Petronas and Sarawak Shell Berhad (SSB) on the B11 offshore gas processing facility, in December 2003, where 12 Twisters are now used (Cottrill, 2012). The unit dehydrates 600 MMcfd of non-associated sour gas fed to the onshore Malaysian LNG plant at Bintulu, to control pipe corrosion, and has at the time of writing been in continuous operation since start-up with over 99% availability. Table 9.2 compares capital expenditure for the Twister and other options for offshore gas development – see Figure 9.9.

The higher capital cost of the Twister, compared to transmitting wet gas, is generally more than balanced by savings in the pipeline and onshore facilities, particularly where corrosion may be otherwise anticipated. The manufacturers of the Twister also believe that it can be used subsea. Possible uses identified by a feasibility study that was supported by the Norwegian Research Council included

Table 9.2 Capital Expenditure on Surface Facilities, Comparing Conventional and Twister Options.

Capital Expenditure (surface) Scope

1000 MMscfd 100m Water Depth	TEG	Wet Gas CRA	Wet Gas CS	Twister	TEG	Wet Gas CRA	Wet Gas CS	Twister
Platform	185	120	120	165	Manned	Unmanned	Unmanned	Unmanned
Pipeline (100km) (diam. inches)	45	120	50	35	3x24" CS	3x24" CRA 6" CS	3x24" CS 6" CS	40" CS 12" CS
Slugcatcher	35	35	35	–	Yes	Yes	Yes	No
Onshore treatment	50	55	55	35	Gas/liquid treatment	Gas/liquid treatment. MEG regen.	Gas/liquid treatment. MEG regen.	Liquid treatment
Total US$M (%)	315	330	260	235	(121%)	(127%)	(100%)	(90%)

Notes: CRA = corrosion resistant alloy; CS = carbon steel; MEG regen.=glycol regeneration.

FIGURE 9.9

The Petronas SSB B11 Platform (lefthand side), and the Twister installation on it.

debottlenecking existing offshore or onshore processing facilities by subsea gas conditioning such as dehydration and dewpointing.

The latest Twister application is to be in the Tunu processing facility in Nigeria, operated by Shell. The six tubes will have a capacity of 4.5 million cubic metres/day.

Pre-treatment of sour gas is also being examined as an application for the Twister. The unit allows a large reduction in natural gas temperature, facilitating the low temperature removal of carbon dioxide and hydrogen sulphide.

9.3.7 RF heating to recover oil from shale

Oil shale is rock that contains oil, and it is estimated that if the oil could be efficiently extracted from the shale in an environmentally friendly manner, a largely untapped reserve of oil could be opened up to extend the life of our oil economy. Originally reported in 2006, but summarised in an aerospace journal in 2008, when high oil prices made oil from shale more attractive, the Raytheon company in the US has introduced radio frequency (r.f.) heating as a way of easing oil removal from the rock (Anon, 2008). The r.f. energy (see Chapter 3) allows the shale to release the deposits of thick oil, which can then be transported to the surface with the help of a supercritical extraction process, developed by another US company, CF Technologies. Supercritical extraction is briefly discussed in Chapter 5. By the time both processes have been employed, the oil is a light product, comparable to kerosene. The company suggests that the techniques could be used in other difficult extraction areas, such as marginal reserves, tar sands and spent wells.

9.4 Floating production, storage and offloading systems (FPSO) activities

FPSOs are increasingly being used for a variety of processes offshore. They are either converted tankers/bulk carriers or new-build purpose-built processing vessels.

Generally, they may be considered as a floating equivalent to an oil or gas platform such as those the UK has in the North Sea, but can be readily moved to other locations and tend to accommodate an increasing variety of unit operations associated with oil and/or gas processing. They are ideal candidates to host intensified unit operations.

The swirl flow devices discussed in Chapter 3 used as tube insets to aid mixing and heat transfer have found ready application in many FPSOs. CalGavin reports that Total has used a combination of shell-side and tube-side enhancement to reduce the size, costs and fouling potential of heat exchangers on the Dalia FPSO in the Gulf of Guinea. The enhancement technology saved around 269 t in wet weight and cut the plot area needed by 75% (78 m^2). Keeping down equipment size and weight while improving reliability is always critical offshore – and never more so than when operators are pushing the limits of current technology. (For example, the Dalia field in the Gulf of Guinea, off the coast of Angola. Thought to contain up to a billion barrels of oil, Dalia extends over 230 km^2 at water depths of up to 1,500 m). The oil is viscous, sour, and hidden in complex geological structures under more than 700m of unstable sediment (Web 1, 2012).

The heat exchangers are employed to transfer heat between dry crude oil (after passing through an oil/water separator) to the incoming wet crude. The size reduction allows nine heat exchangers to be reduced to two, with the approximate wet weight reduced from 400t to 131 t. Cost saving is approximately 65% and it is achieved using the hiTRAN tube inserts in HELIXCHANGER helical baffled shell-and-tube heat exchangers.

Velocys, now part of Oxford Catalysts, is one of the leading players in attempting to extend intensification on to FPSOs. The technology was introduced in the UK by John Brophy (Brophy, 2004), and data were subsequently published by the company (Tonkovich et al., 2008). The Velocys technology (see also Chapter 5 in the context of heat exchanger reactors and micro-reactors) may be used for offshore gas upgrading, producing syncrude from local natural gas reserves. The process involves steam reforming, followed by a Fischer-Tropsch reaction and hydro-cracking to produce, in this case, diesel fuel, as shown in Figure 9.10.

The unique feature of the process, as applied on an FPSO, is the use of microchannel reactors with channel sizes of 0.25 to 5 mm, capable of handling heat fluxes and reaction rates orders of magnitude higher than conventional reactors. The steam reformer has, of course, combustion on one side of the plates and an endothermic reaction on the other side. The overall plant demonstrates flexibility due to modularisation, typical outputs being 100 to 10,000 barrels/day. The largest unit, illustrated in Figure 9.11, is located on a conventional hull.

The paper by Tonkovich et al. cited above deals with methanol production, tailored to an FPSO vessel, and employing plants similar to those designed by Heatric Ltd. One of the principal differences between the plant described earlier and the methanol unit is the need to carry out distillation and compression processes, the flowsheet being shown in Figure 9.12.

The features of this plant that make it compatible with FPSO use are short distillation columns to reduce the impact of vessel tilting, minimum freshwater needs

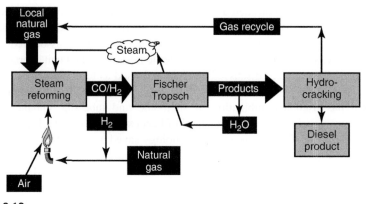

FIGURE 9.10

The processes on the FPSO for diesel production from natural gas.

(courtesy of Velocys)

FIGURE 9.11

The Velocys concept, located on a vessel.

and a minimum number of components, aided by the compactness afforded by the design of the unit operations. Steam reforming is performed, and water from the reformer is recycled, with water from the distillation unit. The compressor delivers dry product gas at 50 bar to the methanol synthesis reactor – a micro-channel unit with minimum recycle needs – a conversion of 70.5% for a single pass through the reactor is claimed. Methanol distillation results in >95% purity. The plant would be made up of nine assemblies, each being 3.9 m long, 3.9 m high and 1 m wide. An

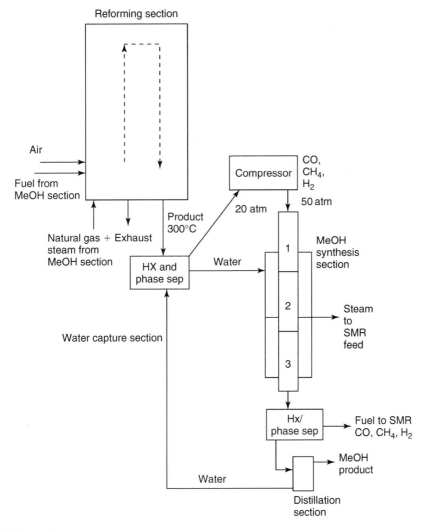

Reforming section

Air

Fuel from
MeOH section

Compressor

CO,
CH$_4$,
H$_2$

50 atm

20 atm

Product
300°C

Natural gas + Exhaust
steam from
MeOH section

HX and
phase sep

Water

1

MeOH
synthesis
section

2

Steam
to
SMR
feed

Water capture section

3

Hx/
phase sep

Fuel to SMR
CO, CH$_4$, H$_2$

MeOH
product

Water

Distillation
section

FIGURE 9.12

Integrated methanol production flow sheet – a simulation was carried out based on a
1,000 tonne/day output.

interesting aspect of the design within micro-channels is the use of a capillary capture structure (a wick – akin to the micro-heat pipe components in Chapter 11) to
take the water in liquid form out of a vapour–liquid stream passing through a channel. This is used to take water, by capillary action, from the synthesis gas and from
the off-gas from the methanol reactor. The Velocys concept also contains a further

FIGURE 9.13

A biomass-to-liquids plant in Austria using Velocys micro-channel reactors.

(Courtesy of Oxford Catalysts Group).

innovation – a micro-channel distillation unit, a possible alternative to the HiGee units described elsewhere.[1]

Although not offshore, but neatly illustrating the ability of the systems to be packaged, the Velocys technology has recently been employed successfully in a biomass-to-liquids (BTL) demonstration plant, jointly operated by the Oxford Catalysts Group and the Portuguese incorporated holding company, SGC Energia (SGCE), at the biomass gasification facility in Güssing, Austria, is now up and running (Figure 9.13). The plant is designed for the small-scale distributed production of biofuels via the Fischer–Tropsch (FT) reaction. The demonstration plant, which is being managed by SGCE, has been fully operational for over a month. Initial results indicate that the equipment – including the Güssing gasifier, a gas conditioning unit supplied by SGCE and an FT micro-channel reactor skid developed by Velocys – is all operating smoothly. The FT micro-channel reactor, comprised of over 900 full-length micro-channels, is working effectively. It is proving to be very efficient at controlling temperatures in highly exothermic (heat-generating) FT reaction and at maintaining isothermal conditions throughout the reactor. The demonstration plant is already producing over 0.75 kg of high quality synthetic FT liquids per litre of catalyst per hour – 4–8 times greater productivity than conventional

---

[1] If we want to overcome the influence of tilt on vessels in distillation or other separation processes, reducing the height of the column by using large numbers of micro-channels is one option. The other, and possibly more stable process, is to apply a continuous high gravity field that dwarfs any perturbations brought about by vessel motion.

systems. The unit is also demonstrating robust responsiveness to shutdowns and start ups. These results confirm the significant process intensification potential of the FT technology. Performance will improve further after the steam superheating section of the plant is debottlenecked at the next scheduled shutdown in the next couple of months. Going forward, the demonstration plant will be operated over a range of conditions to establish and confirm performance. It also will be tested in an extended three-month steady-state run (Anon, 2012).

(The reader interested in a different approach to gas-to-liquids or gas-to-ethylene processing may wish to study the concept proposed by Kenneth R. Hall at Texas A&M University (Hall, 2005) which eliminates the need for syngas production).

9.5 Safety offshore – can PI help?

The offshore industry is particularly sensitive to safety concerns – not surprisingly as there have been major accidents, including the Piper Alpha fire in the UK North Sea – involving platforms handling oil and gas. The HSE Publication 'Improving Inherent Safety' (Mansfield et al., 1996) makes several references to PI within the context of offshore technology. It is useful to read the introduction to the section Approaches to Hazard Management, given below:

Hazard management approaches are often described as being based on three main principles: prevention, control and mitigation. Prevention relates to measures taken to eliminate or reduce the hazard at source, or to reduce the likelihood of it being realised. Control measures are those taken to keep the hazard within the design envelope, either by containment or control systems, or to actively react to events that could result in an accident. Mitigation measures are those intended to deal with the hazard once the accident has occurred, either to bring the hazard back under control or to limit its effects.

…The concept of inherent safety may provide a means to turn the attention of designers towards elimination and reduction, and this in turn could lead to a better integrated combination of prevention, control and mitigation measures.

Inherently safer plant uses basic design measures to achieve hazard elimination, prevention and reduction. The classic definition of an inherently safe plant or activity is one that cannot under any circumstances cause harm to people or the environment. This may be because:

- *It only uses materials that are harmless;*
- *It has such small inventories of hazardous material that these are insufficient to cause significant harm even if released;*
- *The hazardous materials are held in a form or under conditions that render them effectively harmless (diluted, at ambient temperatures and pressures).*
- *The same principles could be applied to hazardous equipment, in that inherently safe equipment would not involve high energy systems such as pressure, vacuum, or high speed rotation.*

Recognising that the attainment of the classical definition is unlikely (and as defined above, would not involve high-speed rotating plant – a gas turbine or a high speed spinning disc reactor, for example) – the emphasis seems to be on *safer*.

Most of the applications of PI offshore have involved one or more of the following:

- Inventory reduction.
- Weight reduction.
- Volume reduction.
- Reducing manpower needs.

Safety is recognised as being of prime importance in offshore environments, and the contribution of PI to hazard reduction is substantial. The HSE report concludes (in part) that:

> *The order of magnitude reductions in inventory needed to significantly reduce the size and duration of fire and explosion hazards, may not be achievable without some step change advances in separation technology.*

It may be said that advances in separation technology during the past 20 years should enable this concern to be overcome – HiGee and the micro-distillation column of Velocys are two examples in support of this. In a positive vein the report suggests that by combining multi-phase metering and pumping capabilities with intensified separations and heat transfer equipment, subsea completions and transfer of products to shore may be feasible.

Ed Terry of Sauf Consulting (now incorporated within Galbraith Consulting) presented views on the offshore industry attitude to PI at a PIN meeting (Terry, 2004). A sticking point for the implementation of PI in some offshore processes might be, he pointed out, the ability of current production facilities to change capacity or throughput rapidly. Platforms are, he said, designed for 20–120% production. Balancing this was the trend to take equipment packages offshore, and PI tends towards being embedded in such packages.

9.6 Summary

The offshore oil and gas sector, be the plants used by the industry necessary to extract and process the fluids located on ships, platforms, down pipes or under the sea bed, was one of the first to embrace process intensification with compact heat exchangers such as the PCHE. The sector has pioneered other PI uses in separations and more recently in reactions, and the potential for growth and further innovation is considerable. More extensive replication onshore, as the demand for these resources continues to grow, is likely.

References

Anon, 1991. High tech exchanger has it all on a plate. Offshore Eng. (May).

Anon, 1998. EEBPP Good practice module: experience in the operation of compact heat exchangers. GPG 198, DETR. Available from: <www.thecarbontrust.co.uk> 2007.

Anon, 2004. Minutes of the Tenth Meeting of the Process Intensification Network. 3 June, Edinburgh. Heriot-Watt University. See <www.pinetwork.org>.

Anon, 2008. Industry outlook: black gold. In: Parmalee, P.J. (Ed.), Aviation Week and Space Technology. January 28, p. 390. See <www.aviationweek.com/awst>.

Anon, 2012. Biomass to liquids plant in Austria. <www.oxfordcatalysts.com/press/egs/epe.1110.p.42-43.pdf> (accessed 20.09.12.).

Banister, J., Rumbold, S., 2005. A compact gas-to-methanol process and its application to improved oil recovery. Heatric web site publication. See <www.heatric.com>.

Betting, M., Epsom, H., 2007. Supersonic separator gains market acceptance. World Oil, 197–200. (April).

Brophy, J., 2004. Modular gas-to-liquids technology. Paper Presented at Tenth Process Intensification Network Meeting. 3 June, Edinburgh. Heriot-Watt University. See <www.pinetwork.org>.

Chen, Jian-Feng, et al., 2003. HiGee Technology and Applications: Novel Technology for Chemical Reaction and Separation. Chemical Industry Press Co.., Chapter 5, ISBN 7-5025-3842-9/TQ.1541, (in Chinese).

Cottrill, A., 2012. Tunu deal is the next leap ahead for twister. Technology Upstream, 27 April, upstreamonline.com (accessed 15.08.12.).

Fowler, R., 1986. Intensified mass transfer with HiGee. Presentation to the Association Francaise Techniciens Petrole, 13 May.

Fowler, R., Khan, A.S., 1987. VOC removal with a rotary air stripper. Session 181, Proceedings of the AIChE Annual Meeting, New York, 15–17 November.

Hall, K.R., 2005. A new gas to liquids (GTL) or gas to ethylene (GTE) technology. Catal. Today 106, 243–246.

Harbold, G., Park, J., October 2008. Single-stage vacuum deaeration technology for achieving low dissolved gases in process water. Paper IWC-08-22, Proceedings of the International Water Conference, San Antonio, Texas.

Knott, T., 1993. PCHEs slice gas treatment costs. Offshore Eng., 14–16. (October).

Mansfield, D., Poulter, L., Kletz, T., 1996. Improving inherent safety. Health and Safety Executive Offshore Technology Report OTH 96 521. HSE Books.

Oswald, B., 1999. Process intensification: an oil and gas perspective, 'More and Less'. Minutes of Inaugural Meeting of the Process Intensification Network. April. Newcastle University. See <www.pinetwork.org>.

Petty, C.A., Parks, S.M., 2004. Flow structures within miniature hydrocyclones. Miner. Eng. 17, 615–624.

Reay, D.A., 1999. Learning from experiences with compact heat exchangers. CADDET Analysis Series No. 25, June, The Netherlands. CADDET.

Tonkovich, A.L., Jarosch, K., Arora, R., Silva, L., Perry, S., McDaniel, J., et al., 2008. Methanol production FPSO plant concept using multiple microchannel unit operations. Chem. Eng. J. 135S, S2–S8.

Van Eck, P., Epsom, H.D., 2006. Supersonic gas conditioning – introduction of the low pressure drop Twister. Proceedings of the European Annual Meeting of the Gas Processors Association, Oslo, 20–22 September.

Waldie, B., 1995. Novel high intensity gas-liquid contactor. Proceedings of the First International Conference on Science, Engineering and Technology of Intensive Processing. September. Nottingham University.

Web 1., 2012. Boosting heat transfer on the Dalia FPSO. <www.calgavin.com> (accessed 01.09.12.).

Weissman, J.G., 1997. Review of processes for downhole catalytic upgrading of heavy crude oil. Fuel Process. Technol. 50, 199–213.

Weissman, J.G., Kessler, R.V., 1996. Downhole heavy crude oil hydroprocessing. Appl. Catal., A: Gen. 140, 1–16.

Zhang, D., Zhang, P.-Y., Zou, H.-K., Chu, G.-W., Wu, W., Zhu, Z.-w., et al., 2010. Application of HIGEE process intensification technology in synthesis of petroleum sulfonate surfactant. Chem. Eng. Process. Process Intensification 49, 508–513.

Application Areas – Miscellaneous Process Industries

10

OBJECTIVES IN THIS CHAPTER

Our discussions of applications in the previous two chapters have centred on chemical processing, the principal application area of process intensification. However, other industrial sectors (and other aspects of our society – see Chapter 11) can benefit from PI, and several successful examples in industry, ranging from nuclear reprocessing to food and drink, exist. Here we explore these, and hope that the reader will employ his or her 'technology transfer' skills to adopt the experiences of PI users in other sectors to his own, if different.

10.1 Introduction

The chemicals (bulk, fine and pharmaceutical) and offshore industrial sectors are not the sole preserves of process intensification, as has already been illustrated in this book. In this chapter, the reader is introduced to other process industry sectors where PI has made significant inroads into unit operations. Some sectors adopted intensification techniques in advance of the chemical industry (e.g. nuclear reprocessing), others pioneered techniques, such as ultrasonics, decades before they became accepted in areas such as chemical reaction intensification (e.g. in metallurgy). The food sector uses microwaves routinely for speeding up cooking processes, and effluent treatment can benefit from oscillatory baffle reactors. Aerospace, although perhaps not regarded as a process industry, is also included in this chapter.

10.2 The nuclear industry

In recent years, there has been a growth of interest in nuclear power, due to the realisation that in countries like France, where nuclear power dominates electricity production (accounting for about 80%), carbon dioxide emissions are low compared to countries that rely on the combustion of fossil fuels such as coal, oil or gas. Other than recent problems in Japan (the failure of the Fukushima nuclear power station) and Germany, (where all nuclear plants are, under current plans, to close by 2022) nuclear power remains a strong option in the UK and elsewhere. Although the use of process intensification is not obvious in nuclear reactors, the development by Heatric of highly compact heat exchangers for reactor use is of considerable interest (see also Chapters 4 and 5).

10.2.1 Highly compact heat exchangers for reactors

Some years ago Heatric presented their thinking and detailed their activities in developing heat exchangers for the new types of reactor (Li et al., 2006). The paper highlighted the concepts that Generation IV and other programmes are contemplating. These reactors vary in type from very high temperature gas-cooled reactors (VHTR) to liquid metal fast reactors (LFR and SFR) with cooling media that include:

- Helium
- Supercritical carbon dioxide
- Sodium
- Lead
- Molten salts

Activity is not just focused on production of electrical power with efficiency greater than that associated with the Rankine Cycle (typically 30–35%). There is now genuine interest in nuclear energy as a heat source for hydrogen production, via the sulphur iodine process (SI) or high temperature electrolysis – see Chapter 6 for a discussion on intensified electrolysis.

The production of electrical power at higher efficiency via a Brayton Cycle using helium as the working fluid, as well as hydrogen production, requires both heat at higher temperatures, up to 1,000°C, and high effectiveness heat exchange to transfer the heat to either the power or process cycle. This presents new challenges for heat exchangers. If plant efficiencies are to be improved there is a need for:

- Highly effective heat exchange at minimal pressure drop.
- Compact heat exchange to improve safety and economics.
- An ability to build coded heat exchangers in a variety of nickel-based alloys, oxide dispersionstrengthened alloys (ODS) and ceramic materials to address the temperature, life and corrosion issues associated with these demanding duties.

Heatric has already given consideration to many of these challenges. Their printed circuit heat exchanger (PCHE) and formed plate heat exchanger (FPHE) technology, which are commercially available, will fulfil all of the duties up to temperatures of 950°C. In addition, products currently under development will further increase the temperature and pressure range, while offering greater corrosion resistance and operational life. Examples are shown in Figures 10.1 and 10.2. In the case of recuperators for the helium Brayton cycle, Heatric suggest that the FPHE is feasible (Southall and Dewson, 2010) as pressures are modest, and at high temperatures 800H or 617, alloys might be used.

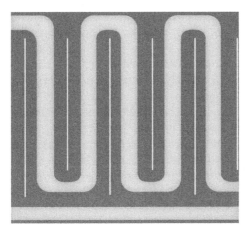

FIGURE 10.1

Chemically machined nickel alloy 617 plate.

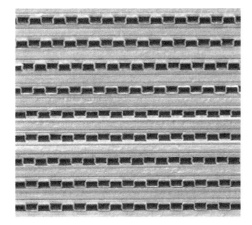

FIGURE 10.2

Diffusion-bonded formed plate heat exchanger, developed within Heatric's reactor programme.

Another type of reactor uses sodium as the coolant, and here Heatric suggest that a hybrid heat exchanger combining the merits of the PCHE and FPHE might be used. This intermediate heat exchanger (IHX) would transfer heat between the liquid sodium and the CO_2. The nature of the two fluids necessitates different considerations on each side of the exchanger – sodium needs larger (finned) channels and exhibits very high heat transfer coefficients, while CO_2 is at a relatively high pressure, in excess of 200 bar but is compatible with smaller etched channel sizes.

In the USA, the PCHE has been identified as a potential contributor to the Next Generation Nuclear Plant. Work at Idaho National Laboratory (Ravindran et al., 2010) on a model of a 600 MW plant using helium as the primary and secondary coolant, involved simulating a loss of coolant accident (LOCA), one of the critical events that might occur, (hopefully, very rarely) in the operation of a nuclear power plant. Starting with a coolant mass flow of just under 300 kg/s, the LCA simulation showed coolant flow decreasing to zero within 45 s. Thermal profiles within the PCHE were predicted as a result of this LCA.

The development of new reactor types (or in many cases upgrading designs that date from the 1950s and 1960s) has suffered in many countries due to embargoes on new construction, in part because of safety concerns. Many do, however, believe that a balanced low carbon future will include a mix of renewable energy sources, clean fossil fuel use and nuclear energy.

10.2.2 Nuclear reprocessing

PI technology is relevant to nuclear reprocessing on at least two counts: life cycle analysis and criticality. While decommissioning costs are not given much attention in the design of conventional (non-nuclear) plant, they are an increasingly dominant consideration when nuclear installations are being planned. It goes without saying that decommissioning costs will be less if the plant mass can be reduced, and this is one of the principal attractions of PI for the industry. The second aspect, criticality, is a consequence of the need for the radioactive contents of any reactor or separator to be held well below the critical limit, above which uncontrolled energy release is triggered. Given the size of conventional separators and reactors, this means that the operational concentrations of reactive species must be held at comparatively low values that are not necessarily optimal for the process in question. Where equipment volumes can be dramatically reduced, this constraint no longer applies and much higher concentrations can be accepted, leading to higher equipment effectiveness.

10.2.3 Uranium enrichment by centrifuge

The centrifuge has become a hot potato in recent years, due to the initiation of programmes in countries such as Pakistan and Iran to enrich uranium ultimately to allow nuclear weapons to be constructed.

The Urenco Group, responsible for uranium enrichment, is a British/German/Dutch organisation with processing facilities in all three countries. The role of enrichment plants is to increase the concentration of U-235 isotope in the fuel to a level acceptable for use in light water reactors or advanced gas-cooled reactors, used in many power stations. Urenco's existence depended upon the reliability of high speed rotating equipment. The uranium enrichment service, which uses gas centrifuges, is a mature business of over 25 years standing with an annual turnover of £300 million. Centrifuge improvements keep the plant competitive.

Many tens of thousands of centrifuges are used in cascade. The feed is uranium hexafluoride with 0.7% U-235, and this is concentrated by the bank of centrifuges to 3–5% U-235. The UF_6 is pumped to containers in cooling chambers. The original centrifuge was the Zippe type which was developed in the Soviet Union by a team of 60 German scientists working in detention, captured after the Second World War. The centrifuge is named after the team's leading experimenter, Gernot Zippe (see Figure 10.3). Data on later types are less accessible. The gas centrifuge creates a very steep pressure ramp across the radius; gas enters at the axis and is removed at the periphery. Novel bearings (pin in a cup of oil) and magnetic bearings are used, and the rotor is under vacuum.

The separation power is proportional to (peripheral speed)^4 and length. R&D is driven by the need to attain higher speeds without proportional increases in manufacturing cost. The original aim was a design life of ten years and a 1% failure rate per annum. However, much older units are still running (second-generation machines, Urenco are now looking at the sixth generation). Essentially, no maintenance is needed; the centrifuges outlive other parts of the plant. Speeds are greater than 1,000 rps.

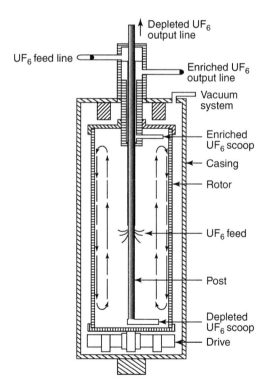

FIGURE 10.3

The Zippe centrifuge.

Table 10.1 A selection of centrifuges used around the world, with outputs.

Type	P1	P2	Russia	Urenco	US
Rotor material	A1	MS	CFRC	CFRC	CFRC
Speed (m/sec)	350	500	700	700	>700
Length (m)	1–2	1	<1	3–4	12
kg SWU/yr	1–3	5	10	40	300

It is historically interesting to note that after the captured scientists were released in 1956, Gernot Zippe was surprised to find that engineers in the West were years behind in their centrifuge technology. He was able to reproduce his design at the University of Virginia in the US, publishing the results, even though the Soviets had confiscated his notes. Later, in Europe during the 1960s, Dr Zippe and his colleagues made the centrifuge more efficient by changing the material of the rotor from aluminum to a stronger alloy called maraging steel, which allowed it to spin even faster. This improved centrifuge design is used by Urenco to produce enriched uranium fuel for nuclear power stations.

The exact details of advanced Zippe-type centrifuges are closely guarded secrets, but the efficiency of the centrifuges is improved by making them longer and increasing their speed of rotation. To do so even stronger materials, such as carbon fibre reinforced composite materials, and various techniques are used to avoid forces causing destructive vibrations, including the use of flexible bellows to allow controlled flexing of the rotor, as well as very careful control of the rotation speed to ensure that the centrifuge does not operate for very long at speeds where resonance is a problem.

Table 10.1 shows the basic parameters of centrifuges, including uranium output, used in Russia, the US and the UK, where carbon fibre reinforced composites are used as wall material. Pakistan's nuclear programme developed the P1 and P2 centrifuges – the first two centrifuges that Pakistan deployed in large numbers. The P1 centrifuge uses a rotor, and the P2 centrifuge uses a maraging steel rotor, which is stronger, spins faster, and therefore enriches more uranium per machine than the P1 centrifuge's rotor. An array of centrifuges is shown in Figure 10.4. Qualification is the key to reliability, and this takes 3–4 years, based on a preproduction run of several hundred units. Consistency of manufacture is important.

Urenco has looked at diversification – using the concept for other isotope enrichment and the production of composite products such as tubes and flywheels for energy storage. One application for the latter would be underground trains, with a 100 kg flywheel of 300 mm diameter × 1 m long storing 10 MJ. The keys to success were:

- Phased development.
- Consistent manufacture to high tolerances.
- Conservative design based upon experience.

FIGURE 10.4

An array of centrifuges at a reprocessing plant (Green, 2003).

An interesting aside concerns the ability of national intelligence agencies to identify who is using ultracentrifuges. The US data source contains the following statement:

The AC motors that drive the centrifuge require higher-than-normal-frequency AC, and the several hundred hertz AC is detectable by other countries' Signals Intelligence (SIGINT) and Electronic Signals Intelligence (ELINT) operations. Otherwise, a cascade plant has the unremarkable real-estate and power signature typical of a large wholesale commercial outlet, and it is monitored by monitoring the acquisition of specialized parts and materials.

10.3 **The food and drink sector**

A useful introduction to PI in the food industry can be made by summarising the presentation of Richard Poynton, of Profit and Planet, at a meeting of the UK Process Intensification Network (Poynton, 2006). He started by comparing the positions of the chemical and food sectors. In the latter, one needed to be very flexible, give products of consistent quality and at low cost. Poynton listed what one must accommodate and what one must deliver in the industry (see PIN website). The PI opportunities were based largely upon the massive use of resources by the sector. Food takes 14% of the total UK energy use, 10% of the water and produces 10% of the waste. It was stated that 10% of the sector is at risk from climate change. If PI can contribute to minimising energy, water use, and waste production, it could be of major importance. Poynton suggested that some demonstration projects should be carried out, involving, for example, mixing plus heating/reaction plus cooling. He cited the continuous oscillatory baffle reactor as an example of equipment appropriate to the sector (see Chapter 5).

Concentrating on separations, which are common in food and drink processing and, of course, waste treatment, Muralidhara (2006) highlighted the removal of impurities, producing new products, and minimising and recovering waste for reuse as opportunities for PI in the sector.

The Food Processing Knowledge Transfer Network – FPKTN, (Food Processing Faraday) in the UK, now merged with a company called Quotec, produced some excellent documents relating to developments in the sector. The FPKTN report on emerging technologies for food processing (Johnson et al., 2008) highlighted many technologies that we would call process intensification that the KTN believed would come to the fore in the period 2013–2018. This sets the scene rather well for this section of the chapter.

A wide variety of electrical technologies were discussed in the context of the preparation and assembly of raw materials. Pulsed electric fields are considered as a way of improving the efficiency of extraction processes (e.g. sucrose from sugar beet) or to preferentially extract a component from a biological source which is made up of a number of similar compounds. The report cites electrokinetic dewatering, electroflotation, electrodialysis and electromigration as having potential being used at the time, outside the food and drink sector.

Packet soup manufacture could benefit from novel dewatering methods, as could jam manufacture, where concentration/water removal (to a certain extent) is needed.

Supercritical fluid extraction has been used for many years in the removal of caffeine from coffee. The authors point out that control of this extraction method via pressure and/or temperature allows the composition of the extractant to be further improved. It is also seen as a way of extracting material from waste streams and was cited as being more economical than steam distillation.

The next section addressed by the FPKTN was component structuring – texture being an important one (as can be seen in the discussion on ice cream later in this section). The spinning disc reactor receives a mention in this context. It is suggested that the SDR, in creating a layer one-particle deep, could continuously produce particles with precise and controlled size distributions – as was done with chemical crystals in the UK and China (see Chapter 6). An example that is gaining increasing importance as we recognise the harm that excessive amounts of salt in food can do to us is the use of the SDR to control reactions where a second ingredient is added at right angles (one which adheres to the layer on the disc). Salt or flavour could be added in much lower quantities than in bulk mixing to give the same flavour. Another established PI technique, the oscillatory baffled reactor (OBR – see Chapter 5) is proposed for sugar crystal size distribution in chocolate and ice crystal size in ice cream, although to date it seems that in the latter case, only the SDR has been tried. Intensive aeration using gas micro-bubbles is suggested as a growth area, where their current use was in separating yeast cells. Supercritical and high-pressure carbon dioxide is also being examined for aeration of baked products as an alternative to yeast or chemically raising agents. (An interesting development in the use of ultrasound in baking, via research at Heriot-Watt University in Scotland, is discussed later in Section 10.3.4).

The section on heating of components is, perhaps surprisingly, rather brief. Of course, microwaves have been established in the sector for decades, as have other electric field heating methods. Nevertheless, the authors do cite the use of lasers for ultra-rapid cooking (a patent only at this stage), and superheated steam as an alternative to frying for the production of frozen chips (and for surface drying other products).

Non-thermal preservation of food components is seen as an important aim, using pressure, electric fields in particular. The preservation of components section encompasses surface decontamination, carbon dioxide processing and biological control strategies. Surface decontamination could benefit from cold atmospheric plasmas to kill microbes on surfaces, and ultraviolet LEDs could also benefit food hygiene.

A number of other technologies had been studied earlier and were excluded from this FPKTN report, but they included acoustic and microwave accelerated thawing of food, infrared surface pasteurisation, concentration membranes for water removal and radio-frequency heating.

For the reader keen to explore novel food processing technologies in more depth, Doona et al. (2010) is recommended as a reference book.

10.3.1 Barriers to PI

The main barrier to PI within the sector was that the food industry was innovative in products, but not innovative in processes. As well as institutional barriers in the sector, a number of technical hazards can be identified. Surface deposits (fouling) are perhaps a greater problem in the food and drink sector than elsewhere, as cleanliness is essential. Fouling, of course, leads to inefficiencies due to increased thermal resistance and/or pressure drop. It is therefore important that, in selecting PI plant for applications here, accessibility for cleaning, preferably CIP (cleaning in place), is available. It is sometimes difficult to incorporate CIP in PI plant, however, and care must be taken in selecting appropriate solutions.

Other technical challenges include process control – a feature that arises in many discussions of intensified plant, particularly where massive volume reductions are anticipated. The talk also identified what are called 'transition' issues. This encompasses areas such as the compatibility of PI plant with service plant and batch plant.

10.3.1.1 Solutions

Poynton's 'solutions to savour' included the following areas:

- Cleaning – A lab-on-a-chip device could be used to monitor cleaning in equipment.
- Process control – in continuous processes more sophisticated control is needed.

10.3.2 Sector characteristics

At a meeting in Cambridge University some years ago, Christina Goodacre (2012), then of DEFRA, the UK government body responsible for the food sector, highlighted unique factors in food processing that would need to be taken into account if PI was to be relevant. Manufacture in this sector often relies on mixing, blending, emulsifying and aeration of physically complex products, by non-reactive processing. This limits the relevance of combined unit operation technologies based on reactors. Appearance, texture and shelf-life stability using minimum stabilisers are important. Achieving controlled emulsification, crystallisation and product structure is a major issue. Products tend to be low added-value, and therefore CAPEX (capital expenditure) reduction is important. Plant monitoring tends to be mainly for product quality, and then only at the minimum level required to operate. There are few hazardous processes, and so safety drivers are rarely encountered.

The prospect for control of fluid mechanics leading to better products is the main area of interest. Product micro-structure, either through novel techniques or obtained in very small and controllable volumes could have application in this sector. There is also interest in distributed product finishing and formulation.

Much more recently in New Zealand in 2006, Fee and Chand (2006) reported work on the use of PI in the on-farm extraction of proteins from the milk of individual cows. Based upon the use of batch cation exchange chromatography, it was found that local (i.e. distributed) extraction using PI could greatly enhance the yields, compared to carrying out the extraction in large, centralised processing plants. By targeting the cows with milk having the highest protein content, it also allows the farmer to maximise his protein yield without adversely affecting the acceptable level of protein in the bulk milk shipped to the central processors.

Comments by DEFRA on texture (and by implication, taste sensations) are even more interesting now, as work was recently completed (2009) on using the spinning disc reactor to produce an ice cream base (see Section 10.3.5.2).

10.3.3 Induction-heated mixers

Many food products require heating and, with viscous liquids, this can be an inefficient and time-consuming task. An elegant solution is to use an induction-heated mixer (see also Chapter 7), which takes heat to the place where it is needed, where most agitation and mixing occurs. This ensures uniform heating of the liquid, rather than via the slow process of conduction from a heated jacket around the walls. With potential uses in other sectors, reduced thermal degradation, rapid response and ease of retrofitting are some of the advantages claimed (Web 2, 2000).

10.3.4 Electric fields for drying and cooking

Drying of porous media, such as pasta and bread, has been made difficult using conventional dryers because of the case hardening phenomenon. In addition to the use of microwaves described in Chapter 3, researchers at Bristol University (in the UK) and Nestlé have suggested that EHD augmentation can help to overcome this problem, in particular, using a targeted Corona wind effect (see also Section 3.4).

Microwaves, as will be obvious to most who cook at home, are proven in the microwave oven as an effective way of rapid heating, cooking and drying. Such processes can be done on an industrial scale, as can radio frequency (r.f.) drying, which operates at over 27 MHz. At this frequency, the water molecules are subject to rapid realignment in direction, causing molecular friction and hence heating. Drying can be effectively controlled to specific moisture contents – often critical in food processing. The ability to moisture profile and to dry to very low moisture contents has led to the use of r.f. drying in making biscuits, processing cereals and vegetables, and in drying spices. Reductions in drying times from 6 hours without r.f. to 2 hours using this method are routine. It has also been reported (Dexter, 2001) that domestic r.f. ovens could be superior to microwave ovens – for example, defrosting and cooking a medium chicken in one hour.

The use of ultrasound in the food and drink industry takes many roles. As a cutting and forming device of pastry batter prior to baking, or as a detector of metal pieces after bread baking, the principle is well established in this industry. Most recently, ultrasound has also been explored as a strategy to modify the organoleptic properties of doughs and batters, in an attempt to make them more appealing both aesthetically and from a taste viewpoint. In particular, the 'free-from' range (e.g. gluten-free and salt-reduced products) benefits the most from this novel PI technique. Traditionally, gluten-free products look denser and with a reduced volume due to the lack of gluten that sustains an expanded matrix at processing.

Research carried out at Heriot-Watt University, Edinburgh, Scotland has demonstrated that treatment with ultrasound at a specific window of operation improves this feature, as well as the product sensory qualities and the processing overall time. This could allow the bakery industry to cater not only for allergy and intolerance sufferers, but also for the growing trend of a more health and diet conscious population. Reducing salt in bakery products has also a major drive from both consumers and from governmental diet guidelines. Salt is an important ingredient in leavening agents, so the reduction of it has an impact on the processing by making handling more difficult and increasing waste, yielding poorer production results. Ultrasonic treatment has been demonstrated to enhance efficiency at the production line by reducing waste, accelerating the process and therefore, reducing costs, and to improve the appearance and taste of these products. The Scottish researchers, in partnership with Macphie of Glenbervie, Nortek Piezo, Mono Bakery Equipment and Fosters Bakery teamed up in 2012 to exploit this PI technique at a factory level with initial deliverables installed on-site by the end of 2013 (Anon, 2012).

10.3.5 Spinning discs in the food sector

Spinning disc technology, as used in the SDR, has been tested on foodstuffs. In particular, it has been used for dehydration of food, and has been proposed for rapid pasteurisation of heat-sensitive liquid foods. In Figure 10.5, the SDR is illustrated heating treacle. The good mixing characteristics and excellent heat transfer are evident in the highly uniform temperature distribution across the disc.

Research at Leeds University in the UK (Dunn, 2007) that studied the pouring of liquids onto heated or cooled spinning discs and collection of them at the

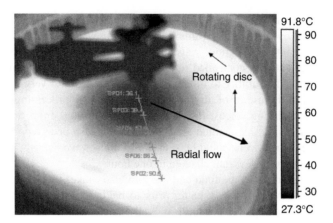

FIGURE 10.5

The spinning disc reactor used for heating treacle. Rapid warm-up and temperature uniformity can be noted.

(Courtesy: www.protensive.co.uk)

periphery, suggested that the phenomenon could provide the food industry with more efficient, and shorter mixing and processing times. It was demonstrated that mixing and heating of foods such as fruit juice, mayonnaise and ketchup could be accelerated, and other tests showed that the process could remove allergens such as those in peanuts. There was a reduced risk of overprocessing of the foods and just-in-time (JIT) manufacture could be contemplated.

The virtues of the SDR extolled in Chapter 5 are as important to food processing as to the chemical and pharmaceuticals sector. It offers intense and uniform mixing, high heat transfer rates and short residence times. It provides the right fluid dynamic environment to keep a viscous liquid well mixed during a heat transfer duty. Far higher operating temperatures can be used, safe in the knowledge that the short residence times (typically 0.5 to 3 seconds) will prevent overcooking or burning of the product. Through control of rotational speed and liquid flow rate, the reaction, duration and intensity of the processing can be tailored to its needs. The end result is a continuous process to compete with current batch technologies. Product quality is improved and processing times are dramatically reduced.

However, there are even more advantages to be realised in terms of business and manufacturing improvements. In less than ten seconds product grade changes can be made or the unit can be cleaned down completely. A wash jet for a few seconds will clean the reactor down upon completion of a process. In manufacturing terms, this has huge implications for the way processes can be run. With instant and incremental capacity of any number of products and grades comes the possibility of JIT manufacture. Truly responsive processing, with no overproduction means warehouses can be drastically reduced in size or eliminated altogether.

As mentioned elsewhere, CIP (cleaning in place) is often demanded in this sector and the ability to wash discs with jets is a strong advantage.

10.3.5.1 SDR for custard manufacture

Hydrolysis of starch requires a combination of good mixing, elevated temperatures and controlled processing times. The raw material is in the form of an aqueous slurry. Heating the starch with water causes the granules to swell. This water is now immobilised in the starch and so the product thickens to a viscous liquid.

The current custard making process involves conventional batch-stirred vessels that use relatively gentle stirring at mild operating temperatures for long processing times. Poor mixing and heat transfer, results in the generation of large temperature gradients and uneven processing of the material. This often leads to an undesirable heterogeneous (lumpy) product and charring of heating surfaces. The temperature of the mixture and the duration of cooking also influence the product with low heating levels, giving insufficient cooking time to fully process the product, while too much heating can degrade or over-thicken the mixture. At saucepan-scale managing these processing issues is simple enough. The SDR provides adequate stirrer power in a responsive mode to manage the process. At industrial scales, issues around mixing and heating viscous liquids present serious well-documented problems.

One key advantage of the spinning disc reactor design, evident in Figure 10.5 above, is that of quick even heating of liquids for short periods of time before cooling on the walls of the reactor. This allows rapid continuous processing of the liquids under conditions not attainable in conventional stirred vessels. This was demonstrated for the simple process of custard manufacture where heat transfer and mixing is a key issue. The system used is illustrated in Figure 10.6 (Anon, 2007b).

This system uses a 100 mm diameter disc heated to 150°C using a 3.5 kW bath which pumps a heat transfer fluid through a chamber below the rotating disc. This processes a slurry containing approximately 200 g of custard powder (predominantly cornstarch granules) per litre at a flow rate of 2 ml/s to 3 ml/s. The slurry is pumped over the disc which rotates at 600 to 1,200 rpm. This produces a thin film of liquid that is rapidly heated before being cooled on the walls, maintained at a temperature of 70°C. Control of the level of cooking, and hence output product viscosity is achieved through the choice of flow rate and rotational speed.

A second set of pumps is used to feed two different dye streams onto the disc surface. These are blended into the cooking slurry by the shearing action of the centrifuged film to produce a well-mixed final product. Due to the low inventory within the system, the choice of colour can be changed quickly with little product wastage between choices. Images of the disc at the start of processing and during the custard manufacture are shown in Figures 10.7 and 10.8. At the end of the process, a simple water wash jet is introduced at the centre of the disc to flush remaining custard out of the system, leaving a clean heating surface.

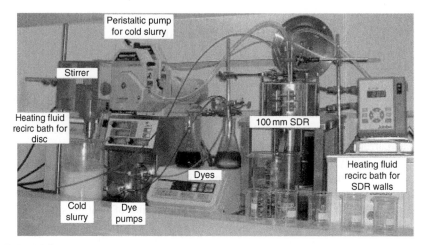

FIGURE 10.6

Set-up for custard making on the SDR showing (left to right): 3.5 kW bath for disc; pump for slurry; custard slurry; pumps for dye; spinning disc reactor and SDR bath for walls.

(Courtesy: Protensive)

FIGURE 10.7

100 mm heated disc used for processing, showing feed (1 × slurry and 2 × dye) and wash water nozzles.

(Courtesy: Protensive)

During the processing of the 1 litre batch, a range of coloured products were made, samples of the material taken from the process are shown in Figure 10.9. Initial plain custard was changed to green custard and then to red custard before being returned to plain custard. A sample of the initial slurry is also shown for comparison.

FIGURE 10.8

Plain custard product exits the SDR.

(Courtesy: Protensive)

FIGURE 10.9

Grade changes can be achieved rapidly using the SDR: (left to right) plain product; with dye #2; with dye #1; plain product; initial slurry.

(Courtesy: Protensive)

10.3.5.2 Ice cream on SDRs

The Department of Food Science at Leeds University in the UK, has investigated the use of the SDR, with its intense mixing benefits, for processing an ice cream base. The ice cream base contains a number of ice cream ingredients, in this case: butterfat, lecithin, xanthan gum, sugar, skimmed milk and double cream. The processing of ice cream involves blending all of these components together, followed by pasteurisation, homogenisation, ageing and freezing of the ice cream base. (Ageing allows time for the proteins and polysaccharides to fully hydrate. (Possibly taking over 4 h, but in this case, 18 h, in order for the fat to cool down and crystallise.) Ageing also improves whipping qualities of mix, and the body and texture of the ice cream.)

The processes on the spinning disc, homogenisation and pasteurisation were carried out at 60–90°C, the disc being heated at 65–95°C. The disc was rotated at 2,900 rpm and the flow rate was 6 ml/s, giving a residence time of 0.45 s.

It was concluded that highly stable ice cream base emulsions could be produced using the SDR. Subsequent processing to make model ice cream resulted in an acceptable product in terms of taste/texture, and one that was microbiologically safe. The time for ageing was much lower than the 18 h normally needed for conventionally made ice cream.

The research may lead to much lower capital cost/energy-intensive ways for making ice cream (Akhtar et al., 2009).

10.3.5.3 Scale-up – so how big a trifle can I make?

With a residence time of the order 1–2 seconds on the disc surface, such a continuous process on a spinning disc lends itself to truly making 'instant' custard on demand; the custard is made at the rate that the cold slurry is pumped to the spinning disc. In the case of a small disc 100 mm in diameter this would be in the region of 10 litres/hr. Increasing the disc area allows for increasing the flow rate, yet being able to maintain the same heat transfer and mixing performance, residence time and film thickness. By scaling up to a 300 mm diameter disc, flow rates in the region of 100 litres/hr are achievable; a 500 mm diameter disc could achieve 250 litres/hr. Since the units are compact, several could be used to match process throughput requirements. (A similar scale-up must of course be addressed for commercial-scale ice cream production.)

Homogenisation is of course a very common process in the food and drink sector. Koehler and Schuchmann (2011) report that milk undergoes high-pressure homogenisation at the rate of about 10 million tonnes/year. Using a microstructured orifice as part of the high-pressure system, these authors have employed a microfluidic valve to simultaneously mix and emulsify the milk, leading to substantial energy and cost reductions as well as an improved product quality – three of the major benefits of PI! (see also Section 10.3.7).

10.3.6 Deaeration systems for beverage packaging

Much has been reported about the HiGee concept in the chemicals and offshore processing sectors, but its use in the food and drink industries has been less evident. Already referred to in Chapters 8 and 9 in the context of oil and gas processing and the chemicals sector, the use of rotating packed beds of the type described in

Chapter 6 has been extended to other application areas by GasTran Systems (see Appendix 3) including deaeration of beverages.

The challenge in the drinks sector is to remove dissolved oxygen from feed water. The presence of dissolved oxygen in the feed water for beverage packaging lines is becoming an increasing concern in the industry. The carbonated soft drink, high gravity beer brewing and fruit juice packaging industries have each recognised the importance of lowering dissolved oxygen levels and achieving consistent CO_2 injection in order to decrease foaming, gain faster and more consistent fill rates, prolong shelf life and produce a quality product.

The GasTran system, the basis of which is shown in Figure 10.10, uses vacuum to remove dissolved oxygen from water to much lower levels than other deaeration methods currently in use. This system has an integrated holding tank and transfer system to reduce floor space requirements and give bottling plants the ability to remove large holding tank systems and transfer pumps. This system reduces overall equipment size, energy requirements and total cost of operation. In the last year, GasTran deaeration systems were installed at PepsiAmericas plants in Austin and Indianapolis, both in Indiana, in the US. The following benefits were noted:

- Increased filling speeds.
- Improved carbonation capability.
- Reduction in low fills.

The Austin plant elected to have the system designed to provide deaerated water to either a can line or a bottle line so that its impact could be evaluated on both

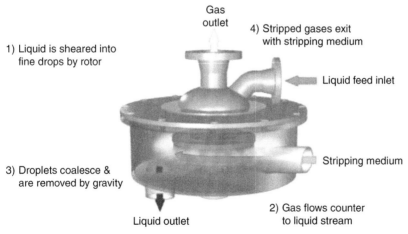

FIGURE 10.10

The rotor used in the GasTran System rotating packed bed deaerator – the rotor diameter is less than 1 m.

(Courtesy: GasTran Systems)

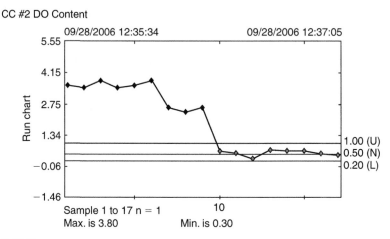

FIGURE 10.11

The dissolved oxygen content at Austin before (left-hand side); and after (right-hand side) installation of the GasTran System. The original system used a CO_2 deaeration column.

(data courtesy: GasTran Systems)

canned and bottled product output. The most significant and noticeable difference achieved by using the GasTran unit was the dissolved oxygen performance levels. Without stripping gases, the system reduced the dissolved oxygen levels in the process water from 8.7 parts per million to 0.3–0.6 parts per million with the water at 9–5.5°C (see Figure 10.11). (By comparison and contrast, current industry deaeration practice targets a much higher level, typically between 1.0 to 2.0 parts per million.) The GasTran technology enabled the Austin plant to produce more volume with existing blending equipment. The line has also experienced more consistent content weights on all carbonated soft drink products.

The systems installed at Indianapolis (shown in Figure 10.12) exhibit the usual features of intensified process plant, including a small footprint (the rotor is on top of the platform – top centre of the photograph), reduced energy requirements, ease of scale-up and the ability to reach equilibrium performance very rapidly. In addition, as the GasTran system does not require CO_2 and nitrogen to facilitate the deaeration process, the use of CO_2 has been much reduced, resulting in a saving of over $87, 000 per annum. The benefits to PepsiAmericas, mirror several of those associated with successful PI installations, as indicated by their director of engineering:

> *The GasTran deaeration system delivered ultra-low dissolved oxygen in the process water, which enabled the bottling line to produce consistent carbonation to our fill lines, increase plant capacity and improve profitability. Rob Schlafer, Director of Engineering, PepsiAmericas East Group.*

Such was the success of the system that GasTran supplied another 5 units up until 2008 (Harbold and Park, 2008) to PepsiAmericas (canning and bottling plants).

FIGURE 10.12

The deaerator installed on the bottling line.

(Courtesy: GasTran Systems)

As well as saving carbon dioxide (a bought-in gas), line speed increases resulted because of lower levels of dissolved oxygen in the fluid, increases being up to 50% (allowing an increase from 400 to 600 units/min) on a 710 ml plastic bottling line.

10.3.7 Intensified refrigeration

Unfortunately, there is no equivalent to microwaves (for heating) that allows us to rapidly cool or freeze foodstuffs.[1] Intensification of heat and mass transfer during cooling processes, if feasible, could help to overcome the inertia of transfer through solidifying masses. The fluidised bed freezer, that can rapidly freeze food such as peas by fluidising them in a freezing air steam, involves a degree of intensification.

Magnetic refrigeration is being developed for commercial applications – this uses the magnetocaloric effect – and it is suggested that mobile refrigeration and freezing duties could be carried out commercially by around 2015, the system requiring further development (Tassou et al., 2010).

However, some work at the renowned Odessa State Academy of Food Technologies in the Ukraine on the use of acoustic fields at around 22 kHz has been

---

[1] Magnetic refrigeration and Peltier devices may employ electricity in a way that does not involve conventional refrigerants to generate cooling, but cannot be categorised as intensified devices in this context.

shown to aid the concentration and fractionation of food solutions by intensifying heat and mass transfer in the system boundary layer, Burdo et al. (2008).

The rate of growth of the ice block, leaving behind grape juice in these particular experiments, was increased by 20–30% on the application of the acoustic field. The greatest effect was observed during the first 20–30 minutes of ice block growth, and further work is needed to see whether commercial gains are possible. The reader may also study the research of Fikiin (1983), who demonstrated rapid cooling of fruit and vegetables by a combination of relatively simple process changes. Tassou et al. (2010) also carries data on thermoacoustic and thermoelectric refrigeration within the food sector.

10.3.8 Pursuit dynamics intensified mixing

A novel form of intensification has been developed by Pursuit Dynamics in the UK, based upon the PDX® Reactor:

> *Which utilises a supersonic vapour flow and condensation shockwave, which is generated from the injection of high velocity steam. Steam is introduced into a special annular "conditioning" chamber that is wrapped around the core of the PDX reactor unit. From here, it is injected into the process fluid at supersonic conditions, generating high levels of shear and turbulence within the process fluid leading to the creation of a controllable cross bore condensation shockwave. The combination of these mechanisms provide unsurpassed homogenous mixing, agitation and heating of the process fluid.(Anon, 2007a)*

One application within the food and drink processing sector involves heating wort in the brewing industry. The system, illustrated in Figure 10.13, is retrofittable, and as far as wort treatment is concerned, it involves low thermal shock homogeneous heating. The nature of the mixing action encourages highly turbulent flow that enhances the chemical reactions being undertaken. As the steam injection takes place under partial vacuum, low temperature boiling can be carried out. This helps to contribute towards the up to 50% energy savings attributed to PDX, as well as reducing process times and lower wort recirculation rates (smaller pipework, etc.).

A more dramatic illustration of the capability of DX to reduce processing time is that at Welcome Food Ingredients, by combining the actions of a pump and a mixer with steam heating, the Pursuit Dynamics PDX Fluid Handling System has reduced cooking times by 95% at the sauces manufacturer. Clean down time between batches has also been cut by some 80%, the companies claim. The PDX system works by using the energy of injected steam to accelerate a stream of fluid through its central bore. It is well suited to food production, says Pursuit Dynamics, because it does not affect taste and texture; it can also handle mixtures of powders and liquids.

Welcome Food was "very impressed with the speed of the PDX process – it mixed and cooked a 60 kg batch of sauce in three minutes, compared with an hour for conventional plant". It has also been used to make stocks, soups and chutneys. "It may enable us to create new sauces, pastes and other products that cannot

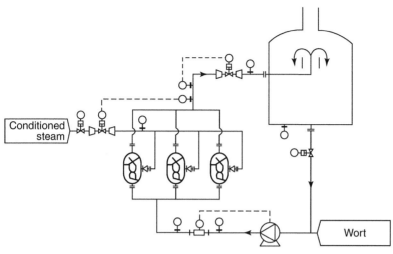

FIGURE 10.13

The PDX wort heater used in brewing.

(Courtesy: Pursuit Dynamics)

be manufactured commercially using conventional processes", says operations director, Alex Sutton. "The faster cooking and mixing, together with short changeover times, will make a significant difference to our business".

Now called the PDX Sonic, generic claims regarding its benefits to the sauce and similar product sectors are:

- Energy reductions – of between 25% and 45% compared with traditional cooking and preparation processes.
- Small footprint – a system to produce 4,000 kg of sauce would utilise 4.5 m^2.
- Increased throughput – the PDX Sonic can produce a range of quantities from small batches to 24 t/h.
- Increased flexibility – batches can be as little as 200 kg, ideal for new product development.
- Increased product quality – the PDX Sonic can improve particulate integrity, compared to traditional manufacturing methods (tomatoes, vegetables, meats), as it has no moving parts.
- Reduced raw materials – due to the unique mixing capabilities, the PDX Sonic can reduce the ingredient use of starches, spices, gum and salt – giving healthier recipes if required.

10.3.9 **The Torbed reactor in food processing**

In the snack food industry, most products available are either fried (visit any potato crisp factory) or cooker extruded before being dusted with oil and flavours.

A trend towards lower fat and healthier eating snacks has allowed the development of new low-fat products, particularly based on expanded half-pellet products. Newer half-pellet products adsorb less oil when fried (20% instead of 40%). The Torftech Torbed process, described in Chapter 5, is capable of expanding these half pellets in hot air thus producing an oil-free base product that can then be flavoured with minimal oil addition (7–10%).

The Torbed process has shown itself to have unique capabilities in this application, in producing higher quality, lighter and more consistent product with lower energy consumption. Indeed some shapes of product can only be air expanded in this type of reactor. High heat and mass transfer and precision of control with the Torbed reactors have provided these advantages. The installation is additionally quiet and compact. A typical unit is illustrated in Figure 10.14.

The process is in successful operation in several major production plants in Europe, Australasia, Central America and the Middle East. Throughputs vary but typically range from 50 to 500 kg/h. In Germany, a 1,000 mm diameter Torbed reactor processes some 1,000 t/year of low-fat snack products to supply just one regional supermarket chain. Many other precision roasting, drying and bloating processes are in commercial use in most major continents of the world.

Intensification was associated with food production and farming long before the chemical process industries became involved. Recent discussions about the growing population and the need to "double food production to feed the 9 billion" (Tomlinson, 2011) will probably ultimately lead to a growing awareness of what our form of intensification can contribute to the need for environmentally and ecologically friendly increases in food production.

FIGURE 10.14

The Torftech Torbed unit in a food processing duty.

10.4 **Textiles**

Textiles and chemicals are closely related. Processes such as dyeing have been subject to intensification over a period of many years, and the production of synthetic fibres involves chemical processes which may be given the same PI opportunities as in the bulk or fine chemical sectors. The case study below gives information on in-line mixers (see Chapter 7) which can reduce processing times by up to 90% (Ryan, 2000).

Effectively undertaking a debottlenecking exercise, Courtelle needed to increase production of Duracol Black (DB) fibre. Downstream facilities were doubled, two spinning lines being commissioned, but the upstream plant for preparing pigmented paste was inadequate for the new demand. Following some development trials, Silverson, a US manufacturer of a range of mixers, was approached to supply a high shear in-line mixer unit. This was situated in a pipeline through which the materials are drawn by high speed rotation of the centrifugal rotor. Shear is imposed on the material in the workhead before it is forced out through the stator at high velocity. In the final variant, multi-stage in-line mixers were used to obtain the desired product quality in as short a time as possible.

10.4.1 **Textile preparation**

The preparation of cotton before it can be woven and subjected to downstream processes such as dyeing, involves, amongst other processes, the removal of impurities from the raw cotton in particular, waxes, fats, pectins and proteins. There are a number of environmentally friendly bioseparation processes that can perform this purification duty, without leading to effluents that do not biodegrade. These can be based upon enzymatic reactions, that are, as pointed out by Yachmenev et al. (2004) rather slow. The aim of their research was to show that the process could be speeded up using active PI methods.

The work carried out in New Orleans in the US extended the use of ultrasound to speed up the biopreparation of the cotton, in particular impacting upon the rate of enzymatic reactions. A critical parameter in the rate at which the reaction takes place is the wettability of the cotton, measured in terms of the wicking time. It was found that the use of ultrasound in trials with treatment times of up to 60 minutes (considered as being relatively short) gave significant improvements in enzyme efficiency, reflected in up to an 80% reduction in wicking time. The authors concluded that a combination of enzymatic bioscouring with applied ultrasound could lead to substantial reductions in wastewater effluent quantities, energy use and total processing costs.

10.4.2 **Textile finishing**

Textile finishing processes comprise washing, bleaching, dyeing and coating, as applied to bulk textiles or garments following weaving and/or production of

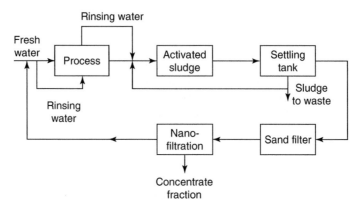

FIGURE 10.15

Current methodology for water reclamation in a textile finishing company (Van der Bruggen et al., 2004).

synthetic materials. These are energy-intensive and use large amounts of water that is generally discharged as effluent. Effluent costs are also an increasing burden on industry. Van der Bruggen et al. (2004) concentrated on minimising the production of waste, such as contaminated effluent, by changing the overall process, rather than just addressing the task of post-process clean-up.

The process proposed by the researchers includes micro-filtration pretreatment of used finishing baths, followed by a nano-filtration process. Membranes feature in both of these, and their ability to operate at process temperatures removes the need for substantial reheating. Figure 10.15 shows the process before modification where, although filtration is used, there is a sludge discharge in spite of the partial reuse of water. The additional features proposed to further improve the process, moving towards zero emissions, are shown in Figure 10.16. The water stream is represented by the solid line, while the dotted line is the energy stream, with energy being recovered from an incinerator that uses the final part of the organic distillate. MD is membrane distillation; NF, nano-filtration; and MC, membrane crystallisation.

The simulation showed that the concept of integrated membrane systems to encourage recycling and/or recovery in the textile finishing sector was feasible. Although the dyes themselves could not be reused, the calorific content of their residues could add to the energy input for heating. The only waste streams generated by the plant are minor amounts of ash water for cleaning the filters and the ashes from the incinerator.

10.4.3 Textile effluent treatment

As implied from the earlier section on textile finishing, the effluent from textile processes – predominantly waste water– can be heavily contaminated with a number of chemicals. As reported by Nachiappan and Muthukumar (2010), of the over 100,000 commercial dyes, producing over 70,000t of dyestuff, 1–15% of this is released to the environment, mainly in waste water.

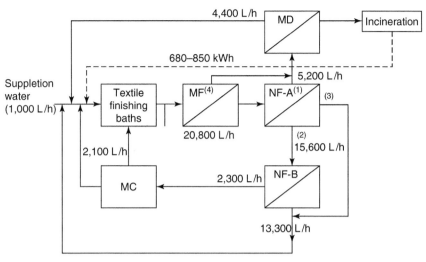

FIGURE 10.16

Proposed integrated water treatment system using membranes. The numbers represent: (1) to be replaced by ultra-filtration in the case of sulphur dyeing, disperse dyeing and pigment dyeing; (2) for reactive dyeing, metal-complex dyeing and mordant dyeing; (3) for vat dyeing and other techniques where no salts are used; (4) operates in dead end. (Van der Bruggen et al., 2004.)

The work in India, where of course the textile finishing sector is very large, concentrated upon three hybrid (combined) intensification methods for effluent clean-up. These were sono-sorption, sono-Fenton and sono-Fenton-sorption. (Note: Fenton's reagent is a solution of hydrogen peroxide and an iron catalyst that is used to oxidise contaminants or waste waters. Fenton's reagent can be used to destroy organic compounds such as trichloroethylene and tetrachloroethylene). These processes rely upon an adsorbent as the basis of the purification procedure, and naturally enough, in the region of India where the research was carried out, tea waste was selected for this role.

It was found that the highest chemical oxygen demand (COD) reduction was 95.5% in sono-Fenton-sorption, aided by the high-mass transfer coefficients and intra-particle diffusion coefficients achieved with this combination of processes. Biodegradation increased from 0.31 to 0.71, and another plus point – it was found that the tea waste could be reused for up to three cycles, then incinerated and the ash used as a another adsorbent – a good example of sustainability.

10.4.4 Laundry processes

Large amounts of energy and water are expended in laundry processes for washing textiles. There have been a number of attempts to improve the energy efficiency of dirt removal from clothes, including the use of low temperature (or room temperature) detergents that allow cold-water washing, and electric fields. The most intensively researched of these has been the use of ultrasound. Warmoeskerken (2002) and colleagues in The Netherlands studied the use of ultrasound to encourage mass transport within textile fibre bundles. It was acknowledged that pulsing flows could improve mass transport by yarn deformation, but it was only through power ultrasound that the limitations of mass transport using this process could be overcome. This research showed that the characteristic mass transport rates in textiles could be increased by a factor of six if ultrasound was used. The ultrasound frequency was in the range above 25 kHz.

The mechanism forming the basis of the phenomenon of soil removal is acoustic cavitation. If the sound field is sufficiently intense, bubbles formed can grow very rapidly and become unstable. This leads to bubble implosion – transient cavitation. When a transient acoustic bubble collapses near a solid boundary, the implosion generates jets of liquid directed towards the solid surface – the fibre. These jets account for the cleaning effect of the ultrasonic waves. The impact is shown in

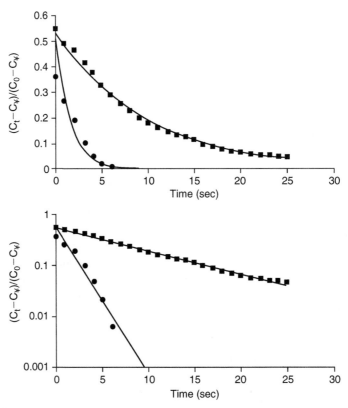

FIGURE 10.17

Experimental salt-rinsing results with, and without, the application of ultrasound. The upper graph shows the relative salt removal, while the lower graph is a semi-log plot of the same data (Warmoeskerken et al., 2002).

Figure 10.17. Another advantage of ultrasound claimed by the Dutch team was that it was likely to lead to much less damage to the textile fabric (see also Moholkar and Warmoeskerken 2004, for further research by the group).

10.4.5 Leather production

Leather production takes place at the opposite end of the textile sector to laundry processes. One of the principal operations in the preparation of leather for use in clothes and elsewhere is tanning. This is carried out to change the leather texture so that it becomes pliable, long-lasting and resistant to external agents that can attack it. Tanning is energy and water intensive, and traditionally takes a long time to complete. Work in Germany at the Fraunhofer Institute and a company has successfully reduced the tanning time by using high pressure CO_2 at pressures of up to 100 bar (Weidner et al., 2007).

The industrial process for tanning, as described by the authors, allowed up to 24 hours for turning skins into raw chrome-tanned leather, this process being carried out in rotating drums incorporating the tanning solution. The authors noted that this did not necessarily lead to fully tanned leather. Experiments on a smaller scale were carried out to assess the time reduction that could be achieved in a high pressure CO_2 environment. A small-scale unit was also used to replicate the conventional tanning process, so a valid comparison could be made – the fully tanned criterion being a chrome content of 3% by weight in the tanned leather. Full process conditions are described in the cited paper, but it was found that CO_2 pressurisation for full tanning could reduce the processing time by a factor of six, to around five hours.

10.5 The metallurgical and glass industries

The use of PI in sectors such as metals and glass has not aroused great interest, although their profile is perhaps not as high as those in the chemicals sector. In both the casting of metals and the solidification of glass, the final product quality is highly dependent upon the solidification process. As well as being used in textiles as described above, ultrasound (see Chapter 3) can positively influence both heat and mass transfer in high temperature process industries, and these features are used to benefit, for example, aluminium castings. Intensification techniques can also be used to improve quality, as well as speeding up production. (Of course if reject rates are reduced, this is analogous to a production increase.)

10.5.1 The metallurgical sector

In the case of casting, the final quality of the product is dependent upon many factors which affect the metal solidification. The mechanical properties of the casting will largely be determined by the grain structure and any defects present. Despite improvements in casting techniques, both gas porosity and shrinkage are still a major cause of rejects in the foundry industry, particularly for aluminium and light alloy castings. High scrap levels result in low yields and additional, unnecessary, energy use, adding to product costs. Any process which reduces defects and improves the metal structure of a casting would therefore benefit the whole industry.

10.5.1.1 Tube drawing

An early project that investigated how ultrasound might benefit metals production was carried out in Bucharest in the 1970s. Atanasiu (1980) reported on experiments to intensify aluminium alloy tube drawing by axial ultrasonic oscillation of the plug inside the die. The ultrasound could be introduced into the deformation zone through either the plug or the die, or both. The experiments undertaken showed that for industrial applications, the application of ultrasound to the plug was most effective. As well as requiring minimum changes to existing equipment, the energy balance was most favourable.

Based on work on an industrial drawing bench with a drawing force of 100 kN, a resonance frequency in the range 18.2 to 20.1 kHz was applied, giving plug amplitude oscillations from 0.7 to 1.5×10^{-4} mm and drawing velocities from 0.2 to 45 m/min. It was found that the effect was to reduce the yield-point of the alloys and diminish the contact friction between the plug-metal and die-metal interfaces. The reduction in drawing force needed was accompanied by the elimination of one or two stages between two intermediate annealings, leading to increased output.

10.5.1.2 Molten metal degassing

The use of ultrasound in the foundry industry for degassing purposes, investigated in the UK as part of a Government-supported energy efficiency programme, was shown to offer the potential to reduce scrap levels and therefore energy costs (Presteigne, 1997). It was also shown that ultrasonic vibrations improve product quality by increasing the mechanical properties of solidified metals. Improvements were seen in yield stress, tensile strength, elongation (increase in length before fracture), reduction of area (change in cross-sectional area before fracture), impact strength and hardness.

Investigations were carried out on two aluminium silicon alloys. It was found possible to degas melts of LM25 rapidly using ultrasound. In tests, a degassing time of twenty minutes by traditional methods was reduced to just one minute using ultrasonic techniques. In industry, this would improve the availability of holding furnaces, increasing throughput and reducing energy consumption.

Degassing by applying ultrasound for short periods minimises disruption of the melt surface. This reduces oxidation, and hence dross production, leading to increased furnace yields. The cavitation induced by ultrasound can be seen in Figure 3.12 in Chapter 3. One of the results of this is intense micro-mixing, in addition to accelerated degassing.

Applying ultrasound over the solidification temperature range also results in grain refinement in straight-sided moulds. The degree of refinement is increased by raising the power density of the applied ultrasound. Intermetallic inclusions are broken up if ultrasound is applied to the melt during the nucleation and growth of the compound.

The mechanical properties of LM25 are enhanced by ultrasonic irradiation. The degree of improvement grows with increasing ultrasonic power. In tests, the hardness of LM25 increased by 10%, while the ultimate tensile strength increased by 11%.

Applying ultrasound reduces the thermal gradient across the solidifying charge resulting in a more uniform grain structure – *a feature likely to benefit cooling processes in many other industrial sectors*. Additionally, aluminium undercooling is substantially decreased, indicating improved nucleation which leads to reduced grain size.

10.5.1.3 Microwave sintering

Another important metallurgical process, sintering, can be intensified by using microwaves. Roy et al. (1999), working at Pennsylvania State University, have indicated that using a microwave sintering machine can offer benefits in terms of speed (reducing time from 5–10 hours to 90 minutes), energy efficiency, simplicity and improved product quality. Finer micro-structures can, it is claimed, also be achieved. The ability of metal powders to absorb microwave energy, rather than reflect it (as solid metals would do) inhibits sparking.

10.5.1.4 Plasma electrolytic coating

Plasmas themselves are 'intense' phenomena, and the use of plasmas to coat materials onto metals for their protection or to impart other properties is long-established. Recent work by Timoshenko and Magurova (2005) showed that the plasma electrolytic oxidation of a soft metal, in this case magnesium alloy, could be intensified by pulsations. The coating, which can protect the magnesium, could be produced at an accelerated rate if pulsed polarisation in positive and negative current intervals was applied. The application of oxide ceramic coatings to metals such as aluminium could also be of benefit, as positive impacts on the coating properties resulted.

10.5.1.5 Copper extraction using emulsion liquid membranes

Solvent extraction has been routinely used for copper refining since its invention in the 1960s, and it is suggested that well over 10% of refined copper production from ores is obtained by solvent extraction, accounting for millions of tonnes of metal. It is even more popular for refining copper from scrap. Research in India (Sengupta et al., 2006) indicates that the use of emulsion liquid membranes (ELM) can help to intensify copper extraction by reducing the solvent and carrier requirements compared to the conventional extraction process. The work showed that one of the concerns about ELM, the stability of the membrane, could be alleviated and the membranes were found to be very stable. Others had suggested that continuous ELM extraction could be 40% cheaper than solvent extraction, but the batch experiments conducted in India had not been extrapolated to a continuous process to allow a further economic analysis to be done. However, the research was believed to be sufficiently encouraging to stimulate the use of ELM in metallurgy.

10.5.2 **The glass and ceramics industry**

10.5.2.1 *Ultrasonic refining of glass*

Under a research project funded within the UK energy efficiency programme some ten years ago, British Glass investigated the development of a glass plasma arc melter and an ultrasonic refiner. The results of these feasibility studies were such that it was decided to fund pilot plants to investigate both technologies (Anon, 1998). As illustrated in Figure 3.12 in Chapter 3, the early demonstrations of ultrasound showed that in a static bath ultrasound cleared the bubbles in a period that was orders of magnitude shorter than without the applied field. Additionally, in a continuous flow of a fluid having the viscosity of molten glass, ultrasound was able to stop the passage of bubbles and agglomerate bubbles rapidly, causing them to rise to the top of the glass more quickly. It was therefore decided to proceed to look at the process using molten glass.

Ultrasound affects bubble interaction in two ways – vibration causes an attraction between two bubbles (Bjerkens force) and the Bernoulli force arises from the flow of the fluid parallel or normal to a bubble. A perpendicular flow also causes inter-bubble attraction. Additionally, if bubbles increase in size, due to the ultrasound, they rise much more quickly, proportional to the (radius).[2] As mentioned above (Section 10.5.1.2), ultrasound enhances cavitation, which is also beneficial.

The results of the research did show that the use of an ultrasonic probe system injecting small amounts of energy into the glass refining stage appears to be an efficient and clean method for accelerating refining and homogenisation. Further research was needed to determine optimum glass flow configurations. The location of the ultrasound inducer is shown in Figure 10.18.

10.5.2.2 *Microwaves in ceramics manufacture*

It is occasionally found that modifications can be made to conventional processes by introducing an intensification method that operates in parallel (or in series) with the conventional process. The research carried out by EA Technology in the UK resulted in a hybrid oven for firing ceramics, bringing together conventional gas firing and 'intensified' microwave heating of the ceramic ware (Anon, 2000).

Microwaves used to implement dielectric heating of the ceramics ensure that uniform heating takes place throughout the body, allowing faster firing of the ceramic materials. It was found that the microwave-assisted gas firing (MAGF) gave the following benefits:

- Increase in kiln throughput of 191%.
- Reduction in delivered energy of up to 50%.
- More consistent materials properties.
- Improved strength or equivalent properties achieved at lower kiln temperatures.

---

[2] Tolerant of pressure, in this case high pressure.

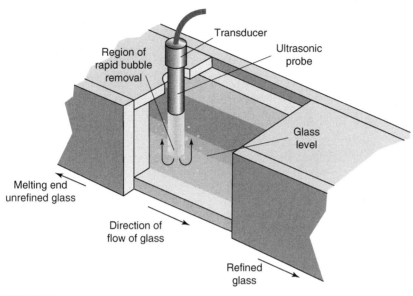

FIGURE 10.18

Cut-away schematic of the ultrasonic refiner.

- Reduced pollution (e.g. hydrogen fluoride).
- Improved yield.
- Reduction in the formation of undesirable phases, e.g. in silica refractories.
- Potential to combine microwave firing and drying.
- Enhanced binder removal.

10.5.2.3 Vermiculite processing using the Torbed reactor

A Torbed installed at Cape Industrial Products, in Scotland, generated energy savings sufficient to cover the cost of the investment in 16 months. This was used for exfoliation (principally water removal) of crude vermiculite ore as a replacement for five rotary furnaces. At the same production level as the five units it replaced (4 tonnes/h) use of a single Torbed reactor resulted in less vermiculite wastage and substantially reduced energy use. The unit is shown in Figure 10.19.

The fluidising agent was combustion gas at 1,100–1,300°C. This allowed the vermiculite to be suspended above the blade ring in a well-packed bed, and gave the bed the toroidal motion to give excellent mixing and heat transfer properties. In this case, atmospheric emissions from the furnaces had created local environmental problems, necessitating Environment Agency intervention, but as a result of the Torbed system being installed, working conditions around the exfoliation process were now the same as those in the rest of the factory, as the unacceptable emissions had been eliminated.

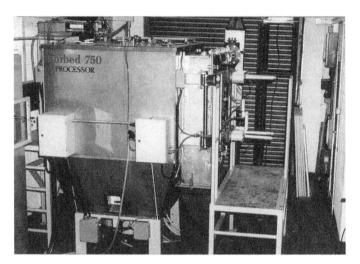

FIGURE 10.19

The Torbed unit at Cape Industrial Products.

10.6 **Aerospace**

Infra-red detector thermal control (cooling) using the Joule-Thompson effect is but one way of cooling detectors to improve their sensitivity. As such devices have to be small – they fit in the noses of missiles, for example – intensification is second nature to the aerospace specialists in this field. The example shown in Figure 10.20 is a heat exchanger tube manufactured in stainless steel by Honeywell Hymatic of Redditch, in the UK. This milli-engineered component of the infra-red detector cooling system has a tube of approximately 0.3 mm diameter and 30 fins/cm of tube length.

The UK leads in the technology of nanosatellites. Such earth-orbiting or deep space mission devices, with a weight of less than 10 kg, dramatically reduce the cost of access to space because they are cheaper to construct and require much less power to launch. Surrey Satellite Technology Limited (SSTL) recently moved from micro- to nano-satellites with the launch, in June 2000, of SNAP-1, weighing 6 kg and containing micro GPS navigation systems, computing, propulsion and attitude control technologies, all developed in the UK (Anon, 2008).

Another 'nano' area that is gaining ground in aviation is energy harvesting (sometimes called energy scavenging). This involves thermoelectrics or conversion of vibrations into electrical energy, and it has been suggested that engine exhaust ducts and passengers might even provide sources for conversion to electricity (Scott, 2012). This is discussed further in Chapter 11 in the context of power generation.

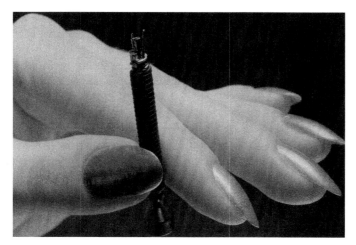

FIGURE 10.20

A micro-heat exchanger for cooling infrared detectors in missile noses.

(Courtesy: of Honeywell Hymatic)

10.7 **Biotechnology**

A wide range of intensification methods can be used in bioprocessing, ranging from high gravity fields, electric fields and ultrasound, to membrane processes and some reactors. Intensive bioprocessing has been the subject of many contributions at PI conferences since 1995. Applications discussed have ranged from bioreactors to ultrasonic enhancement of micro-filtration. Several workers have associated fish farming with biointensification techniques. Indeed, the word 'intensification' is perhaps used more widely in agriculture, horticulture and aquaculture than in the chemical and related process industries!

The reactor is normally associated with chemical processes, but work in the US is directed at developing enzymatic microreactors. It is believed that these could be used as sensors, for energy production, chemical synthesis and environmental clean-up. The inherent advantage of enzymes in reactors is that they operate at ambient temperatures and pressures. This, of course, potentially leads to less complex engineering than that necessary for most chemical micro-reactors.

In the UK, Heriot-Watt University has been studying marine bioprocess intensification, in particular using novel bioreactors (Wright et al., 1999). With optimisation and conversion of current technologies these have the potential to yield more efficient units.

A limiting factor in the widespread commercial acceptance of a large range of marine metabolites is the efficient production of, for example, sufficient quantities of antibiotics and nutraceuticals to allow for structural analysis and clinical testing. Conventional methods utilised for physical and chemical process intensification require careful analysis of their potential application to shear-sensitive bioprocess systems. Stress induction, for example, provides one route to marine bioprocess intensification due to the expression of metabolites not otherwise possible. Use of high pressure as a stressing agent and/or intensification tool is discussed, and its potential, demonstrated by showing the existence of barotolerant[2] (at 120 MPa) marine microorganisms obtained from shallow surface waters (<1.5 m deep). Microorganisms associated with the surface of, for example, seaweed, show a greater likelihood of being barotolerant.

Research work in Singapore and Korea has investigated the potential for biomolecular process intensification in the 'cellular biofactory' based upon *E. coli* (*Escherichia coli*) that is used for producing many biomolecules (Choe et al., 2006). The processes involved in the biofactory, in particular extraction, solubilisation and refolding techniques have, suggest the authors, reached the stage when the linking of several unit operations, with intensified upstream and downstream processing, should soon become feasible. The main target is improved process economics.

The leader in the use of electric fields in many areas of contacting and bioprocessing is Professor Laurence Weatherley, currently working in the University of Kansas. Recently he has, in conjunction with a researcher at The Queens University of Belfast, examined how high voltage electric fields might affect enzymatic processes. Their recent work examined the enzymatic hydrolysis of sunflower oil, comparing the performance of mechanical agitation and electrostatic spraying of the aqueous enzyme with the oil. Examining batch and continuous operation, it was generally observed that the reaction rate was significantly enhanced as the magnitude of the applied electric field (to 40 kV) was raised (Weatherley and Rooney, 2008).

10.7.1 Biodiesel production

Prof Adam Harvey at England's Newcastle University (Harvey, 2006) examined the use of oscillatory baffle reactors as components to intensify the production of biodiesel fuels. The research used rapeseed oil as the feedstock, the attraction being generally that it is a renewable energy source, it reduces CO_2 emissions and pollution; and it could attract tax relief in the UK. The range of PI projects in this area include a portable plant, solid catalysts (which allow a reduced number of process steps compared to liquid catalysts), the development of a reactive extraction process direct from the oilseeds, examination of cold flow properties and the production of biodiesel from algae.

The oscillatory baffle reactor/oscillatory flow reactor (OBR/OFR) types are seen as for niche applications, where one wants to convert a long residence time batch process to a continuous one. In the case of biodiesel, Prof Harvey indicates that a conversion could be carried out in 10 minutes, compared to 1–6 hours in continuous industrial processes. One variant is shown, by means of a flow diagram, in Figure 10.21, while Figure 10.22 shows components of the OFR. The aim is to

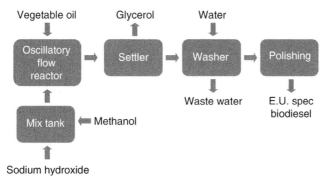

FIGURE 10.21

The flow sheet of the intensified biodiesel plant at Newcastle University.

FIGURE 10.22

The co-author, Prof Harvey, beside an OFR for the biodiesel plant.

make the plant portable so that it will fit into a standard shipping container. The unit could be sold worldwide to, for example, farmers to produce their own fuel locally.

Biodiesel research has also been undertaken in the UK at Cranfield. Richard Jackson, of BHR Biofuels Ltd, introduced a relatively new activity within BHR Group to a recent meeting: the development of biodiesel manufacturing plant using PI technology. After giving the background on BHR and highlighting reasons for making biodiesel (reduced viscosity, a tax rebate, the reduction of harmful acid species, better cold pour point and clouding issues), Jackson said that the starting point is vegetable oil which is made of triglycerides. Continuing a brief description of the chemistry involved, Jackson said that this was a natural product and was not clean in processing. The process is an ester–ester conversion and there is not a lot of heat involved. The target is 96.5% ester content, which is difficult to achieve because of some impurities in the feed, for example, fish oils. The presence of water also causes soap formation and any free fatty acids increase the catalyst needs.

With regard to the physics, BHR believe that a large-scale adiabatic reactor could be used with high mass transfer and near plug flow. Temperature is around 100°C and the residence time can be seconds to a few minutes. The reactor pressure should be sufficient to suppress methanol boiling at the desired temperature. Challenges are associated with the mixing duty and the fact that operation is currently mainly batch mode.

The experimental work started with a small batch reactor and then moved to the BHR Flex-reactor, where temperatures were 70–150°C and residence times 1–10 minutes. Trials were used to optimise operating characteristics with production of biodiesel at 1 tonne/day. Scale up has now reached 75 tonnes/day and the target fuel specification was achieved first time. The plant can actually process 92 tonnes/day, 20% more than the design value. BHR Biofuels is now working on a 150 tonne/day unit (Jackson, 2007).

10.7.2 Waste/effluent treatment

Effluent is associated not just with many process industries, as has been illustrated above in the case of the textile finishing industry, but also with most aspects of our urban existence. The need to treat effluent arises for several reasons. Some effluents can be transformed into useful products, such as feedstock for animals or even pharmaceutical products; the ability to concentrate effluent, particularly in an energy-efficient manner, can reduce disposal costs; and the ability to recycle effluents, once separated from less desirable components, can also be economically beneficial. The link between effluent treatment and PI is strong and growing, as the above influences become the subject of legislation or economic necessity. The range of techniques that can be used to aid clean-up covers both active and passive intensification methods. Included below are several examples including the use of ultrasound already dealt with in a number of process uses in this chapter.

As we have already emphasised, the sensitivity of PI unit operations to contaminants within process steams – such as particulates causing fouling, blockage or corrosive fluids – so the use of PI for effluent treatment has to take into account the fact that many effluents are, by their nature, 'foul'. Obviously, care can be taken in the selection of appropriate components for effluent treatment, and a major opportunity for PI is that it can be used as the basis for compact, reduced capital cost and energy-efficient effluent treatment plant. An early UK case study undertaken in the 1990s illustrates the benefits that could accrue to the use of PI in effluent treatment.

10.7.2.1 Wet oxidation plant conceptual study

Wet air oxidation is a well-established treatment process for dealing with spent pulping liquors, sewage sludge and process waste water. The basic principle is to combine the aqueous waste stream containing organic and/or inorganic contaminants with air under suitable temperature and pressure conditions so that the aqueous effluent remains as water. The high solubility and diffusivity of oxygen in aqueous solutions at elevated temperatures (150–325°C) enables a strong driving force for oxidation, via an exothermic reaction at pressures of up to 200 bar. Commercial systems vary from single units handling about 1.5 m³/h, to multiple units handling up to 1,000 m³/h.

The study, one of the first to examine a complete plant from the PI viewpoint, looked at wet air oxidation of phenol, a contaminant regularly found in industrial effluent streams. The plant considered handles 10 m³/h of effluent with a phenol concentration of 10 ppm. A flow sheet model was used to produce material and energy balances, allowing individual equipment items to be sized and costed (Anon, 1994; Reay, 1999). Two intensified plant layouts were considered, one of which was skid-mounted. (The latter concept is feasible if the oxygen storage and effluent holding facilities are located separately.)

The potential for PI in this application can be both by the use of compact plant and through changes to the process conditions. The equipment has been intensified in three ways:

- A fluidic mixer has been installed to shorten reactor mixing time.
- Two compact plate and shell heat exchangers have replaced four shell and tube units.
- Use of a vortex mixer combined with improved process control has eliminated the need for the stirred pH control vessel.

The main process alteration was to change from air to pure oxygen. As can be seen from a comparison of the cost and component breakdown illustrated in the pie chart in Figure 10.23, this eliminates the air compressors (and also reduces reactor size), although necessitates oxygen storage facilities (assuming on-site O_2 generation).

The benefits of the intensified plant are:

- Capital cost reduction – from £1, 600, 000 to £600, 000.
- Primary energy reduction – from 682 kW to 380 kW.
- Reduced operating cost when O_2 generated on site – from £62, 000 to £43, 000.
- Reduced space requirements.

10.7.2.2 Treatment of waste water and contaminated soil

An excellent review of hybrid methods (including ultrasonics) to treat effluents and groundwater containing phenolic compounds has recently been published (Gogate, 2008). Phenols arise from many processes, including petroleum refining, paint manufacture and pharmaceuticals production, and effluent treatment is necessary to

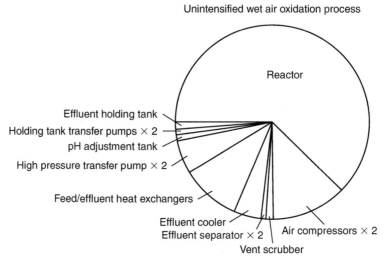

Total cost £1,600,000

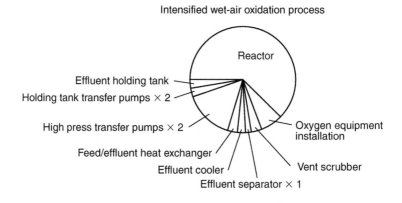

Total cost £600,000

FIGURE 10.23

Features of a standard (top), and intensified (bottom), effluent treatment plant, illustrated by means of a pie chart.

eliminate them. The emphasis of the review of over 60 papers in the field is to high-light the hybrid technologies, many of which involve ultrasound as one component, that appear to be the best way forward for clean-up. It is generally concluded that combinations of methods, rather than the use of ultrasound alone (as sonochemistry with, for example, hydrogen peroxide added to the stream or sonication), are the best solutions.

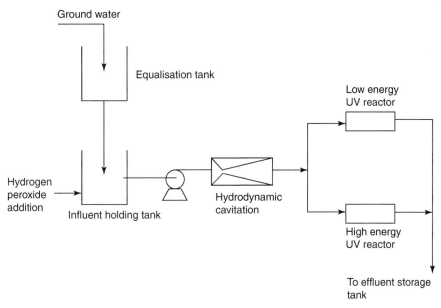

FIGURE 10.24

The CAV-OX process that uses an alternative to ultrasound to create cavitation (Gogate, 2008).

The hybrid techniques (all of which involve ultrasound), in addition to the use of ultrasound plus hydrogen peroxide or ozone, include the use of different catalysts/additives, sonophotocatalytic oxidation, sonoelectrochemistry and the use of cavitation (brought about by ultrasound, as in the textiles example above) in conjunction with enzymatic treatment. However, as illustrated below, cavitation can be introduced by other means (as marine engineers will well know).

The CAV-OX process employs ultraviolet radiation, hydrodynamic cavitation and hydrogen peroxide to oxidise organic compounds present in water at ppm level concentrations to non-detectable levels. It does not need ultrasound to generate cavitation. The system has been used with success for the effective degradation of volatile organic compounds, primarily trichloroethane, benzene, toluene, ethyl benzene and xylene. In the system, hydrogen peroxide is added to the effluent, in this case contaminated ground water, which is then pumped through a cavitation nozzle, followed by ultraviolet radiation, as shown in Figure 10.24. Hydrodynamic cavitation produced in a nozzle is used to generate additional hydroxyl radicals, which help in increasing the rates of degradation. Moreover, the author claims that cavitating conditions along with UV irradiation also result in an increased number of free radicals due to the dissociation of hydrogen peroxide.

Miladinovic and Weatherley (2006) have reported on the intensification of the removal of ammonia ions from waste water. By combining ion-exchange with a nitrification column in a packed bed, some of the costs associated with using ion-exchange alone are removed and the approach is, as the research shows, more robust in terms of accommodating fluctuations in ammonia concentrations in waste water. The system does need good aeration in order to provide the large oxygen demand.

Work in the Ukraine (Mishchuk et al., 2007) on the electro-remediation of soil for the removal of undesirable contaminants, such as cobalt and uranium, showed that controlling the pH, in particular reducing it, could give substantial intensification to the removal process. Use of the method was able to speed up the removal of radio-nuclides and heavy metals to less than 24 hours, compared to only modest improvements seen after many days without pH regulation. The methods for moderating the pH are the subject of further investigation, but results suggested that occasional pH control in the cathode chamber, together with appropriate acid concentration changes, was required to produce the highly promising results. Readers may also be interested in the research in Singapore by Dutta and Ray (2004) using a Taylor-Couette reactor (see Chapter 6) as the basis of a photocatalytic reactor for water purification. Work in Japan on this concept has been reported recently (Jia et al., 2011) but further work is needed before any commercial possibilities could be realised.

10.7.2.3 Waste water treatment – an industrial viewpoint

To the technology champion, process intensification must often seem to be a no-brainer as far as its selection is concerned. However, the practising process engineer has to weigh up all the options, taking into account negative as well as positive attributes of potential solutions to a treatment challenge. Gary Howard, presenting at the European Process Intensification Conference in 2011 (Howard, 2011), reviewed drivers for PI in this area and presented examples of work in his company where PI had been considered.

He discussed an example of a new greenfield refinery design in Asia, where the national regulator insisted on substantial water recovery and flow limits regarding discharge into the sea. While conventional effluent treatment could probably meet all Standards, the requirement to recover clean water led to high concentrations of contaminants in the reject streams. With an added problem of an excessive COD, a stage was needed to take out COD residuals.

The oil removal system used American Petroleum Institute (API) separators, tilted plate interceptors and dissolved air flotation in series. This was influenced by the company's experience of these systems, at the expense of any PI opportunities. Similarly, the biological treatment system used a trickling filter followed by activated sludge. Options involving PI, including a membrane bioreactor, were rejected, again in part to their lack of experience of this system.

According to Howard, in asking 'should we use PI?' Foster Wheeler has addressed the increasing need that the company has seen for PI solutions for both effluent treatment and water by developing an iterative process that should lead to successful implementation. The majority of the steps are those recommended in PI implementation in Chapter 12, adapted for the specific area of effluent treatment/water recovery. The emphasis, however, was on the ability to manage the constraints of the intensified process.

10.8 Summary

A wide range of process industries have applied process intensification – obvious candidates include the food sector (microwaves) and the nuclear processing industry using high gravity centrifuges. However, less high profile examples are coming to the fore, including biodiesel production, currently of interest as a renewable energy resource, and the perhaps less glamorous but equally important area of waste destruction and remediation of contaminated sites. It is interesting to note that active enhancement methods play a strong role across all these sectors.

References

Akhtar, M., Blakemore, I., Clayton, G., Knapper, S., 2009. The use of spinning disc reactor for processing ice cream base – effect of ageing in making model ice cream. Int. J. Food Sci. Technol. 44, 1139–1145.

Anon, 1994. Process intensification: its potential in effluent treatment processes. Energy Efficiency Best Practice Programme R&D Profile 44, May.

Anon, 1998. Enhanced energy efficiency through the ultrasonic refining of glass. Future Practice Report No. 68, Energy Efficiency Best Practice Programme, UK, March.

Anon, 2000. The microwave-assisted gas firing of ceramics. Energy Efficiency Best Practice Programme Future Practice Profile 57, March.

Anon, 2008. Surrey Satellite Technology Limited, <http://www.sstl.co.uk>.

Anon, 2007a. Pursuit Dynamics brochure. See <www.pursuitdynamics.com>.

Anon, 2007b. <http://www.protensive.co.uk/pages/technologies/single/categoryid=food/rowid=15>.

Anon, 2012. Patent Application No. PCT/GB2012/051365 (filling date14th June 2012).

Atanasiu, N.E., 1980. Intensification of tube drawing by axial ultrasonic oscillation of the plug. Ultrasonics, 255–260. (November).

Burdo, O.G., Kovalenko, E.A., Kharenko, D.A., 2008. Intensification of the processes of low-temperature separation of food solutions. Appl. Therm. Eng. 28, 311–316.

Choe, W.-S., Nian, R., Lai, W.-B., 2006. Recent advances in biomolecular process intensification. Chem. Eng. Sci., 886–906.

Dexter, A., 2001. Process intensification in the food industry. Proceedings of Welsh Meeting on Process Intensification, Energy Efficiency Office, December.

Doona, C.J., Kustin, K., Feeherry, F.E., 2010. Case studies in novel food processing technologies: innovations in processing, packaging and predictive modelling. Woodhead Publishing Series in Food Science, Technology and Nutrition No. 197. ISBN 1 84569 551 8, October 2010.

Dunn, J., 2007. New spin on mixing and heating. Food Manufacture See <http://www.foodmanufacture.co.uk/> news..

Dutta, P.K., Ray, A.K., 2004. Experimental investigation of Taylor vortex photocatalytic reactor for water purification. Chem. Eng. Sci. 59, 5249–5259.

Fee, C.J., Chand, A., 2006. Intensification of minor milk protein purification processes. Proceedings of the Second Conference on Process Intensification and Innovation, Christchurch, New Zealand, 24–29 September.

Fikiin, A.G., 1983. Etude des facteurs d'intensification de l'echange de chaleur dans le refroidissement des fruits et legumes. Int. J. Refrig. 6 (3), 176–181.

Gogate, P.R., 2008. Treatment of wastewater streams containing phenolic compounds using hybrid techniques based on cavitation: a review of the current status and the way forward. Ultrason. Sonochem. 15 (1), 1–15.

Goodacre, C., 2001. Process Intensification: relevance to the food industry. Proceedings of the Sixth Process Intensification Network Meeting, Jesus College, Cambridge University, 27 November. See <www.pinetwork.org>.

Green, R., 2003. Back to the future. Nucl. Eng. Int., 36–39. September.

Harvey, A.P., 2006. Biodiesel process intensification projects at Newcastle University. Proc. PIN Meeting, 16 November. See <www.pinetwork.org>.

Harbold, G., Park, J., Using the GasTran deaeration system to achieve low dissolved oxygen levels for superior line speed and product quality: a case study in carbonated soft drink bottling. Presented at the Seventh International Conference on Process Intensification, Albany, NY, 2008. Available from GasTran Systems.

Howard, G., 2011. Process intensification in industrial wastewater treatment. Proceedings of the European Process Intensification Conference (EPIC 2011), IChemE Symposium Series No. 157, Manchester.

Jackson, R., 2007. The BHR Biofuels intensified biodiesel production route. Proceedings of the Fifteenth PIN Meeting, Cranfield University, October. See <www.pinetwork.org>.

Jia, N., Horie, T., Ohmura, N. 2011. Process intensification of water purification using Taylor vortex photocatalytic reactor. Proceedings of the European Process Intensification Conference (EPIC 2011), IChemE Symposium Series No. 157, Manchester.

Johnson, K.A., Marsh, R.A., Meacock, G., 2008. Emerging technologies for food processing. Food Process. Knowl. Transf. Netw., 9. October.

Koehler, K., Schuchmann, H.P., 2011. Homogenisation in the dairy process – conventional processes and novel techniques. Procedia Food Sci. 1, 1367–1373.

Li, X., Le Pierres, R., Dewson, S.J., 2006. Heat exchangers for the next generation of nuclear reactors. Proceedings of ICAPP'06, Paper 6105, Reno, NV, USA, 4–8 June 4–8.

Miladinovic, N., Weatherley, L.R., 2006. Intensification of ammonia removal in a combined ion-exchange and nitrification column. Proceedings of the Second Conference on Process Intensification and Innovation, Christchurch, New Zealand, 24–29 September.

Mishchuk, N., Kornilovich, B., Klishchenko, R., 2007. pH regulation as a method of intensification of soil electroremediation. Colloids and Surfaces A: Physicochem. Eng. Aspects 306, 171–179.

Moholkar, V.S., Warmoeskerken, M.M.C.G., 2004. Investigations in mass transfer enhancement in textiles with ultrasound. Chem. Eng. Sci. 59, 299–311.

Muralidhara, H.S., 2006. Role of process intensification in food and bioprocessing for the 21st century: challenges and opportunities. Proceedings of the Second Conference on Process Intensification and Innovation, Christchurch, New Zealand, 24–29 September.

Nachiappan, S., Muthukumar, K., 2010. Intensification of textile effluent chemical oxygen demand reduction by innovative hybrid methods. Chem. Eng. J. 163, 344–354.

Poynton, T.R., 2006. PI in food and drink processing – opportunities, barriers and some solutions to savour. Presentation at PIN Meeting, Newcastle University, November. See <www.pinetwork.org> for presentation.

Presteigne, K., 1997. Improved castings and enhanced energy efficiency with power ultrasound. EEBPP Future Practice R&D Profile 55, March.

Ravindran, P., Sabharwall, P., Anderson, N.A., 2010. Modelling a Printed Circuit Heat Exchanger with RELAP5-3D for the Next Generation Nuclear Plant. Report INL/EXT-10-20619, Idaho National Laboratory, December 2010.

Reay, D.A., 1999. Learning from experiences with compact heat exchangers. CADDET Analysis Series No. 25, Sittard, The Netherlands.

Roy, R., Agrawal, D., Cheng, J., Gedevanishvili, S., 1999. Full sintering of powdered-metal bodies in a microwave field. Nature 399, 670–688. 17 June.

Ryan, C., 2000. Mixer doubles Courtelle output. Plant Control Eng. (9), 27. June.

Scott, M., 2012. Current thinking. Flight Int. pp. 26, 27, 24–30 July.

Sengupta, B., Sengupta, R., Subrahmanyam, N., 2006. Process intensification of copper extraction using emulsion liquid membranes: experimental search for optimal conditions. Hydrometallurgy 84 (1–2), 43–53.

Southall, D., Dewson, S.J., 2010. Innovative compact heat exchangers. Proceedings of ICAPP '10, San Diego, CA, USA, Paper 10300, 13–17 June 2010.

Tassou, S.A., Lewis, J.S., Ge, Y.T., Hadawey, A., Chaer, I., 2010. A review of emerging technologies for food refrigeration applications. Appl. Therm. Eng. 30, 263–276.

Timoshenko, A.V., Magurova, Yu.V., 2005. Investigation of plasma electrolytic oxidation processes of magnesium alloy MA2-1 under pulse polarisation modes. Surf. Coatings Technol. 199, 135–140.

Tomlinson, I., 2011. Doubling food production to feed the 9 billion: a critical perspective on a key discourse of food security in the UK. J. Rural Stud. doi: 10.1016/j.jrurstud.2011.09.001.

Van der Bruggen, B., Curcio, E., Drioli, E., 2004. Process intensification in the textile industry: the role of membrane technology. J. Environ. Manage. 73, 267–274.

Warmoeskerken, M.M.C.G., van der Vlist, P., Moholkar, V.S., Nierstrasz, V.A., 2002. Laundry process intensification by ultrasound. Colloids and Surfaces A: Physicochem. Eng. Aspects 210, 277–285.

Weatherley, L.R., Rooney, D., 2008. Enzymatic catalysis and electrostatic process intensification for processing of natural oils. Chem. Eng. J. 135, 25–32.

Weidner, E., Renner, M., Geihsler, H., Brandin, G., 2007. Production of leather under the influence of compressed carbon dioxide. Proceedings of the European Congress of Chemical Engineering (ECCE-6), Copenhagen, 16–20 September.

Yachmenev, V.G., Blanchard, E.J., Lambert, A.H., 2004. Use of ultrasonic energy for intensification of the bio-preparation of greige cotton. Ultrasonics 42, 87–91.

Application Areas – the Built Environment, Electronics, and the Home

OBJECTIVES IN THIS CHAPTER

As in Chapter 10, examples of the use of process intensification outside the chemicals sector will aid technology transfer among those considering PI as a technology in any industry, and it is also believed that this will encourage innovation in areas where energy efficiency and the use of renewable energies are becoming increasingly important, such as absorption and adsorption cooling/air conditioning, micro-power generation and the plethora of electronics-based equipment in commerce and the home.

Another point worth making here, illustrated later in the chapter, is the time between concept and practical application of some intensified units – although the area of application may be one in which progress is seen to be made very rapidly by consumers. The Rotex (latterly Rotartica) absorption heat pump/chiller is a case in point and is an object lesson in patience – and, like the spinning disc reactor company, Protensive, has subsequently succumbed to market forces beyond its control. One only needs to look to China to see exploitation of PI on a large-scale, however.

The chapter concludes with our third Key Table, summarising PI applications (Table 11.2).

11.1 Introduction

Most of those not active in PI will have encountered applications of process intensification in their domestic environment, rather than in their work. The microwave oven uses electric fields to intensify the energy input to foodstuffs and liquids, to cook and/or raise their temperature. We may find it difficult to identify other orders-of-magnitude intensification processes within our living spaces but an examination of consumer electronics and the power devices needed to drive such systems can reveal existing and potential applications of PI. Even the heat pump/chiller, used to maintain comfort conditions, benefited from the rotating intensification methods invented by Professor Ramshaw.

Since the first edition, there has been an accelerated trend towards developing micro-systems for cooling, particularly for applications in the electronics sector, and domestic combined heat and power (CHP) has become a reality – although the latter may not strictly involve process intensification. This has in all cases, however, necessitated compact heat exchangers and prime moves (e.g. Stirling engines and their associated regenerators), as has the growing demand worldwide for domestic heat pumps. An area not mentioned before is the use of piezoelectric devices for energy scavenging. The development of nano-materials has accelerated investigation of miniature power sources that can convert mechanical or thermal energy into electricity – see section 11.3.5.

The use of PI in such areas will hopefully continue to make a positive contribution to the larger scale applications discussed in earlier chapters. It should be noted, and will become evident, that the problems with thermal plant of the type discussed below relate directly to the need to reduce size while increasing, by a disproportionate amount, the output and/or heat fluxes that need to be catered for. Enhancement (or intensification) is the main route leading towards overcoming this bottleneck. In this chapter, a number of such devices and their applications are described.

11.2 Refrigeration/heat pumping

There is a raft of technologies that are integral parts of conventional processes in a wide range of industrial sectors. Refrigeration, steam generation, the provision of compressed air and the use of prime movers are four principal ones. These may be classified as 'services', but in an intensified plant, where many services are integral (or could become so), they must be compatible with other compact unit operations.

It is fortunate that, because of the high energy use of these service providers and the vast number in use in industry, they have received attention in the context of compact heat exchangers, combustion systems, compressors, etc. Thus, while most may not yet be compatible with 'desktop' plant, there would certainly be appropriate units available for plant of, say, 20% of the size of current practice. One of the original applications of the PCHE was in refrigeration, and the classic example of the opportunity offered by repackaging a water chiller using CHEs, as shown in Figure 11.1, gives a volume reduction of just this order.[1] The point to emphasise here is that some pieces of intensified plant *can* help in allowing a complete redesign of a commonplace unit.

Also, these examples are not all related to process plant – the Rotex chiller/heat pump discussed below was destined for the domestic heating and cooling market, but incorporates several examples of PI, such as highly compact heat exchangers and rotating absorption and heat transfer enhancement. The micro-sized chiller for cooling micro-electronics components – the electro-adsorption chiller – brings adsorption down to the micro-scale (Gordon et al., 2002).

---

[1] Note that the PCHE is too expensive for such an application in current circumstances, but PHEs can replace shell and tube units to give a reasonable volume reduction. Much can also be done with plate-fin heat exchangers – see Chapter 4.

Package employing heatric   Conventional package
heat exchangers

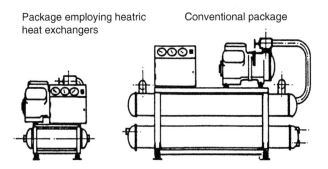

FIGURE 11.1

The PCHE-based chiller compared with a conventional unit of identical performance.

11.2.1 The Rotex chiller/heat pump

Absorption cycle heat pumps and chillers tend to be physically large (they are essentially a collection of heat exchangers), and in the process industry applications these are invariably shell and tube units. While pressure is on the manufacturers of units for such applications to reduce their physical size, the importance of compactness to the small commercial or domestic chiller/heat pump user is even greater. This is often made difficult by the demands for increased performance. Interotex Ltd (now disbanded) based in Cheltenham, UK, followed up the work of Cross and Ramshaw in their ICI patent (see Chapter 1) by applying process intensification to the whole of the absorption heat pump process in order to achieve high thermal performance within compact dimensions (Gilchrist et al., 2002).

Shown in cycle form in Figure 11.2, the key components in the double-effect unit are enhanced using rotation (see the schematic in Chapter 1, Figure 1.6). All heat and mass transfer, apart from combustion, take place within a hermetically welded spheroid container of about 500 mm diameter and 500 mm long. This rotates at 550 rpm, and thus generates high 'g' on liquids flowing over the several surfaces within the casing. The faster moving and thinner films allow much more effective mass transfer (e.g. absorption) and evaporation. The machine is rotated using a small electric motor, but the primary energy input is via a radiant burner located to the left of the generator, visible in the figure.

A feature of absorption cycle machines is a solution heat exchanger, which is a heat recovery unit between two liquid streams running between the absorbers and generators. In Rotex, highly compact plate heat exchangers are used in this role (see Chapter 4), although they do not rotate with other components. Additionally, pumping power within the machine is derived from rotation, using pitot tubes that convert fluid kinetic energy into pressure energy. The performance of the laboratory machine (cooling coefficient of performance [COP] versus ambient temperature) is shown in Figure 11.3.

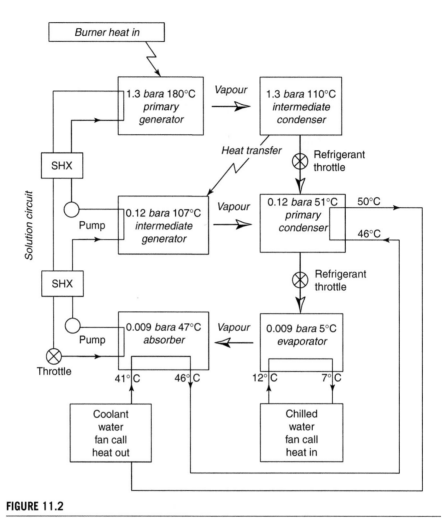

FIGURE 11.2

The Interotex double effect absorption cycle (Gilchrist et al., 2002).

The early work on Rotex was directed at heat pump use, for example, for heating in the UK. Later interest in the US and Spain dictated that a chiller capability was introduced, and latterly the cooling duty has dominated. The Spanish manufacturer has introduced a solar capability as well. Rotartica, a company established in 2001 and owned by Fagor Electrodomesticos and Gas Natural in Spain, has now begun to produce air-conditioning equipment to use thermal solar energy – based on Rotex. The company expects to install equipment both in the residential and industrial sector. The firm will separate production, producing to stock the core of the system (the rotating components), and carrying out final assembly of the air

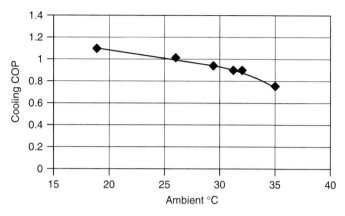

FIGURE 11.3

The performance of the double effect Rotex absorption machine in cooling mode
(Gilchrist et al., 2002).

FIGURE 11.4

Rotartica single and double effect absorption chillers.

conditioners, heat exchangers, etc., on demand. Rotartica is targeting the tertiary
commercial sector, i.e. houses with environmentally conscious owners and office
buildings, using the machines illustrated in Figure 11.4

At the same time in the 'non-comfort' world, they expect industrial applications
where an air conditioned environment is specified to improve productivity. Rotartica
has tested its equipment in various pilot installations (Demolab of Ikerlan-Enerlan,
Edesa and Gas Natural). The details of a trial on a 4.5 kW air-cooled, single effect
Rotartica machine in Madrid have recently been reported (Izquierdo et al., 2008).

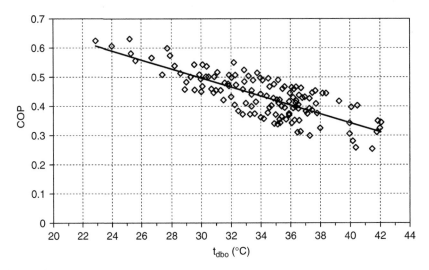

FIGURE 11.5

Data from the Madrid trials showing the COP as it varies with outdoor dry bulb temperature – this is a single effect machine (Izquierdo et al., 2008).

The trials took place at the Eduardo Torroja Institute Heat Pump and Absorption Chiller Laboratory at La Poveda, Arganda del Rey, Madrid, in August 2005. Measurements were recorded over a 20-day period. The hot water inlet temperature in the generator varied throughout the day from 80°C to 107°C. Daily and period energy flows were recorded, along with the energy balance for the facility.

The total energy supplied to the chiller generator came to 1,085.5 kWh and the heat removed in the evaporator to 534.5 kWh. The average COP for the period as a whole was 0.49, based on the ratio of the coolth delivered to the heat input. When the electric power used by auxiliary equipment, such as circulating pumps and the motor to rotate the components was factored into the equation, primary energy based COP came to 0.37. The results obtained for Madrid can be extrapolated to other regions of Spain with t_{dbo} lower than 35°C (see Figure 11.5).

As pointed out by Professor Ramshaw (Ramshaw, 2005), this development was founded on an ICI patent filed in the 1980s in which it was suggested that elevated acceleration fields could intensify the operation of the multi-phase processes involved in an absorption heat pump cycle. He goes on to say:

The development work was initially spearheaded at Cheltenham by Caradon Mira, ICI and British Gas, with a significant cash input being provided by the Energy Technology Support Unit (the forerunner of Future Energy Solutions). Originally, the target was a gas "boiler" having an efficiency of 140%. However, with the prevailing low gas prices, the pay-back time was over five years and interest switched to air conditioning duties. Subsequently, Fagor (a Spanish white goods manufacturer) and Gas Natural (a major Spanish gas distributor) joined the consortium, because of their air conditioning interests. Development work switched to Spain several years ago and it is highly gratifying that the technology is now being commercialised – albeit after twenty years. This saga illustrates how long it can take for radically new energy-saving technology to be accepted.

A message here for those developing radical new processes is patience!

11.2.2 Compact heat exchangers in heat pumps

It has been argued elsewhere (Reay and Kew, 1999) that the fundamental differences between heat and work transfer imply that the greatest advances in intensifying the heat pumping process are likely to come from CHE developments, rather than from improvements in compressor technology. This is because heat exchangers having a high heat transfer duty can be made extremely small if suitable fabrication techniques are available. On the other hand, it was shown by Kew in the above paper that turbines, pumps and compressors do not benefit from miniaturisation, at least in efficiency terms. In practice, the efficiency of a turbo-machine will be decreased if it is scaled-down. Clearly, miniature turbines and compressors, even with extremely high rotational speeds, will necessitate small power outputs or inputs. Similar reasoning applies to positive displacement machines. However, as can be readily observed in photos of heat exchangers such as the PCHE and Marbond, compact heat exchangers become surrounded by not-so-compact headers and interconnecting pipework. These latter components can become a significant proportion of the volume of the equipment. What can be done to eliminate these and reap the full advantage of compactness?

Figure 11.6 shows one concept proposed by Heriot-Watt University, Edinburgh (Reay et al., 1999). This is a water–water heat pump employing a rotary vane compressor embedded in a printed circuit-type heat exchanger. The unit shown was designed to operate using R134a with evaporating and condensing temperatures of −4°C and 60°C respectively, an output of 3 kW and a COP of 3. The unit (excluding water-side headers, water pumps and compressor motor) would occupy a block 25 mm×700 mm×200 mm, or 0.0035 m^3. Interestingly, the unit here has a similar duty per unit volume to that of an absorption cycle unit (ammonia/water) that was investigated by Heriot-Watt University and Absotech in Germany, based upon the Hesselgreaves compact heat exchanger surface (see Chapter 4, Figure 4.1). This design had a 100 kW cooling duty using ammonia and water working fluid pair, and a core size of 1,000 mm×550 mm×200 mm, or 0.11 m^3 (Reay et al., 1998).

More recently, the work at Garimella's Georgia Institute of Technology's laboratory (Determan and Garimella, 2012) has led to the successful testing of a 300 W nominal cooling capacity ammonia–water absorption heat pump with overall dimensions of 200×200×34 mm^3, operating over a range of heat sink temperatures from 20°C to 35°C with 500–800 W of desorber heat input. This was directed at mobile and portable uses but with a volume of 0.00136 m^3 it would approach the

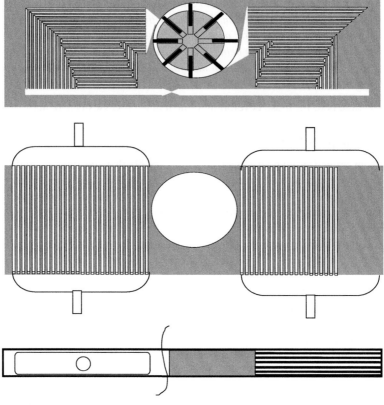

FIGURE 11.6

Concept of a heat pump with the compressor integral with the heat exchanger(s).

compactness of the Hesselgreaves unit (for a 100 kW unit, all other things being equal, the Georgia unit would scale to 0.45 m³). The shims incorporating the several components are shown in Figure 11.7. The performance of the unit is illustrated in Figure 11.8 – the COP is given for a range of absorber inlet temperatures (x-axis) for several desorber heat inputs. As the heat sink temperature increases, the COP decreases. The authors stated that at a standard chiller rating of 35°C heat sink temperature, the system delivered a cooling duty of 230 W, and was able to deliver 300 W cooling with a heat sink at 20°C.

Although of more conventional construction, the transcritical CO_2 heat pump developed by the same group, in conjunction with the United Technologies Research Center, also in the USA, represents a growing trend towards the use of CO_2 as a working fluid in heat pump and power cycles. Goodman et al. (2011) report on transcritical carbon dioxide micro-channel heat pump water heaters (see also Chapter 4 for data on the Japanese supercritical work in this vein – Tsuzuki et al., 2009).

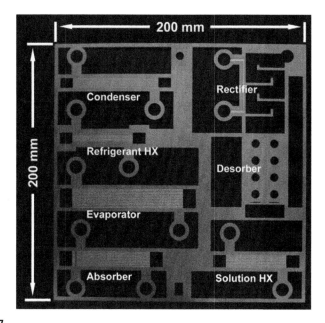

FIGURE 11.7

Individual components of the absorption unit on an etched shim (Determan and Garimella, 2012).

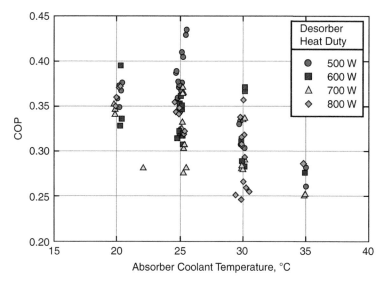

FIGURE 11.8

Micro-absorption system cooling mode coefficient of performance (Determan and Garimella, 2012).

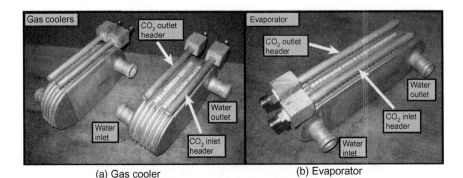

FIGURES 11.9

(a) Gas coolers, and (b) evaporator. Both are aluminium micro-channel units (Goodman et al., 2011).

The heat pump consisted of three heat exchangers (as well as two hermetic reciprocating compressors). The heat exchangers, illustrated in Figure 11.9, were (Figure 11.9(a)) a cross-flow water-to-CO_2 aluminium brazed-plate, micro-channel gas cooler, a cross-counterflow water-coupled aluminium micro-channel brazed-plate, evaporator (Figure 11.9(b)) and a counterflow brazed micro-channel suction line heat exchanger. With regard to the gas coolers, which have five or seven plates, each with offset strip fins, between which the water flows, the CO_2 flows inside 16 aluminium micro-channel tubes that were wrapped around the plates. The gas coolers are multi-pass on the water side, while the evaporator, of similar construction, is single pass. With regard to size, the evaporator, as an example, measured 254×84×85 mm^3 (length×width×height), with over 13,000 mm^2 surface on the working fluid side.

An important international initiative in encouraging the use of CHEs in heat pumping systems was started in 2006. Annex 33, one of around 40 Annexes supported by the IEA Heat Pump Centre, was established with Brunel University, Uxbridge, UK as the operating agent, or annex manager. With a three-year timescale, the goals were as follows:

(a) To identify compact heat exchangers, either existing or under development, that may be applied in heat pumping equipment, including those using vapour compression, mechanical vapour recompression and absorption cycles. This had the aims of decreasing the working fluid inventory, minimising the environmental impact of system manufacture and disposal, and/or increasing the system performance during the equipment life, thereby reducing the possible direct and indirect effects of the systems on the global and local environments.

(b) To identify, where necessary propose, and document reasonably accurate methods of predicting heat transfer, pressure drop and void fractions in these types of heat exchangers, thereby promoting or simplifying their commercial use by heat pump manufacturers. Integral with these activities was proposed to be an examination of manifolding/flow distribution in compact/micro-heat exchangers, in particular in evaporators.

(c) To present listings of operating limits for the different types of compact heat exchangers, e.g. maximum pressures, maximum temperatures, material compatibility, and minimum diameters, and of estimated manufacturing costs or possible market prices in large-scale production. It was intended within this context that opportunities, for technology transfer from sectors where mass-produced CHEs are used (e.g. automotive), would be examined and recommendations made.

The outcomes of the Annex (Anon, 2010) were several, ranging from fundamental research on boiling in narrow channels to guidelines for selecting and using CHEs in heat pumping systems. Considerable market data were collected (including potential industrial uses of heat pumps in the UK), and a number of novel heat exchanger concepts, including the use of new materials and the application of process intensification methods, will, it is believed, allow equipment manufacturers in the future to achieve the Annex aims.

Particular aspects that it is considered worth highlighting here are:

1. The increasing interest in and use of CO_2 as a working fluid. This has implications in terms of the equipment used and the concepts for heat pumping that might be applied – see above and Chapter 4.
2. The vast portfolio of research on heat transfer and fluid dynamics in narrow channels in CHEs. The research highlighted in Sweden, Japan and the USA are of particular note. While much of this has been conducted as fundamental research, or for applications such as electronics thermal control, technology transfer to the heat pump sphere is occurring – such as the Japanese PCHE illustrated in Chapter 4 earlier.
3. There is a need to educate the heat pump industry in the use of CHEs, their merits and limitations and the types that are available. The use of materials other than metal, as indicated in some of the research in the USA, could reveal new opportunities.

Annex 33, together with other recent research on compact and micro-heat pump systems, brought together many experts in the heat pump/CHE field and the Annex Report (Anon, 2010) has proved to be a major and constructive source of data for those interested in using CHEs in heat pumping equipment. The efficiency of heat pumps is improving, thanks in part to effective heat exchangers. The use of compact and micro-heat exchangers is also leading to cost and size reductions that will hopefully accelerate their penetration into space conditioning and process heating applications. When this occurs on a large-scale, the heat pump will be seen to be one of the most effective energy-saving devices in applications ranging from the home to petrochemical plant.

One of the most active research centres in the USA, where CHEs are under development for a variety of heat pump/refrigeration duties, is based at Georgia

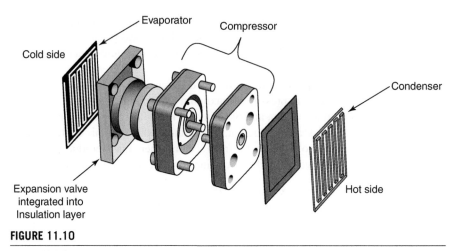

FIGURE 11.10

The ARCTIC system expanded to show the components.

Institute of Technology where Srinivas Garimella is studying a variety of concepts. Some of these are discussed above and in later sections, as is research at another US centre of expertise, the Cooling Technologies Research Center at Purdue University.

11.2.3 Micro-refrigerator for chip cooling

Since the proposals reported in 1999, others have adopted similar strategies in order to 'compress' heat pumping/refrigeration cycles. The ARCTIC project at the University of California is directed at chip cooling in microelectronics, and necessarily adopts a rotary compressor, in this case a Wankel type that is projected to give a compression ratio of 4.7:1 (Heppner et al., 2007). The schematic of ARCTIC is shown in Figure 11.10. By using MEMS components within the unit, and a compressor with a footprint about the size of an Intel Pentium 4 chip (25 mm×30 mm) the researchers indicate that a theoretical COP of 4.6 is achievable, with a cooling capacity of 45 W, a temperature difference of 40°K and a compressor speed of 1,000 rpm.

Purdue University in the USA has modelled a linear compressor for electronics chip cooling, based upon the vapour compression cycle. The prototype compressor illustrated in Figure 11.11 has a displacement of 3 cm^3 and an average stroke of 0.3 cm (Bradshaw et al., 2011).

Several aspects of micro-electro mechanical system (MEMS) technology have been investigated for refrigerator and heat pump technology, extending to compressor concepts, many of which (unlike the Purdue University device), are as yet untried. A few of the fundamental principles for converting electricity into mechanical displacement/forces to create displacement could be envisaged as the basis of

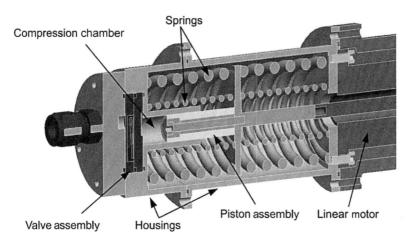

FIGURE 11.11

Section view of a prototype linear compressor (Bradshaw et al., 2011).

a compressor – some of the aspects of micro-fluidics involving electric fields may also come into play – see Chapter 3 and later in this chapter. Those interested in extending PI to heating and cooling systems may well consider a number of technologies previously associated with micro-electronics and micro-actuators being applied in energy-efficient heat pumping and refrigeration systems – if only for cooling of the lab-on-a-chip.

One of the most active European groups in chip cooling is at EFPL (Ecole Polytechinque Federale de Lausanne) in Switzerland. Over several years the group has worked on the thermal management of data centres, often large buildings housing massive computing power with challenging heat dissipation duties. Thome and Marcinichen (2011), in their introduction to their on-chip micro-evaporation work, explore the benefits of heat recovery from such centres, which exhibit a growth rate in electrical energy use of 10–20% per annum, and in the USA alone, costing $7.4billion in electricity charges annually. Citing units consuming 50 MWe each, the authors argue that it would be economically (and environmentally) much more sensible to employ on-chip two-phase cooling which can operate at a temperature sufficiently high to allow reuse of the waste heat at the condenser – which could be in excess of 70°C.

In another option discussed, (Marcinichen et al., 2012) the team examined the possibility that heat captured in the data centre can be reused by a power plant. Since the waste heat of the data centre is of a relatively low quality, it can only be delivered downstream of the condenser of the power plant, as illustrated in Figure 11.12. This would then increase the temperature of the water leaving the condenser (46°C typically, in the case examined) to a maximum temperature dependent upon the condensing temperature of the data centre cycle – this being of the order of 70°C. Therefore,

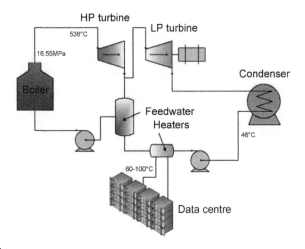

FIGURE 11.12

Integration of heat recovered from a data centre into a power utility (Marcinichen et al., 2012).

any additional heat added to the power plant's cycle will result in less fuel needing to be burnt, thus saving fuel and reducing the CO_2 footprint of the power plant. It was estimated that a 2.2% rise in power plant thermal efficiency might be the outcome in this case – a significant improvement.

11.2.4 Absorption and adsorption cycles

Both absorption and adsorption are inhibited in their intensification potential by heat and mass transfer limitations. The method applied in Rotex above has been successful in intensifying the absorption cycle, but the adsorption refrigerator or heat pump uses a solid and a gas as working media and active enhancement of the solid can be difficult.

One of the leading experts in adsorption cycle enhancement or intensification is Professor Bob Critoph of Warwick University (UK), who has been developing innovative adsorption cycle chillers, in particular, over many years. He points out that with fluids used such as silica gel and water, where one needed 1 kg of adsorbent for a 30 W cooling duty, there is a need for some intensification. The basic cycle does have some advantages however, it is rugged, not sensitive to orientation and the regenerative cycle has a good COP.

Several units have been developed at Warwick that help to reduce size and/or improve performance, based upon a number of different refrigerants and adsorbants, the latter including activated carbon and zeolites. A critical problem area that needs addressing if size reduction can be achieved is poor heat transfer through activated carbon, which tends to be granular in form. Thus, inherently, the units are large and expensive. The use of ammonia as the refrigerant allows more compact

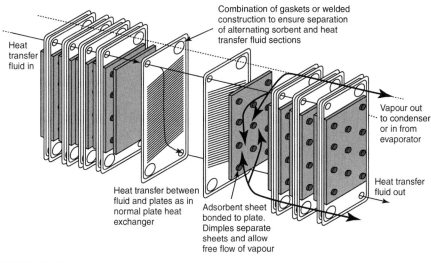

Combination of gaskets or welded construction to ensure separation of alternating sorbent and heat transfer fluid sections

Heat transfer fluid in

Vapour out to condenser or in from evaporator

Heat transfer fluid out

Heat transfer between fluid and plates as in normal plate heat exchanger

Adsorbent sheet bonded to plate. Dimples separate sheets and allow free flow of vapour

FIGURE 11.13

The plate adsorber configuration developed at Warwick University.

units, by 2–3 times, but the challenge remains one of heat transfer into and out of the granular bed. One of the systems developed at Warwick was the convective wave machine. This achieved a cooling COP of 0.8, while a second machine achieved 0.9. The challenge here was the ammonia gas circulator, and the work is now part of a Carbon Trust project. A second variant uses monolithic carbon – solid blocks give a higher packing density. The bed thermal conductivity is higher, as is the heat transfer coefficient. Tubes can be embedded in the carbon and fins allow a laminated generator. The cycle time (the adsorption cycle is a regenerative cycle) is 10 minutes. Multi-bed regenerative cycles are used as the third option, and this is based upon low-cost modules. The system was used in an EU-funded project on tri-generation of heat, power and cooling – SOCOOL.

The fourth concept uses plate heat exchangers (PHEs), with monolithic carbon bonded into thin sheets of 1 mm thickness. Compared to the 30 W/kg for conventional machines, the PHE-based unit can reach 6 kW/kg adsorbent – a substantial intensification of capacity (see Critoph and Metcalf, 2004; and Figure 11.13). A derivative uses the plate and shell concept, with circular plates that have been tested at 1 kW cooling duty. Industrial collaboration is bringing in diffusion bonding/Ni brazing expertise. Other work includes a unit for car air conditioning (TOPMACS), using the activated carbon–ammonia pair, and work with Fiat on truck refrigeration. Tamainot-Telto et al. (2009) describe the car air-conditioning work at Warwick University, where a plate heat exchanger adsorption generator is a key component. Illustrated in Figure 11.14, this arrangement was selected because of its good heat transfer and low thermal mass. The exchanger uses nickel-brazed shims and

FIGURE 11.14

Plate heat exchanger adsorption generator – size 150 × 150 × 150 mm^3 (Tamainot-Telto et al., 2009).

spacers, creating adsorbent layers only 4 mm thick between pairs of liquid flow channels. The performance (evaluated under European Union car air-conditioning test conditions) based upon the use of engine coolant waste heat at 90°C revealed that a pair of the generators, containing about 1 kg of activated carbon, produced an average cooling duty of 1.6 kW, peaking at around 2 kW. Coefficient of performance was 0.22.

For those interested in the optimisation of such cycles, see Metcalf et al. (2012).

Heat pipes (see Section 11.4.2) can be used to improve the conductivity of the adsorber and work at Warwick University has been directed to this end. Work in China (Li et al., 2007) has used the principle of heat pipes in an adsorption ice-making machine. Heat pipes fulfil the roles of heating and cooling the beds, as well as recovering heat between them. Other variants are discussed by Dutour et al. (2005). The work is relevant to chemical reactors as well (see Chapter 5), as the incorporation of these super thermal conductors into reactor vessels can aid iso-thermalisation, remove hot spots and add or extract heat, for example, handling an exotherm.

Although proposed to aid the regeneration in chemical processes, a little lateral thinking directs us to the possibilities of using the ideas of Cherbanski and Molga (2009) in heat pumps and refrigeration systems that use adsorbers. The Polish group has proposed microwaves as a means for desorbing solvents from the adsorbent, based upon a survey of work of others over almost ten years. They found

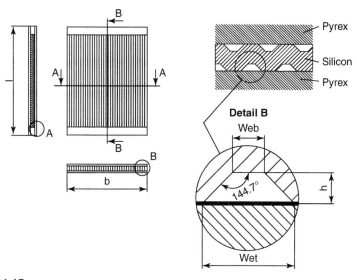

FIGURE 11.15

Details of the SINTEF micro-condenser. Refrigerant-side pressure drop was 180 kPa/m with a mass flux of 165 kg/m^{2}s (Munkejord et al., 2002).

research that suggested that desorption using microwaves was particularly effective when activated carbon adsorbents were used – the most common type in adsorption refrigeration/heat pumping, see Cherbanski and Molga (2009).

For more recent work, specifically on the use of microwaves to regenerate heat pump adsorbent beds, see Demir (2012). Demir found that the period of the cycle was improved by about 20% and the COP improved by 61% compared to the conventionally heated cycle. Of course some applications would find the use of microwave enhancement an inconvenience.

There are a number of instances where microtechnology has been studied for heat pumping duties. This is relevant to the broader field of PI for at least two reasons – firstly, it serves as an application example with substantial practical application potential, and secondly, it allows an energy-efficient concept to be used for heating and/or cooling micro-engineered processes. It is interesting to note (see section 11.4) that the passive and active cooling methods for computer chips are the subject of intense patent activity and the successful techniques coming to the market auger well for the lab-on-a-chip concept and its thermal control. These go beyond the relatively common thermoelectric devices.

Leading laboratories for micro-heat pumps include Battelle in the US (see Appendix 5) and SINTEF in Norway (Munkejord et al., 1999). The work at SINTEF is on devices with main dimensions of the order of a few centimetres with trapezoidal channels of 0.5 mm width concentrated on the heat exchangers and their performance. Illustrated in Figure 11.15, the micro-condenser so constructed showed that a heat flux of $135\,kW/m^2$, based on the refrigerant-side area, was possible. The overall 'U' value was $10\,kW/m^2K$.

More recently, a group of laboratories and universities in the USA, ranging from Georgia Institute of Technology and the University of Nevada to Cummins Inc. and OpenCell Technologies, undertook a wide-ranging study of micro-channel component technology in ammonia–water absorption cycle heat pumps. Concentrating initially upon the absorber (Garimella et al., 2011), micro-channel tubes were placed in an array – see Figure 11.16 – with the arrays stacked vertically. The ammonia–water solution flows down the outside of the tubes as a film/droplets, giving good heat and mass transfer. The tube o.d. was $1.575\,mm$ and the authors claim that the construction/layout is suitable for the other absorption cycle components, as well. (For other concepts, see section 11.2.2.)

For an excellent review of the technologies used in thermally activated systems such as adsorption and absorption for heating and cooling, see Deng et al. (2011).

b Microchannel tube array

a Absorber outer shell and viewports

c Solution drip tray

FIGURE 11.16

Micro-channel tubular units for an absorption cycle heat pump (Garimella et al., 2011).

11.3 **Power generation**

There are several power generation processes where PI has a role to play. Currently the most significant is the fuel cell, where HEX-reactors can be used as reformers on either a relatively large or micro scale. The fuel cell is being developed as a power pack for laptop computers, mobile phones and micro-aerial vehicles. However, as may be seen from section 11.3.2, the gas turbine engineers have recognised a similar opportunity.

11.3.1 **Miniature fuel cells**

The consumer electronics industry currently relies on batteries for power in devices such as laptop computers, mobile phones and video cameras. There are several approaches to alternative, more powerful, lighter and longer lasting energy sources, including hybrid fuel cell/battery systems and fuel cells operating in isolation. Motorola, some years ago, worked on the former solution, using a miniature direct methanol fuel cell (DMFC) and a rechargeable battery. One of the key technologies to crack is the fuel delivery system and micro-fluidic devices were studied (Anon, 2000).

Metal foam (see, for example, Figure 3.5) has already been discussed in the context of heat exchangers. Micro-reactors, highly relevant to the subject of small fuel cells, have also been introduced in earlier chapters. The construction of metal foam-based methanol steam micro-reformers to generate hydrogen for polymer electrolyte membrane fuel cells (PEMFCs) has been reported and in Guangzhou, Chinese researchers have looked at laminated micro-reactors in which copper-based catalysts have been supported by metal foams (see Figure 11.17; Yu et al., 2007).

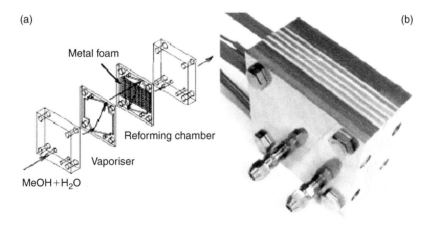

(a)

Metal foam

Reforming chamber

Vaporiser

MeOH + H$_2$O

(b)

FIGURE 11.17

The micro-reactor configuration in expanded form (a), and constructed (b). The reformer size is 40 × 70 × 2 mm deep.

FIGURE 11.18

The use of micro-fuel cells for laptop power supplies necessitates refuelling. Here we can see the fuel storage vessel connect to top up the reformer supply.

Interestingly, this research illustrated the care needed in selecting materials for catalyst supports. Eventually, high activity H_2 production was achieved with 1–2% CO concentration, but it was found that the foam itself played a critical role in the success of the reaction. To avoid the possibility of early catalyst poisoning, CuZn foam with a Cu/Zn/Al/Zr catalyst was found to be optimum, giving excellent activity and stability, as well as selectivity. An alternative Ni-doped catalyst on FeCr alloy was not so successful in operation. Although other research in China (Chen et al., 2007) suggests that the fuel processor for a PEMFC, based upon a monolithic reactor for methanol oxidation, can work well (in this case for a power output of 70W), the integration with other equipment remains challenging. In the case of the Dalian research, other processes/components included CO preferential oxidation, combustion, vaporisation and the micro-heat exchanger.

A useful review of micro-fuel cells (Kundu et al., 2007) with duties of 1–50W, but concentrating on those delivering less than 5W, divide applications into three categories – portable electronics, about which we are all familiar; military portable; and healthcare. The latter category might include micro-biofuel cells for implantable power sources for pacemakers and the like. Kundu et al. list about 17 micro-fuel cells within these categories. One of these, the NEC laptop with a DMFC unit, is illustrated in Figure 11.18. This shows the methanol refuelling cartridge (temporarily fitted during filling only) on the back of the laptop (Anon, 2007).

Kundu and colleagues, working at Samsung Electromechanics in Korea, developed the silicon-based micro-channel reactor shown in Figure 11.19. Here a

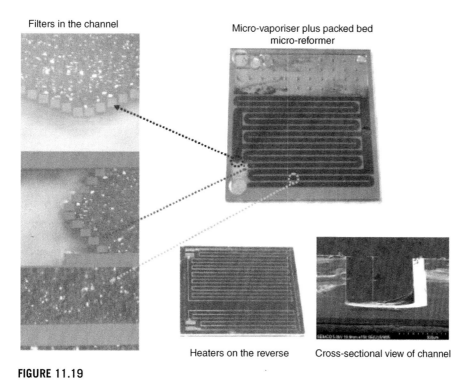

Filters in the channel

Micro-vaporiser plus packed bed micro-reformer

Heaters on the reverse   Cross-sectional view of channel

FIGURE 11.19

The micro-reformer developed at Samsung.

$Cu/ZnO/Al_2O_3$ Johnson Matthey catalyst was used. In the experiments, a serpentine patterned micro-reformer proved superior in terms of activity to a parallel-patterned unit. Nevertheless, dates for commercialisation, conclude the authors in the review, remain uncertain.

Few reports address the challenges of the mass production of fuel processors. Work at IMM in Mainz, Germany, has resulted in an assessment of costs associated with production runs of 100,000 units per annum for a heat exchanger reactor for a fuel processor (Kolb et al., 2007). The structure is based upon micro-structured plate heat exchanger technology (see Chapter 4). The plate heat exchanger unit has dimensions of approximately 60 mm×25 mm with a thickness of about 10 mm. Based upon a production process involving embossing of plates and laser welding them together, the costs breakdown was as follows:

Cutting	27%
Catalyst	18%
Embossing	4%
Material (steel)	9%
Catalyst coating	9%
Joining	9%

Joining (headers, etc.) 9%
Stacking 5%

An analysis based upon 3D printing as an alternative would be an interesting exercise today!

For those wishing to delve more deeply into micro-fuel cells, Zhao (2009) covers many types in his comprehensive text.

11.3.2 Micro turbines

The current trend towards decentralised power generation has generated interest in small-scale, or micro-combined heat and power plant – micro-CHP. While these units can have power outputs of a few kW, the other micro concept relates to the gas turbine alone. While the typical micro-CHP unit is not 'intensified' in our recognition of the term, the micro-turbines proposed (with outputs of a few tens of Watts) enter the arena currently inhabited by micro-fuel cells. Thus, uses in portable electronic equipment, micro-aerial vehicles, nano-satellites and other similar-sized devices are put forward as reasons for the development of such small turbines.

In a study in the US, (Waitz et al., 1998) analysed later by Ribaud at ONERA (2001), the combustor for one such turbine 500 times smaller than conventional gas turbine engines was designed. The authors stated that the motivation for this work was the development of micro-gas turbine generators capable of producing 10–100 W of electrical power while having a volume of less than $1\,cm^3$ and consuming approximately 7 g of fuel per hour. The attraction would be a device with 10 times the energy density of the best current lithium batteries. The micro-turbine is illustrated in Figure 11.20.

Data on other micro-turbines are included in a most useful review by Weiss (2011). Weiss also gives data on novel thermodynamic cycles that are used exclusively in MEMS-type devices. Of particular current interest is the micro-Rankine cycle machine. One unit, illustrated in Figure 11.21, was built using MEMS technology, and had a four-stage turbine and a spiral groove micro-pump. The size was approximately $1.5\,cm^2$ with a thickness of a few mm. It was tested at speeds to 300,000 rpm using compressed air expansion and the output was predicted to be 0.38 W.

A recent entrant to the heat exchanger field, Hiflux Ltd (see Appendix 4), has developed a plate and pin heat exchanger (commonly called pin-fin) where pins replace conventional fins. A market targeted by the company is the micro-CHP market driven by gas turbines (the turbines are not as small as those discussed earlier), where the unit would be used as a recuperator (Reade, 2008). The company involved reported in 2011 that the heat exchanger was a component of two systems under trial in Europe.

11.3.3 Batteries

Like fuel cells, batteries are undergoing a transformation in terms of their specific energy (stored energy/unit weight). This has been spurred for many years by the

FIGURE 11.20

The micro-turbine of MIT, overall diameter 12 mm (Ribaud, 2001).

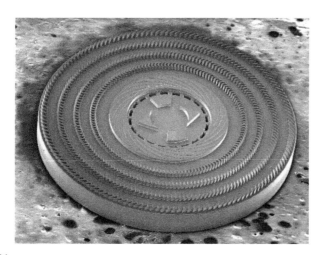

FIGURE 11.21

Four-stage micro-turbine rotor with groove seal (Weiss, 2011).

demands of mobile equipment (including the mobile phone, portable radios and music players) and the increasing use of so-called hybrid vehicles, where batteries and internal combustion engines can combine to reduce emissions. With regard to hybrid or fully electric vehicles, the choice of battery type ranges from the well-established lead–acid unit with low specific energy, to lithium–ion and

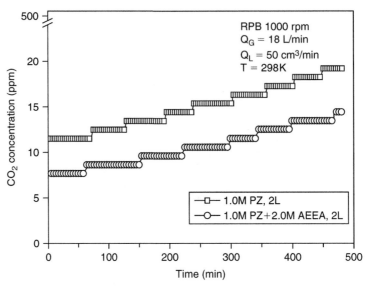

FIGURE 11.22

CO_2 concentration in the discharged gas using various absorbents (PZ = piperazine and AEEA = 2-(2-aminoethylamino) ethanol).

sodium–sulphur batteries, with superior storage capabilities. The zinc–air battery has a theoretical specific energy of over 1,000 Wh/kg, better than competing types.

The zinc–air battery has one significant drawback, in that the life and performance can be reduced due to CO_2 in the air causing carbonate formation in the electrode. Researchers at the National Tsing Hua University in Taiwan (Cheng and Tan, 2006) have developed a means to reduce the CO_2 levels at the electrode from 500 ppm to less than 20 ppm, using a high gravity rotating packed bed (RPB) covered with an amine-based absorbent. The absorbent, introduced as a liquid at the centre of the RPB (see also Chapter 6) absorbs the gaseous CO_2 fed in at the outer circumference of the bed. The bed was rotated at 1,000 rpm, this being found to be the optimum speed. Other variables examined by the group included amine concentration, solution and gas flow rates. Regeneration of the amine was also examined, as this would be necessary in practical use. With regard to size, the RPB had a volume of just over $300 \, cm^3$ and was deemed sufficiently compact for motorcycles and cars. Data in Figure 11.22 shows that 8 hours of operation is feasible before the concentration of CO_2 reaches 20 ppm.

Reference has been made elsewhere to the benefits of PI by combining unit operations – the HEX-reactor is an obvious example. Incorporation of heat exchangers in, or as, structural members has been proposed in a number of areas, such as in military aircraft. Now BAE Systems (Anon, 2012) has patented a 'structural battery' that it is currently developing to lighten the load of soldiers who currently operate carrying up to 76 kg of kit each, mostly made up of electrical equipment for communicating and signalling, which can adversely affect the soldiers' mobility and speed of movement. This new technology stores the electrical energy within the physical structure of a device – reducing or eliminating the need for traditional batteries. This could represent a significant reduction in weight and bulk, as well as minimising the burden and cost of carrying spares.

While the benefits to the defence sector have already been demonstrated in a high-tech micro-unmanned air vehicle and a rudimentary torch, the technology is also being applied beyond the battlefield. Through a partnership with leading race car manufacturer Lola, the Lola-Drayson B12/69EV, zero emission 850 horsepower Le Mans Prototype car will incorporate structural batteries to power some of the on-board electronic systems. Upon completion, the vehicle aims to become the world's fastest electric racing car.

To develop the technology, BAE Systems scientists merged battery chemistries into composite materials that can be moulded into complex 3D shapes and form the structure of the device itself. This structure can then be plugged in when it needs recharging or can utilise renewable power sources, such as solar energy.

Current developments have demonstrated the ability to store useful energy in composites such as carbon fibre and glass reinforced plastic, but in the future, energy storage could also be incorporated into the fabric for a wide range of lightweight applications, from tents with their own power supply to electric blankets becoming a reality.

11.3.4 Pumps

Associated with much of the above when designed at the small-scale, is a micropump. As stated by Liu and Chan (2011), such pumps have applications across a range of sectors, from pharmaceuticals to electronics cooling and medical appliances. For those interested in designing miniature centrifugal impellers, this work at Nanyang Technological University is useful – dealing with flow rates of the order of 1 l/min and blade heights and inner radii of 0.84–1.72 and 1.5–3.00 mm, respectively.

Other pump types are discussed in Chapter 3, in the context of electric fields, and in section 11.4.1 below, where micro-fluidics is considered, again in the context of electric fields (e.g. electro-osmosis pumping).

11.3.5 Energy scavenging

As mentioned in the introduction the area of energy harvesting or energy scavenging is attracting rapidly growing attention in sectors such as aerospace and even energy recovery from the heat of the human body. In the area of aeronautical engineering the demand for energy scavenging has been driven by the trend towards the all-electric aircraft. (Not an aircraft that has electric propulsion units – at least not yet – but one in which most of the on-board systems are electric-powered, rather than via engine bleed air). Scott (2012) states that the Airbus A380 uses 1.2 MWe, while the later Boeing 787 has a demand for 1.5 MWe. This, coupled with the need to improve engine thermal efficiency to meet environmental constraints, suggests that means for power generation other than those that have a detrimental effect on propulsion efficiency, should be examined. As with fuel cells, energy scavenging is a potential player.

Harb (2011), in a most useful review of technology, lists the potential micro-energy sources that might be used for electricity generation. These include motion (vibration, floors, engine vibration); radio-frequency (RF) energy from base stations; TVs; satellite communications; thermal energy; pressure gradients; and micro-scale water flow. With regard to harvesting from heat sources, Harb indicates that pyroelectric cells may be employed (others investigate thermoelectric devices). The difference between the two is that the former can develop a temporary voltage when they are heated or cooled, while thermoelectric devices sustain a constant voltage when one side is heated and/or the other side cooled. In discussing the pyroelectric cell, Harb indicates that applying consecutive heating and cooling cycles to two PZT (lead zirconate titanate) cells in parallel yielded 0.5 mJ and a maximum voltage of 31 V. The thermoelectric converter, in this case 1 cm^2 in size, yielded about 18 μW.

At the nano-scale Chang et al. (2012) suggest that piezoelectric nanofibres are a suitable way forward for energy scavenging. It appears that the production of such fibres can be undertaken (electro-spinning is proposed), while materials used include PVDF and PZT. Their flexibility suggests clothing applications, but the authors state that real interest from potential users will come about when the power from a single nano-fibre nano-generator increases from the present 10^{-11} W to the order of 10^{-6} W.

11.4 Microelectronics

The whole concept of microelectronics is highly relevant to PI and perhaps the ultimate 'local' chemical plant. The disposable credit card-sized blood analysers that are proposed for rapid diagnosis lean on microelectronics technology to some extent for their development (see Figure 11.23). There are many analogies between microelectronics technology and process intensification. The printed circuit heat exchanger was one of the first examples of this, but concepts such as the lab-on-a-chip (see for example Gleichmann et al., 2011) or the desktop chemical plant (as in Figure 11.23) can be related directly to computer components or concepts. It is therefore important to develop the synergy between the two fields, so that the vast progress made in microelectronics, in terms of manufacture, micro-miniaturisation and thermal control, to name three areas, can be used in PI technology.

It is the manufacturing technology – exemplified by the micro-turbines discussed above – that is important to both microelectronics and micro-intensified processes, and those involved in the latter can learn much from the former. See, for example, Helvajian, 1999. The links between refrigeration cycles and electronics thermal control has also spawned some micro-refrigerators that can incorporate PI technology (see section 11.2).

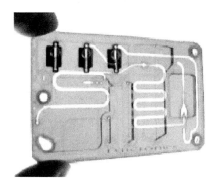

FIGURE 11.23

Slightly larger than the lab-on-a-chip, the credit card-sized analyser from Micronics Inc. is a prime example of intensification.

11.4.1 Micro-fluidics

Before proceeding with the discussion on microelectronics-related topics, it is useful to remind ourselves that micro-fluidics (introduced in Chapter 3) is an important subject that allows us to appreciate flow phenomena at the micro- and nano-scale. It is of growing importance in intensified unit operations, where characteristic lengths may be well below 1 mm, for example. The paragraph below, from a quote by US commentators on the impact that micro-fluidics might have on the process industries, sets the scene.

> *Will micro-fluidics change the process industries forever? That certainly seems possible. At the very least, this new area of technology will change the way chemical engineers think and learn. Coming to grips with phenomena like low Reynolds numbers and electrokinesis will effectively invert the traditional skill of scale-up. Tomorrow's chemical engineers may well be more concerned with radical scale-down.*

Micro-fluidics, and the development of components and systems that involve fluid flow at the micro-scale, are areas of growing scientific and commercial interest (Reay, 2005). The reduction in system size offered by the successful integration of micro-fluidic components has opened up new opportunities in medicine, aerospace, power generation, chemical engineering and a number of other fields. In his monograph (Zajtchuk, 1999) on the impact of bio- and nano-technology on medicine, the US surgeon Russ Zajtchuk made a number of forecasts.[2] One related to

---

[2] Some of which have already been proved accurate, such as disposable diagnostic chips at US$1 each! His review is well worth reading.

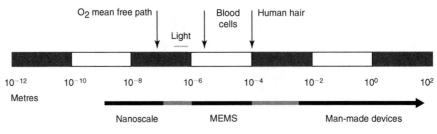

FIGURE 11.24

Length scales in micro-fluidic systems.

(Courtesy: Daresbury Laboratory data)

nano-technology, an area relevant to a number of facets of micro-fluidics, being as follows:

> *If a breakthrough to a universal assembler occurs during the next 10 to 15 years, an entirely new field of 'nano-medicine' will emerge by 2020. Initial applications will be focused outside of the body, in areas such as diagnostics and pharmaceutical manufacturing, but eventually the most powerful uses will be within the body. Possible applications include programmable immune machines that travel through the bloodstream and supplement the natural immune system, cell-herding machines that stimulate rapid healing and tissue reconstruction and cell repair machines that perform genetic surgery.*

However, such developments have created a number of scientific questions that require addressing. The characteristic dimensions of micro-fluidic systems, identified by some as being in the range of 1 to 100 microns (as shown in Figure 11.24, from Daresbury Laboratory (Anon, a), between a blood cell and a human hair – do encourage us to look at flow and other physical phenomena at scales below those we regard as conventional. Those working in fields such as micro-heat exchangers will recognise dimensions greater than 100 microns as having the prefix 'micro' attached to them, and may quite happily be using well established expressions for Reynolds numbers and Prandtl numbers. However, many papers, usefully summarised in the review by Mehendale et al. (1999), suggest that we should be looking at deviations from Navier-Stokes equations in channels with a characteristic dimension such as width greater than 100 microns.

When we wish to predict the performance of unit operations within what is called by some the 'micro-fluidic regime', it is important to have reliable correlations for fluid behaviour within the small flow passages. As a first step in introducing the subject, it is worthwhile reviewing some of the data, much of which may be inconsistent, that illustrate deviations from that, in solely 'compact' equipment.

Next, critical to the successful application of micro-fluidic devices is the satisfactory control of heat and mass transfer. Having worked for many years in a field where quite complex and numerous heat and mass transfer phenomena are taking

Table 11.1 Characteristics of Capillary Structures.

Material	Porosity (%)	Effective pore radius (μm)	Permeability ($\times 10^{13}$ m^2)	Thermal conductivity (W/mK)
Nickel	60–75	0.7–10	0.2–20	5–10
Titanium	55–70	3–10	4–18	0.6–1.5
Copper	55–75	3–15	–	–

place – that of heat pipes and thermosyphons, including micro units – David Reay, co-author, has been intrigued by the emergence in micro-fluidics of a number of relatively well known active and passive techniques for fluid stream manipulation. Most of these were examined perhaps 30–40 years ago for enhancing flow in micro-channels that were components of macrosystems or for overcoming the limitations of purely passive capillary forces. They included osmotic, electro-osmotic and electro-hydrodynamic flow pumping (see also Chapter 3) mechanisms that increasingly influence, or can be used to influence, flow at the micro-scale. Table 11.1 gives some of the characteristics of the porous structures which might be said to represent micro-fluidics in the 1960s (relevant to heat pipes, discussed below). Interestingly, while capillary forces perhaps cannot dominate larger scale systems, they can have a positive or negative impact on micro-fluidic devices, and interface phenomena in general can become important players in micro-systems where fluid flow is a feature.

The demand for improved heat transport capabilities and a coincidental reduction in mass, as dictated by spacecraft thermal control, is effectively illustrated by Astrium, the Toulouse-based (France) space organisation, in Figure 11.25 (Figus et al., 2003). Thales Alenia Space in Cannes, France, recently examined the performance of micro-condensers for thermal control systems in telecommunications satellites. The increase in the power dissipated in these satellites has increased to such a level that the overcoming of problems associated with thermal control is becoming critical (El Achkar et al., 2011). It is proposed by the research team that one solution could be to increase the dissipation to space of the heat via radiation. This could be implemented using a vapour compression cycle cooling unit, but compactness of course is important, and one component being examined by Thales Alenia Space is the micro-condenser. Using n-pentane as the working fluid, and mass velocities of 1–10 kg/m^{2}s, the presence of any bubbles in the square channels in the borosilicate condenser (inner hydraulic diameter 0.49 mm) was examined. This was necessary so that space-qualified units can be designed with a high degree of confidence using micro-channel heat exchangers at much smaller mass velocities than are common in condensers.

The heat transport capability for spacecraft is commonly expressed in Wm, so hardware under development needs to have access to thermal control systems (most likely micro-fluidic) that can transfer 1 W 100 m and weigh 10 g, or 10 W 1 m and weigh just in excess of 1 g. Devices using two-phase flow, such as capillary pumped

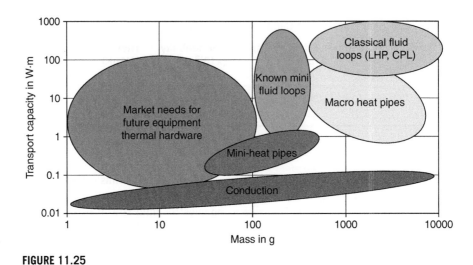

FIGURE 11.25

Performance demands for spacecraft thermal hardware (Figus et al., 2003).

loops, are seen as one option and are currently used where weight is not the most critical feature (Maydanik, 2005). (Notwithstanding this, four of these devices are used on the Hubble Space Telescope.) Space technology is essentially process intensification under another name!

11.4.2 Micro-heat pipes – electronics thermal control

A technology which impinges on several of the areas discussed in this book is the heat pipe. In this section we are concerned with micro-heat pipes which are very small thermal superconductors. Effective thermal control that can be offered using micro-heat pipes will be needed in both microelectronics and micro-chemical processing, such as Pentium chips and the 'lab-on-a-chip'. As heat pipes in the micro-electronics application area are orders of magnitude more numerous than in any other use and the challenges for the future of heat pipes and their derivatives are considerable, this section concentrates on this application.

As pointed out by Thermacore Inc. (McGlen et al., 2003), the major manufacturer of heat pipes for electronics thermal control, the microelectronics, telecommunications, power electronics and, to some extent, the electrical power industries, are constantly striving towards miniaturisation of devices that inevitably result in greater power densities. Therefore, there is a challenge to develop efficient management of heat removal from these high flux devices.

However, this miniaturisation, together with increasing processing speeds, decreases the heat transfer surface area and increases power. This generates very high heat fluxes resulting in large temperature rises. Therefore to maintain the chips within operating conditions more heat must be removed and traditional methods, such as

Table 11.2 A Selection of PI Applications.

Sector	Sub-Sector	PI Process
Bulk chemicals	Steam reforming of natural gas	Catalytic plate reactor
	Crude oil processing	Electric fields – electrostatic phase separation
	Reforming of naptha to gasoline	Catalytic plate reactor
	Dehydrogenation of alkanes and ethyl benzene	Catalytic plate reactor
	Olefin hydroformylation	HEX-reactor
	Terephthalic acid production	Rotation (mop fan)
	Acid stripping – sulphuric acid	Rotation (RPB)
	Formaldehyde from methanol	Micro-reactor
	Polymerisation of ethylene	Laser-induced reaction
	Methyl acetate production	Reactive distillation (NL)
	Ethyl acetate production	Reactive distillation (Spain)
	Ethanol/propanol distillation	Rotation (HiGee) (original ICI use)
	Polycondensation	Rotation (SDR) (Newcastle University)
	Hydrogen peroxide distillation	Ejector and/or heat pump (Sulzer)
	Methanation	Heat pipe reactor
	Styrene polymerisation	Spinning disc reactor (Newcastle University)
	Steam reforming	HEX-reactor (multiple adiabatic bed PCR)
	Methyl acetate production	Combined unit ops – reactive distillation
	Benzene nitration	Micro-reactor (Newcastle University)
	Suspension polymerisation	OBR
	Transesterification	Sonochemical reactor
	Ethanol from bioethanol	Micro-reactor
	Polyolefins production	Rotor stator reactor
Fine chemicals/ pharmaceuticals	Diagnostics contrast agents	Ultrasound
	Paint formulation with TiO_2	Ultrasound to enhance emulsion polymerisation
	Solvent extraction	Electric fields – electrostatic emulsion phase contactor
	Organic synthesis electrochemical reactions	Selective pulse electrical heating

(*Continued*)

Table 11.2 (Continued)

Sector	Sub-Sector	PI Process
	Emulsion manufacture	Crossflow membrane emulsification (Exeter University/Unilever)
	Hydrogenation of fats and oils	Supercritical CO_2 processing (De Gussa)
	Inkjet printer ink	Intensified reactor series
	Organic synthesis	Impinging jet reactor
	Titanium dioxide processing	Photocatalysis in a micro-reactor
	Pharma drug intermediate	Rotation – SDR
	Paracetamol crystallisation	OBR – mixing
	Fragrance intermediate	Rotation – SDR
	Aroma production	OBR
Offshore processing	Gas stripping	Centrifugal field – tangential injection into static vessel
	Gas stripping and deaeration	Centrifugal field – rotation
	Liquid–liquid separators	Centrifugal field – hydrocyclone
	Absorption – CO_2 sequestration	Centrifugal field – HiGee
	Natural gas conditioning	Membranes (Shell)
	Gas conditioning	Supersonic hydrocyclone 'Twister' (NL)
	Crude oil processing	Down hole catalytic processing
	Gas conditioning	Compact heat exchangers
	Enhanced oil recovery	Compact reactors
	Water deoxygenation	Centrifugal contactor
	FPSO and diesel production	Compact reactors
	Separations	Micro-distillation units
Nuclear industry	Reactors	Compact heat exchangers
	Fuel processing	Rotation by ultra-centrifuge
	Fuel processing	Pulsations – OBR for liquid– liquid extraction
	Solvent extraction	Electric fields/membrane (BNFL/ Southampton University)
	Sludge dewatering	Centrifugal transportable unit. (BNFL)
Other energy sectors	Coal gasification	Membrane air separation
	Biodiesel production	Mixing – OBR
	Biodiesel production	Reactive extraction
	Oil from shale	Electric fields (r.f.) and supercritical extraction
Food and drink	Breweries/wort boiling	PDX intensified mixer
	Custard manufacture	Spinning disc reactor

(Continued)

Table 11.2 (Continued)

Sector	Sub-Sector	PI Process
	Several	Electrical enhancement – induction mixers
	Pasta manufacture	Electric fields – microwaves
	Vegetable oil processing	Corrugated membranes in crossflow microfiltration. (Newcastle University)
	Oligosaccharide production	Centrifugal fields (Aston University)
	Baking and cooking	Electric fields – microwaves Ultrasound
	Various	SDR – dehydration
	Liquid foods	SDR – mixing
	Fruit juices, etc.	Electric fields – ultrasound enhancement in refrigeration
	Beverage packaging/gas stripping	Rotation – HiGee
	Sauce manufacture	PDX intensified mixer
	Food preservation and cooking	High pressure (to 800 MPa) processing
	Starch production	Rotary vacuum filter (NL)
	Sterilisation	Ohmic heating (Birmingham University)
	Yeast culture growth	OBR
Textiles	Fibre production	In-line mixers
	Textile finishing/effluent	Membranes/ultrafiltration
	Leather tanning	High pressure CO_2 processing
	Washing	Electric fields – ultrasound
Metals	Tube drawing	Electric fields – ultrasound
	Degassing of melt	Electric fields – ultrasound
	Sintering	Electric fields – microwaves
	Copper extraction	Membranes
	Coating titanium	Pulsed UV laser sputtering
	Sheet metal annealing	Electric fields – induction heating
	Gold extraction	Nozzle mixers for oxygen injection
Polymer processing	Resin curing for pipe manufacture	Ultrasound
	Biodegradable polymer production	OBR
Glass and ceramics	Glass refining	Electric fields – ultrasound
	Ceramics manufacture	Electric fields – microwave hybrid oven
	Gypsum/ceramics drying	Superheated steam dryers (TNO)

(Continued)

Table 11.2 (Continued)

Sector	Sub-Sector	PI Process
Aerospace	Infra-red detector cooling	Compact heat exchangers
Effluent treatment	Gas separation/product recovery	Electric fields – electrochemical separation
	Water purification	Electric fields – electro-acoustic filtration
	Waste remediation	Plasma reactor
	Waste water purification	Active enhancement – jet loop reactor (Venturi)
	Waste water purification	OBR with photocatalysis
	Water treatment – phenolic compound removal	Electric fields plus cavitation
	Dewatering slurries	Electric field (r.f.) and pressure
	Water purification	Rotation plus photocatalysis – Taylor-Couette system
Refrigeration/ heat pumps	All cycles	Compact heat exchangers
		Micro-heat exchangers (shims/tubes)
	Absorption chiller	Rotation
	Adsorption cycles	Heat pipes/foams
	Adsorption cycles	Microwave regeneration
	Vapour compression cycle units	Micro-heat exchangers
	Vapour compression cycle units	Rotation
Power generation	Fuel cells	Metal foams
	Fuel cells	Micro-reactors/heat exchangers
	Gas turbines	MEMS micro-combustors and components
		Micro-surface recuperators
	Batteries (CO2 removal)	Rotation – RPB
	Boilers	Carbon capture
Microelectronics	Chip thermal control	Micro-heat pipes
		Micro-refrigerators
	Chip thermal control	Micro-vapour chamber isothermalisation.
Miscellaneous	Desalination	Thin film heat transfer enhancement
	Boilers	Rotation
	Gas clean up	Rotation (mop fan)
	Target cooling	Rotation
	Fouling control	Mixing – tube inserts
		Surface modification
	Heat exchangers	Nano-particles
	Heat exchangers	Rotating disc
	Mixers	EHD

simple forced air convection, become inadequate. Thermal management is now the limiting factor in the development of higher power electronic devices and new, preferably low cost, methods of cooling are required. The nature of the heat concentration in very small volumes/areas of surface, challenges innovative thermal engineers, with solutions such as jet impingement and micro-refrigerators being considered.

There are some mitigating steps than can alleviate the thermal control problem. For example, the move from bipolar to CMOS and the change in voltage from 5V to a voltage lower than 1.5V have extended the lifetimes of standard cooling systems, at least for low power electronics (PC, phone chip, etc.). However, the International Technology Roadmap for Semiconductors (ITRS) indicates no slowdown in the rate of increase in cooling demands for the next generations of chips. Predictions for 2006 were of peak heat fluxes around $160\,W/cm^2$, and in spite of alternative solutions appearing regularly, further increases are highly likely.

As mentioned earlier, the challenges of miniaturisation have similar implications there. The current status in that sector is neatly summarised by Swanson and Birur at NASA (2003) as:

> *While advanced two-phase technology such as capillary pumped loops and loop heat pipes offer major advantages over traditional thermal control technology, it is clear that this technology alone will not meet the needs of all future scientific spacecraft....*

The need to handle much higher heat fluxes has prompted much research in two-phase heat transfer, much of it directly or indirectly related to heat pipes. For example, Thermacore and its parent, Modine, were industrial supporters of a major UK university-based project on boiling and condensation in micro-channels that started in 2006, involving one of the authors. This project, one of the largest in the area funded in Europe (approximately £1million), was supported partially on the basis of the demands of chip cooling.

Much of the research will be relevant to micro-heat pipes, because, as highlighted earlier, as power ratings are increased and sizes are decreased, multi-phase heat transfer in channels of very small cross-section becomes increasingly attractive, compared to single-phase cooling. Multi-phase pressure drop imposes a constraint on the lower limit of cross-sectional dimension of,50μm and micro-channels in the range of practical interest of 50–1,000 microns may either be components of heat pipe capillary structures, or at the extreme, the whole heat pipe vapour and liquid cross-section. As an example, cooling by evaporation in micro-channels embedded in a silicon chip is a strong contender for cooling new electronic devices at heat fluxes exceeding $100\,W/cm^2$ – and these micro-channels may, of course, be an array of micro-heat pipes. Figure 11.26 illustrates some of the micro-heat pipe configurations available.

In the interim, work by researchers at UES and the US Air Force Research Laboratory have reported heat flux capabilities greater than $140\,W/cm^2$ using novel capillary fin-type structures (Lin et al., 2002). However, at high heat fluxes (some exceeding $280\,W/cm^2$) the temperature difference between the evaporator wall and the vapour temperature was rather high – around 70°C. For the reader who wishes

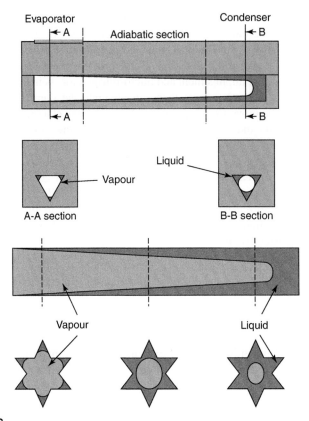

FIGURE 11.26

The shapes of micro-heat pipes. The capillary action to move the liquid from the condenser to the evaporator is generated in the corners.

to consult an overview of the many methods for managing heat in microelectronics, including systems that compete with heat pipes, the review by Suresh Garimella (Garimella, 2006) is interesting and covers micro-channels, micro-pumps, jet impingement and phase-change energy storage, as well as micro-heat pipes.

The notebook computer is the most popular heat pipe application for CPU cooling. Many millions of heat pipes are made per month for this use, and they are now also used in memory chips. Some desktops have five or more heat pipes in their cabinet. The units in Figure 11.27 are of triangular cross-section and are used in a Korean subnotebook.

Furukawa Electric in Japan has developed an innovative sheet-type heat sink, a vapour chamber claimed to be the thinnest in the world. This is directed at the cooling of semiconductor chips in mobile electronic equipment such as notebook PCs, cellular phones and PDAs (Personal Digital Assistants) as well as CCDs (charge-coupled devices, imaging units used in digital cameras). The heat sink can also be

FIGURE 11.27

Location of two triangular cross-section micro-heat pipes in a sub-notebook personal computer. Water is the working fluid (Moon et al., 2004).

FIGURE 11.28

The Furukawa electric vapour chamber.

used for equalising the heat distribution within the casings. The Furukawa argument is that as equipment (such as laptops and phones) gets lighter, thinner and smaller, conventional rigid heat pipes, even at the miniature scale, become difficult to install due to the limited space allowed for mounting. The perceived need was to develop an efficient heat conductor that is as thin and as flexible as paper, while providing heat-dissipating and isothermalising functions. Furukawa points out that although heat-dissipating sheets, called thermal sheets, using carbon graphite and the like having high heat conductivity, have recently been used in practical applications, their effectiveness is not adequate for current demands. The company addressed this challenge by developing a sheet-type heat sink with a thickness of less than 1 mm, as shown in Figure 11.28. The width, length and thickness can be selected to match the product to the design requirements of individual electronic equipment with regard to heat dissipating performance and shape.

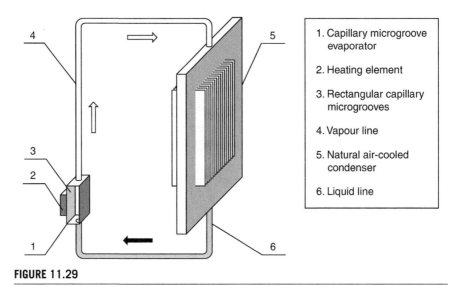

FIGURE 11.29

Schematic of the passive micro-cooling system.

With regard to performance, a unit 0.6 mm in thickness, 20 mm in width and 150 mm in length can dissipate up to 10 W and several tens of W if the thickness is increased to 1 mm. Furukawa estimated two years ago that production would rapidly reach five million units per annum. Before we leave this interesting area of PI, there are two examples that illustrate features of heat pipes that can offer benefits in other areas of intensification. Firstly, Figure 11.29 shows a capillary-pumped loop with an evaporator (1) containing micro-grooves where heat is taken in from the chip, and a condenser (5) where it is rejected. The system can transport over 130 W and has a cooling performance of up to 1.68×10^6 W/m². Most importantly, it is passive – needing no external power input. It is suitable for remote cooling of micro-systems with high powers and thermal sensitivity (Hu and Tang, 2007). Of course, the micro-system does not need to be an electronic component, but it could be a micro-reactor. It is also worth pointing out, and this is relevant to the second example below, that the evaporator operates to even out heat fluxes and the feedback to the microchip surface is analogous to the heat exchanger side of a HEX-reactor acting to even out the reaction process and aid high product quality.

The second example that has implications for reactions, but is more commonly associated with electronics thermal control, is the flat plate heat pipe. Two variants are shown in Figures 11.30 and 11.31. It is often used for heat spreading or temperature flattening and is highly effective in this role. The applications of the flat plate unit include, but are not limited to:

- Multi-component array temperature flattening.
- Multi-component array cooling.
- Doubling as a wall of a module, or mounting plate.
- Single component temperature flattening.

FIGURE 11.30

Embedded heat pipes in a flat plate.

(Courtesy: Thermacore Ltd)

FIGURE 11.31

A heat pipe vapour chamber.

The ability of flat plate heat pipes (and vapour chambers) to double as structural components opens up many possibilities, and not just related to electronics thermal management. The example shown in Figure 11.30, although not strictly a flat plate heat pipe (rather heat pipes in a flat plate), is the first step towards producing a near-isothermal surface. These are embedded heat pipes; it is very common to take the standard heat pipes and embed them into the base of a heat sink. This allows the use of high volume, lower cost heat pipes in a heat sink design. The heat pipes are flattened, soldered or epoxied into the heat sink and then machined to create a flat mounting surface to the electronics being cooled. This could also form the base of a flat plate reactor, where temperature uniformity is important.

The heat pipe vapour chamber is a completely hollow flat plate in which evaporation, and condensation of the working fluid in it, can take place. Again it can be used to isothermalise a surface on which a variable heat distribution is occurring. The variable heat input/output could be a catalysed reaction surface.

11.5 Summary

In this final applications chapter, we have examined a number of applications where PI has been applied. Some, such as the Rotex heat pump, have arisen out of the work of chemical engineers and others have been spawned out of electronics or propulsion technologies. The message from all of the examples is that technology transfer can be extremely valuable and an appreciation of a broad range of engineering fields is essential if the full potential of PI is to be exploited – even when those in some of the sectors may not use the term 'process intensification'!

References

Anon, (a). <www.cse.clrc.ac.uk/c4m/c4m.shtml>.

Anon, 2000. Beyond the battery. Ind. Eur. Mar. 10 (9), 2000.

Anon, 2007. <http://www.nec.co.jp/press/en/0309/1701.html>.

Anon, Annex 33 Report (2010). Available free to IEA Heat Pump Centre members on <http://www.heatpumpcentre.org/en/projects/completedprojects/annex%2033/publications/Sidor/default.aspx> (accessed 09.05.12).

Anon, 2012. Front line military technology promises battery revolution. BAE Systems Press Release, 13 February 2012. <http://www.baesystems.com/article/BAES_035647/frontline-military-technology-promises-battery-revolution> (accessed 26.07.12).

Bradshaw, C.R., Groll, E.A., Garimella, S.A., 2011. A comprehensive model of a miniature-scale linear compressor for electronics cooling. Int. J. Refrig. 34, 63–73.

Chang, J., Dommer, M., Chang, C., Lin, L., 2012. Piezoelectric nanofibres for energy scavenging applications. Nano Energy 1, 356–371.

Chen, G., Li, S., Li, H., Jiao, F., Yuan, Q., 2007. Methanol oxidation reforming over a $ZnOCr_2O_3/CeO_2ZrO_2/Al_2O_3$ catalyst in a monolithic reactor. Catalysis Today 125, 97–102.

Cheng, H.-H., Tan, C.-S., 2006. Reduction of CO_2 concentration in a zinc/air battery by absorption in a rotating packed bed. J. Power Sources 162, 1431–1436.

Cherbanski, R., Molga, E., 2009. Intensification of desorption processes by use of microwaves – an overview of possible applications and industrial perspectives. Chem. Eng. Process. : Process Intensification 48, 48–58.

Critoph, R.E., Metcalf, S.J., 2004. Specific cooling power intensification limits in ammonia-carbon adsorption refrigeration systems. Appl. Therm. Eng. 24, 661–678.

Demir, H., 2012. The effect of microwave regenerated adsorbent bed on the performance of an adsorption heat pump. Appl. Therm. Eng. doi: 10.1016/j.applthermaleng.2012.06.022.

Deng, J., Wang, R.Z., Han, G.Y., 2011. A review of thermally activated cooling technologies for combined cooling, heating and power systems. Prog. Energy Combust. Sci. 37, 172–203.

Determan, M.D., Garimella, S., 2012. Design, Fabrication, and Experimental Demonstration of a Microscale Monolithic Modular Absorption Heat Pump. Appl. Therm. Eng. 47, 119–125.

Dutour, S., Mazet, N., Joly, J.L., Platel, V., 2005. Modeling of heat and mass transfer coupling with gas-solid reaction in a sorption heat pump cooled by a two-phase closed thermosyphon. Chem. Eng. Sci. 60, 4093–4104.

El Achkar, G., Miscevic, M., Lavieille, P., Lluc, J., Hugon, J., 2011. Flow patterns and heat transfer in a square cross-section micro condenser working at low mass fluxes. Proc. 3rd Micro and Nano Flows Conference, Thessaloniki, Greece, 22–24 August.

Figus, C., et al., 2003. Capillary fluid loop developments in Astrium. Appl. Therm. Eng. 23, 1085–1098.

Garimella, S.V., 2006. Advances in mesoscale thermal management technologies for micro-electronics. Microelectron. J. 37, 1165–1185.

Garimella, S., Determan, M.D., Meacham, J.M., Lee, S., Ernst, T.C., 2011. Microchannel component technology for system-wide application in ammonia/water absorption heat pumps. Int. J. Refrig. 34, 1184–1196.

Gilchrist, K., Lorton, R., Green, R.J., 2002. Process intensification applied to an aqueous LiBr rotating absorption chiller with dry heat rejection. Appl. Therm. Eng. 22, 847–854.

Gleichmann, N., Malsch, D., Horbert, P., Henkel, T, 2011. Simulation of droplet-based microfluidic lab-on-a-chip applications. Proc. 3rd Micro and Nano Flows Conference, Thessaloniki, Greece, 22–24 August.

Goodman, C., Fronk, B.M., Garimella, S., 2011. Transcritical carbon dioxide microchannel heat pump water heaters: Part 1 – validated component simulation modules. Int. J. Refrig. 34, 859–869.

Gordon, J.M., Ng, K.C., Chua, H.T., Chakraborty, A., 2002. The electro-adsorption chiller: a miniaturised cooling cycle with applications to micro-electronics. Int. J. Refrig. 25, 1025–1033.

Harb, A., 2011. Energy harvesting: state-of-the-art. Renewable Energy 36, 2641–2654.

Helvajian, H. (Ed.), 1999. Microengineering Aerospace Systems The Aerospace Press/AIAA, USA.

Heppner, J.D., Walther, D.C., Pisano, A.P., 2007. The design of ARCTIC: a rotary compressor thermally insulated micro-cooler. Sens. Actuators A 134, 47–56.

Hu, X., Tang, D., 2007. Experimental investigation on flow and thermal characteristics of a micro phase-change cooling system with a microgroove evaporator. Int. J. Therm. Sci. 36, 1163–1171.

Izquierdo, M., Lizarte, R., Marcos, J.D., Gutierrez, G., 2008. Air conditioning using an air-cooled single effect lithium bromide absorption chiller: results of a trial conducted in Madrid in August 2005. Appl. Therm. Eng. 28, 1074–1081.

Kolb, G., Men, Y., Schurer, J., Tiemann, D., Wichert, M., Zapf, R., et al., 2007. Mass-production of miniaturised microstructured fuel processors for distributed energy generation. Proc. European Congress of Chemical Engineering (ECCE-6), Copenhagen, 16–20 September.

Kundu, A., Jang, J.H., Gil, J.H., Jung, C.R., Lee, H.R., Kim, S.-H., et al., 2007. Review paper. Micro-fuel cells: current development and applications. J. Power Sources 170, 67–78.

Li, T.X., Wang, R.Z., Wang, L.W., Lu, Z.S., Chen, C.J., 2007. Performance study of a high efficient multifunctional heat pipe type adsorption ice making system with novel mass and heat recovery processes. Int. J. Therm. Sci. 46, 1267–1274.

Lin, L., et al., 2002. High performance miniature heat pipe. Int. J. Heat Mass Transfer 45, 3131–3142.

Liu, G., Chan, W.-K., 2011. The effect of flow coefficient on the design of miniature centrifugal impeller. Proceedings of 3rd Micro and Nano Flows Conference, Thessaloniki, Greece, 22–24 August.

Marcinichen, J.B., Olivier, J.A., Thome, J.R., 2012. On-chip two-phase cooling of datacenters: cooling system and energy recovery evaluation. Appl. Therm. Eng. 41, 36–51.

Maydanik, Yu.F., 2005. Loop heat pipes. Review article. Appl. Therm. Eng. 25 (5–6), 635–657. April.

McGlen, R., Jachuck, R., Lin, S., 2003. Integrated thermal management techniques for high power electronic devices. Proc. 8th UK National Heat Transfer Conf., Oxford University.

Mehendale, S.S. et al., 1999. Heat exchangers at micro- and meso-scales. In: Compact Heat Exchangers and Enhancement Technology for the Process Industries. Proc. Int. Conf., Banff, Canada, pp. 55–74, 18–23 July.

Metcalf, S.J., Critoph, R.E., Tamainot-Telto, Z., 2012. Optimal cycle selection in carbon–ammonia adsorption cycles. Int. J. Refrig. 35, 571–580.

Moon, S.H., et al., 2004. Improving thermal performance of miniature heat pipe for notebook PC cooling. Microelectron. Reliab. 44, 315–321.

Munkejord, S.T., Maehlum, H.S., Zakeri, G.R., Neksa, P., Pettersen, J., 2002. Micro technology in heat pumping systems. Int. J. Refrig. 25, 471–478.

Ramshaw, C., Editorial. PIN News: The Newsletter of the Process Intensification Network, Issue 12, March 2005. See <www.pinetwork.org>.

Reade, L., 2008. Hot Stuff. Eureka Magazine, 12. 13, December.

Reay, D.A., 2005. Microfluidics overview. Proc. Seminar on Microfluidics, organised by TUV-NEL, East Midlands Airport, 13 April.

Reay, D.A., Hesselgreaves, J.E., Sizmann, R., 1998. Novel compact heat exchanger/absorber technology for cost-effective absorption cycle machines. Exploratory Award Contract JOE3-CT97-1016, Final report, Commission of the European Communities, March.

Reay, D.A., Kew, P.A., 1999. The contribution of compact heat exchangers to reducing capital cost. IMechE seminar on recent developments in refrigeration and heat pump technologies, London, 20 April 1999. IMechE Seminar Publication 1999–10, 37–46.

Reay, D.A., Kew, P.A., 2006. Heat Pipes, fifth ed. Butterworth-Heinemann/Elsevier, Oxford.

Ribaud, Y., 2001. La micro turbine: L'example du MIT. Mec. Ind. 2, 411–420. in French.

Swanson, T.D., Birur, G.C., 2003. NASA thermal control technologies for robotic spacecraft. Appl. Therm. Eng. 23, 1055–1065.

Scott, M., 2012. Current thinking. Flight International, 26. 27, 24–30 July.

Tamainot-Telto, Z., Metcalf, S.J., Critoph, R.E., 2009. Novel compact sorption generators for car air conditioning. Int. J. Refrig. 32, 727–733.

Tsuzuki, N., Utamura, M., Ngo, T.L., 2009. Nusselt number correlations for a microchannel heat exchanger hot water supplier with S-shaped fins. Appl. Therm. Eng. 29, 3209–3308.

Thome, J.R., Marcinichen, J.B., 2011. On-chip micro-evaporation: experimental evaluation of liquid pumping and vapour compression cooling systems. Proceedings 3rd Micro and Nano Flows Conference, Thessaloniki, Greece, 22–24 August.

Waitz, I.A., Gauba, G., Tzeng, Y.-S., 1998. Combustors for micro-gas turbine engines. ASME J. Fluids Eng. 120, 109–117. March.

Weiss, L., 2011. Review: power production from phase change in MEMS and micro devices, a review. Int. J. Therm. Sci. 50, 639–647.

Yu, H., Chen, H., Pan, M., Tang, Y., Zeng, K., Peng, F., et al., 2007. Effect of the metal foam materials on the performance of methanol-steam micro-reformer for fuel cells. Appl. Catalysis A: General 327, 106–113.

Zajtchuk, R., 1999. New technologies in medicine: biotechnology and nanotechnology: a review. Disease-a-Month 45 (11), 453–495. November.

Zhao, T.S., 2009. Micro Fuel Cells: Principles and Applications. Academic Press, London, ISBN 9768-0-12-374713-6.

Specifying, Manufacturing and Operating PI Plant

12

OBJECTIVES IN THIS CHAPTER

In the context of process intensification, the task of process engineers is to implement novel and possibly unique unit operations in either an existing plant or on a greenfield site in such a way that management will be convinced that the investment is worthwhile, and will ultimately see such a unit operation working successfully. The routes to doing this involve a range of enabling technologies, software and strategies, aspects of which are discussed in this chapter.

By example, based upon efforts of research engineers at university, the methodology developed by Britest and BHR Group is adapted to allow logical procedures to be adopted to enable PI designs to replace conventional processes. Other approaches adopted and/or proposed in Europe are also briefly discussed.

As an appendix at the end of this chapter, four case studies using the PI assessment method taught to students at Heriot-Watt University, Edinburgh are given. The methodology closely follows those of BHRG and Britest.

12.1 Introduction

Here, we effectively take the reader out into the field, where he or she may be asked to consider applying PI technology to a greenfield site, either to an existing unit operation or to a whole plant which is due for upgrading or replacement. A number of stages of assessment have to be made before one can proceed to select a PI technology, not least being the achievement of a full understanding of the existing process that is necessary before one can consider changing or replacing it.

The first half of the chapter deals with broad issues leading up to installation, commissioning and operating plant, while the second part presents an example of a process intensification methodology, in this case based upon Britest and BHR Group procedures (see below). In this case, it is used to select an intensified chemical reactor.

Some of the factors to be taken into account have already been introduced in Chapter 3, where it was stressed that an in-depth understanding of the process was one prerequisite to starting a PI exercise. This is so important that it is well worth repeating here. The corollary, concerning how far the operating parameters might be modified to accommodate PI, is equally significant.

12.2 **Various approaches to adopting PI**

The approach to the adoption of PI plant varies from country to country, in part due to different strategic requirements, but also due to the different pace of development of methodologies. The initiative in the UK was taken by BHR Group at Cranfield and Britest, a large consortium representing companies and research institutes, based in Manchester. In the Netherlands, Henk van den Berg, latterly chairman of the Novem process intensification taskforce, produced a tool for PI implementation based upon process systems engineering (van den Berg, 2001). In another article in the same journal, van den Berg and colleagues (Capel et al., 2001) stress the importance of systematically carrying out the conceptual design of PI plant. The thrust of the approach was summarised in four statements:

- Analyse existing flowsheet (of the process) and technology – collect know-how.
- Decompose flowsheet into functional sections – task prevails over equipment.
- A black box approach is the end point of the analysis of the process.
- Rebuild the process systematically and include known technologies and break-throughs, document alternatives and choices.

The Dutch group stressed that the main difference and advantage of their process synthesis approach, compared to conventional design procedures, was the emphasis on process *functions* rather than *unit operations*. While some may argue that this is solely semantics, the constraint on lateral thinking, which might be self-imposed if one considers unit operations as key to the process, is removed. An early PI example quoted illustrates this, in relation to a methyl acetate plant. Here the process function required concurrent reaction and distillation tasks, which, by removing constraints that might be imposed by thinking of separate unit operations, could be implemented in a single reactive distillation column. Readers are also recommended to consult the proceedings of a Dutch PI conference (Anon, 2000a) for a detailed analysis of approaches to PI implementation. (See Chapter 5 for a discussion of reactive distillation, and Appendix 3 for data on a useful review paper.)

Bakker of DSM (2004) identified steps used in his company in implementing PI solutions and a number of examples from DSM are used to illustrate a methodology similar in procedure to that given in this chapter. A specific methodology for the scale-up of micro-reactors has recently been developed by a Finnish team working at Lappeenranta University of Technology (Kolehmainen et al., 2007). The researchers point out that one of the scale-up problems is the uniform distribution of fluids into possibly thousands of channels.

Recently, Dutch and German PI workers have built upon the experiences of implementing combined unit operations to prepare a discourse in which intensification and process systems engineering are discussed in concert (Moulijn et al., 2008).

At a meeting of the Dutch PI Network in 2012, PQ Corporation (Philadelphia Quartz) presented their approach to considering PI within their processes, (Beckers, 2012). The company produces silicates (waterglass) and silicas that are used in a range of products from catalysts to toothpaste. The company had carried out two quick scans and one in-depth scan. The first quick scan involved an autoclave process (exothermic reaction) for making water glass – a batch process with steam addition and carried out at 20 bar. Questions were asked such as whether the process could be made continuous, or whether the exothermic heat could be employed as an energy source. The outcome of this exercise was a focus on speed by increasing reaction speed and operating semi-continuously. The in-depth scan focussed on the manufacture of silica gel, resulting from another quick-scan exercise. Greater in-depth analyses were carried out within groups, and employing specialists, leading to long-term innovations such as the use of a spinning disc unit to produce silica gel.

12.2.1 Process integration

It will not be surprising to chemical engineers that mention of process integration – the other 'PI' – arises when considering the optimum placing of intensified unit operations within a complex site. It is therefore to be expected that process integration methodologies, developed at ICI and elsewhere and now effectively disseminated by the Centre for Process Integration at Manchester University, UK, have been studied with a view to using the software to select and optimise unit operations, such as compact heat exchangers and HEX-reactors (Efthimeros et al., 2000; Kalitventzeff et al., 2000). Although the work to allow integration software to accept all but the simplest intensified unit operations (e.g. compact heat exchangers) has yet to be fully implemented, the European Union-supported project discussed in these two papers were ground-breaking; and additionally were the first examples of the use of expert systems to select optimal intensified process technologies. Some of the steps are those adopted in the BHR and Britest methodologies, and of course are common to any sensible process analysis. These steps, many of which will be recognisable from later discussions in this chapter, were:

- Define process requirements.
- Identify candidate generic energy-saving technologies (note that the thrust of the research was to save energy).
- Compute the technology requirements to fit the process requirements.
- Select the candidate technologies from those in the marketplace.
- Carry out optimal insertion of the technologies (the process integration step).
- Evaluate the solutions.

12.2.1.1 The Linnhoff March early approach

Process integration methodologies started with heat exchangers (generally trains of shell and tube types in, for example, a petroleum refinery) and early work by Linnhoff March in the UK stressed that process integration could be used when intensified unit operations were being considered. In these early days (late 1980s) the emphasis was on compact heat exchangers and tube inserts (see Chapters 3 and 4), so it was logical that there was an emphasis on these components. Linnhoff March stated:

> *Because compact heat exchangers or shell and tube exchangers containing turbulators both have different heat transfer characteristics and cost/area relationships to conventional exchangers, it is essential to take this into account when*

optimising a design. New process intensive designed heat exchangers should only be used if there are good economic reasons or if safety constraints dictate their use. Pinch technology (process integration) allows the trade-offs between operating and capital costs of a mix of heat exchangers to be explored and thus the justification for using new devices can be made...

Today this approach may seem comparatively unsophisticated, but the parallels with the Dutch approach some ten years later can be seen. Inevitably, there will be a greater integration of intensification technologies and process simulation software that will allow routine use of PI over the coming decades, particularly as constraints on energy use and the need for sustainability become imperative.

This trend can be seen with the work within the European Commission FP7 project, INTHEAT. Pan et al. (2012) at the University of Manchester (in the UK) have been extending the group's expertise in process integration to examine computer models for heat exchanger network (HEN) retrofitting with intensified heat transfer techniques, such as tube inserts (see Chapter 4) and shell-side helical baffles. In shell-and-tube heat exchanger trains these will combine to improve energy recovery in many processes. Examples quoted include 4–6% energy savings with around 22% saving in capital cost of the exchangers, as opposed to adding more surface.

12.2.2 Britest process innovation

One of the most successful pioneers in helping companies to realise the benefits that we attribute to PI (without necessarily spelling out that PI technologies are used) is Britest (see Appendix 6). Britest points out that:

since 1998, the application of Britest's innovative tools and methodologies has enabled members to realise capital and operating cost savings in excess of £50million, as well as delivering other substantial benefits including:

- *Higher yields of better quality product.*
- *Environmental benefits through waste reduction.*
- *Better understanding of processes and complex operational issues.*
- *Better, more efficient plant design.*
- *Capturing and retention of process knowledge.*

Application of Britest's whole process design thinking can achieve:

- *Reactions in minutes, not hours.*
- *Smaller, more flexible plant.*
- *Sustainable chemical processing.*

The reader will recognise that many of the above outcomes relate specifically to PI, and the brief case study from the Britest website (where there are several similar interesting case summaries) is typical of the benefits accruing to the successful use of PI, backed up by appropriate process simulation tools and flowsheeting expertise.

Jeremy Double of Britest presented an update on the approach to PI implementation within the context of choosing appropriate targets at the European Process Intensification Conference in 2011 (Double, 2011). The emphasis on a fundamental understanding of the needs of the process is reiterated, and he correctly points out that many of the specific heat and mass transfer characteristics of PI equipment and mechanisms are not readily available to chemical engineers.

12.2.2.1 Case study

Within NPIL Pharma (formerly Avecia Pharmaceuticals), the use of Britest tools facilitated an opportunity to move from large scale batch processing (with expensive raw materials and requiring a complex solvent mixture with no practical opportunity for solvent recycling) to small scale continuous processing with lower cost materials and integrated solvent recycle.

The identified benefits in this programme included a 20% reduction in product cost, a 40% reduction in capital and a 75% reduction in organic waste compared to the original batch facility.

12.2.3 Process analysis and development – a German approach

The Institut fur Chemische Verfahrenstechnik at Stuttgart University has reported at length on the topic of process design and intensification – in particular as they impact on several examples of chemical reaction engineering (Eigenberger et al., 2007). A number of processes were examined, ranging from steam reformers to diesel exhaust gas treatment (an interesting non-chemical plant application of PI which readers of the paper may find valuable).

The guidelines adopted are similar to approaches already described and to the more detailed exercise presented later in this chapter. Modelling had a strong role to play, but there was an emphasis on simplicity. The guidelines were divided into four steps as follows:

- Obtain an understanding of the process features by physical modelling, numerical simulation, parameter analysis and (if necessary) experimentation.
- Based on these results, attempt to apply simple models to the most typical process patterns.
- Recognising that excursions out with these patterns – instabilities – could occur, process control would be applied.
- Confine the process to a geometrically simple structure – a one-dimensional geometry, state the authors. They point out that 1D flow in the absence of lateral non-uniformities is one prerequisite for uniform product quality – a major benefit of PI.

It is now appropriate to consider the several stages in an approach to implementing PI. (This is not the only approach possible, but it is a valid one and also one that is relatively straightforward to present in a book such as this introductory text.)

12.3 Initial assessment

It will be obvious from the discussion above that the implementation of a full PI exercise, associated with an existing plant/process or for a new plant on a greenfield site, is a complex procedure. Before rushing ahead to specifying plant, there are a number of aspects of the existing plant which are relevant to any plans involving PI. These are discussed below. The PI approach should not run counter to any of the possible corporate concerns of the host organisation. These may include:

- Energy
- Environment
- Legislation
- Quality
- Throughput
- Economic survival

As discussed in Chapter 2, and as should be evident from the examples of applications given in Chapters 8 to 11, these can all be positively addressed by PI, but need consideration on a case-by-case basis.

12.3.1 Know your current process

Obviously, if starting with a greenfield site, manufacturing a totally new product and envisaging a new intensified plant, understanding the current process may be irrelevant. However, certainly in the short-term, a considerable proportion of process engineers who are considering adopting PI will be interested in intensifying a part of a process, perhaps where a bottleneck occurs or where energy and/or materials wastage are high. Although this may not bring the full benefits of PI that have been explained in earlier chapters, it can allow the concept to be introduced gently into an organisation, and possibly with minimum risk. In this context, it is worth highlighting one of the important benefits of PI to a company, insofar as risk minimisation is concerned, as identified by Ian Henderson of Protensive Ltd (others are identified in Chapter 2):

> Opportunities arising through a corporate strategy, including minimising risks, e.g. by adding a product made by PI rather than replacing output of the core business.

Where one is examining the possibility of adding PI unit operations to existing plant, it is vital to have fully quantified data and a full knowledge of the process into which the piece of PI plant will be incorporated – this is exactly the same exercise one would undertake if, for example, the aim was to reduce the energy consumption in the process by fitting a waste heat recovery system. If a current process is being examined with a view to replacing it completely with a PI plant, the same necessity exists. Errors and omissions are likely to have a profound effect on the success of later engineering activities.

One would not be thinking of moving to PI equipment unless the current plant was not meeting one's requirements in at least one of several important respects – for example, it has excessive energy consumption or retains a large inventory of nasties that could cause severe pollution if released. It is therefore important, also, to ensure that the comparison between the existing process and the one which is proposed is done on a fair basis. In order to give one's case the greatest credence, the original plant should be examined to see if it is running under optimum conditions. If neglect has caused increased energy consumption, compromised safety or limited production rates, one may find that relatively low-cost modifications will improve performance significantly. This is called good housekeeping. If this allows you to optimise the plant operations to such an extent that a PI modification becomes unnecessary, as may be the case in rare situations, so be it!

As an example of this, again from the Britest case study portfolio on their website (see the output highlighted in italics), at AstraZeneca, the organisation helped to deliver both financial and technical results by providing:

- Improved technical review and understanding of a process (resulting in an increased process yield of 20% in one example).
- Provision of a predictive capability to understand the technical risks associated with scaling-up a process.
- Improved documentation, allowing projects to be transferred between sites and project teams more easily.

Less relevant to the current discussion, but of interest to PI exercises in general, the Britest exercise also permitted:

- More rapid identification of opportunities for the exploitation of alternative manufacturing technologies.
- Increased contribution and closer working between process chemistry and process engineering by establishing a common language.

If one is determined to incorporate new processes within the plant, the knowledge gathering exercise will contribute towards gaining sufficient data about the plant and processes to help take the next step – an initial assessment of the process-limiting factors. What one does not want in any new or modified process is a change which could detrimentally affect the production rate, product quality, plant safety or energy use. Visit any production plant and talk to process engineers and one will readily find out that the perceived ideal operating conditions are jealously guarded by operators, even in these days of computer control. One of the co-authors of this book, with several decades of experience in assessing new technologies for retrofitting to process plant, recalls the feelings of a baking oven superintendent when there was a hint that fitting a process heat recovery unit on its exhaust might slightly alter the colour of the loaf crust! Some of the most important analyses which can be carried out specific to the chemicals sector are:

1. Ensure that a process integration analysis has been performed in order to minimise the number of process elements (e.g. heat exchangers), as well as making certain that there is no heat transfer pinch.[1]
2. Can higher pressure be used to increase either reaction rates or mass transfer? Clearly the higher pressure must be paid for in terms of compression costs, but the overall system benefits may justify this – high pressure is briefly discussed as an intensification process in Chapter 3.
3. Can oxygen be substituted for air? In this case we can have a double potential benefit, not only because the oxygen partial pressure and reactivity is raised, but also because the volume flow associated with nitrogen is eliminated. Again, these advantages must be balanced against the increased hazards associated with neat oxygen.
4. Can solution concentrations or operating temperatures be raised (by raising pressure?) in order to enhance the reaction kinetics? The downside may be increased by-product and/or lower selectivity.
5. Can electric heating via microwaves or resistors be justified because of its easy controllability despite its relatively high cost? Section 3.4 in Chapter 3 is devoted to electrically-enhanced PI mechanisms.

It is interesting to read a paper written by Infineum, a global manufacturer of additives for engines and lubricants, in which the company discusses the merits of PI for medium-scale (defined as >10ktonnes/year) processes, (Hobin, 2011). The paper points out the necessity of evaluating the appropriateness of the chemistry and the processes. In stating that the likelihood of building a whole new site based on PI is remote (in Infineum), Hobin does say that a new plant could be built based on PI, or equipment in an existing plant might be replaced using PI unit operations. It is suggested that costs of PI plant could be lower, due to reduced process equipment capital cost, but even with a 50% reduction in the cost of the process equipment, it is suggested that total capital cost will reduce by only 35%. This is wholly attributed to the fact that piping and tanking costs remain the same as in a conventional plant. This runs contrary to the arguments put forward in Chapter 2, (section 2.3), where it is suggested that Lang actors can be reduced across the board. Hobin suggests that other benefits, such as improved product quality or lower operating costs, might be the more attractive features making PI saleable. The points do make a valid lead in to the procedures continued below.

---

[1] There is significant work on using process integration methodologies to identify the best places to site process intensification unit operations. Much of this work is co-ordinated at the Centre for Process Integration, University of Manchester, UK (see Appendix 4).

12.3.2 **Identify process limiting factors**

For the next stage in an assessment, it is imperative that the process limiting factors are identified. The likely contenders are:

1. Intrinsic kinetics (chemical or phase change, etc.)
2. Heat transfer
3. Mass transfer
4. Mixing rates

Note: The mention of kinetics will immediately strike a chord with chemical engineers, and most likely represent an association with chemical reactions – PI plant is often used to accommodate fast exothermic reactions, where other process plant (e.g. a stirred pot) cannot match the fast kinetics that are desirable. The heat transfer engineer will recognise phase-change as a kinetics-dominated process, so the refrigeration engineer looking to intensify a solid–gas adsorption process or the kinetics of a phase change material, such as a heat storage salt, will also be able to associate him – or herself with kinetics. The example given below, however, is related to reactions in a chemical process.

12.3.2.1 Kinetics

If there is a fundamental kinetic limitation at the standard process temperature, it may be possible to adopt more severe process conditions, bearing in mind that intensified equipment will entail much shorter residence times. Thus, the balance between the desired and undesired reactions may be influenced favourably. Certain processes may spuriously appear to be kinetically limited, especially if the reaction involves intermediates which are only present in small concentrations, e.g. dissolved oxygen in fermentations or organic oxidations. The key factor is the half-life of the relevant reacting species. If this is short, compared with the mixing rate or circulation time in the reactor, then the system is limited by the prevailing fluid dynamics rather than the kinetics. Obviously, when the kinetics are not limiting, the fluid environment must be intensified to relax the other restrictions.

While continuous processing is standard for most products manufactured on a large-scale, batch production tends to be the norm in the fine chemical and pharmaceutical industries. Although batch manufacture may be superficially flexible, it runs counter to the whole intensification ethos. This is because the heat and mass transfer duty for the entire batch is concentrated in a restricted period and the heat/mass transfer equipment must be sized to accept the peak batch load. If this is not done, then reactor runaway can ensue, with disastrous consequences. On the other hand, with continuous production, the respective steady state heat and mass transfer rates are much less and automatically involve smaller equipment.

It is therefore well worth considering the switch from batch stirred vessels to some form of continuous equivalent, even if the chemical kinetics are slow and therefore limiting. An annual output of 200 tonnes/year corresponds to a volume flow <20 ml/s. Two simple 100 m tubular reactors in parallel, each 3.6 cm diameter, could provide 2.7 hours residence time and have a total reactor volume of $0.25\,m^3$ when coiled within a vessel. Thus, even for lengthy reaction times, continuous operation is worth considering. An important aspect of this is an assessment of the reactions one might consider intensifying.

The foregoing indicates that the ideal process characteristic for the application of intensification is that the reaction or phase change kinetics should be as fast as possible. If we regard the H_2/O_2 rocket motor as a chemical reactor, its performance is very definitely limited by heat and mass transfer considerations. Nevertheless the very fast combustion kinetics allows residence times of several milliseconds and a production rate of 1 tonne/m3s! In the process industries these rates are, unsurprisingly, beyond reach. However, a careful evaluation of the factors outlined above can lead to significant intensification opportunities, even with long established processes. The example below, based upon spinning disc reactor polymerisation (see also Chapter 8), illustrates how PI can allow these very significant increases in the reaction rate.

In the SDR, it was shown that the average increase in conversion in one disc pass for the 40% prepolymer feed was of the order of 10–15% and about 16% for the 60% prepolymer feed. With residence times of the order of 1–2 seconds, the polymerisation rate in the SDR is estimated to be between 12,000% per hour to 64,000% per hour, as opposed to 10–50% per hour in conventional reactors operating at more extreme conditions. Molecular weights and polydispersity indices were almost unchanged in comparison to the feed.

Another example, outside reactors, relates to the crystallisation of sodium chloride in a vacuum evaporator, which has long been performed in very large vessels. The boiling (and hence the zone wherein supersaturation is generated) is confined by hydrostatic pressure considerations to within about 50cm of the gas/liquid surface. From knowledge of the crystal growth rates and the crystal surface area in the circulating slurry, it can be established that the supersaturation half-life is roughly one second, whereas the circulation time is about one minute. Therefore, crystals only grow in a small proportion of the volume provided, indicating an opportunity for innovation.

In most of the attractive examples of intensified process equipment (e.g. in-line mixers, spinning disc reactors, rotating packed-beds, micro-reactors, etc.) the fluid residence time is measured in seconds. Therefore, a process designer should consider the use of these devices, provided that the reactions are (or can be made to be) completed in this time frame. If this is not the case, then the fluid intensity should be 'detuned' to match the relatively relaxed kinetic environment. In this event, with a continuous process, a simple tubular reactor with very modest flow velocities could provide adequate plug flow and residence times up to several hours. In the case of a number of biological processes, as an alternative example, substantial intensification can be achieved using a continuous oscillatory baffled reactor (COBR, see Chapter 5), where residence times may be at best minutes and could extend to hours.

12.3.2.2 Heat and mass transfer

Heat and mass transfer has already been included in the discussion of kinetics, but it is worth mentioning some of the areas where intensified heat and mass transfer are attractive, but no reactions are taking place. Several of the applications of PI discussed in Chapters 9 and 10 are of this type, and highly compact heat exchangers, electrical enhancement methods, such as ultrasound, and induction-heated mixers are examples of where a need to improve heat and/or mass transfer has been recognised. In some of these examples, the heat transfer may be limiting because of the nature of the material being heated or cooled. As discussed in Chapter 9, the rapid processing of foodstuffs can, in some instances, affect the taste or consistency and such changes may be anathema to the food scientist and others in the sector, not least the customer. Therefore, more rapid heat and mass transfer are not always desirable. For similar reasons, excessive reaction rates may result in an unacceptable product. The phrase 'horses for courses' springs to mind (one often used by a leading specifier of waste heat recovery systems in the process industries).

12.3.3 Some key questions to address

The preliminary assessment necessitates the process engineer asking a number of questions, some of which relate to his or her ability or that of his or her organisation to handle intensified unit operation design and specification. The questions that it is worth addressing include those listed in Table 12.1. Examples of responses to these questions in specific cases are given later in this chapter.

12.4 Equipment specification

Specification of PI plant can lead the unwary into corners, particularly when building up a case for support. The best advice is to approach the contractor or equipment supplier for answers but, as listed in Appendices 4–6, there are others able to assist. In addition to the questions posed in Table 12.1, there are some critical concerns that affect a company's decision to proceed with new projects. These include:

- Are you likely to be dealing with a single source supplier of equipment?
- What are the health and safety implications?
- Is the PI equipment constructed to current standards (e.g. small pressure vessels, etc.)?
- What about fouling?
- Can you control your PI plant?

If it is company policy not to deal with single source suppliers, the relationship with many companies attempting to introduce new technology into the marketplace may be strained. As technologies develop and the stranglehold of some intellectual property may not be as total as at first thought, competitors will enter the market and the reassurance of competition and multiple sources of new plant will become evident. Nevertheless, the observation that 'the rush to be second', (highlighted many times by Colin Ramshaw in lectures on PI) is a characteristic of many companies looking at new technologies. However, the business case for proceeding with an innovative PI

Table 12.1 Checklist for Process Engineers Considering Implementing PI.

Is my plant/process currently running under optimum conditions? If not, why not?

Is it cost-effective to reach optimum running conditions with my current plant/process?

If my plant/process is optimised, is there still a case for looking at PI?

Do I know all the current operating parameters on my plant/process?

Can I obtain these data?

Do I have information on the operating parameters of any replacement plant?

Will any replacement plant/process have significantly different requirements to the existing plant (throughput, be continuous or intermittent [batch], product type, operating temperatures, pressures, quality control, etc.)?

Are these requirements realistic for a conventional plant?

If they are, why do I want to study a PI plant?

If they are not realistic for a conventional plant, how will a PI plant help?

Will a radical change of process/plant affect my product?

Is my knowledge of PI sufficient to answer the above?

If not, are there others in my organisation who could assist me?

Should I consider employing outside consultants?

Do I have sufficient time and resources within my own team?

Do my timescales and potential budget allow for any R&D work?

If not, am I limited to equipment that is already fully developed?

If this is the case, is such equipment available?

Do I then have the information to carry out:

   A risk assessment (health & safety)?

   A technical feasibility study?

   A rigorous financial appraisal?

Have I considered the availability of grants?

Can I now make a comprehensive case for examining a PI solution to my plant/process?

process can be overwhelming (see Chapter 2) and with technology such as spinning disc reactors (SDRs, see Chapter 5) the benefits can make the risk seem bearable.

As one reduces the size of plant by intensification, an area that has recently received attention is the impact of manufacturing tolerances and measurement uncertainties on performance, (Brandner, 2011). Brandner points out that at the microscale for example, manufacturing tolerances can have a much greater influence on processes than at the microscale. Even surface finishes that would be negligible at large-scale may affect surface behaviour in either heat transfer or mass transfer.

Brandner gives as an example the flow distributor system for microchannel heat exchangers. As highlighted elsewhere in the book, it is important to ensure uniform feed of fluid into each channel, (more particularly so in the case of microreactors), and the tree-like branch flow distributor system frequently discussed by Professor Adrian Bejan is one that should give uniform residence times in the channels, resulting in and evaporator, states Brandner, a phase transition line normal to the flow direction. However, due to manufacturing tolerances and small steps in the structure of the branches, the phase transition region is less than perfect compared to the normal shape that would be the ideal.

Safety, also discussed in Chapter 2, is an integral part of process plant design and specification, whether it is PI-related or not. Etchells (2001) discusses fast exothermic reactions and factors affecting runaway. She points out that exothermic reactions may cause particular problems, due to thermally initiated decomposition, self-accelerating exothermic reactions or rapid gas evolution. The time constants of many PI reactions are very low, and therefore control may be critical. Hazard identification techniques are routine (e.g. HAZOP) in most organisations now, particularly in countries where legislation is driving health and safety at work.

There are a number of publications discussing safety and PI. The paper by Ebrahimi et al. (2012) and introduced in Chapter 2, discusses a safety checklist for intensified processes. In one example, the impact of replacing a bubble column with a tubular reactor is discussed in the context of positive and negative effects on safety, while the switch from auto-oxidation to a microreactor for hydrogen peroxide synthesis is treated in a similar manner. Drawbacks of the microreactor route include the use of methanol (more toxic) and higher pressures (to 30 bar). It is suggested that those carrying out the procedure for PI assessment discussed later in this chapter might use the checklist from this paper, too.

It should not be forgotten, however, that in the majority of PI applications involving 'nasty' substances, safety benefits dominate due to smaller liquid, gas or solid inventories!

With regard to construction standards, these will be dictated by law and/or safety regulations in most instances.

12.4.1 **Concerns about fouling**

Concerns about fouling are high on the agenda of anyone dealing with plant that may comprise small channels – such as compact heat exchangers (CHEs, see Chapter 4) and micro-reactors (see Chapter 5). Operators of CHEs do not have the answer to all the fouling problems which might be met with equipment and it is premature to expect the users of PI plant, with its greater range of types, to know all the answers to fouling minimisation. Some techniques are standard practice and others are at the ideas stage, as can be seen from the outcome of a recent workshop at one of the Process Intensification Network meetings, given below (Anon, 2000b):

- PI is ultimately limited by the fouling characteristics of the system.
- Process incidents/excursions are triggers for fouling/channel blocking.
- How do you clean a fully blocked unit, or do you replace it?
- Membrane operators, e.g. on marine desalination, have learned to cope with sea water fouling. Can we learn from this?
- Install pressure drop sensors across a unit where fouling may occur, with cut-out and operator warnings.

- Intensified units can be very attractive for debottlenecking in some cases.
- Scale removal using magnetic fields could be investigated for PI plant.

Mayer et al (2012) have recognised that fouling in micro-systems can be a problem, and have examined in detail crystallisation fouling in micro-heat exchangers, obtaining data on fouling resistances. Currently, the group in Germany is investigating cleaning procedures.

A message relevant to process engineers addressing fouling problems in any plant is that ... optimum solutions to equipment specification, and all subsequent stages, are only to be found when the equipment supplier and the end user fully exploit their know-how together.

There are instances where the host plant may have concerns about confidentiality that inhibit a full dialogue between both parties, but these are the exception.

12.4.2 Factors affecting control and their relevance to PI plant

Specification of the control system, with knowledge of the parameters which need to be controlled and their ranges, go hand-in-hand with specification of the process equipment components themselves. In any process, one has to examine a number of parameters before deciding upon a suitable control system and associated instrumentation. These include the following main features, described by Dr Ming Tham of Newcastle University, in the UK (Tham, 1999):

- Dead-time to time-constant ratio.
- Linearity.
- Constraints.
- Time variations.
- Interactions.

The instrumentation itself is also subject to similar features and constraints, in particular:

- Response times.
- Availability.
- Dead-time.
- Accuracy and precision.

Some of these features may be identified at an early stage as potential problem areas, which you may find difficult to address with off-the-shelf control systems and, in particular, sensors. Response time is obviously a critical feature of control and instrumentation in PI, where one may be reducing reaction times, for example, close to their minimum. A comparison of response times of different types of loop element is given in Table 12.2.

If it becomes necessary to address the above limitations and others which may be identified by moving outside what is generally available commercially, the PI plant designer may consider the following solutions, as outlined in Table 12.3.

Table 12.2 Response Times of Loop Elements.

Loop Element	Typical Response Time
Signal transmitter	1–5 s (pneumatic)
	Instantaneous (electrical)
Signal converter	0.5–1.0 s (electronic to pneumatic)
Final control element	1–4 s (0–100% valve open)
Sensors	
Thermocouple	Almost instantaneous (bare)
	5–20 s (in thermowell)
Flow, pressure, level	Several seconds
Analysers	5–30 minutes or longer (usually discrete)

Table 12.3 Possible Problems and Potential Solutions.

Possible Problems	Potential Solutions
Instrumentation response times	Cascade strategies
Sensor problems	New sensors; sensor location. Inferential measurement and control
Time-delays	Predictive control. Robust controller designs
Interactions between process states	Selection of control loop pairings. Decoupling control
Interactions between process units	Feedforward strategies

There are a number of questions which need addressing, either on a project-by-project basis, or across the PI field. The immediate needs are to address:

- Whether the existing control strategies and algorithms are able to cope with the new conditions.
- Whether problems can be alleviated by system redesign.

Readers wishing to study control in more detail can read about experiences with a reactor supervisory control system in the paper by Turunen et al. (1999). More recently, a number of papers have started to address control of PI systems in more depth. Of particular interest in this context is the work of Kothare (2006) who studied the dynamics and control of micro-fuel cell components.

Kothare looked at embedded feedback control in the regulation of thermal transients in micro-reactor systems where it is desirable to maintain a specific temperature profile in the micro-reactor. The proposed control scheme, illustrated in Figure 12.1, is a '…receding horizon boundary control scheme using empirical eigenvalues in a constrained optimisation procedure…' to track the desired profile through the micro-reactor, in terms of space and time. The unit was able to maintain a desired temperature profile using resistance heaters to carry out necessary

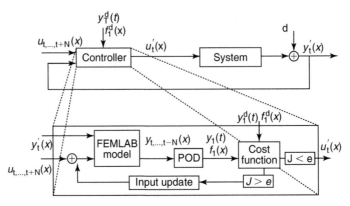

FIGURE 12.1

Receding horizon controller block diagram (Kothare, 2006).

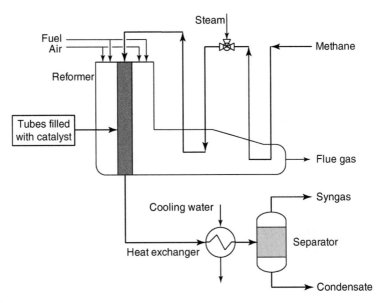

FIGURE 12.2

Typical flow sheet for a conventional steam reformer.

heat input changes. The system has also been applied to another problem in microsystems where fluids are involved, the regulation of channel flows. More recently, Nikacevic et al. (2012) have reviewed control of intensified processes with a view to putting forward a new process synthesis concept that facilitates operation, design and control integration.

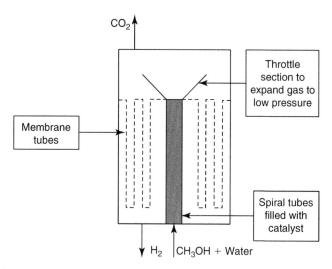

FIGURE 12.3

Possible compact heat exchanger reactor design (based on Buxbaum, 1997).

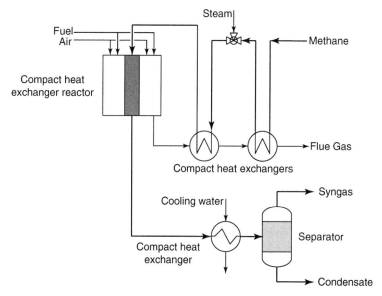

FIGURE 12.4

Flow sheet for steam reforming using compact heat exchangers and reactors.

In an application of control to a spinning disc reactor (SDR), Ghiasy et al (2012) at Newcastle University, in the UK, pH control of a neutralisation reaction was attempted. PH control was pointed out by the authors as being particularly challenging due to the nonlinear and time-varying characteristics, and initial attempts using a proportional integral controller were less than ideal. Including a pH characteriser and a disturbance observer compensated for the pH process non-linearity. In the future, it was suggested that using disc rotational speed as the manipulated variable (rather than reagent flow rate) could be usefully studied.

12.4.3 Try it out!

A number of potential pitfalls that could inhibit the process engineer proceeding to fully implement a PI assessment have been identified above. These can be overcome using expertise available inside or out of the host companies. However, increasingly, one will find that an academic institute, an equipment manufacturer or an independent body will be able to offer a trials service. This allows the customer to examine, on a PI plant, whether his or her reaction, separation or heat transfer process (as examples) can be satisfactorily carried out. One company offering such a service in the UK is the Centre for Process Innovation (CPI) on Teesside (see Appendix 5). CPI provides services ranging from small consultation projects to full pilot plant design, build, operation and optimisation. At the time of writing (mid-2007) the following PI products were available for customers to trial and test:

- Structured reactor.
- Micro-reactor.
- Spinning disc reactor.
- Continuous oscillatory baffled reactor demonstrator.

12.5 Installation features of PI plant

It is important to realise that many intensified plant and processes need treating carefully. In particular, do not expect them to be as tolerant to excursions in conditions as large-scale processes. Listed below are a number of considerations that are easy to overlook when planning the introduction of a new unit operation.

- Ensure that all relevant staff is trained in dealing with PI plant.
- All pipes, ducts and equipment headers should have been sealed at the factory and not opened until ready for connection.
- Ensure that metering equipment upstream is compatible with your PI plant, not the original plant.
- PI equipment may require clean-room assembly facilities.

12.6 Pointers to the successful operation of PI plant

Based upon selected case studies from elsewhere in this book and discussions with operators of PI plant, a list of hints for the safe and successful operation of PI plant is given below. Although not to be regarded as exhaustive, it should give the reader a strong feel for the considerations that lead to reliable operation. Sensors, accurate measurement of pressure drop changes in a compact process unit, etc., are covered. Many practising plant engineers will recognise that a proportion of the recommended actions apply equally to other types of unit operation – in fact, much of the list was derived from recommendations relating to compact heat exchangers (Anon, 1998):

- Have the equipment suppliers present during commissioning, if possible – the reputation of the suppliers, as well as the user, is at stake if commissioning does not proceed well.
- Ensure that all rotating plant is shielded from personnel during operation – a normal safety procedure.
- Check the kinetics of reactions where intensified reactors are used, particularly if product changes are undertaken.
- Continuously monitor pressure drops through compact plant – a good indication of blockages, etc.
- If one is changing the product, check materials compatibility – one of the advantages of some PI unit operations is their flexibility, but normal corrosion considerations come into play if the nature of the throughput is changing.
- Ensure all filters are regularly cleaned and/or replaced.
- Have in place a cleaning strategy should any blockages occur in PI plant – companies operating in fine chemicals and the food and drink sector have long recognised the importance of cleaning procedures, and companies such as Alfa Laval have pioneered processes for this.
- If the plant or unit operation is critical to production, have a replacement unit available – this may not be as expensive as it seems, as PI plant can be highly compact and use a lot less exotic metal in its construction.
- Use water treatment as and when necessary. Scaling may not be important in a conventionally-sized plant, but could bock a PI one. See, for example, scaling tests on highly compact heat exchangers that could represent the thermal control side of a HEX-reactor (Anon, 1993).
- Consider cleaning-in-place (CIP) facilities.
- PI plant may be designed to very tight limits – be aware in operating the units that this can also limit flexibility. Very high product quality may be one outcome of a particular PI unit operation but, in order to achieve this, the feed rates and temperatures will most likely need precise controlling.
- In case of any failure, contact the equipment manufacturer before trying to rectify matters.
- Make sure that maintenance procedures are followed to the letter. Some plant will need to be returned to the manufacturer for maintenance.

It will be unlikely that the leap from conventional process plant to an intensified site will take place in one step. Although this is highly desirable, in order to maximise the benefits of PI, it is probable that most users will have a mix of conventional and intensified unit operations, at least to start with. The second major part of Chapter 12 deals with the selection procedure for PI plant, with a specific application in mind.

12.7 The systematic approach to selecting PI technology

As already illustrated at the beginning of this chapter, a number of organisations have developed, or are in the process of developing, methodologies to assist companies implement PI projects. The information given below is an edited version of the methodology developed at BHR Group Ltd and published on the Britest website (see Appendix 6) and in BHRG PI Conference Proceedings.

12.7.1 A process intensification methodology

Applications for PI equipment with a technology driven approach have regularly been sought. The implications of this are that the novel technology that does exist is developed by organisations which have knowledge of one or maybe two types of equipment, and when the organisations are confronted with a problem or are looking for applications their approach will be on the basis of trying to make the process operate in their particular equipment allowing the procedure to run at its most efficient rate.

The principal suppliers of intensified unit operations increasingly realise that the ideal approach is driven by business requirements, although, of course, process drivers will retain an important place. Process drivers are those where the physical and chemical requirements of the process are determined and then used to select the equipment which best suits the process. Business drivers are financial issues relating to the operability and profitability of the process, and companies such as Protensive rightly see the massive business opportunities afforded by the radical PI approach.

One way of increasing the use of PI lies in producing methodologies capable of promoting a different approach to process development. Below is a development methodology in which process intensification is brought into consideration for upgrades of existing processes. Ideally the methodology would be applied right from the initial stages of a new process, including involvement in the selection of the synthesis route, as this is where opportunities for the application of PI can be missed.

For companies who rely heavily on simulation software for process examination, there will be a gradual transition to incorporate the capabilities of PI plant in these. As an example, ANSYS Fluent is involved in a major project that is developing correlations for two-phase heat transfer at the microscale, suitable for microreactor and micro-heat exchanger design and sizing, see Web 1 (2007).

12.7.1.1 Background
The old route to process development
Taking, as an example, a new process in the chemicals manufacturing (or some similar) sector, a typical approach would be for the chemist to identify a number of

chemical synthesis route options. The most promising ones would then be chosen for further analysis. By the time the engineer had become involved, the synthesis may already have been developed to the stage where it can only be done in conventional plant, usually batchwise or semi-batchwise in stirred tank reactors (STRs). Very often engineers are not given sufficient data on the kinetics of the synthesis in order to consider whether the operation could be intensified, even if they do have knowledge of PI.

Bring in the engineer at the beginning

The engineer should get involved earlier in the project and encourage the chemist to develop faster reactions, or not slow down intrinsically fast reactions. Slow reactions are preferred in STRs as they are easier to control. Inhibitors can be added to slow reactions down and allow the study of the reaction steps, but then continue to be used for full-scale operation if the reaction scheme has been shown to produce acceptable results. Some synthesis routes could be discarded which would have been ideal for PI, but are not possible with conventional equipment, as intensified equipment is capable of handling very fast, exothermic reactions.

Use a PI methodology

The methodology sets out structured procedures to follow for considering PI *during process development*. The overall methodology, known here as the framework, consists of a number of protocols detailing the information needed to ensure the potential for PI is fully examined. The framework is formatted to apply to situations where an existing chemical plant[2] is to be replaced or upgraded. Each of the methodological steps is described below:

- Overview the whole process.
- Examine the chemistry and the unit operations.
- Identify business and process drivers.
- Identify rate-limiting steps.
- Generate design concepts.
- Analyse the design concepts.
- Select the equipment.
- Compare the PI solution(s) with conventional equipment.
- Make the final choice.

The order of these may vary, and some activities are continuous throughout the project. More data are given below and an example of the approach used by chemical engineering postgraduates and final year students is included.

---

[2] Although some of the terminology is related to chemical plant, the essential steps in the process are also applicable to other processes. In general, however, chemicals and petrochemicals have received most of the attention to date in this area. There are parallels with process integration (discussed earlier in this chapter), where it took some time before the technique developed for chemical plant was used outside the sector.

12.7.1.2 The approach

(Note: The steps below relate to chemical processes involving reactors, but the principle is the same.)

Step 1: Analyse the business drivers

Determine why it is desirable to change the plant. The term 'business drivers' relates to the reasons why PI might be considered, and these are usually economic in nature as these reasons are normally of an economic nature. While safety, health and environmental concerns are increasingly becoming important factors, as discussed elsewhere in this book, it is the business driver as represented by the new opportunity that dictates much of the real interest in applying PI. Another major business driver may be to have a higher and more efficient production rate. These drivers are required to set targets for the plant design to meet. Where increased production rate is not a priority, it might be found that plant flexibility – the ability to handle different products and to change from one to another rapidly – is important. A detailed analysis of the business drivers may, of course, reveal that the original enthusiasm for a full PI plant was misplaced and partial revamping (or making efforts in another direction) may prove more fruitful.

Step 2: Knowledge elicitation (process understanding)

An understanding of the whole process is required which is gained through the knowledge elicitation stage. The approach is split into separate chemistry and plant audits, though there will be interaction between them. (Note that the chemistry audit will also be relevant to sectors such as food and drink, natural gas processing, etc.)

The chemistry audit examines the whole reaction scheme. The potential to use different solvents, catalysts or operating conditions should be considered. Ideal operating conditions and those conditions that promote by-product formation should be determined, such as temperature of operation or residence time. Check if the chemical reaction rate is inhibited in any way. Some knowledge of the kinetics and thermodynamics of the reaction is essential.

The plant audit examines what the existing plant currently does. The audit should include all physical aspects of the reactor, including mixing and heat transfer capabilities, feed rate and position of feed addition. It is necessary to have a fundamental understanding of the reactor to determine where and how the reaction occurs. If the intention is to run a new chemical reaction scheme in existing equipment, as is the case in many fine and speciality chemicals processes, the equipment should be audited as if it were already running the new process. It may be necessary to go back to the original process plant contractors to obtain full data.

Step 3: Examine PI blockers

Blockers are those properties or conditions of a process which may prevent the application of PI. Many are to do with the nature of the chemicals themselves, such as the presence of solids – literal 'blockers'. PI equipment often has narrow channels, which large solids would foul or cut-off completely. Fine solids can be handled (Anon, 1991). There may be some business blockers which relate to practical problems of running PI plants, such as flexibility or continuous operation versus batch production. Batch production is preferred in some sectors of the chemicals industry, such as pharmaceutical manufacture, where there is a requirement for batch identification. At this stage the process engineer will be looking for ways of overcoming business blockers, having been told that the main motivation for considering PI is a new business opportunity.

Step 4: Identify rate limiting factors

In the most radical PI plant, one is trying to run the process as fast as possible – in a reactor dealing with a rapid exotheric process one tries to approach the kinetic limit. Rate limiting factors are conditions preventing the overall process running at a faster rate. These may be mechanical limitations, such as low heat transfer area, poor mixing or limited supply of feedstock to the reactor from an upstream operation. Chemical rate limiting factors, for example slow kinetics or mass transfer into a solid reactant, may occur. Rate limiting factors and blockers are considered in parallel as there can be common elements, such as slow reactions which are both a PI blocker and a rate limiting factor. PI should aim to remove, or at least alleviate, rate limiting factors.

   An excellent example of the identification of the rate-limiting step in a hydrogenation reaction in a trickle-bed reactor and how it was identified as mass transfer and overcome using a compact liquid phase reactor is provided by Akzo Nobel (Kooijman et al., 2000).

Step 5: Assess PI viability

The potential for intensifying a process is determined by pulling together the results of the audits, blockers and rate limiting factors into a mid-methodology assessment. This will ensure all the required information has been gathered and properly considered. Even if it is determined that full PI is not possible, it is worth continuing with the methodology as improvements to the conventional plant could be found that partially intensify it, concentrating on some individual unit operations, for example. This may not satisfy the business drivers, however, where a full radical solution may deliver vastly superior outputs.

Step 6: Review the business and process drivers

Business and process drivers are required to set targets to be met by the plant design. The business drivers identified at the start of the methodology, which are the economic reasons why it is desirable to intensify the process, should be reviewed to keep a clear idea of the overall aims of the project. Process drivers are those characteristics of, in this particular example, the chemical reaction scheme that determine the required operating conditions within, and performance of, reactor equipment to allow the process to run at its most efficient rate. A process driver example is the rate of heat release from a reaction determining the heat transfer capability required of the equipment.

Step 7: List initial concepts (continuous activity)

Throughout the methodology, ideas or concepts will become evident on how to intensify the process, which will tend to be equipment-driven concepts for applying familiar equipment. These ideas should be documented for discussion in the proper manner at the appropriate methodological stage. Accepting an initial concept early on could introduce bias into the rest of the methodology, preventing further, possibly superior, plant concepts being suggested.

Step 8: Generate design concepts

A creative problem-solving session should be held, in which plant concepts are suggested for meeting the process and business drivers. Include the initial concepts in this session. A database of available PI equipment and capabilities would be useful here, so that no possibilities are overlooked, but concepts should not be restricted to familiar plant items. The success of the concepts generation stage depends on thinking laterally to come up with possibly novel solutions to a problem. If this is run in the form of a brainstorming session, possibly extending over a day or more at an appropriate venue, significant progress can be anticipated. If confidentiality concerns permit, outside experts could be introduced to participate in such an exercise. One of the authors of this book was involved in such a session extending over a weekend at a major cement manufacturing site. The outcome comprised a number of new cement process routes involving fluid bed technology.

Step 9: Select best concept

All the concepts suggested must be analysed to study how each of them matches the business and process drivers. There may be factors which limit or rule out the use of a particular piece of equipment, such as it not being available in the required material of construction for corrosion resistance purposes. The best concept must now be chosen. Some economic analysis may be required if there is more than one feasible choice.

Step 10: Implement laboratory scale PI protocols

It will be necessary to prove that the selected concept will work with actual process chemicals. BHR Group has expended considerable effort in designing PI laboratory protocols to demonstrate the performance of continuous, intensified operation without the need for a pilot plant. This will allow the quantification of any potential benefits of intensification, such as improvement in product quality, shorter reactor residence time and lower reacting inventory. The background to this approach is described by Reynolds (2002).

Step 11: Compare intensified concept with conventional (existing) plant

List the strong and weak points of the existing and conceptual plant. Showing that the conventional plant is not fully suitable for a process, due to mechanical rate limiting features, could be just as important as showing the benefits achievable by PI when trying to justify its use.

Step 12: Make the final choice of plant

The person or team responsible for making the ultimate choice of equipment should have an open mind to the use of PI. This final decision process involves factors currently outside the scope of this methodology, such as the risk of using novel equipment, legislation and time span to commissioning of the plant. A high risk factor and long lead time to commissioning may rule out the use of PI, even if significant financial and operability benefits have been shown to exist. It may be concluded that the risk associated with the use of novel equipment, perhaps yet to be tested, justifies R&D effort.

Examples of the use of the methodology are given in the appendix to this chapter – these were based upon exercises given within the Process Intensification MEng and MSc modules at Heriot-Watt University, carried out principally, but not exclusively, by chemical engineering students.

In some cases, the early order of the methodology steps may differ from the above order, but the outcome is consistent.

12.8 The ultimate goal – whole plant intensification

It has been pointed out in this chapter that parts of a plant can be intensified, for example for overcoming a bottleneck, or one can build a new, fully intensified, plant. The example used to illustrate the Britest and BHRG methodology above takes a reactor unit – a single operation – as an example.

Ideally, in order to realise the full potential of intensification, all the plant operations should be involved rather than just the reactor, heat exchanger or separator. At ICI in around 1980, S.F. Kelham developed a multifunctional element for demisting, drying and condensing wet chlorine gas, leaving a high pressure chlor alkali cell. The device consisted of 1 cm layers of polymer slabs containing appropriate channels for the individual process duties, the whole unit being assembled like the plates of a plate/frame filter. This general concept is now being extended in a number of research centres as microprocessor technology, with the channel dimensions being reduced to around 100 μm or less, with many more process functions potentially available in one multi-plate unit. The obvious advantage of this approach is that expensive pipework is eliminated so that high processing complexity may be cheaply available. Thus the equipment installation factor, which routinely absorbs most of the system costs, can be dramatically reduced – as illustrated in Chapter 2.

The principle of multifunctionality can be extended with considerable benefit to those operations performed within a rotor. For example, the HiGee contactor described in Chapter 6 currently only comprises a mass transfer element. However, a full distillation operation requires both a condenser and a reboiler, each of which is a multiphase duty capable of benefiting from a high acceleration environment. Therefore, it is logical to include a peripheral reboiler and an inner condenser to enhance the original concept, along the lines of some of the devices described in Chapter 1. This strategy for plant design has already been followed in the Rotex absorption heat pump which was described in Chapter 11.

It is generally acknowledged that the key element in a process is the reactor, since its performance has a dominating influence on the downstream separation/purification equipment and the overall profitability of the operation. The reactor also dictates the greenness of the process in terms of the environmental acceptability of the feedstock, and the selectivity in converting this to the desired product, with the minimum of undesired by-product. As already pointed out, we can learn significant lessons from the way nature constructs the innards of a typical mammalian body, with its reactor (stomach) and downstream separator (intestines) suffused and modulated by a heat transfer fluid (blood) and pump (heart). Even thermoregulation can be aided by micro-heat pipes. If we are to eliminate most or all of the system costs associated with piping, then a high degree of intensification and multi-functionality needs to be incorporated into the process components. Nature has done a very good job in this respect!

A further driver for multi-functionality is provided by the need for process integration, whereby the minimum high-grade heat is supplied to the process and cascaded rationally down through the temperature spectrum before rejection to the environment. For cryogenic gas separation, of course, low-grade heat is allowed to rise to the environmental rejection temperature. In the latter context, the Linde double distillation column is a good example whereby the condenser of one column operating at high pressure acts as the reboiler for a second at lower pressure. This is elegant both thermodynamically (because the ΔT's associated with one condenser coolant and one reboiler heating medium are eliminated) and because pipework/pumps are avoided.

A good example where process intensification has the potential to transform a complete plant operation centres on the manufacture of sulphuric acid. This can be seen from a study of the SO_3/H_2SO_4 contact process. In order to appreciate the potential impact this approach could have on a well-established process, it is worth discussing the flow sheet for the manufacture of sulphuric acid/oleum. A sulphur burner produces SO_2 which is then reacted over vanadium pentoxide catalyst at 1 bar to produce gaseous SO_3. This is then absorbed in recycling sulphuric acid to give product oleum. As is often the case, the reactor is the heart of the process. Here, its pressure capability is limited by its brick construction, and its ability to withstand higher temperatures is restricted by the low melting point of the V_2O_5 catalyst. If an alternative reactor, for example a catalytic plate unit, could be devised to overcome these limitations then the use of oxygen at elevated pressure would be well worth consideration. The attractive features of this approach are:

1. Large reduction in reactor size due to lower volume flows and improved reactivity.
2. Improved heat transfer should raise the temperature at which reaction heat can be rejected to raise superheated steam/electric power.
3. Inter-reactor coolers are eliminated.
4. The higher operating pressure allows the absorption units to be reduced in size and will also permit the SO_2 off-gases to be reduced.

The postulated alternative reactor could well be based on a catalyst comprising platinum on an alumina support. This was rejected when the contact process was originally developed because the SO_2 gas streams of the day contained too many catalyst poisons. This resulted in the more robust, though less active, V_2O_5 catalyst being adopted. This example is a good illustration of the broad impact that process intensification can have on a complete process, provided that the entire process is reviewed creatively to exploit the strengths of the new technology. We must not be hidebound by the existing temperature or pressure trajectories within a conventional process.

Another case which illustrates this point involves the manufacture of a polymer on a spinning disc reactor (SDR). In this case, the thin polymer film is subject to intense heat/mass transfer and mixing as it flows to the disc periphery in about 1–5 seconds. Although typically a polyesterification will be performed at about 280°C and occupy several hours in conventional stirred vessels, this temperature has been chosen bearing in mind the balance between the desired and degradation reactions *during the reactor residence time*. If that time is significantly reduced, then all the process variables must be re-optimised. In the case of the SDR polymeriser, this allows the consideration of higher reaction temperatures which will accelerate the kinetics and reduce the viscosity, possibly making a good reactor even better.

12.9 Learning from experience

It is interesting to carry out a post–mortem on projects, be they successful or unsuccessful. The risk associated with process intensification and other technologies that may be new to those wanting to develop them or incorporate them in their plants can be rather high and much can be gleaned from following the stories associated with previous successes and failures. This can be particularly true when analyses are carried out of government-funded programmes, where strategic decisions to support particular areas of technology (including PI) can be identified. The experiences may be rather late in being reported, but it is generally a case of better late than never.

A case in point was the analysis of investment in and subsequent commercialisation of energy-efficient technologies (including compact heat exchangers) in Japan in the 1970s and 1980s. The programmes were managed by the New Energy and Industrial Technology Development Organisation (NEDO). Kimura (2010) compiled a most interesting analysis of the reasons for success or failure of the technologies during the R&D stage or later in commercialisation, where the market either developed as predicted or in some cases did not materialise. For those setting out on long-term development programmes with the ultimate aim of commercialisation, this paper is recommended reading. It also stresses the important role government can play in the many stages up to product launch and beyond.

Of the technologies bearing some relationship to PI, Sumitomo's development of a stainless steel plate fin heat exchanger (also used with catalysts on one side) was rated successful. It was adopted for fuel cells and micro-gas turbines. Less successful was an electrohydrodynamically enhanced heat exchanger. While EHD was used successfully to improve performance, the development came when the ozone layer problem was beginning to be tackled. When CFCs were banned, the heat exchanger proved less attractive, because the CFC was an ideal EHD fluid. There was no immediate replacement. Thus, regulatory measures, after the event, can lead to downfall!

Over-stating target efficiencies is nothing new. The failure of a thermoelectric generating system using low temperature waste heat was due to disappointing outcomes compared to performance predictions earlier in the R&D.

12.10 Summary

By explanation and example, this chapter illustrates that it is essential to prepare all stages of the decision-making process in assessing applications of PI (as with other processes) with great care. There is a good phrase which originated from Dr Alan Deakin when he was a senior heat transfer engineer at BP:

Innovation needs cogent argument to back it up.

Initially directed at engineers wishing to incorporate compact heat exchangers in their processes, the phrase is even more pertinent to those attempting to promote process intensification within their organisations.

References

Anon, 1991. Testing of printed circuit heat exchangers. Future practice R&D profile 12, energy efficiency best practice programme. Available from The Carbon Trust (UK). <www.thecarbontrust.co.uk>.

Anon, 1993. Investigation into the fouling and scaling of a printed circuit heat exchanger. Future practice R&D profile 39, energy efficiency best practice programme. Available from The Carbon Trust (UK). <www.thecarbontrust.co.uk>.

Anon, 1998. Experience in the operation of compact heat exchangers. Good practice guide 198, energy efficiency best practice programme. Available from the Carbon Trust. <www.thecarbontrust.co.uk>.

Anon, 2000a. Procesintensificatie: een uitdaging voor de procesindustrie, Novem, Rotterdam, 10 October, (principally in English).

Anon, 2000b. Minutes of the Third PIN meeting, UMIST, 27 April. <www.pinetwork.org>.

Bakker, R.A., 2004. Process intensification in industrial practice: methodology and application. In: Stankiewicz, A., Moulijn., J.A. (Eds.), Re-Engineering the Chemical Processing Plant – Process Intensification Marcel Dekker, New York. Chapter 12.

Beckers, J., 2012. Process Intensification: Greener Processes. Proc. Dutch PIN Meeting, 19 April 2012. See <www/.senternovem.nl/knowledge_networks/pin>.

Brandner, J.J., 2011. Microstructure devices for process intensification: Influence of manufacturing tolerances and measurement uncertainties. Proc. Third Micro and Nano Flows Conference, Thessaloniki, Greece, 22–24 August 2011.

Capel, R., de Waard, W., van den Berg, H., 2001. Process intensification: a result of systematic design. Npt procestechnologie No. 2, 26–27. March–April.

Double, J.M., 2011. Process intensification: Choosing appropriate targets. Proc. European Process Intensification Conference, EPIC 2011, Manchester. IChemE Symposium Series No. 157. pp. 1–7.

Ebrahimi, F., Virkki-Hatakka, T., Turunen, I., 2012. Safety analysis of intensified processes. Chem. Eng. Process. 52, 28–33.

Efthimeros, G.A., Tsahalis, D.T., 2000. Intensified energy-saving technologies developed in EU-funded research – a review. Appl. Therm. Eng. vol. 20, 1607–1613.

Eigenberger, G., Kolios, G., Nieken, U., 2007. Thermal pattern formation and process intensification in chemical reaction engineering. *Chem. Eng.* Sci vol. 62, 4825–4841.

Etchells, J., 2001. Prevention and control of exothermic runaway: an HSE update. Loss. Prev. Bull. vol. 57 (February), 4–8.

Ghiasy, D., Boodhoo, K.V.K., Tham, M.T., 2012. Control of intensified equipment: a simulation study for pH control in a spinning disc reactor. Chem. Eng. Process. 55, 1–7.

Hobin, P.J., 2011. Issues that need to be addressed when trying to capture process intensification credits. Proc. European Process Intensification Conference, EPIC 2011, Manchester. IChemE Symposium Series No. 157. pp. 23–27.

Kalitventzeff, B., Marechal, F., 2000. Optimal insertion of energy saving technologies in industrial processes: a web-based tool helps in developments and co-ordination of a European R&D project. Appl. Therm. Eng. vol. 20, 1347–1364.

Kimura, O., 2010. Public R&D and commercialisation of energy-efficient technology: a case study of Japanese projects. Energy Policy 38, 7358–7369.

Kolehmainen, E., Rong, B.-G. and Turunen, I. (2007). Methodology for process intensification applied to the scale-up of microreactors. European Congress of Chemical Engineering (ECCE-6), Copenhagen, 16–20 September.

Kooijman, C. and Vos, H.J., 2000. Liquid phase hydrogenation: An energy saving alternative. In: Procesintensificatie een uitdaging voor de procesindustrie. Novem, Rotterdam, 10 October, (in English).

Kothare, M.V., 2006. Dynamics and control of integrated microchemical systems with application to micro-scale fuel processing. Comput. Chem. Eng. vol. 30, 1725–1734.

Mayer, M., Bucko, J., Benzinger, W., Dittmeyer, R., Augustin, W., Scholl, S., 2012. The impact of crystallisation fouling on a microscale heat exchanger. Exp. Therm. Fluid. Sci. 40, 126–131.

Moulijn, J.A., Stankiewicz, A., Grievink, J., Gorak, A., 2008. Process intensification and process systems engineering: a friendly symbiosis. Comput. Chem. Eng. vol. 32, 3–11.

Myers, D.B., Ariff, G.D., James, B.D., Lettow, J.S., Thomas, C.E., Kuhn, R.C., 2002. Cost and Performance Comparisons of Stations Hydrogen Fuelling Appliances. Directed Technologies Inc – prepared for the Hydrogen Program Office.

Nikacevic, N.M., Huesman, A.E.M., Van den Hof, P.M.J., Stankiewicz, A.J., 2012. Opportunities and challenges for process control in process intensification. Chem. Eng. Process. 52, 1–15.

Pan, M., Bulatov, I., Smith, R., Kim, J-K., 2012. Optimisation for the retrofit of large-scale heat exchanger networks with comprising different intensified heat transfer techniques. Appl. Therm. Eng. doi: 10.1016/j.applthermaleng.2012.04.038.

Reynolds, I. (2002). Laboratory protocol of process intensification. Proceedings of the Eighth PIN Meeting, Cranfield University, 14 November. Powerpoint talk available on <www.pinetwork.org>.

Shu, J., Lakshmanan, V., Dodson, C., 2000. Hydrodynamic study of a toroidal fluidized bed reactor. Chemical Eng and Proc. 39, 499–506.

Turunen, I., Haario, H. and Piironen, M. (1999). Control system of an intensified gas–liquid reactor. Proceedings of the Third International Conference on Process Intensification, BHR Group.

Tham, M. (1999). Presentation on control and Process Intensification. Proceedings of the Second PIN Meeting, DTI Conference Centre, London, 16 November, see <www.pinetwork.org>.

Van den Berg, H., 2001. Process systems engineering, an effective tool for process intensification? Npt procestechnologie vol. 2, 23–25. (March–April).

Web 1., 2007. Website of the project: boiling and condensation in microchannels. <www.microchannels.org/>.

Appendix: Applications of the PI Methodology
Case Studies 1–4

The four case studies below are for processes that students at Heriot-Watt University selected and to which they applied the PI methodology given in the main part of Chapter 2 (section 2.7).

Case Study 1: The Toray Process
Introduction

This case study deals with the production of caprolactam from cyclohexane. Caprolactam is a monomer used to manufacture nylon-6. The most established method used to manufacture caprolactam (which for our purpose will be referred to as the Raschig route) begins by oxidation of cyclohexane to form cyclohexanone. Responsible for the Flixborough disaster, this reaction involves high recycle flows and high fluid inventory; it is inherently unsafe. Conversion of cyclohexanone to caprolactam by the Raschig route involves is a complicated operation, resulting in by-product ammonium sulphate production of 4.4 kg/kg caprolactam. This volume of by-product (which is normally sold as fertiliser) is so great, caprolactam production is often determined by the price of ammonium sulphate rather than caprolactam itself. In addition, the Raschig process introduces by-products into the caprolactam stream which must be separated later – caprolactam must be produced as a very high purity in order to polymerise correctly, as the polymerisation occurs through step growth. Nylon from impure caprolactam is often unfit for purpose; for example, unable to be spun into fibres. A simpler process will reduce the safety risk, downstream processing costs and economic sensitivity to ammonium sulphate for the process.

Numerous alternatives to the Raschig process exist. One example is the Toray process, which uses actinic light to process cyclohexane, eliminating a step in the reaction. This study applies the PI methodology to the intensification of the Toray process.

A full process description of both the Raschig and Toray process is given in the appendix to this case study. This work is taken from the design project undertaken by this student at Heriot-Watt University. The work in the appendix constitutes the entirety of the research performed on the Toray process prior to undertaking the PI module. (Heterogeneous catalysis of the Beckmann arrangement was the responsibility of another member of the group, for this reason it is not considered in detail in this work.)

Step 1: Analyse business drivers

Business drivers towards the intensification of the Toray process are:

1. *Lower electricity costs*: The main setback of the Toray process is in the electrical requirement of the high pressure mercury vapour immersion lamps used to supply light to the reaction. This makes the process economical only near a cheap source of electricity such as a nuclear power station. A PI plant may be able to decrease the residence time significantly and improve the operating conditions of the lamps; thereby improving their lifespan.
2. *Lower by-product production*: Temperatures above 20°C cause formation of by-products in the Toray process. Light of a low wavelength causes by-products

467

in the form of a tar on the lamp surface. It is well known that continuous stirred-tank reactors (CSTRs) have a high residence time for part of the fluid (i.e. tail of the E curve in residence time distribution) which can lead to product undergoing further reactions before it leaves the reactor, for example.

3. *Lower capital cost of corrosion resistant materials*: In addition to concentrated acids used in the process, NOCl is also highly corrosive, requiring special materials of construction. Decreasing plant size will therefore mitigate capital cost for specialist construction materials.

4. Miniaturisation could allow processing of cyclohexane (manufactured by hydrogenation of benzene in refineries) to caprolactam near to the where it is produced.

5. Full PI or partial revamp? Intensifying the entire plant to one unit operation may be possible in the future; however, at this stage a partial revamp will be considered, in addition to the combination of some unit operations in a single item of plant equipment.

Step 2: Knowledge elicitation
Chemistry audit
Reaction: NH₃ oxidation

$$2NH_3 + 3O_2 \rightarrow NO + NO_2 + 3H_2O$$

1. *Solvents*: This can be performed over catalysts in an aqueous solution.
2. *Catalysts*: Catalyst is a 90% platinum, 10% rhodium metal gauze. This gives a selectivity of 97% under optimised conditions. Cobaltous cobaltic oxide CO_3O_4 is the most effective alternative, giving a selectivity of 95% (Temkin, 1979).
3. *Process conditions*: The process conditions are already heavily intensified, running at 700–850°C and pressures of 1–10 atm. This reaction step is also part of the Ostwald process for nitric acid production, it is likely the optimum conditions for this reaction have previously been pursued a great deal. An alternative route to nitrosylsulphuric acid may be the reaction of sodium nitrite with sulphuric acid. (There are lots of routes to nitrosylchloride production to discuss.)

Reaction: Photonitrosation

Cyclohexane   Nitrosyl chloride   Cyclohexanone Oxime Hydrochloride

1. *Solvents*: The reaction produces a two-phase mixture as cyclohexanone oxime is insoluble in cyclohexane. A solvent is optional but not necessary. Carbon

tetrachloride is one example; however, the ozone depleting effects of this solvent are well known. Toluene may also be used. Solvents require recovery downstream.

2. *Catalysts*: No catalyst is required. Cutting edge developments allow selective hydrogenation of cyclohex*ane* to cyclohex*ene* which is useful in other routes to caprolactam. However, this is outside the scope of this study.

3. *Process conditions*: By-products are reduced if the temperature is maintained below 20°C. NOCl is bubbled through the reactor conventionally; however, NOCl has a vapour pressure of 267 kPa at 20°C (see appendix) increasing the operating pressure could eliminate mass transfer through the bubble.

Reaction: Beckmann rearrangement

1. *Solvents*: Water can be detrimental to the reaction, and oleum serves as a desiccating agent in the conventional process.

2. *Catalysts*: It is possible to perform the reaction in the gas phase, using a fluidised bed of B-ZSM-5 catalyst. This reaction is performed at 300°C. Previous studies have illustrated the upper limit on flow rate imposed by a fluidised bed is a limitation; a rotating packed bed (RPB) may offer one solution to this problem.

Plant audit

Photonitrosation reactor

– Currently mixed by circulation using an external pump.
– Heat transfer through a jacketed vessel is inherently poor due to low surface area.
– Feed position: CSTR operation is likely to produce by-products due to high residence times for a minority of fluid elements. Plug flow could increase selectivity/conversion.
– Possibility that the fluid is effectively opaque to the reacting photons, as in the photopolymerisation of Boodhoo (2002). Photons react with fluid near the lamp and are absorbed, as a result they do not reach the bulk of the fluid in the reactor and reaction occurs mostly near the lamps.

Overall, Corrosive components may include:

- HCl
- H_2SO_4
- NOCl

Nitrogen derivatives are likely to be more corrosive.

Step 3: Analyse PI blockers

Nature of chemicals: Corrosivity could limit the application of moving parts. Below is a table of qualitative corrosivity data taken from Sinott (1999). Aqua Regia is a mix of nitric and hydrochloric acids, which gives off NOCl gas, used here in substitution.

	Al	Brass	Cast Iron	Copper	Mild Steel	Stainless Steel	Molyb-denum	Aust.	Glass
Anhydrous ammonia	R	X	X	R	R	R	R	R	R
Aqueous ammonia	R	X	X	X	X	R	R	R	R
Aqua Regia	X	X	X	X	X	X	X	X	R
Hydrochloric acid	X	X	X	X	X	X	X	X	R
Oleum	R	X	R	X	X	X	X	X	R
Sulphuric acid 70%	X	X	X	X	X	X	X	X	R

Mild Steel: Mild steel BSS 15; OK at T <60°C: Al: Aluminium; X: Corrosive at 100°C: Molyb.: Molybdenum stainless steel 18/8; R: Resistant at 100°C: Aust.: Austernitic ferric stainless steel.

Presence of solids: Cyclohexane will be obtained from a refinery and is likely to contain a small amount of particulate (such as oxidised iron, for example) as a result of flow through pipes and storage. A filtration could be employed immediately before any sensitive equipment as a precaution although this will have associated pressure drop. Tar build-up can be reduced but not eliminated by reducing short-wave emissions and will have to be accounted for in the photonitrosation reactor.

Business blockers

Electricity consumption: The intensified process should reduce but not eliminate the electrical requirement involved in light emitting. Toray originally positioned their plant in close proximity to a nuclear power plant, ensuring supply; a similar approach may be necessary. Large petrochemical and refineries usually use combined heat and power to produce both electricity and utility steam; it may be possible to obtain cheap electricity in this manner.

Alternative catalyst-based solutions: State-of-the-art routes to caprolactam are currently emerging in heterogeneous catalysis, such as an ammoximation process, producing oxime from cyclohexanone. Hazards associated with cyclohexanone production can in turn be avoided through selective hydrogenation of benzene using supported ruthenium catalysts – cyclohexene oxidation is much safer in comparison to (saturated) cyclohexane.

Sales of ammonium sulphate: Sales of this by-product will be reduced, which may be detrimental to sales.

Step 4: Identify rate limiting factors

Inside the photonitrosation reactor, mechanical rate limiting factors are:

- *Poor heat transfer area*: A jacketed vessel is severely limited in this area. In addition, it may be possible to improve heat transfer from the lamp into the lamp coolant, or reduce this load.

- *Poor mixing*: In the conventional reactor, mixing is achieved through circulation using an external pump. As illustrated by Boodhoo (2002) light in photochemical reactions often does not penetrate beyond a few millimetres into the reaction mixture, as light reacts, photons are consumed in the reaction. It is possible that reaction inside the Toray reactor occurs only near to the surface of the lamps.

Chemical rate limiting factors may include:

- *Lamp efficiency*: The high pressure mercury vapour lamps in the Toray process produce a large amount of heat and low wavelength light; this is essentially waste. Low wavelength light can cause tar build-up, this can be mitigated by adding absorbent additives to the lamp cooling water.
- *Water*: NOCl degrades to HCl in water. This makes it unable to form the NO radical necessary in the reaction to cyclohexanone oxime.
- *Selectivity and concentration*: The concentration of NOCl must remain low in order to promote selectivity towards the desired product. This may impose lower limits on residence time, for example, increasing the size of the reactor.

Step 5: Assess PI viability
PI is viable due to:
- Poor transmission of actinic light.
- *Improved temperature control*: Keeping the reaction below 20°C is important to prevent by-product formation. This is difficult to achieve using a jacketed vessel as both the lamp and the (exothermic) reaction produce heat.
- *Decreased materials cost*: Corrosive reactants will require expensive construction, a result of PI is decrease in equipment size.
- Reduced inventory of hazardous chemicals will allow faster start-up and shutdown.
- Improved energy usage through enhanced process integration.

Step 6: Review the business and process drivers

Step 7: Initial concepts
Step 7 should be built upon as the methodology is followed; new ideas will come to mind and should be recorded for consideration.

Replacing the photonitrosation step:
- Spinning disc reactor
- Continuous oscillatory baffle reactor
- Optical fibre
- Glass plate reactor

Replacing Beckmann rearrangement:
- Torbed
- Magnetically stabilised bed reactor
- HEX reactor

Replacing unit operations upstream:

- Combine absorber and HCl absorption/reactor into one column, multiple outlets? Divided wall?
- RPB absorbers
- HiGee columns
- Turbo expansion upstream/downstream of ammonia oxidation

Separation of reactant and product from photoreactor:

- Centrifugation
- Hydrocyclone

Step 8: Generate design concepts
Replacing the photonitrosation step

Spinning disc reactor: Using a spinning disc reactor (SDR) for photoreactions has been explored previously (Boodhoo, 2002). Flow is well mixed with no backmixing on an SDR, it will also increase exposure to the UV lamps. However, the residence time has an upper limit of seconds at best. To combat this, the temperature could be highered, increasing reaction rate according to the Arrhenius equation. By-product formation may be limited in this case by the short residence time. However, laboratory scale experiments would need to be conducted before the applicablity could be assessed. The fast reaction rate will also need a proportionate increase in light flux as the required moles of photons will increase.

Continuous oscillatory baffle reactor: A continuous oscillatory baffle reactor (COBR) benefits from a capacity to deal with the three-phase reaction mixture (gas, NOCl; liquid, cyclohexane; liquid, oxime). It is commonly applied to situations where a long residence time is required along with plug flow, and operation of the system approaches the multiple stirred tanks model for mixing. Temperature control is also very easy with a COBR. Moving parts may pose an issue with the corrosive environment; however, there are a number of ways around this. For example, stationary baffles with oscillatory flow is possible and this dramatically reduces number of moving parts, or a magnetic coupling is used in high pressure COBRs.

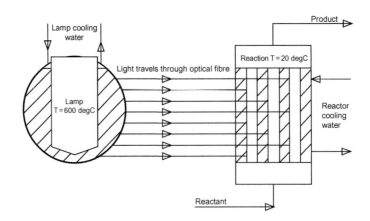

Optical fibre: It may be possible to supply light to the reactor using optical fibre. This would segregate the lamp from the reactor; unwanted transfer of heat from the lamps would be eliminated. This has the following benefits:

- Elimination of safety risk due to over-temperature of the hydrocarbon by lamp heat – lamp coolant is no longer critical to hazard prevention.
- Increased reaction control. It will become possible to engineer the distribution of the lamp light around the reactor through where the ends of the optical fibre are positioned. It may also be possible to control the brightness of individual fibres to control the reaction rate in localised areas within the reactor.
- There is the possibility that the lamps could be substituted for other sources. For example, a laser might be used. These alternatives might prove more attractive in comparison to the lamps; for example, lasers are monochromatic, reducing tar build-up.
- Most importantly with respect to the previous point, the application of solar light may be possible. Using solar light in the Toray process has previously been researched. These authors used a parabolic trough set-up (similar to solar thermal collectors) and reported an increase in conversion. This may be due to the effects of plug flow. Large dish-like solar collectors can be used to collect sunlight – this is used in a commercial setting with optical fibres to supply natural lighting to offices. A dish would be able to follow the sun, making it a more effective collector than a parabolic trough. Optical fibre sources could switch, using sunlight during the day and artificial light during the night or at low light levels. The proposed design would offer improvements such as: 24 h operation; reduced pipework; and reduced safety implications of over-temperature by parabolic mirrors.
- Lamps could be replaced with less downtime as the reactor is segregated physically from the lamps by the fibres – effectively, solid glass.
- The lamps working life may be extended as the lamp conditions can be run at more suitable conditions. It may be possible to utilise the heat from the lamps in other applications as it could be recovered at a higher temperature.

Although a tubular reactor is shown in the diagram, many reactor geometries may be suitable for use with fibre optics, for example a COBR could benefit as its cylindrical shape would normally make supply of uniform distribution of light to the reaction difficult. A pilot plant may be necessary to access the practicality of this idea, and whether sunlight can be collected economically and effectively remains to be seen; $1 \, kW/m^2$ of light falls on the earth from the sun as a rough maximum; the area of the solar collectors may be substantial, particularly if only part of the light is ultraviolet. If this is the case, it might be possible to combine the reactor with concentrated solar thermal power, where steam is raised using the unwanted components of the light.

Glass plate reactor: The main benefit of using a glass plate reactor is that there is proof of it working on the laboratory scale. However, the 'rush to be second' is a major PI blocker. Companies want to keep up with competition, but often will be reluctant to be the first to try a new process. The results of research and development funding can be unpredictable and this often puts companies off investing in the area.

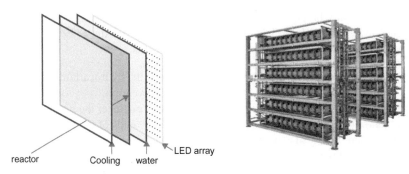

reactor   Cooling   water   LED array

Corning glass plate reactor, incorporating LEDs.

Beckmann rearrangement

Gas phase heterogeneous catalysis: Current technology in Beckmann rearrangement replaces oleum with a solid acid catalyst (usually a zeolite, B-ZSM-5 is most effective in a practical setting). The reaction is performed in the gas phase in a fluidised bed reactor. Torbed reactors are the logical option for intensification of fluidised beds. Inside a torbed, gas enters at an angle, causing the fluidised bed to rotate. Rotation causes higher heat/mass transfer and allows for higher throughput, as a result the size of the reactor is reduced (Shu et al., 2000). A normal fluidised bed has an upper limit on flow due to the effects of entrainment, where particles are blown out of the top of the bed. A magnetically stabilised bed (MSB) reactor uses a magnetic field to trap catalyst particles and hold them in the bed. This allows flow rates above entrainment, as well as flows below the minimum flow rate (where the pressure drop across the bed is not sufficient to support its weight). However, the catalyst must be ferromagnetic; in practice, this limits choice severely (iron and nickel metals). Coating these particles with the desired catalyst has been suggested but is likely to be inefficient on a full-scale plant. The reactor has been tested for downstream purification of caprolactam at laboratory scale (Meng et al., 2003).

HEX reactors: Within the Toray process the Beckmann rearrangement consists of two reactions. In the first, hydrochloric acid is replaced by sulphuric acid bonded to the oxime, this produces hydrogen chloride gas. During the second, the true Beckmann rearrangement reaction occurs and caprolactam is produced. Again, plug flow improves the reaction in comparison to a CSTR by segregating the two reactions. Coupled with the highly exothermic heat of reaction and corrosive nature of the reactants, a glass plate reactor is again put forward by Aubert (2011). However, if some degree of corrosion is acceptable, other reactors may be suitable; for example, PCHE reactors are available in corrosion resistance metals such as titanium. Heterogeneous catalysis, with reaction in the gas phase, may be possible through catalyst coating.

A DeanHex reactor may also be applicable to this process-ceramic (silicon carbide) reactor of similar design to a glass plate reactor, but also offering high heat transfer rates; specially suited to highly exothermic, corrosive reactions (Cybulski et al., 2010).

Replacing unit operations upstream.

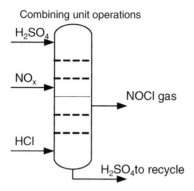

Combining unit operations

Combining unit operations: The design does not differ significantly from the original design; the units have not been integrated, they are segregated by a liquid seal of $HNOSO_4$ on the middle plate. This design is not an intensified process. Divided wall columns are not applicable here.

Use of centrifugal force: An RPB could dramatically reduce the size of both the NO_x absorption and the NOCl production absorber. This will make use of corrosion-resistant materials economical. From the above table we can see most metals are corrosive in 70% H_2SO_4 – this may limit applications in practice. It might be possible to mitigate this using a glass-lined vessel and corrosion-resistant packing. Packing inside an RPB tends to be of a fibrous nature; structured packing is less common and random packing is unsuitable. It may be possible to use either a fibrous ceramic packing or a corrosion-resistant structured packing made of a polymer gauze (Sulzer, 2011). Structured packing has an advantage over others in that it can be engineered to account for differences in flow conditions through the radius of the bed. (The mass velocity of the gas will increase as it travels through the packing to the centre, as the cross-sectional area available for flow is reduced. The opposite will occur with the liquid; mass velocity will fall as it travels out from the centre. Flow conditions are heavily dependent on these mass velocities, and the operation is in turn impacted by mass velocities.) One common technique is to use two concentric cylinders of two types of packing, for example.

Step 10: Implement laboratory-scale PI protocol

As aforementioned, COBR and SDRs could be tested in the laboratory for applications as photonitrosation reactors. These established intensified reactors are comparatively easy to obtain and could prove effective in the application. For example, the oscillations in a COBR processing a liquid/liquid-phase mixture can be 'tuned' to either enhance mixing or separation of the two phases, depending upon frequency of oscillations – towards the outlet of the reactor, it might be possible to use this effect to assist the phase separation afterwards, which may reduce the separator size dramatically.

The chosen reactor may be tested at pilot plant scale. The modular construction aids scale-up as data obtained for one plate is more directly applicable to multiple plates in the reactor.

Step 11: Compare intensified to conventional plant

Replacing lamps with LEDs will give substantial benefits in reliability, eliminating the costs associated with downtime. The 2,400 h quoted below is reduced due to the harsh operating conditions for the lamp. Emission spectra cause spikes in the low wavelength region which cause tar build-up.

Chromacity	Emission Spectra	Monochromatic
Efficiency	35–65 lumens/W	29–60 lumens/W
Power	60 kW	3.5 W
Lifespan	24,000 h reduced under operating condition	>50,000 h redundancy

Step 12: Make final choice of PI plant

The final choice is as follows:

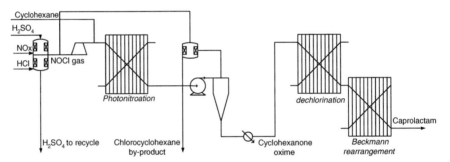

The two absorbers have been combined in one unit and will operate on the same drive shaft. This will simplify the design; however, it remains to be seen whether the two unit operations are entirely compatible. Compression liquefies the NOCl product which is then reacted in the photonitrosation reactor – the heart of the process. A glass hydrocyclone is used for separation of the two phases as it incorporates no moving parts. The lighter phase is distilled in a HiGee column and the excess NOCl is returned to the reactor. Dechlorination and Beckmann rearrangement are carried out in two separate stages; however, the conditions in reality will probably make it impossible to segregate the two reactions entirely.

Overall modest increases in selectivity and yield can be expected. A smaller plant footprint and more efficient reactor are the most substantial gains.

Checklist for PI implementation

Is my plant currently operating at optimum conditions?

No. The plant consumes electrical energy, the by-products produced must be removed downstream.

If not, why not?

The light generated by the lamps is not utilised effectively. High residence times and shortwave emissions produce by-products; iron materials of construction also contribute through catalysis of caprolactam breakdown by iron oxides during distillation downstream.

Is it cost-effective to reach optimum conditions with my current plant/process?

No. Use of lamps and the design of the reactor itself is the root cause of the problem. Surrounding equipment could also be improved.

If my plant/process is optimised, is there still a case for looking at PI?

Yes. The reactor should definitely be intensified. Surrounding equipment may undergo debottlenecking or retrofitting refurbishments also. For example, absorber packing could be replaced.

Do I know all the current operating parameters on my plant/process?

No. Data is missing. Design parameters for absorption such as mass transfer rates prevent a full design of these units. The stoichiometric number of photons required in the photonitrosation reactor places a lower limit on the power requirements of the light source.

Can I obtain these data?

Yes. Some of the data will be available in literature. Even though the photonitrosation process is propriety, manufacture of NOCl in industry is commonly done on-site as the compound decomposes on storage. There should therefore be some information available on this aspect of the process. In practice, testing on stream at the site may be one route to data. As aforementioned, PI is easier to scale-up from laboratory/pilot scale plant, and experiments may be another route to source data.

Will any replacement plant/process have significantly different requirements to the existing plant?

Yes. The product should be of higher purity, reducing downstream processing significantly. Operating pressures will be higher. The photonitrosation reactor itself will require significantly less maintenance due to the replacement of the lamps with LEDs; however, it will also be more difficult to mechanically clean tars from the surface.

Because the light is monochromatic and the temperature is controlled much more rigidly in the reactor, tar build-up will be reduced. However, it is unlikely to be eliminated completely. In the event mechanical cleaning is impossible, chemical cleaning may be assisted by the glass construction. It remains to be seen what chemicals could be used; a strong alkali may be applicable given the tars form under strongly acidic conditions.

Are these requirements realistic for a conventional plant?

The electricity requirement is the main setback of the conventional Toray process. Along with the cost of replacing the purpose-built lamps, failing economics have forced Toray to cease using this route to manufacture.

If they are, why do I want to study PI plant?

In reality it is likely that an intensified Toray process may still be uneconomical. The most sustainable route to caprolactam is through heterogeneous catalysis routes. However, the Toray process does not need a catalyst (which would have to be regenerated, incurring costs) and may be applicable where these catalysts cannot be used, for example, due to licensing disagreements or the scarcity of noble metals. The argument for intensifying the process is covered in detail under the heading Step 1: Analyse business drivers.

If they are not realistic for a conventional plant, how will a PI plant help?

The main benefits are a reduction in electrical energy consumption through more effective use of the light. Also, more corrosion resistant materials will have a longer lifespan, becoming economical due to the minimisation effects of intensification.

Will a radical change in process/plant affect my product?

Yes. The effects will be positive. Reduction in impurities means a better quality of end product is possible. The polymerisation step produces longer chain lengths when it is more pure because there is less chain termination by monofunctional groups within impurities. Discolouration due to impurities will be reduced, making the product more attractive to customers. Discolouration is measured in Hazen units and often forms a product specification for caprolactam as low discolouration is required in some applications.

Is my knowledge of PI sufficient to answer the above?

No. My knowledge of PI is not sufficient to perform the full mechanical design and costings necessary to determine indefinitely whether such a product is practically and commercially sound. Full design may involve more direct work with the manufacturers (such as Corning) as the knowledge base required is often present only within these specialist companies.

If not, are there others in my organisation who could assist me?

Heriot-Watt University has specialist facilities for research into oscillatory baffle reactors. However, in order to research other intensified equipment it will be necessary to go further afield. For example, Newcastle University in the UK may be better equipped to research SDRs. A blend of in-house work and sub-contracting will be necessary.

Should I consider employing outside consultants?

Outside consultancy will be required at different levels through the course of development. In the beginning, consultant specialists may be brought in to brainstorm ideas for design improvements. Towards the end they may be required to ensure the unit meets technical standards.

Do I have sufficient time and resources within my own team?

Yes. It is reasonable to assume a design team will have the necessary skills to organise the management framework necessary to research process intensification of the process. In particular, since the conventional Toray process is uneconomical the option of a textbook design is not available. However, outside contribution may be necessary in particular areas as discussed above.

Do my timescales and potential budget allow for any research and development work?

Yes. Routes for possible research and development have been explored from the outset. It should be possible to budget for these, and the results of experiments should be conclusive to enable a decision for the plant design.

If not, am I limited to equipment that is already available?

No. Intensified equipment such as the Corning glass plate reactor are established and available commercially. However, most PI equipment is assembled on a case-by-case basis in applications such as PCHE heat exchangers and unique design elements will be easy to incorporate with relatively little increased cost.

If this is the case, is such equipment available?

The reactors should be available. RPBs are also available commercially; however, glass-lined vessels may require attention from a specialist company.

Do I then have the information to carry out the following:

1. Risk assessment (health and safety)? Yes. There is enough information to perform a risk assessment for laboratory scale research.
2. Technical feasibility study? No. Research on the laboratory scale is necessary to determine the practicality of the proposed design.
3. Rigorous financial appraisal? Possibly. Costs of the equipment and installation should be easy to come by. However, there may be a potential for development costs to increase if any flaws must be overcome or if a decision is not immediately clear.

Have I considered the availability of grants?

The Smart scheme, offers a maximum of 35% towards eligible costs in the area of research and development. However, this grant applies to SMEs in Scotland, UK (http://www.innovateuk.org/deliveringinnovation/smart.ashx) and may not apply to Toray, who are a Japanese company. An additional requirement for a Smart grant, is that the business must be independent of a university; if research was being carried out on a Heriot-Watt campus, it would not be applicable.

Can I now make a comprehensive case for examining a PI solution to my plant/process?

More work may be required before investment can be sought indefinitely. In particular, more quantitative information will be needed on aspects such as absorption design in order to assess how effective the intensified equipment will be.

Annex to Case Study 1
Vapour pressure of NOCl

Antoine Constants taken from Landolt and Bornstien (2001).

$T = 293.15°C$
$A = 6.48644$
$B = 1094.73$
$C = -23.45$

$$logP = A - \frac{B}{C + T}$$

$$logP = 6.48644 - \frac{1094.73}{-23.45 + 293.15}$$

$$logP = 2.42737$$

$$P = 10^{2.42737} = 267.53\,kPa$$

Raschig Process
Chemistry

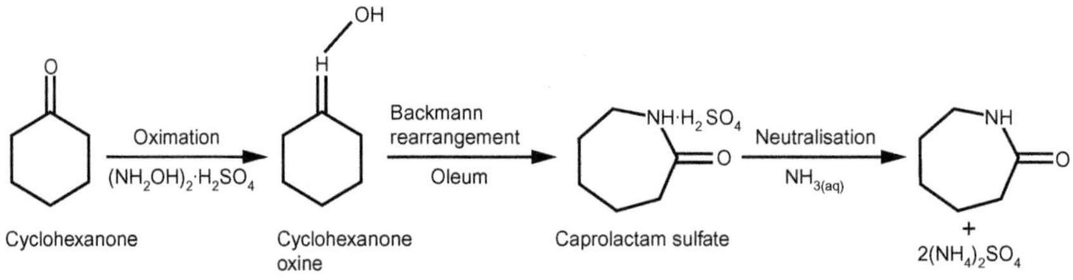

Cyclohexanone — Oximation — $(NH_2OH)_2 \cdot H_2SO_4$ — Cyclohexanone oxine — Backmann rearrangement Oleum — Caprolactam sulfate — Neutralisation $NH_{3(aq)}$ — NH — $2(NH_4)_2SO_4$

The first step of the conventional method to obtain caprolactam from cyclohexanone consists of oximation, by reacting cyclohexanone with hydroxylammonium sulphate. The hydroxylammonium sulphate is neutralised with ammonia according to the following reaction (Ohm and Stien, 1982).

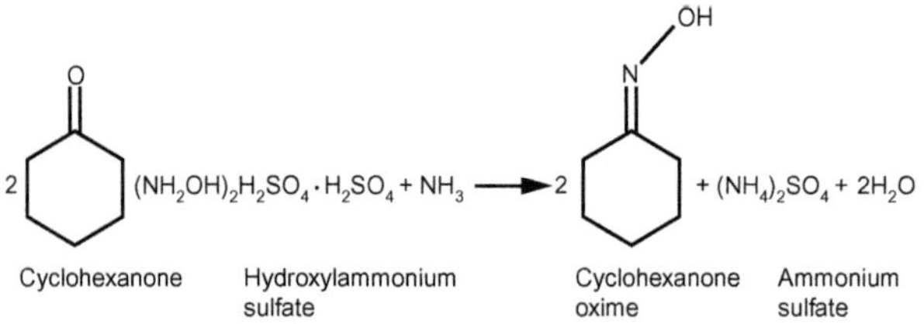

$$2\ \text{Cyclohexanone} + (NH_2OH)_2H_2SO_4 \cdot H_2SO_4 + NH_3 \longrightarrow 2\ \text{Cyclohexanone oxime} + (NH_4)_2SO_4 + 2H_2O$$

Cyclohexanone   Hydroxylammonium sulfate   Cyclohexanone oxime   Ammonium sulfate

480

Neutralisation with NH_3 is necessary to drive the equilibrium forward by maintaining a pH of 4.5 which is necessary for the oximation (Schuit and Geus, 1979).

After the reaction, the oxime and aqueous phases can be separated. The Beckmann rearrangement method requires a strong acid; Oleum (H_2SO_4 $xSO_{3(l)}$ where $x \approx 0.2$) being the most common, also serves as a desiccating agent (Lowenheim and Moran, 1975, Ohm and Stien, 1982).

The resulting caprolactam is freed from the acid salt by neutralisation with aqueous ammonia, yielding crude caprolactam.

Hydroxylamine sulphate is usually manufactured on-site. Conventionally, this is through the Raschig process. NO_x gas (a mixture of NO and NO_2) is manufactured by catalytic air oxidation of ammonia. Ammonium carbonate is made by dissolving CO_2 and ammonia in water. The NO_x gasses are absorbed in ammonium carbonate solution to give ammonium nitrite:

$$NO + NO_2 + (NH_4)_2CO_3 \rightarrow 2NH_4NO_2 + CO_2$$

(NO_x gasses + ammonium carbonate $\rightarrow$ ammonium nitrite)

Sulphur is burned to produce SO_2, which reacts with ammonium nitrite to produce hydroxylamine disulfonate $HON(SO_3NH_4)_2$ (Ohm and Stien, 1982, Chevaul and Lefebvre, 1989):

$$2NH_4NO_2 + (NH_4)_2CO_3 + 4SO_2 + H_2O \rightarrow 2HON(SO_3NH_4)_2 + CO_2$$

Which is then hydrolysed to produce hydroxylammonium sulfate: (Ohm and Stien, 1982)

$$2HON(SO_3NH_4)_2 + 4H_2O \rightarrow (NH_2OH)_2 \cdot H_2SO_4 + 2(NH_4)_2SO_4 + H_2SO_4$$

The BASF process (also known as the Badishe process) and the Inventa process are two very similar alternatives to the Raschig process. In the first step, concentrated nitric oxide is obtained by catalytic oxidation of ammonia and oxygen; Steam is used as a diluting agent to avoid working in the explosive range (Ohm and Stien, 1982).

$$4NH_3 + 5O_2 \rightarrow 4NO + 6H_2O$$

The steam is then condensed; nitric oxide can then be reduced to hydroxylammonium sulphate using hydrogen and a catalyst of either palladium or platinum, on a carbon substrate, suspended in dilute sulphuric acid. The reaction is as follows:

$$2NO + 3H_2 + H_2SO_4 \rightarrow (NH_2OH)_2 \cdot H_2SO_4$$

Ammonium hydrogen sulphate solution can be used in place of diluted sulphuric acid (Maxwell, 2004). The result is ammonium hydroxylamonium sulphate $(NH_3OH)(NH_4)SO_4$ which performs the same oximation reaction in a specially designed column, resulting in ammonium hydrogen sulphate $(NH_4)HSO_4$ as a by-product.

Process description

Conventionally, a Raschig process is used to manufacture the hydroxylammonium sulphate as described above. Ammonia oxidation is performed isothermally over a platinum foam catalyst at 700 to 850°C (Chevaul and Lefebvre, 1989). Ammonium carbonate solution is contacted with NO_x gas in an absorption column producing ammonium nitrite, which is then contacted with SO_2 in a second column to produce hydroxylammonium sulphate.

Oximation occurs at around 85°C (Maxwell, 2004). Exothermic effects of the reaction are largely due to neutralisation by ammonia. An organic solvent can be introduced into the reactor to extract oxime from the aqueous phase (Leudeke, 1978). Alternatively, molten oxime can be removed as a separate organic phase (Ohm and Stien, 1982); despite having an OH group, the oxime is insoluble in water (<1 g/l at 20°C).

The Beckmann rearrangement is highly exothermic. Oxime concentration is kept low to avoid violent reaction, a temperature of 100 to120°C (Lowenheim and Moran, 1975) can be maintained by an external heat exchanger set-up (shown).

After neutralisation the crude caprolactam usually undergoes some form of solvent extraction. Residual caprolactam in the aqueous phase is also extracted before the ammonium sulphate solution is crystallised to be sold as fertiliser. High levels of caprolactam purity are necessary, vacuum distillation at 2-20mbar avoids temperature decomposition of the caprolactam (Sulzer, 2010). Several crystallisation stages can also be used to purify the caprolactam.

The main disadvantage of the conventional process lies in excessive production of ammonium sulphate. Per kg of caprolcatam manufactured, 1.6kg $(NH_4)_2SO_4$ occurs in Hydroxylamine manufacture by the Raschig process, 1.1kg during oximation, and 1.7 during the Beckmann rearrangement implies a total of 4.4kg/kg caprolactam. Impurities can dramatically impact the quality of nylon produced using caprolactam, for example making it impossible to spin into fibres – for this reason, purification is performed to exceptionally low tolerances. The conventional process produces a lot of by-products that are costly to remove. Handling of concentrated mineral acid required for the Beckmann rearrangement increases risks to health & safety and the environment, and requires corrosion resistant materials of construction.

The figure below shoes the authors' interpretation of the Raschig process.

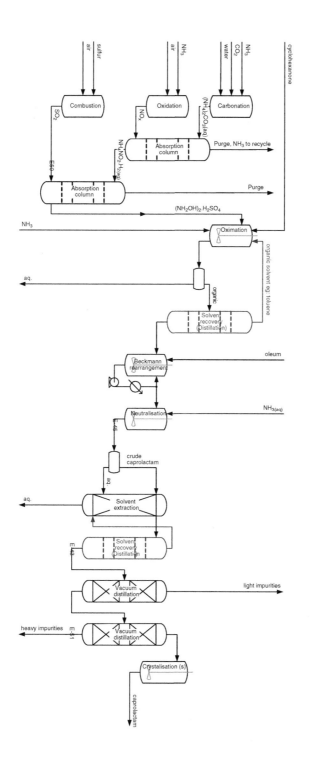

Photonitrosation of cyclohexane (Toray process)
Chemistry

Photonitrisation of cyclohexane (Maxwell, 2004).

Photonitrosation of cyclohexane (PNC) produces cyclohexanone oxime from cyclohexane in one reaction, bypassing the cyclohexanone precursor. The cyclohexanone oxime hydrochloride produced then undergoes a Beckmann rearrangement (not shown) in the conventional manner (oleum or H_2SO_4, followed by neutralisation with NH_3). HCl is added to the reaction to increase selectivity by inhibiting side reactions. HCl in the oxime hydrochloride is replaced by H_2SO_4 during the Beckmann rearrangement (Maxwell, 2004).

Nitrosyl chloride is normally manufactured on-site, in a three-step process. Ammonia is oxidised in air to yield nitrogen oxides:

$$2NH_3 + 3O_2 \rightarrow NO + NO_2^* + 3H_2O$$

Nitrogen oxides are then contacted with sulphuric acid in an absorption column, reacting to from nitrosylsulphuric acid:

$$2H_2SO_4 + NO + NO_2 \rightarrow 2HNOSO_4 + H_2O$$

Gaseous HCl is then added to the nitrosylsulphuric acid to produce nitrosylchloride:

$$HNOSO_4 + HCl \rightarrow NOCl + H_2SO_4$$

Process description

NOCl manufacture follows the chemistry – the three reactions taking place in three reactors. Ammonia oxidation can be performed by combustion in air or catalytically in aqueous solution. This is contacted with H_2SO_4 in an absorber, as aforementioned, to yield $HNOSO_4$. In the third reactor, NOCl leaves as a gas, along with unreacted HCl, necessary in the photonitrosation. H_2SO_4 leaves as liquid which can be purified and recycled.

Photonitrosation is performed in a specially designed reactor; high-pressure mercury immersion lamps satisfy the need for actinic (chemically active) light (wavelength 500 nm). The reaction is carried out at 20°C, however, the lamps operate at temperatures in excess of 600°C (Funken and Becker, 2001) so they are immersed in glass cooling jackets. Shortwave (<365 nm) emissions cause tar build-up on the lamps; to address this problem the lamps can be fitted with filters or absorbing additives can be added to the coolant. NOCl is bubbled through the liquid cyclohexane phase; any unreacted gas is directly recycled and bubbled through again. The resulting cyclohexanone oxime hydrochloride is insoluble in the cyclohexane phase and can be separated out. Chlorocyclohexane produced in a side reaction is removed at this stage by distillation, and again unreacted NOCl is recycled.

The two phases of the Beckmann rearrangement are performed in two separate stirred tank reactors. Crude caprolactam can be purified by, for example, extractive distillation to obtain caprolactam of a purity suitable for polymerisation. As a result of using conventional Beckmann rearrangement the process still produces some ammonium sulphate $(NH_4)_2SO_4$ at approximately 1.7 tonnes per tonne of caprolactam. However, Toray look to tackle this problem by performing the Beckmann rearrangement with a gas phase fluidised bed reactor and solid acid catalyst.

Despite operating through a free radical reaction mechanism, a selectivity of 86% is achieved in the photoreactor. Overall yield of caprolactam is 81% (stiocheometric). The process uses significantly less material in comparison to other caprolactam routes, however, electricity demand is significantly higher due to the mercury lamps. These make the process economical near a cheap source of electricity (Maxwell, 2004, Seongbuk-Gu 2010, Wittcoff et al., 2004).

Alternatives to NOCl

The main disadvantage of NOCl is its high corrosivity. Tert-Butyl nitrite has been suggested as an alternative NO source. Optimum reaction conditions are very similar: low concentration of tert-Butyl nitrite; temperatures between ambient and 0°C and light of similar wavelength. The reaction produces mostly an azodioxy dimer (azocyclohexane N,N'-dioxide) rather than the oxime. However, this derivative can easily be converted catalytically to the desired product (Gilbert, 1971).

Azocyclohexane N,N'-dioxide.

Advantages and disadvantages

The photonitrosation process was not pursued further as the route to caprolactam, due to the following associated disadvantages:

- *Electrical requirement of lamps*: High power demands make the process economical only near a cheap source of electricity such as a nuclear power station.

Each lamp requires 60 kW, each 10,000 t/a reactor requires 50 lamps – the 160,000 t/a Toray plant lamps required 48 GW of electricity (t/a = tonnes of caprolactam per year).

- *Cost of frequent replacement of lamps*: Cost due to being custom-built for the specific purpose, frequency of replacement due to operating conditions. For example, linear extrapolation of the lamp replacement costs gives $65.4 mil/a for a 160,000 t/a plant.
- *Corrosivity of NOCl*: Requires corrosion resistant materials of construction, implying increased capital costs.
- *Failure of coolant is a potential safety risk*: Reactor contains flammable hydrocarbon, potential for over-temperature from lamps on failure of coolant.
- *Sunlight requirements of solar alternative*: Solar alternative unlikely in China due to overcast climate.

As a proposed alternative to immersion lamps, parabolic mirrors are used to focus light through glass tubes lying along their focal lines. Reactant flowing through the tubes is exposed to the concentrated light necessary for the reaction. In the proposed design, excess cyclohexanone oxime hydrochloride is produced and stored so that the rest of the plant can operate continuously when there is no daylight. Experiments performed by DLR (German aerospace centre) on laboratory plant scale showed a selectivity of 92.4% compared to 86% for mercury lamps. They have also shown reactor effluent can be stored for several months with no change in composition.

Funken et al. (1999) investigated a computer simulation in Aspenplus of two pilot plants, differing only in the reactor configuration (immersion lamp vs. solar). Solar operations required 4 times less electrical energy, and 8 times less cooling energy than its counterpart.

Although offering 85% higher capital cost initially, a solar photoreactor offers substantial savings in operating and maintenance costs, as well as their consumption of electrical power, the purpose-built lamps frequently need to be replaced. (Funken et al., 1999, Funken and Becker, 2001)

The authors' interpretation of the PNC process is given below.

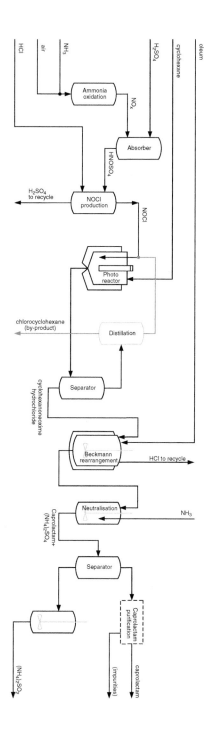

References

Aubert, T., 2011. Method for preparing lactames, comprising a photonitrosation step, followed by Beckmann transition step. U S. Pat. Appl., 0137027.

Boodhoo, K., 2002. Photopolymerisation in spinning disc reactor. In: Reay, D.A. (Ed.), Seventh PIN Meeting. 23rd May 2002. Process Intensification Network.

Chevaul, A., Lefebvre, G., 1989. Petrochemical Processes: Magor Oxygenated, Chlorinated and Nitrated Derivatives. Editions Technip, France.

Coker, A.K., 2010. Ludwig's Applied Process Design for Chemical and Pertochemical Plants. Elsevier, Oxford.

Cybulski, A., Moulijn, J.A., Stankiewicz, A., 2010. Novel Concepts in Catalysis and Chemical Reactors. Wiley-VCH, WeinHiem.

Funken, K., Becker, M., 2001. Solar chemical engineering and solar materials research into the 21st century. Renew. Energ., 469–474.

Funken, K., Muller, F., Ortner, J., Riffelmann, K., Sattler, C., 1999. Solar collectors versus lamps - a comparison of the energy demand of industrial photochemical processes as exemplified by the production of epsilon-caprolactam. Energy, 681–687.

Gilbert, A., 1971. Photochemistry. RSC.

Landolt, H., Bornstien, R., 2001. Vapour Pressure and Antoine Constants of Nitrogen Containing Organic Compounds. Springer.

Leudeke, V., 1978. Caprolactam. In: Mcketta, J.J. (Ed.), Encyclopedia of Chemical Processing and Design Marcell Dekker Inc.

Lowenheim, F.A., Moran, M.K., 1975. Faith, Keyes and Clark's Industrial Chemicals. John Wiley & sons Inc.

Maxwell, G.R., 2004. Synthetic Nitrogen Products. Kluver Acedemic.

Meng, X., Mu, X., Zong, B., Min, E., Zhu, Z., Fu, S., et al., 2003. Purification of caprolactam in magnetically stabilized bed reactor. Catal. Today 79, 21–27.

Ohm, R., Stien, C., 1982.. In: Grayson, M. (Ed.), Encyclopedia of Chemical Technology, third ed. John Wiley & sons

Schuit, G. & Geus, I., 1979. The catalytic reduction of nitrate and nitric oxide to hydroxylamine: kinetics and mechanism.

Seongbuk-Gu, S., 2010. Chemical Engineering Research Information Centre [Online]. Available.

Sinott, R., 1999. Chemical Engineering Design. Butterworth-Heinemann, Oxford.

Sulzer, 2010. Separation Technology for the Chemical Process Industry. In: Chemtech, S. (Ed.).

Sulzer, 2011. Sulzer Chemtech website [Online]. Denmark. Available: <http://www.sulzerchemtech.com/DesktopDefault.aspx/tabid-98/> (accessed 20.10.2011).

Temkin, M., 1979. Ammonia oxidation. In: Eley, D. (Ed.), Advances in Catalysis Academic Press Inc., New York.

Wittcoff, H.A.R., Plotkin, B.G., Jeffery, S., 2004. Industrial Organic Chemicals. John Wiley & Sons Inc.

Case Study 2: Monoethanolamine (MEA) Production

Overview

Ethanolamines are used in a wide variety of places such as in the manufacture of detergents, gas sweetening, lubricants, cutting oils and cyanide-free electroplating. There is more than one method to produce MEA due to its simple aliphatic chemical composition, methods such as reduction of glycine using sodium borohydride/lithium aluminium hydride in acidic conditions or using molybdenum trioxide or nickel chloride with water to carry out the reduction of glycine are all possible routes. The production of monoethanolamine can also be achieved efficiently by reacting liquid or gaseous ammonia with liquid ethylene oxide in an acid catalysed system. The benefit of the use of ethylene oxide is clear as it is highly reactive and will react until it is all converted to a less stressed molecular form which is why it is used in many industrial systems.

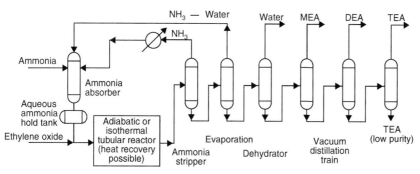

Typical flow scheme of mono-, di- and triethanolamine using ethylene oxide and ammonia injection.

The industry-favoured reaction between ethylene oxide and 20–30% aqueous ammonia is an exothermic process carried out in a tubular reactor at 30–160 bar(a) and 60–150°C. Selectivity towards one of the three alkanolamines is controlled through the ratio of ammonia to ethylene oxide with ratios of 1:1 giving a distribution of 12% (mono-), 23% (di-) and 65% (tri-); a ratio of 10:1 will give a distribution of 75% (mono-), 21% (di-) and 4% (tri-); other higher ratios up to 40:1 have been noted but no data on effects could be located. It is anticipated that further increases in exposure to ammonia, hence increased segregation of ethylene oxide and ethanolamine, would further increase the MEA production rate. The benefits of this process are that there is near full conversion of the ethylene oxide, the most expensive reactant, and the excess unreacted ammonia can be extracted and recycled. To use a higher NH_3:EO of around 20:1 in an intensified unit with enhanced mixing properties, such as a spinning disc reactor (SDR), may allow for ethylene oxide and ammonia to encounter one another in such an excess that the formation of MEA is favoured with DEA and TEA production minimised.

Overall Chemical Reaction

$\Delta H_f = -507.5\,\text{kJ/mol (MEA)}$
$\Delta H_f = -493.8\,\text{kJ/mol (DEA)}$
$\Delta H_f = -664.2\,\text{kJ/mol (TEA)}$

Primary and Secondary Reaction Mechanisms

The aspect of the process that will be focused on in this analysis is the reactor section because intensification at this point in the process will enable for, at the very least, simplification/elimination of downstream units. In this typical commercial process, an adiabatic tubular reactor is used to process the fluid. As you can see from the heats of formation, it is imperative that ethylene oxide is reacted in exclusion from the already formed MEA, as it will seek to form DEA and thus TEA as well.

PESTLE analysis
Political
- There is a political drive for cost-efficient, post-combustion carbon capture.
- Governmental energy departments all have policies aimed at increasing energy efficiency in industries and in particular the heavy bulk manufacturing sector.
- There is ever-increasing emphasis on low-to-no impact processing which intensification aims to address.
- Specifically in the UK at the moment the government are pushing an agenda of technology development and manufacturing and have recently announced the Queen Elizabeth Prize for Engineering of £1m which is open to all with team limits of three people.[3]

Economical
- Size reductions allow for capital cost reductions and thus increase viability.
- Increased efficiency will lead to increased income and reduced effluent processing costs.
- Operational and maintenance costs are reduced through smaller plant.
- There is an Enhanced Capital Allowance scheme operated by the UK Department of Energy and Climate Change which covers new business first year expenses for compact heat exchangers (CHEs).

Social
- Visual impact is a common complaint from local communities and intensification can aid in addressing their concerns.
- The requirements for construction of new process technologies can lead to job increases in the area.

Technological
- The improvements in technology allow for improved efficiency utilising enhanced heat and mass transfer aiding in providing the same conditions for reaction for all molecules.
- Lower inventory systems will have less impact on the local area should there be a major safety failure.

Legal
- The size reductions that increase safety enhance compliance with HSE regulations in terms of the phrases 'as long as reasonably practicable' and 'so far is reasonably practicable'.

Environmental
- Lower energy demand from plant reduces related emissions although MEA has a better than average associated emission given its use in CCS.

---

[3] http://webbook.nist.gov/cgi/cbook.cgi?ID=C102716&Units=SI&Mask=2#Thermo-Condensed, TEA condensed phase thermochemistry heat of formation, accessed 11/11/2011

  - Higher conversions, possible with PI, solutions towards desired products reduce the environmental impact, overall plant process requirements through ancillary services and energy consumption and hence emissions caused are also reduced.

Business drivers

  - Reduction of plant energy demand improves economical and environmental efficiency – using rotating systems reduce energy demand of reactors and separators.
  - Reductions in size of plant – typical plants reach substantial elevations which cause environmental and visual impacts, rotational systems are intrinsically smaller and minimise the visual impact of processing.
  - Decrease in plant size can facilitate the application of on-site production allowing just-in-time outputs – in the case of MEA production it could allow for the design of micro-plants on the post-combustion capture site.
  - Lowering of operational costs – a decrease in plant energy demand, scale of maintenance and operational staff requirements will be possible.
  - Improved safety due to lower inventory and storage of hazardous materials.
  - Reductions in plant size per output clearly have benefits in reducing initial capital costs, although it may slightly increase the overall cost of project development.

Process drivers

  - Higher conversion of ethylene oxide to monoethanolamine will be beneficial to the business case and achievable through better mixing.
  - Increased production rates – allows for reductions in sale price to generate more sales and hence more profit. This also allows for the CCS plants to become more viable.
  - Reductions in production of lower value DEA are preferable.

In terms of business drivers, it is the high cost of these specific amines that have hampered the economics of amine-based post-combustion carbon capture. Should intensification of the production process present significant capital savings, then the cost of amine can be dropped to enhance the viability of this well-defined and highly desired process which could be retrofitted to power stations. The current average cost of MEA in the European market is €1,455/t, DEA is €1,000/t and TEA is €1,435/t which is given in the current Euro/Sterling exchange rates is roughly equivalent to the pound-per-ton price and has been rising through strong demand as well as because of production material cost increases. Although PI can address the majority of process improvement issues, raw material cost is not one that can be affected by the decisions of the engineer to the same extent. Ethylene oxide prices are at the lowest in Europe at €1,301/t and €900/t on the US market, and ammonia prices are at their lowest at €340/t in the EU but are only slightly more expensive in the US at €390/t. This clearly illustrates the margins that producers are working with and moves to promote production of MEA and any reduction in plant capital expense, operational energy demand, safety costs and environmental maintenance requirements will be a welcome improvement to the business case. This also highlights the economic incentive to operate out of the EU region and in the US where tax rates are generally lower as well.

In terms of process drivers there is the desire to increase production of the mono-substituted amine as it is of the highest market value of the three. MEA is going to see rises in demand as carbon capture systems and research pilots become more common and then commercially operational. Mixing is one key issue that can be addressed by PI as the need to encase the ethylene oxide in ammonium hydroxide in order to maximise conversion. PI can also address the issue of the highly exothermic reaction by making use of high efficiency CHEs to maintain the reaction and to pre-heat the feed or provide downstream/external heat as well as the high efficiency in most of the intensified reactor types.

Rate limiting steps
- The primary reaction is highly exothermic and will require highly efficient cooling systems to maintain the optimum reaction temperature.
- The traditional system involves relatively high pressures of up to 160 bar which may be lowered slightly but will need to remain well above ambient to maintain ethylene oxide in the liquid phase.
- If ethylene oxide molecules do not have sufficient access to free ammonia molecules, they will react with higher substituted amines, forming the mostly undesired DEA and TEA.

Generation and analysis of design concepts
- The amine heat exchanger could be intensified, it could be integrated to supply some heat to the feed streams for the reactor or to some of the downstream separations such as the water evaporator.
- The last three columns are traditionally plated distillation columns which could quite easily be switched for the rotating distillation column that would provide the same efficiency, increase safeness and reduce size, environmental impact and capital costs.
- It will also be the case that using CHEs in the reboilers and condensers on separation units to give greater control of heat flows and recovery can increase the economic and thermal efficiency of these operations.

Reactors
Comparison of the three main types of reaction system including ideal conditions, kinetic characteristics and other features, are given in the table below.

Reactor / Comparison item		Batch	Continuous flow	
			Tubular type	Tank type
Ideal conditions	1. Temperature, pressure, composition in reactor	Uniform at each moment	• Concentration changes in the direction of flow • No gradient of reaction rate and temperature in radial direction • No mixing and diffusion in axial direction	• Complete mixing • Uniform composition (equal to that at the outlet)
	2. Residence time distribution of reactant	None	None	Yes
Kinetic characteristics	1. Required reaction volume (Equal conversion basis)	Relatively small	Relatively small	Large
	2. Distribution of products (Consecutive reaction)	Large yield of intermediate product	Large yield of intermediate product	Small yield of intermediate product
	3. Probability of reaction with specific composition ratio	Impossible	Impossible	Possible
Features	1. Flexibility	Large	Small	Medium
	2. Application	Multi-purpose and small-scale production	Mass production	Medium

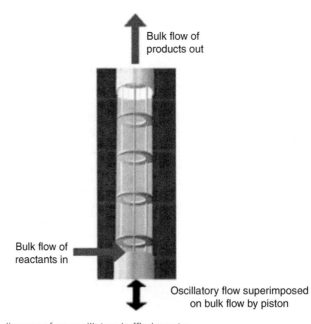

Schematic diagram of an oscillatory baffled reactor.

- The continuous oscillatory baffled reactor could provide for significant size reductions.
- The ability to inject reactants at various points to maintain the excess throughout the tubes, and thus production rates, as well as offering reduced retention times and lower cost.
- The reduced size of the OBR compared to a tubular reactor would be beneficial for just-in-time manufacturing at a CCS site.
- The downside to this is that it is designed to accommodate long reaction times and for this process a fast reaction is desired.

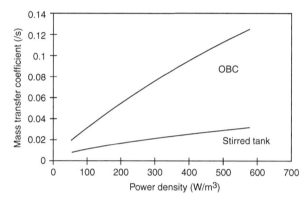

Comparison between the mass transfer rates in a continuous stirred tank and an oscillatory baffled reactor.

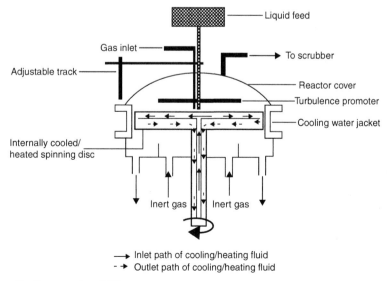

Schematic diagram of an SDR.

- The SDR can handle the pressure range and provide effective mixing and allow for the use of gaseous ammonia if desired, in a high ratio but overall low inventory, and limit the volume of water passing through the process.
  - Given that this is a fast reaction occurring in residence times of 30 s maximum for MEA favoured processes.
  - Using this system, the ammonia solution could be injected into the ethylene oxide on the plate, allowing for optimal mixing to ensure ethylene oxide is exhausted and occurs to allow a reduction in the levels of di- and triethanolamine aided by saturation with ammonia.
  - After this reaction, the fluids can be passed to a rotating column or a membrane/molecular sieve filter to extract any unreacted ammonia or both the ammonia and the water leaving the ethanolamine to be purified. It may be possible to use a second spinning disc to extract ammonia from the liquors but a suitable chemical to carry out the capture which has low-to-no effect on the product, would first need to be identified.
  - The spinning disc can be effectively cooled using the greater surface contact with the thin film mixture which will allow for effective extraction of heat for use elsewhere in the plant.
  - This reaction is highly exothermic and it would follow that HEX reactors would be a suitable option.
  - The corrosive nature of ethylene oxide would more than likely eliminate the implementation of printed circuits as it would eat away at the etching.
  - The HEX reactor could also be integrated with the amine cooler and possibly the feed streams to provide pre-heat for the reactor. This may be a complicated operation to manage but similar systems have carried out in this form of heat transfer.
  - Micro-reactors can minimise hotspots on the vessels and are not generally limited by pressure.
  - Micro-reactors allow for precise control of the reaction which is required in this case due to the exothermic nature of the reaction and the use of ethylene oxide also present significant control requirements on the process.
  - The rotating packed bed (RPB) provides a good option for this reaction to be carried out in a two-phase system.
  - Ammonia gas and liquid ethylene oxide could be contacted over a zeolite packed rotating bed reactor thus providing the protons necessary for the reaction while reducing the inventory through the elimination of the water in the reactor.
  - This process would make ammonia recovery simpler as it is essentially carried out in the vessel and there would be little to no water formation thus eliminating the need for first three columns in the process flow sheet.
  - MEA would be favoured in this process as the ethylene oxide would have to pass through the proton and amine rich packing where it would react to MEA and leave through the liquid exit as a relatively pure product stream.
  - The RPB provide similar control of heat and effective mass transfer as the SDR.
  - A gas or liquid phase system may be used and it will be possible with a little cleaver design to have an integrated process capable of switching from an

RPB using gaseous ammonia to a spinning disc utilising aqueous ammonia by using some innovative pipe junctions in a similar fashion to train lines.

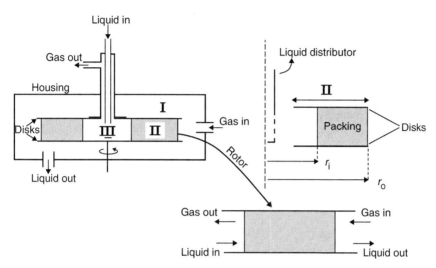

Schematic illustration of the RPB reactor.

Identification of the optimum process

Three processes have been identified as suitable for this system: the first involves the use of an SDR, aqueous ammonia and CHEs, the second utilises an RPB, gaseous ammonia and similar downstream equipment to the SDR process and the final system uses an RPB in place of the SDR with zeolite catalysis to enhance yield.

The most suitable piece of equipment for the reaction of ethylene oxide and ammonia in a bid to maximise MEA production is the SDR because it integrates the mixing point and reactor as well as providing the desired mixing environment to give the fast reaction.

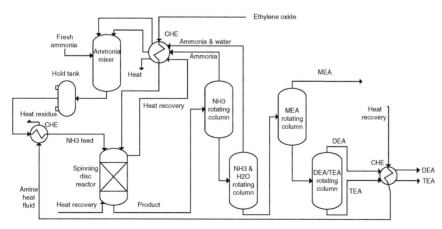

MEA manufacturing process using aqueous ammonia and liquid ethylene oxide with SDR, rotating columns and compact integrated heat exchangers.

This process follows a similar configuration as the first flow scheme presented in this analysis but the final column has been eliminated and the other operations have been intensified. The tubular reactor has been replaced with an SDR and the columns have been switched from plate distillation to rotating distillation columns. There is a CHE system that is used to pre-heat the ammonia feed using the heat from the DEA and TEA residue streams out of the final column and there is a similar system for the ethylene oxide where heat is recovered from the ammonia and water recovery line, and from the SDR to enhance plant efficiency and to reduce inherent associated emissions through heat sourcing. This plant, while very similar to the first one on paper, will be vastly smaller than the former due to the size reductions achievable using these units. This process will have a much smaller visual, physical and chemical impact on the environment and will be much simpler to operate to a high safety standard as well as being easier to clean and maintain.

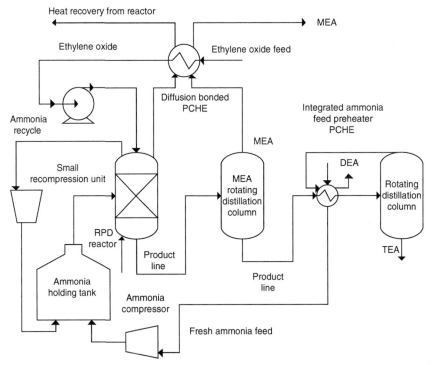

MEA production using gaseous ammonia and liquid ethylene oxide in an RPB reactor, rotating distillation columns and integrated heat systems for capital and energy efficient processing.

In this process, liquid ethylene oxide is fed through a heat recovery exchanger and then into an RPB reactor where it is reacted with gaseous ammonia in a molar ratio of greater than $10(NH_3):1$ over a zeolite catalyst, to provide the proton needed to facilitate the reaction. The excess ammonia gas is then extracted through the gas outlet and recompressed and mixed with the fresh stream in a mixing/holding tank. The product line is passed on to a rotating distillation column where MEA is extracted and the bottom line is passed on to a cross exchanger which heats the fresh ammonia stream and is then itself separated into DEA and TEA.

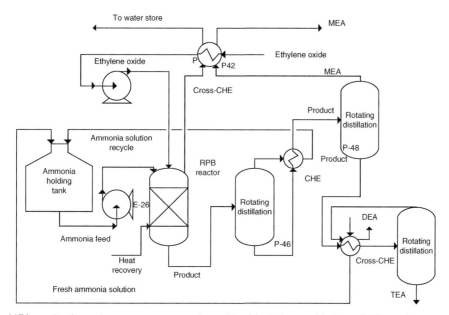

MEA production using aqueous ammonia and liquid ethylene oxide in an RPB reactor, rotating distillation columns and integrated heat systems for capital and energy efficient processing.

This process follows a similar configuration as the first flow scheme presented in this analysis but the final column has been eliminated and the other operations have been intensified. The tubular reactor has been replaced with a rotating bed and the columns have been switched from plate distillation to rotating distillation columns. There is a CHE system that is used to pre-heat the ammonia feed using the heat from the DEA and TEA residue streams out of the final column and there is a similar system for the ethylene oxide where heat is recovered from the ammonia and water recovery line and from the rotating bed to enhance plant efficiency and to reduce inherent associated emissions through heat sourcing. This plant while very similar to the first one on paper will be vastly smaller than the former due to the size reductions achievable using these units. This process will have a much smaller visual, physical and chemical impact on the environment and will be much simpler to operate to a high safety standard as well as being easier to clean and maintain.

Final choice

In order to achieve the largest size reduction and greatest level of facility integration, process two is chosen from the previous section; the RPB utilising gaseous ammonia and CHEs with rotating distillation columns. The benefits of this process are:

- It eliminates the need for huge volumes of water which allows for a significant capital cost reduction and the elimination of one of the downstream units.
- This system also eliminates the need for an ammonia stripper as it will be in the gas phase and easily separable from within the RPB.
- Downstream of the reactor only two or three rotating columns and one integrated heat recovery system are needed.
- There can be a significant integration of the heating system making use of diffusion bonded plate heat exchangers to recover the maximum level of heat from the product streams in order to heat the reactor feed streams.
- The zeolite catalyst will need to be replaced at regular intervals but given the massive cost reduction achieved it would more than likely remain economic to operate this system.
- Using gaseous ammonia in a pressurised system does present significant safety risks but the system is designed to be as low inventory as possible at every operation in a bid to mitigate these risks.
- In this process, the heat from the reactor is recovered and used in conjunction with the heat from the MEA out of the first rotating distillation column to heat the ethylene oxide feed and the heat from the heavy stream out of the first column and from both streams out of the final column to pre-heat the fresh ammonia gas before it is compressed to reduce the work requirement of the compressor.

It has therefore been decided that the best process to go for in terms of size, output, capital and operational cost and overall environmental and visual impact. The RPB system eliminates the need for a dedicated ammonia separation system and only has three large operations in place with some additional compression and heat exchange which is a much smaller operation that presented at the beginning of this particular analysis and would be just as or much more efficient in terms of production rates. The reduction in inventory would help ease any safety considerations in terms of the pressurised ammonia gas which can be severely hazardous if released but in this process would be in a low inventory and could be effectively monitored. This process offers the best overall reduction in size through the elimination of the aqueous ammonia and the use of a gas and zeolite system to provide the protons needed to catalyse the reaction process. The effective mixing and fast reaction time will also be used to optimise the production of MEA over the other two forms which was one of the main drivers in carrying out this analysis.

Checklist

Is my plant/process currently running under optimum conditions?

- No, it is quite energy intensive in the recovery of ammonia out of the product liquors and there is no integration of the heat systems which causes much of the resource to be wasted and increases the carbon emissions associated with the whole process. The reactor and column sizes are large leading to a high visible impact and increased risk of mass devastation should a catastrophic safety failure occur on plant. It also has high capital costs which have hampered the use of the product as a carbon capture medium. The reactor in the conventional process encourages the production of DEA and TEA which are undesired in this analysis and it also has a relatively long residence time further promoting such formations.

If my plant/process is optimised, is there still a case for looking at PI?

- Optimisation is a continual process in which there is always potential for improvement and looking at PI options or already optimised systems is a smart move. The potential efficiency increases, space savings and possible cost savings as well as reduced inventory and just-in-time processing.

Do I know all the current operating parameters on my plant/process?

- I have presented some of the most pertinent information and should a further study be required it would not be all that difficult to obtain or analyse to gain such data.

Do I have information on the operating parameters of any replacement plant?

- Not at this moment in time but a detailed analysis of this or a similar system will provide it, the reactants in this process are relatively common and can be gained on the open market at somewhat reasonable prices, as outlined before.

Will any replacement plant/process have significantly different requirements to the existing plant (throughput, be continuous or intermittent [batch], product type, operating temperatures, pressures, quality control)?

- There will be similar requirements in the reaction stages and there may not be such a temperature requirement in the distillation columns but that will require a greater level of analysis. The process will be continuous and could be reduced down with PI to supply a capture process with a just-in-time processing mantra.

Are these requirements realistic for a conventional plant?

- Yes.

If they are, why do I want to study a PI plant?

- Even at optimum conditions, the plant treated with a PI methodology still offers extensive benefits over a conventional system, such as:
  - Higher purity product and desired product formation.
  - Reduced environmental impact and unit inventory.
  - Lower operating and capital costs compared to that of conventional systems at optimum conditions.
  - Inherent safety increases and reduced risk of catastrophic explosions.

If they are not realistic for a conventional plant, how will a PI plant help?

- I believe that this process would run under optimum conditions as the maximum number of units have been intensified or integrated to provide for greater utilisation of resources and the most efficient production of the target product.
- The use of zeolites for the proton donation instead of using copious amounts of water and thus reduces the size of a significant section of the upstream reaction process and simplifies the recovery of ammonia by the reduction in the number of possible hydrogen bond formations in the product stream.

Will a radical change of process/plant affect my product?

- Yes it will, but in a desired fashion. The desired product forms in a fast reaction and is ideally to be isolated from exposure to any more of the highly reactive ethylene oxide which will then form one of the less desired DEA or TEA products.

Is my knowledge of PI sufficient to answer the above?

- I would like to believe so at this point in time but it may not be the case and it may be beneficial to seek advice from an expert on the systems intended to be implemented and in the reactants and products used.

If not, are there others in my organisation that could assist me?

- Possibly within my university but I would have to enquire within the school.

Should I consider employing outside consultants?

- If it is feasible and if they have sufficient knowledge to bring to the table.

Do I have sufficient time and resources within my own team?

- At present I am significantly pressed for time and have no formal team so I would have to say no, not at this moment in time.

Do my timescales and potential budget allow for any R&D work?

- No

If not, am I limited to equipment that is already fully developed?

- No, I would seek funding and assistance rather than aim for a potential conventional system given its inefficiency and cost overrun.

If this is the case, is such equipment available?

- Yes, equipment is available from suppliers and it may be possible to secure agreements with universities and research organisations to use their equipment to test the theories and assumptions made within the analysis.

Do I then have the information to carry out?

A risk assessment (health and safety)?

- Yes, the information on the majority of the issues that would arise in the theoretical analysis of this system and the proposed PI system is available to facilitate the study.

A technical feasibility study?

- Not at the moment; access to equipment for testing would be required, although there is enough information to carry out somewhat detailed technical analyses.

A rigorous financial appraisal?

- Not a rigorous one, a general indicative analysis would be possible with data available at the moment.

Have I considered the availability of grants?

- Yes, there are grants available from the Department of Energy and Climate Change, the Enhanced Capital Allowance, which can cover the first year expenses on equipment such as CHEs and intensified processing equipment as well as other loans and help available for systems aiding in the viability of the post combustion carbon capture systems.

Can I now make a comprehensive case for examining a PI solution to my plant/process?

- Given some time to carry out more analyses on both technical and commercial grounds and then to secure use of the equipment to test the hypothesis and analytical outcomes.

References

Ammonia ICIS EU Pricing Report 06/05/2011, Edited by Rebecca Clarke. <http://www.icispricing.com/il_shared/Samples/SubPage75.asp>. (accessed 11.11.2011).

Ammonia ICIS US Pricing Report 06/05/2011, Edited by Rebecca Clarke. <http://www.icispricing.com/il_shared/Samples/SubPage149.asp>. (accessed 11.11.2011).

Baroody & Carpenter, 1972. MEA condensed phase thermochemistry. <http://webbook.nist.gov/cgi/cbook.cgi?ID=C141435&Units=SI&Mask=1EFF> (accessed 11.11.2011).

J. Coulson et al., Coulson & Richardson's Chemical Engineering, vol. 2, Fifth ed. 2002, P. 1129.

DEA condensed phase thermochemistry heat of formation. <http://webbook.nist.gov/cgi/cbook.cgi?ID=C111422&Units=SI&Mask=2#Thermo-Condensed> (accessed 11.11.2011.).

Enhanced Capital allowance 2009. <http://etl.decc.gov.uk/NR/rdonlyres/FA9E5D2B-BE76-49CB-BAE4-722316E5599F/0/09CompactHeatExchangers.p> (accessed 18.11.2011).

Ethanolamines ICIS Pricing Report 11/05/2011, Edited by Amandeep Parmar, http://www.icispricing.com/il_shared/Samples/SubPage10100082.asp (accessed 11.11.2011).

Ethylene oxide ICIS EU Pricing Report 06/05/2011, Edited by Heidi Finch, <http://www.icispricing.com/il_shared/Samples/SubPage77.asp> (accessed 11.11.2011).

Ethylene oxide ICIS US Pricing Report 06/05/2011, Edited by Heidi Finch, <http://www.icispricing.com/il_shared/Samples/SubPage151.asp> (accessed 11.11.2011).

Harvey, A.P., et al. 2003. Process intensification of biodiesel production using an oscillatory flow reactor. Journal of Chemical Technology & Biotechnology 78, 338.

Kent, J.A., 2006., Eleventh ed. Synthetic Nitrogen Products, Handbook of Chemistry & Biotechnology, vol. 1. Springer, (P. 1059, Table 22.26)

Llerena-Chavez, H., Larachi, F., 2009. Analysis of flow in rotating packed beds via CFD simulations. Chem. Eng. Sci. 36, 2114.

Reay, D.A., 2011. Process Intensification Module Document. Heriot Watt University, (P. 102)

TEA condensed phase thermochemistry heat of formation. <http://webbook.nist.gov/cgi/cbook.cgi?ID=C102716&Units=SI&Mask=2#Thermo-Condensed> (accessed 11.11.2011.).

Vicevic, M., Catalytic rearrangement of alpha pinene oxide using spinning disc reactor technology, School of Chemical Engineering and Advanced Materials, Newcastle University, July 2004, P. 26.

Queen Elizabeth Prize for Engineering. <http://www.number10.gov.uk/news/queen-elizabeth-prize-for-engineering/> (accessed 18.11.2011).

Zahedi, G., et al. 2009. Simulation and optimisation of ethanolamine production plant. Korean J. Chem. Eng. 26 (6), 1505.

Case Study 3: Catalytic Hydrodesulphurisation Process

Step 1: Overview of whole process

Chemistry

Distillate hydrodesulphurisation (HDS) is one of the catalytic hydrogenation chemical processes utilised to saturate various hydrocarbons streams in order to remove or reduce the sulphur content of refinery unit products. The increasing demand for sulphur removal and product stabilisation has resulted in additional separation units to work alongside HDS; such as hydrogen plant, amine absorbers and additional sulphur units. The increased drive by government to reduce sulphur content of transport fuel in order to reduce direct pollution from impurities is requiring refineries to put in place more stringent removal processes. Along with environmental concerns, society demands that liquid refinery unit products be as clean as possible.

The general HDS process involves treating the hydrocarbon feed with hydrogen under elevated temperatures, pressures and palladium catalysts.

The process takes the following reaction form:

$$X - S + H_2 \rightarrow X - H + H_2S \qquad (12.1)$$

(where X is the hydrocarbon and S is sulphur).

Within refineries, the following refinery unit products are subject to treatment: natural gas, kerosene, diesel amongst others. Transport fuels such as unleaded motor spirit (ULMS), super unleaded motor spirit (SUMS) and diesel are facing tighter sulphur restrictions due to increasing concentrations of carbon emissions.

General HDS of middle distillate – mercaptan (between C9 and C16):

$$C_{12}H_{22}SH + H_2 \rightarrow C_{12}H_{23} + H_2S \qquad (12.2)$$

HDS of lower organo-sulphur containing compounds (dibenzothiophene):

$$C_{12}H_8S + H_2 \rightarrow C_{12}H_8 + H_2S \qquad (12.3)$$

For this process intensification assignment, middle distillates have been identified as the feedstock, as these are normally processed through hydrotreating (jet fuel, kerosene, diesel).

Conventional process

The route used to undertake the removal of hydrogen sulphide is currently being carried out within a conventional co-current trickle bed reactor (TBR). With up to 50% of all refinery unit products potentially being processed through a hydrotreater (ranging from naphtha feedstock to heavy vacuum residue) the reaction can be carried out over a range of potential temperatures and pressures.

Current removal of sulphur via HDS utilises a large volume of catalyst, due to the scale of material processed through daily production within a refinery. In the table above, the LHSV provides an indication of the volume of refinery unit product processed as a ratio of time. At present, the lower levels, indicated for heavy residues (can be up to 90% by volume H_2S concentration) illustrate higher residence times due to the increase in volume of catalyst required, in order to achieve sufficient desulphurisation.

Feedstock	Temperature (°C)	Hydrogen pressure (atm)	LHSV[a] (hr⁻¹)
Naphtha	320	15–30	3–8
Kerosene	330	30–45	2–5
Atmospheric gas oil	340	38–60	1.5–4
Vacuum gas oil	360	75–135	1–2
Atmospheric residue	370–410	120–195	0.2–0.5
Vacuum residue	400–440	150–225	0.2–0.5

HDS process conditions. Chunshan et al., (2011)
[a]LHSV, liquid hourly space velocity.

RED circle: HDS section

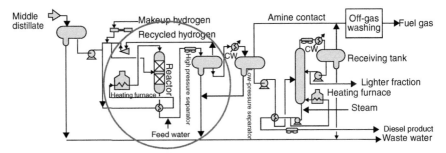

Conventional HDS refinery process. Packinox (2011).

The middle distillates stream (containing the organic sulphur impurities) from the fluid catalytic cracking (FCC) unit is send downstream for hydrotreating. The hydrocarbon stream is mixed downstream with rich hydrogen recycle gas, after which it is passed through a pre-heater and on to the combustion chamber where the mixture is vaporised. The vaporised mixture is then fed through a fix bed catalytic reactor with a gas/liquid co-current flow distribution to drive the desulphurisation.

The catalysts to be utilised are cobalt molybdenum or nickel molybdenum supported on an alumina base. Catalyst changes will not provide PI improvements, but only performance enhancement.

Within a conventional fixed bed HDS unit, the catalytic reaction and sulphide separation are being undertaken around 20–30 barg and at elevated temperatures between 330°C and 370°C (Delgado et al., 2002) where the feed is gaseous. The reactor effluent mixture is passed through a heat exchanger where it is cooled down (with water) and then initially passed through a hot high pressure separator and cold high pressure separator to remove the hydrogen vapour feed which is condensed and separated into rich hydrogen gas, middle distillate and waste hydrogen sulphide. This stage reduces the octane number by saturating some olifinic compounds in the mix. This will reduce the potential use of the refinery unit product due to the lower quality.

The rich hydrogen feed is then circulated via a compressor and fed back through a recycle loop back into the reactor to restart the process. The outgoing product from the reactive column will be redistributed back through the heat exchanger to provide reheating utility.

Other than the initial fixed catalytic reactor and gas separator, both the hydrogen recycle and the reacted liquid feed is sent downstream for further processing for H_2S removal via an amine gas treating unit and distillation column; where the final desulphurised liquid stream will be produced.

Chemistry audit

Reaction mechanism: The reaction kinetics of the HDS process is dependent upon factors related to surface adsorption/desorption of the catalyst in relation to the actions of the carrier H_2 gas. PI identified herein will assess the ability to enhance diffusional effects whilst improving on the current resistances to the mass transfer in terms of:

- Hydrogen gas across the gas/liquid interface.
- Within the internal structure of the catalyst in order to improve reaction rate.
- Alter internals of reaction or flow direction, in order to reduce resistance from co-current liquid stream to and from the catalyst.

Based on the middle distillates feedstock and the existing process, HDS of diesel-mercaptan is based on the following reaction:

$$C_{12}H_{22}SH + H_2 \rightarrow C_{12}H_{23} + H_2S \tag{12.4}$$

$$\mathbf{H_R} = -251 \text{ kJ/kmol}$$

The current process is based on an increase in temperature in order to sustain the reaction; however, this is very energy intensive at present. Based on Le Chatelier's principle, increasing the pressure within the system will also increase the conversion.

Solvents: With operational costs associated with hydrogen proportionally increased with the extent of HDS required; an alternative could be the use of caustic solution downstream of the gas separator to remove mercaptans for lower molecular weight thiols. The alternative is to react the sulphur containing alkane within a heterogeneous reaction with a metal oxide, resulting in the production of insoluble hydrocarbons that can be separated with a membrane. A lower pressure unit could be encouraged through the use of a cobalt molybdenum catalyst as they indicate lower hydrogen requirements and hydrogenation activity over the nickel form. However, this could potentially lead to less nitrogen inhibition.

Operating conditions: One of the major variables in current TBR design is the use of temperature control. As the process is extremely exothermic, the temperature rise of the reaction is proportionally related to the length of the reactor. From a PI view point this is not ideal for a middle distillate HDS process, as the reactor size is already large. Only with increasing temperature does the H_2S removal in the current process become efficient; however, at a reduced fuel octane rating.

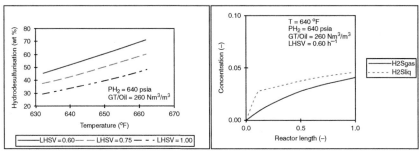

(a) Reactor temperature over HDS reaction   (b) HDS Concentration profile Vs. Reactor Length

Changing operating conditions. (a) Reactor temperature over HDS reaction. (b) HDS concentration profile vs. reactor length. Robinson (2011).

Without the introduction of a quench to limit the hydrogen consumption and a runaway temperature increase, the catalyst bed has been found to be 10–20 m in order to achieve up to 90–95% sulphur removal as required by regulations (Filho et al., 2008). The introduction of a quench and directional changes to the hydrogen feed will ensure less corrosive conditions within the reactor. In turn, this will allow less costly catalysts to be used over the current nickel form.

Equipment

The three main pieces of equipment within this HDS process are:

- Fixed bed catalytic reactor; more specifically, a TBR.
- Amine contact absorption column and regeneration unit.
- Hydrotreating to aid distillate desulphurisation.

These are visibly summarised in the figure below as the distillate HDS unit:

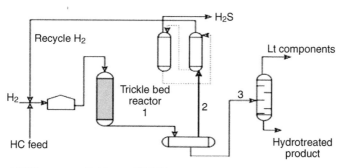

Summarised HDS process. Robinson (2011).

The process intensification will be focused around these three main areas, as this is where the initial separation requires the most energy conservation. As undertaking PI of the multi-functional TBR reactor will allow plant size reductions to be made to both refineries and offshore platforms. At present, TBRs utilise hydrogen gas in the continuous phase, with minimal liquid hold-up.

The trickle bed operates by inducing a distributed liquid phase in interaction with a continuous hydrogen feed, where the bulk mass transfer resistance is known to occur. The adiabatic TBR is typically constructed as a three-phase fixed bed reactor that contains catalysts ranging between 1 and 3 mm and can be a very high column (15–30 m in length) and therefore needs to be intensified (Ancheyta et al., 2011). A typical TBR HDS unit's dimensions are 10 m in height by 2 m in diameter, with a superficial gas/liquid velocity up to 0.3 m/s.

It is proposed that the TBR contains a quench system that is fed at midpoint between the two catalyst beds, as visible in the figure below.

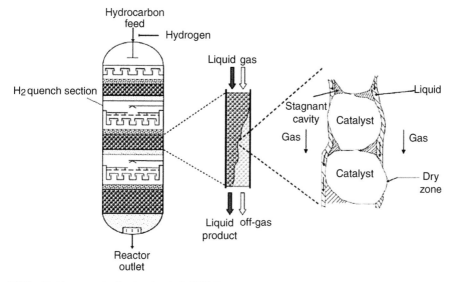

TBR with H_2 quench. Ancheyta et al. (2011).

The presence of hydrogen at high pressures and elevated temperatures for catalysis, along with the joint production of hydrogen sulphide and some ammonium chlorides require the reactor to be constructed from stainless steel to prevent any embrittlement and reductions to any coking that may occur with high carbon chain feeds.

Step 2: Business and process drivers

Business drivers

- *Regulations*: Environmental factors demand the production of ultra-low-sulphur motor spirits.
- *Energy intensive*: Current TBR carries out an extremely exothermic reaction, which prohibits potential scaling up due to the heat effects. The uses of multi-quench streams or carrying the reaction out over smaller units, where the heat is distributed, will help improve inherent safety along with less waste production.
- *Reduce energy consumption*: as the current fixed bed TBR requires large heat to ensure the catalyst reaction is maintained at the high temperature.
- *Utilises hydrogen*: HDS exhibits high operational costs associated with compressing hydrogen contributes to 60–70% of the total hydrotreating operational cost. Minimise operating costs and hydrogen consumption by either implementing a membrane unit or utilising pressure swing adsorption.
- *High capital investment*: Fixed bed TBR design reduces catalyst life and results in added catalyst costs.
- *High conversion cost*: Increased hydrogen consumption with increasing reactor feed to promote conversion results in greater catalyst costs, due to reduced activity. Increased consumption could potentially de-bottleneck plant.
- *Decrease HDS plant footprint*: In order to progress HDS in both refineries and offshore platforms; caustic treatment (NaOH) could be used as this could also reduce environmental impact.
- *Health and safety*: New technology will reduce the required number of operating personnel which will not result in downtime of the existing plant.
- *Operational integrity*: Trying to gain better conversion rates, space time velocity and product selectivity will require greater observations surrounding the reactor's charging with a catalyst bed.

Process drivers:

- *Space velocity*: Middle distillates and above require a lower space velocity, thus reducing the quantity of material processed per hour and weight of catalyst. This will affect the product quality; more specifically reduce the octane number, which will need to remain high throughout this intensive treatment process.
- Poor liquid distribution within current TBR, resulting in the development of hotspots and a potential reactor runaway.
- Filtration and bed plugging in the TBR will result in a lengthened reaction residence time. This will lead to plant de-bottlenecking and loss of production time.

Step 3: Identifying rate limiting steps

- Hydrogen mass transfer to the surface of the catalyst limited by surface deposits.
- Reaction kinetics of hydrotreating is inhibited by the presence of aromatic groups/compounds within the kerosene to diesel feed range.
- Hydrogen mass transfer limitations within the conventional liquid phase TBR (gas limiting reaction).
- Catalyst regeneration affected by the presence of ash-forming constituents present within residual fuels (diesel) resulting in catalyst contamination.
- The conversion will decrease if the flow rate of petroleum feed into TBR is increased whilst the temperature remains constant.
- High pressures and elevated temperatures are required for the catalyst regeneration to be an active process; thus resulting in the need for additional hydrogen quench sections.
- Failure to remove H_2S from the recycle gas reduces the extent of the catalyst HDS reaction resulting in a greater required catalyst volume.

Poor radial heat and mass transfer in a conventional fixed bed TBR means current reactors are treated as adiabatic. The PI illustrated herein, discusses the use of a compact liquid phase reactor, that will alleviate this problem and increase the reaction rate.

Step 4: Review business and process drivers

From the business and process drivers mentioned earlier, the PI intensions are to ensure plant design targets are met by fundamentally reducing hydrogen and catalyst costs and improving downstream processes. The achievable targets from the initial drivers are to ensure economic prosperity due to additional pollution refining costs instated by the government. Through introducing an integrated compact heat exchangers and liquid phase reactors will ensure reduced reactive content at a reduced plant footprint and increased manufacturing competitiveness. However, constraints still exist in identifying new environmentally friendly solvents for separation of mercaptans from the gas separator.

The driving force for this PI still remains overcoming economical and environmental consequences of not removing sulphides from refinery pipeline networks (e.g. corrosion).

Step 5: Assess PI viability and generate design concepts

Based on the initial audits carried out, it is evident that PI will not only be possible across the fixed TBR, but improvements will also be possible to the distillation column in order to alleviate the demand on the TBR by reducing the need for excessive temperatures, pressures along with reduced hydrogen consumption.

To achieve cost reductions and energy gains, alternatives could include improving the hydrogen gas recycle purity, or increasing the partial pressure of the hydrogen feed entering the reactor in order to remove any aromatic compounds present. With the maintenance budget of large refineries being constrained, the reactor will need be of smaller volume but with higher surface area for catalyst activity.

Step 6: Analyse design concepts
DETAILED PFD:

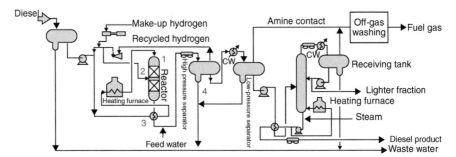

PFD with possible intensifications.

Step 7: List of potential areas for process intensification
The possible applications of PI have been numbered in the PFD diagram above and are referred to in the subsequent list that shows where direct modifications could be made to the conventional TBR:

- Convert TBR into a HiGee reactor for extractive and reactive purposes (**1**).
- Utilise rotating TBR (**1**).
- Utilise a catalytic plate reactor (CPR) (**1**).
- Integrate a non-reacting gas/liquid contact zone which could be used to strip any sulphide from the hydrogen gas as it is passed out of the column (extraction and adsorption technology) – this replaces the amine scrubbing stage (**2**).
- Utilise a pressure control valve to reduce power consumption by the hydrogen compressor.
- Utilise Heatric's printed circuit board integrated between the catalyst beds (**3**).
- Convert the TBR from a co-current downflow to a countercurrent flow distribution.
- Induce pulsated flow into the column.
- Spiral heat exchanger (**4**).

As will be demonstrated below, the high pressure separator will be able to be removed (4).

Step 8: Generate design concepts
Essential PI Focus Areas: Reduced catalyst consumption, increased hydrogen partial pressure, elevated temperatures/pressures and increased space velocity.

TBR replacements
HiGee Separator
With the aim of reducing the reactor length, it is crucial that the multi-functional reactor is optimised. As seen in the detailed PFB above, the weight % of sulphide removed is proportional to the length of the catalyst bed. In order to achieve precise temperature control and to prevent the potential of a runaway reaction is to

implement: the HiGee rotating packed bed (RPB), a compact liquid phase reactor or creating a counter current TBR. Conventional TBRs utilise high pressures to increase the solubility of the hydrogen into the middle distillate then onto the catalyst surface.

With high levels of flow non-uniformity in the conventional concurrent TBR caused by poor radial distribution, the HiGee will intensify mass transfer by inducing centrifugal forces which will enable reductions to residence time, improved energy consumption and increased selectivity of the refined diesel product by improving interphase transfer rates. As an intensive hydrogenation reaction, the use of this modified RTB with integrated catalysts has been found from experiment to increase conventional TBR reaction rate by 60-fold (Sivakuma et al., 2011). By altering the contact of the middle distillate as it passes through, the torus packed with catalysts will improve the wetting ability of the catalyst and reduce the risk of overall hotspots. Extraction and adsorption technology integrated into the separator leads to further energy reductions and reduced carbon emissions from columns.

Countercurrent TBR
Induced Pulsing
As a conventional TBR is operated as close to steady-state operation, research undertaken into the effects of enforcing non-steady flow operation without being fluidised in operation, as this will disturb and potentially deactivate the catalyst bed and prevent efficient interfacial mass transfer. As residence time is restricted in packed beds up to 1 m/s, PI of the TBR can be achieved by manipulating the wetting efficiency of the catalyst with respect to time and distributing short fast pulses of gas oil through the bed and column.

Countercurrent TBR/Rotating TBR
To improve the resistance across the gas-liquid boundary, countercurrent feed distribution will be introduced in order to replicate, as close to fluidised bed operations as possible, without exhibiting the increase in size of the reactor vessel and increased pumping requirements, as a result of pressure drop which will reduce surface area of the catalyst bed, as the heat transfer will be constrained.

The velocity required to expand the catalyst through the bed efficiently will be possible through adopting a counter-current flow through the packed catalyst bed over a co-current configuration. With the middle distillate adding the bulk liquid resistance, hydrogen consumption can be undertaken sparingly, unlike in a conventional TBR. Conventional TBRs are of a large size due to the poor interfacial mass and heat transfer. Counter-current TBRs overcome this by creating higher accelerated forces.

Recent investigations undertaken have demonstrated that reactor improvements are possible whilst keeping the bed rotating and feeding the liquid feed from the top, and allowing it to come into contact with a countercurrent continuous hydrogen feed from the bottom. The HiGee concept can be applied to a limited extent, with the catalysts being held within monolithic catalyst bed where the corrugated pathways will enhance heat transfer. As residence time is restricted in packed bed, the rate will be increased with centrifugal force whilst minimising the turbulence (unlike in a HiGee) as this will lead the liquid film to become thinner, thus increasing the ability of the H_2S to be stripped from the liquid phase and be replaced with hydrogen.

Plant Heat Exchangers

Compact heat exchangers could be implemented when reconditioning the compressor hydrogen recycle gas are integrated between the catalyst beds as a microstructured heat exchanger in order to attain precise control of reactor conditions in achieving high selectivity and preventing a runaway reaction. A printed circuit heat exchanger could be added before and after each catalyst bed in the column in order to provide efficient cooling and heating the feed as it passes through the column. This will prevent the need for extremely high reaction temperatures whilst maintaining a liquid/gaseous product leaving the column for further processing.

Monolithic Catalyst Bed

Monolithic reactors are becoming increasing more popular in applications involving liquid/gas catalytic reactions over conventional pellet-packed TBRs. To this process they could provide benefits ranging from: ability to increase liquid flow at reduced pressure drops, improved catalyst reaction surface area, reduced back mixing and fouling, along with overcoming internal mass transfer limitations which heavily influence the large hydrogen consumption in this HDS process. By utilising washcoated monoliths, along with the higher catalytic effectiveness, it will result in PI in the form of significant reactor size reductions and power requirements due to the improved transport properties through reduced diffusion paths.

Step 9: Select most suitable equipment

Countercurrent TBR with pulsated

When we encountered business drivers (in Step 4), the aim was to improve heat transfer. As the countercurrent operation provides an axial temperature profile across the bed of greater stability, this will create a more intensified reactor of smaller volume, due to the increased activity distributed across the catalyst bed. The micro-channels created, will allow for laminar flow; thus allowing both mass and heat transfer through diffusion, along with maintaining moderate column operating conditions whilst increasing the space time velocity due to reactor residence.

Within a countercurrent TBR, the reaction ratio between middle distillates and cobalt molybdenum will be low and therefore will suit induced pulsing. Flushing liquid into the column at a set frequency and amplitude will not only provide improved distillate to catalyst activity but also demonstrate a flow motion closer to plug flow and thus improve reactor size and performance for a hydrogen gas limiting process (Ancheyta et al., 2011). This will enable more hydrogen-to-catalyst surface coverage (increased catalyst wetability); resulting in increased reaction kinetics for the catalytic hydrogenation. With heat and products being continuously flushed down the column with this pulse strategy, the recycled hydrogen will have greater activity with the catalyst; thus reducing the need for a large compressor and column and improving the overall space time velocity.

Printed Circuit Heat Exchanger (PCHE)

With the PCHE, the large surface created for hydrogen/middle distillate mass and heat transfer through inducing a pulsation and countercurrent flow configuration will allow the gas to increase thermal conductivity within the channels, thus enabling the removal of excessive heating of the catalyst bed. The PCHE section of the catalyst bed could be utilised as pre-heat section for feed coming to reactor in order to reduce the initial heating load. According to Heatric, this CHE is useful for gases/liquids/two-phase systems; thus making it suitable here. The ability to utilise stainless steel will help minimise the corrosion long term.

With the advantage of utilising multiple pass and multiple fluid streams, it will allow for the hot liquid from one of the lower trays (not bottoms) to be partially recycled through the PCHE where it will preheat the feed as it is fed into the column, as it will provide a closer approach temperature and less energy consumption.

This will enable the column to operate with a lower hydrogen consumption duty as the pre-heat temperature will enable the H_2 to become more soluble before application with the catalyst as the heating will closely moderated; thus making the process more energy efficient. Due to the interstage cooling and heating, the hydraulic concerns will be reduced. This PCHE integration will intensify the process by removing the additional heat exchanger being used to heat the fresh feed entering the boiler. Removing this external preheat exchanger will reduce exposure to hydrogen sulphide in the fuel, in the event of a leak.

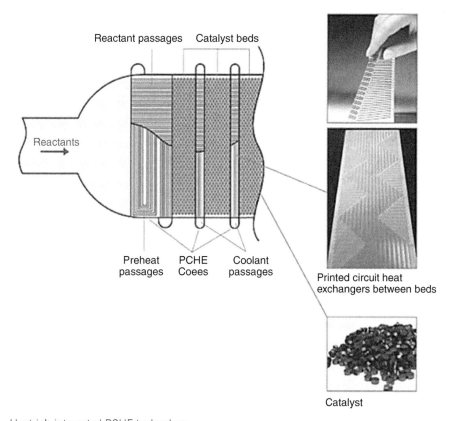

Heatric's integrated PCHE technology.

This may lead to a greater duty on the fired heater; however, this PCHE integration provides containment, process control and hazard mitigation in the form of another leak point. With the feed coming in at high temperatures, the level of impurities will be reduced in comparison to when it is cooled. The concern however, will be that there will be a great pressure drop incurred by the fluid.

With the viscosity of refinery middle distillates (e.g. Ineos Grangemouth kerosene/gas oil, 100°C) circa 1.64–6.3 cP and densities between 770 and 800 kg/m³; these values are within the PCHE requirements. The initial concern may be surrounding high pressure drops; however, with the pulsated flow in the column there will be greater flexibility in design along with high pressure/strength qualities of the PCHE.

Benefits of PCHE

- High heat transfer effectiveness at very low pressure drops.
- Greater resistance to corrosion.
- Reduces the need for reactor turbulence/agitation.
- Reduces the need for increasing catalyst loading, which would have been the normal route to improve performance.
- Improves the overall economics and safety of the column.
- Column integration allows for a smaller plant footprint.

Additional PI Benefits from Countercurrent TBR with Pulsated Flow

Recycle Gas Treatment
To intensify the process and ensure high scalability, the quenching with hydrogen in conventional processes could be replaced with recycling the liquid product stream within the reactor in order to reduce the heat effects or by utilising cool methane. With respect to recycling light cycle oils[4] such as kerosene, these can potentially increase the conversion of H_2S as the system will have greater activity at shorter residence times; thus helping to reduce the size of the vessel.

Feedstock
Meeting hydrogen needs for the HDS process will need to become more cost-effective as a utility. By adopting the countercurrent TBR, there will be an increased hydrogen partial pressure towards the end of the catalyst bed which will result in much lower H_2S impurities towards the end; thus allowing for reduced hydrogen consumption whilst driving the sulphide removal upon the catalyst reaction. The initial H_2S formed will limit the reaction rate, as it will operate in H2S lean mode; however, the combination of the PCHE initially with the pulsation force will significantly reduce this effect.

---

[4] Light Cycle Oils – C8 to C12 carbon chains.

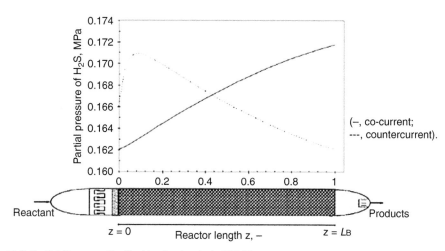

H$_2$S Partial Pressure Profile (Ancheyta et al., 2011).

Step 10: Comparison with conventional HDS unit

With final product quality and control becoming more tough and stringent in terms of sulphide removal, the current fixed TBR co-current configuration is not the most efficient, as the HDS is already known for being product inhibited. The current arrangement leads the HDS reaction rate through the catalyst bed to decrease as it descends proportionally through the packing. Conventional stripping of H$_2$S proceeds with high hydrogen partial pressure will help vaporise the sulphide in the gaseous stream and allow for hydrogenation.

The rate at which hydrogen sulphide is being produced is causing the partial pressure of hydrogen being supplied to decrease, due to the rate of consumption surpassing it. The high pressure drop across the bed has been the cause of reduced heat and mass transfer. However, the countercurrent approach will enable the catalyst bed to operate more effectively. In a conventional TBR the reactor maintains a high hydrogen partial pressure throughout by injecting regular cold recycle quench gas.

Step 11: Make the final choice of plant

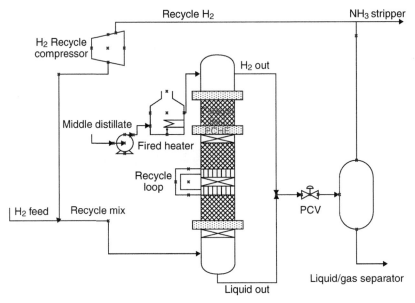

PI modifications to the plant.

Chosen PI: Countercurrent TBR with Pulsated Flow and Integrated PCHE

For refinery hydrotreating applications, the countercurrent pulse induced TBR with the PCHE has been selected as the most ideal process intensification which can be undertaken. The main piece of equipment is the use of PCHE, as the pulse flow can be used in addition to it. Due to the extent of current research being limited to small RPB reactors with pulse induced flows, there is a potential for this to be implemented into a petrochemical site.

Economic Gains

- The pulse flow configuration will help to obtain hydrotreating energy savings up to 83 trillion Btu/year.
- Pulsation will enable large gas-to-liquid ratio within the column, thus enabling it to operate at lower velocities without the risk of entrainment, the reactants being passed over and maintaining the pressure along the column, unlike in the conventional form.
- A pressure control valve in place of the hot high pressure gas liquid separator will significantly reduce recycle compressor requirements.
- This counter flow arrangement and improved axial temperature distribution will help protect against catalyst bed plugging as experienced in the conventional TBR.
- Reduced capital expenditure with a smaller volume H_2 recycle compressor.

- With column intensity being reduced, the hydrogen partial pressure can be refined to around 15 bar instead of the values of 40 bar; thus intensifying the reactor.
- Due to high pressures and temperatures in the column, the integrated PCHE and product recycle will help control and reduce the accumulation of H_2S within the column.
- The constraints surrounding using methane as a quench stream over hydrogen are currently out of the scope of this methodology. For this to be acceptable, research will have to be undertaken more closely into the thermodynamics of the process.
- Will prevent catalyst bed from slumping as the gas feed flow will be maintained across all of bed.
- As the product leaving the column will be a gas/liquid mix as opposed to just gas (due to introduction of the PCHE cooling and heating), only one medium pressure liquid/gas separator will be required downstream due to the hydrogen not being supplied in excess, as the old procedure required. This will reduce the total energy consumed across the separation, along with emissions.
- PCHE will result in a reduction in size of shell and tube heat exchanger by a magnitude of 4-6.

Environmental Gains
- The heat integrated process within the column with the PCHE has the potential to provide reactor energy savings of up to 20–25%. Maximising heat recovery is not only crucial financially, but also assists in minimising environmental impact.
- The removal of a hydrogen quench will not only promote lower cost production, but also reduce harmful water discharge.
- The closer temperature control gained between catalysts beds by the PCHE will increase catalyst life thus reducing the consumption and energy demands associated with treatment with steam to clean them.

Safety and Corrosion Concerns
- Catalyst coking is a major risk when excessive contact time takes place between fuel and bed. Countercurrent flow will reduce this liquid hold-up.
- PCHE have been made in accordance to the ASME regulations.
- The figure on page 517, above, showing the H_2S partial pressure profile, demonstrates the improvement in reducing the concentration of hydrogen sulphide across the length of the reactor. The intensification and simplification of the reactor makes the design inherently safer through reducing both the hydrogen consumption and improvements to axial temperature distribution.
- Hot product recycle helps moderate use of hydrogen gas at high temperatures.
- The high concentration of H_2S over the initial section of the reactor will be controlled by the pulsating flow, thus minimising corrosion affects.
  In countercurrent flow this favours the conversion of a wider range of more difficult heavy distillates; in particular, reactions that in a concurrent reactor are limited by equilibrium.

Step 12: Checklist

1. *Is my plant/process currently running under optimum conditions?*

At present the refinery hydrotreater is running close to optimum operating conditions required for the hydrogenation reaction to take place. The only way performance could be improved would be by improving the catalyst which does not lead to direct process intensification; only performance improvements. Hydrogen gas is being supplied from the catalytic reformer.

2. *If not, Why not?*

The distillation column that proceeds downstream of the TBR could be used to provide heat from the high pressure saturated stream re-boiler back into the column's non-reactive zone to provide both heating and cooling quench instead of recycling the hydrogen gas. If it is believed that a hotter high pressure/temperature separator is required to refine the liquid from the recycle gas, re-boiler steam could be recirculated to provide heat and reduce wasted energy.

- Efficient steam recycles.
- Improve heat recovery system by passing the recycle gas plate through plate fin heat exchanger before compressor to extract heat.

3. *Is it cost-effective to reach optimum running conditions with my current plant/process?*

The current approach of adopting a catalyst bed that accommodated the majority of the column is not cost-effective. The issue here is that with the hydrogen sulphide concentration increasing with the length of the conventional TBR, the residence required for the reaction will be significantly increased. As a result, current processes are struggling with high hydrogen consumptions to overcome this. With this increase residence comes an increase in the overall reactor size and continuous attrition of the bed, which already promotes environmental concerns.

4. *If my plant is optimised is there still a case for looking at PI?*

Yes, the purpose of this PI as described in Step 8 was to have reduced catalyst consumption, reduce hydrogen partial pressure, reduced temperatures/increased pressures and increased space velocity. Only with increasing temperature, does the H_2S removal in current process become efficient; however, at a reduced fuel octane rating.

High pressures and elevated temperatures are required for the catalyst regeneration to be an active process due to the high pressure drop across the continuous catalyst bed; thus resulting in the need for additional hydrogen quench sections.

5. *Do I know all the current operating parameters on my plant/process?*

For the conventional TBR all operating parameters are known apart from the quantity of hydrogen that is to be utilised within the reaction, wetability of catalyst, gas/liquid interfacial area and gas-side mass transfer co-efficient.

6. *Can I obtain these data?*

Based upon the stoichiometric equations, the hydrogen consumption can be estimated. A bench scale laboratory could be used to assess reaction times along with the factors affecting the internal mass transfer. As data regarding the PI of packed beds is limited at the moment, an improved understanding will be required of this gas/liquid interaction with the packing. The experiment could involve analysing liquid/gas distribution, flow ratio between pulsated and normal flow. Comparison should also be drawn between the experimental and theoretical values for the separation effectiveness of the packing when pulsation flow is introduced. The resultant HTUs should be compared as this will indicate the extent of intensification, which will allow for scaling up.

7. *Do I have information on the operating parameters of any replacement plant?*
In terms of operating conditions, the replacement plant will be similar in parts to the conventional process. The catalyst contact time and pressure drop across the bed will, however, be unknown. This change in reactor configuration will result in radical PI effects. However, the PCHE has high pressure capabilities along with enhanced safety performance, which not only reduces inventory but also tube vibrations that are regular within shell and tube heat exchangers. Through calculations undertaken during development, plant equipment will be troubleshooted to find ideal operating conditions.

**8. *Will any replacement plant/process have significantly different requirements
to the existing plant (Throughput, be continuous or intermittent [batch], product type, operating temperatures, pressures, quality control)?***
The degree of compactness of the Heatric PCHE could potentially result in laminar flow, due to the channel sizes which could reduce the refinery treatment rate. Heatric have advised that the channel size utilised can range between 0.5 and 5 mm; however, there is still potential to achieve both turbulent liquid (Reynolds of 5,000) and gas (Reynolds up to 100,000). In terms of quality control, this may cause the process to be intermittent and reduce the sulphide conversion, but in those conditions the internal diffusion will be improved and the gas consumption will be reduced, which currently is 60% of the total cost.

9. *Are these requirements realistic for a conventional plant?*
The utilisation of pulsation flow is realistic within a current process. This form of active PI enhancement will allow for the utilisation of lower velocities, which will allow for reductions in the energy utilised currently through pumping.

10. *If they are, why do I want to study a PI plant?*
Cost reductions; efficiency enhancement and greater final product purity; inherent safety; and a smaller plant footprint, along with reduced environmental impact by decreasing unit hardware sizes.

11. *If they are not realistic for a conventional plant, how will a PI plant help?*
The countercurrent flow distribution, along with PCHE, will create micro-channels which will create better gas-to-liquid flow distribution. The induced pulsation flow will allow for more distributed bubbling of middle distillate and hydrogen within the TBR, resulting in a compressor of reduced volume and size that is

fundamentally required to recycle undissolved hydrogen gas whilst reducing the need to pressurise hydrogen to such high pressures. Removing the need to dilute the incoming feed with hydrogen, will also contribute towards a lower compressor load and hydrogen consumption, which ensures that the column operates in an economical, safe and intensified manner.

The pulsation strategy will reduce the pressure drop across the vessel in comparison to normal conditions and therefore extend the life of TBRs thus providing a financial and maintenance incentive. The pulsations could also act as heat exchanger enhancement through fluid vibration. With respect to cleaning, the ultrasound would not only remove surface fouling within the PCHE, but will also improve mixing and reaction kinetics.

12. *Will a radical change of process/plant affect my product?*

This process change does run the risk of not performing like the conventional TBR but could potentially improve the quality by increasing the efficiency of the column. With pulsated flow being introduced along with a PCHE, the instrumentation and control could be significantly affected between catalyst reaction zones, which would affect response time, residence time and accuracy.

13. *Is my knowledge of PI sufficient to answer the above?*

My knowledge of PI has allowed me to identify potential areas of improvement; however, the feasibility of the selections may come into question.

For example, with the gas distributer being fed from the bottom, the concern is of plugging the distributor with small catalysts from the bed as they are close to each other. To minimise this a porous plate could be utilised to separate the hydrogen feed point from the bed or a monolithic catalyst bed could be used. However, monoliths in large-scale petrochemical applications are relatively new. The counter flow would correct this; however, the addition of pulse will counter this, resulting in lower reactor effectiveness due to poor fluid distribution. More research will be required in this area in order to make an assured decision.

In order to make more detailed PI analysis, outside assistance will be required in relation to the integration of PCHEs upstream and downstream of catalyst beds.

14. *If not, are there others in my organisation who could assist me?*

At present there are no individuals who could advise about how the PCHE will integrate between the catalyst bed and how this will affect the temperature and pressure control within the column. In terms of implementing PI in general, there are sufficient resources available to put together a justifiable PI proposal.

15. *Should I consider employing outside consultants?*

For this PI to be successfully implemented, consultants from Heatric and other heat exchanger in-house design specialists (Foster Wheeler) will be engaged. The channel diameters and depth could potentially be increased; however, this will require greater technical input.

16. *Do I have sufficient time and resources within my own team?*

At present this process is lacking crucial research data; in particular, experimental theory on the affect on the HTU when the following are used: counter flow, angled liquid distribution and manipulated gas distributor hole sizes. In order to generate a valid data set, significant research will be required into the actual internal mass transfer characteristics and how that relates to flooding behaviour. With research on packed TBR very limited, this is crucial to the viability of the project once it is scaled up.

17. *Do my timescales and potential budget allow for any research and development work?*

In order to run these test effectively at a batch scale, a significant level of funding will be required to erect a column and have a large supply of hydrogen present for the reaction. Results will be used to justify whether this process is feasible at a large-scale. Environmental demands for cleaner fuel are getting much stronger. Whilst a TBR integrated with PCHE is being developed, research could be conducted into the role of pulsation in a conventional reactor.

18. *If not, am I limited to equipment that is already fully developed?*

At present there is no piece of equipment developed other than the PCHE to undertake intensive product testing. The single source supplier of this product, currently, is Heatric, who have effectively established this CHE technology.

19. *If this is the case, is such equipment available?*

Other compact heat exchangers do exist, however, for two-phase or liquid gas systems the best option is the PCHE. Presently, research identifies the use of PCHE as not being fully developed, unlike plate fin heat exchangers. For the purpose of research, a brazed plate fin heat exchanger could be used. However, it may be more difficult to clean due to the passive enhancement considerations of thin channels. When an integrated PCHE reactor is available, trials should be repeated.

20. *Do I then have the information to carry out:*

   a. *A risk assessment*

A process risk assessment and associated programme risks will need to be undertaken thoroughly. With the removal of hydrogen quench streams, there is the concern that the product recycles and feed preheating through the PCHE will not provide adequate cooling between catalyst reaction zones. The remainder of the chemicals and processes undertaken will be similar to those operating under the conventional form. In terms of time of PI implementation, programme risks (execution/completion) will be associated with the environmental constraints and timeframe.

   b. *A technical feasibility study*

A technical feasibility will be undertaken by both internal process engineers along with assistance from external contractors and consultants (Foster Wheeler, Heatric).

   c. *A rigorous financial appraisal*

Due to the current unavailability of this type of integrated heat exchanger and reactor configuration, the initial technical costs will be high. To confirm this, a detailed capital expenditure proposal will be generated that details capital, future repair and key indicators such as payback time of this unit.

21. *Have I considered the availability of grants?*
At present, no grants are available.

22. *Can I now make a comprehensive case for examining a PI solution to my plant?*
Having initially demonstrated the existence of high hydrogen operation cost alongside poor interfacial mass transfer with the current process, the relevant information accumulated throughout on technical feasibility, associate project risks, social and environmental implications along with external assistance, make this process a comprehensive case for PI.

References

Ancheyta, G., et al., 2011. Modeling and Simulation of Catalytic Reactors for Petroleum Refining. John Wiley & Sons, New Jersey, (pp. 53-62).

Chunshan, S., et al., 2011. Desulfurization Clean Fuels and Catalysis Program. The Pennsylvania State University, Pennsylvania, USA, (pp. 651–659).

Delgado, S., et al., 2002. Organometallic Modeling of the Hydrodesulfurization and Hydrodenitrogenation Reactions. Kluwer Academic Publishers, New York, (pp. 1–32).

Filho, R. M. et al., 2008. Multi-Feed HDT Reactor High Performance Operation. Brazil. pp. 1–8.

Packinox. (2011). PACKINOX for gasoil hydrodesulfurization - Hydrotreater. Available: <http://195.200.108.28/pkfxpp02.htm> (Last accessed 29.11.2011).

Robinson, K., 2011. Reaction Engineering. Mega-Carbon Company, Illinois, (pp. 2557-2567).

Sivakuma, S.E., et al., 2011. Innovations for Process Intensification in the Process Industry. Indian Institute of Technology Kanpur, India, (pp. 1–4).

Case Study 4: Alkylation

Step 1- Analyse the business and process drivers

The business drivers to intensify the alkylation process are:

- Increased efficiency and reduction in overall process time via optimisation of the reaction and settling stage. This allows increased production rate.
- Reduction in plant size. Some units could potentially be widespread and may visually impact the surrounding environment due to excessive height.
- Reduction in energy consumption. This can be achieved by incorporating more efficient methods of heat transfer or reaction mechanisms, which in turn can lead to lower operation costs and a reduction in CO_2 emissions.

Process drivers which could lead to a change in process may be:

- Flexibility in operation, which may be achieved by the use of advanced reactive and hybrid separation techniques.
- Potential to allow an inherently more safe operation by changing the processing route or reducing the volume of harmful materials.
- Greater conversion of olefin feed to higher octane compounds.
- Potential to utilise the heat produced in the reaction sequence in a synergetic relationship with other parts of the refinery.

Step 2: Overview the whole process knowledge elicitation (process understanding)

Alkylation is used in refining to produce high-octane petrol or gasoline from the combination of light olefins. These olefins are in the C_3–C_5 range and the reaction occurs in the presence of a hydrofluoric or sulphuric acid catalyst. The purpose of this process is to produce a hydrocarbon which is advantageous for gasoline blending, allowing octane and antiknock specifications to be met.

The chemistry

The chemistry involved in the alkylation process is relatively simple; however, there are several reactions which may occur either consecutively or in parallel. The main two-phase reaction mechanism involves the production of carbon cation intermediaries and has several reaction stages (Kranz, 2003). The general reaction is outlined by:[5]

$$C_4H_{10} + C_4H_8 + H_2SO_4 \rightarrow C_8H_{18} + HSO_{4^-} \rightarrow \Delta H_r = -259 \text{ Btu/lb}$$

This equation details the basic reaction mechanism for the production of isooctane from isobutane and butylene.

---

[5] Taken from a hydrogen transfer reaction in catalytic reforming (De Lasa and Doğu, 1991).

In order to fully understand the alkylation process, the reaction is given in more detail. The first stage of the alkylation process involves the reaction of an olefin, in this case, isobutylene with sulphuric acid, represented by:

$$
\begin{array}{ccc}
\text{Olefin} & \text{Alkyl sulfate} & \text{Cation}
\end{array}
$$

$$
\underset{\displaystyle \overset{\text{C}}{|}}{\text{C}-\text{C}=\text{C}} + \text{HX} \rightleftarrows \underset{\displaystyle \overset{\text{C}}{|}}{\text{C}-\underset{\overset{|}{\text{X}}}{\text{C}}-\text{C}} \rightleftarrows \underset{\displaystyle \overset{\text{C}}{|}}{\text{C}-\underset{+}{\text{C}}-\text{C}} + \text{X}^-
$$

This initial reaction takes place via the production of an intermediary cation. Following this reaction with the hydrogen ion from the acid, a cation is produced. This then reacts further with other molecules of isobutylene to produce a longer 8-chained hydrocarbon cation.

$$
\begin{array}{ccc}
\text{buty1 cation} & \text{i-butene} & \text{2,2,4-TMP}^+
\end{array}
$$

$$
\underset{\displaystyle \overset{\text{C}}{|}}{\underset{+}{\text{C}}-\text{C}-\text{C}} + \underset{\displaystyle \overset{\text{C}}{|}}{\text{C}-\text{C}=\text{C}} \longrightarrow \underset{\displaystyle \overset{\text{C}\quad\text{C}}{|\quad|}}{\underset{\displaystyle \overset{}{|}}{\text{C}-\text{C}-\text{C}-\underset{+}{\text{C}}-\text{C}}}
$$

The longer chained cation then undergoes isomerisation and methyl shift to form a more stable cation:

$$
\underset{\displaystyle \overset{\text{C}}{|}}{\underset{\displaystyle \overset{\text{C}\quad\text{C}}{|\quad|}}{\text{C}-\text{C}-\text{C}-\underset{+}{\text{C}}-\text{C}}} \xrightarrow{\sim\text{CH}_3} \underset{\displaystyle \overset{\text{C}\quad\text{C}\quad\text{C}}{|\quad|\quad|}}{\text{C}-\text{C}-\text{C}-\underset{+}{\text{C}}-\text{C}} \quad \text{2,3,4-TMP}^+
$$

The larger cation then undergoes hydrogen transfer with a smaller molecule (in this case, isobutane) which allows the continuation of the sequence as a butyl cation is again produced. This can then react with an alkene, as shown previously, to produce isooctane. This is generally regarded as the main alkylation reaction:

$$
\underset{\displaystyle \overset{\text{C}\quad\text{C}\quad\text{C}}{|\quad|\quad|}}{\text{C}-\text{C}-\text{C}-\underset{+}{\text{C}}-\text{C}} + \underset{\displaystyle \overset{\text{C}}{|}}{\text{C}-\text{C}-\text{C}} \longrightarrow \underset{\displaystyle \overset{\text{C}\quad\text{C}\quad\text{C}}{|\quad|\quad|}}{\text{C}-\text{C}-\text{C}-\text{C}-\text{C}} + \underset{\displaystyle \overset{\text{C}}{|}}{\underset{+}{\text{C}}-\text{C}-\text{C}}
$$

The sequence shown above is, however, not the only set of reactions which take place in the alkylation process, as there are a number of side reactions which occur in the form of polymerisation. These side reactions have been known to reduce the quality of the alkylate produced, as heavier molecules are produced which reduce the octane number. It is thought, however, that some of these longer molecules will be retained in the acid phase and are thus responsible for the acid consumption throughout the alkylation process.

Overall, the product consists of approximately 20% C_5–C_7, 60–65% C_8 and the remaining 15–20% is made up of heavier compounds (Lee and Harriott, 1977). The timescale of industrial alkylation reactions are generally documented to take several minutes.

The alkylation process

An overview of the conventional equipment utilised within the refining industry is detailed in the below figure. The sulphuric acid catalysed reaction is carried out at temperatures between 2°C and 20°C and pressures of 15–85 psig (Johnson, 1994). The reaction is exothermic and thus requires refrigeration.

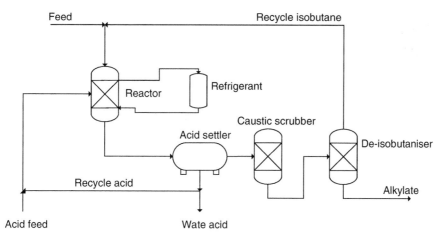

Drawn in reference to Alkylation, SET laboratories Inc. (2011).

In the alkylation process, olefin feedstock is fed to a reactor where it comes in contact with concentrated sulphuric acid (85–95%). The figure below outlines a typical alkylation reactor set-up, with fresh acid and hydrocarbon feed being fed to an acid reactor. The different phases are then allowed to settle on the basis of their different densities, allowing the lighter hydrocarbon to be removed as the top alkylate product. The alkylate is then typically depropanised, debutanised and deisobutanised. It should be noted that only the deisobutaniser has been shown above, in fitting with the context of the example. The acid can then be recycled and reused.

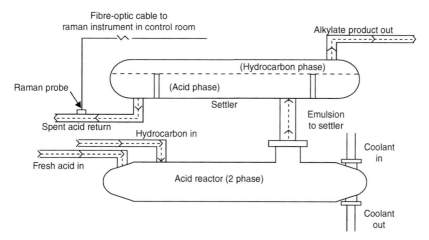

A typical dual alkylation reactor (Process Instruments Inc. Acid Alkylation, 2011).

Generally, there are five important variables which affect the operations in terms of product quality and operating expenditure of an alkylation unit. These are:

1. Isobutane concentration.
2. Olefin space velocity, otherwise known as the volume of olefin per unit volume of sulphuric acid in the contactor.
3. Temperature.
4. Acid concentration.
5. Mixing.

Step 3: Examine the PI blockers

In order to ensure the appropriate level of PI implementation, the conditions and parameters of the process need to be thoroughly considered. As the olefin or alkene feed consists of light hydrocarbons, it follows that they will be present in the gas phase unless chilled below 273 K. This means that fouling and particle size is unlikely to be an issue when considering PI.

One potential PI blocker is the strong acid catalyst required for the alkylation reactions to proceed. This may limit the equipment selection due the highly corrosive nature inside the reactor vessel.

A further possible reason not to implement process intensification within alkylation may be backed by the range of side reactions which can occur. This may lead to a situation where an intensified system could not produce or control the ratio of products required, leading to inefficiencies and resulting in the opposite of the desired effect associated with PI.

Further blockers may arise when the new PI equipment is implemented and tested. If a high feed consumption with poor conversion occurs, then the set-up may need to be revised. This, however, should not deter the use and further consideration of PI implementation.

Step 4: Identify rate limiting factors

Within the alkylation process there are several potentially rate limiting factors which arise from considering the five main variables listed in *Step 2*.

Isobutane concentration: Maintaining the isobutane concentration is necessary within the alkylation process to ensure that a high-octane molecule is produced. If a low concentration occurs then the resultant reaction involves olefin-olefin polymerisation, decreasing the octane number as well as increasing the acid usage. Typical isobutane-olefin ratios are in the range of 5/1 or 10/1, with the below figure illustrating the effect on acid consumption.

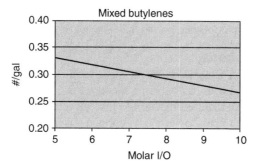

The effect of increase isobutane/olefin concentration on acid consumption in gallons (Caton et al., 2008).

Olefin space velocity: Further rate limiting factors may arise when considering the required contact time or olefin space velocity as mentioned previously. The space velocity refers to a specific contact time which is required to allow a suitable conversion to be reached. The space time is usually described as a ratio of the volume of olefin provided per hour, against the volume of sulphuric acid inside the reactor. Typical values range from 0.25 to 0.5/h (Kranz, 2003). If the contact time is not high enough then the reaction will not proceed as planned and efficiency will be dramatically reduced.

Temperature: As previously stated, the reaction is exothermic and it therefore follows that the heat produced from the aforementioned reactions has to be removed as it is otherwise detrimental to acid consumption and octane production (Caton et al., 2008). This typically takes place by the utilisation of a refrigerant, with operating temperatures of around 4.5–20°C.

Acid concentration: The acid concentration of the reaction should typically continue to stay above 85% (wt), as below this concentration polymerisation persists resulting in the conditions being too adverse to maintain acid concentration. Furthermore, a ratio of between 45% and 65% (vol/vol) (Kranz, 2003) acid-to-hydrocarbon should be achieved to produce high-quality alkylate. It should be noted, however, that acid runaway can occur if the acid strength stems below the 85% outlined previously. The runaway reaction happens when the acid is not strong enough to catalyse the alkylation reaction, resulting in olefin–olefin polymerisation reactions. The polymers produced are soluble in acid, allowing further dilution and thus propagating the runaway reaction. (Liolios, 2001).

Mixing: There needs to be sufficient contact between phases in order to allow the reaction to proceed fully, resulting in a high conversion of olefin-to-high octane compound. If the phases are not adequately mixed it will indefinitely be rate limiting.

Further to these factors, which relate directly to the process variables, it should be noted that a typical alkylation reactor leads one to assume the process would require a certain degree of time for phases to separate. This therefore promotes a batch operation with one reactor, or the requirement for a number of reactors and separators in series; an important point when considering PI in this context.

Step 5: Assess PI viability

If the previous four steps are considered, a good indication of process intensification viability can be gained. If, firstly, the chemistry is considered, there needs to be adequate scope for the required reaction to take place to allow the alkylation process to proceed. There may, however, be scope to enhance the rate of this reaction based on the contact time with the sulphuric acid catalyst. It should be noted that this point focuses more upon the equipment than the chemistry.

There is a wide scope for process intensification to be applied within the concept of the main alkylation reactor within the alkylation process, with many patents in place already.

Step 6: Review the business and process drivers

If the business drivers are once again considered, it can be seen that intensification in the alkylation process would contribute to savings in energy and emissions, alongside efficiency and potentially dramatically reduce process time.

Process drivers bring to light the operating conditions which need to be met to allow the intensified process to be at least as effective as the original scheme. In the situation of alkylation, the main concerns/areas for potential intensification are effective heat removal and promotion of the correct reaction sequence.

Further process drivers arise when considering some aspects of research, which detail that there is a direct link to agitation within the alkylation reactor and an increased octane number of the product (Lee and Harriott, 1977). Zeolites have also been detailed to aid the sulphuric acid alkylation process. Both of these processes may therefore provide or contribute to the necessary incentives to incorporate PI into the existing alkylation process, alongside offering an alternative route to intensification.

Step 7: List the initial concepts

The alkylation process has several different process intensification concepts which can be applied:

- The use of reactive and hybrid distillation to perform the olefin acid contacting, whilst allowing separation to occur without the requirement for a settling vessel. This would reduce process time, equipment size and potentially increase reaction efficiency.
- The use of reactive distillation with a novel, solid catalyst such as a zeolite. Such catalysts are, however, expensive and require regeneration (de Jong et al., 2004-2012).
- Introduction of a compact heat exchanger to utilise the waste heat from the exothermic reaction which takes place, providing efficiency gains elsewhere in the process plant. This may be in the form of a compact reactive heat exchanger, otherwise known as a HEX-reactor. Efficient heat extraction also ensures the reaction proceeds as intended, minimising acid consumption.

- An adapted version of the rotating packed bed could be incorporated to replace the reaction vessel, allowing greater gas–liquid contacting. This would provide a trade-off with increased energy usage against improved reaction and potentially decreased reaction time.
- There may be some potential to utilise a liquid cooled rotating anode to extract the heat produced in the exothermic reaction. This, however, would require careful consideration of the alkylation reaction conditions.

Step 8: Generate design concepts
Concept 1 – A reactive hybrid separations column

As the alkylation process involves two consecutive unit operations, a reaction in a reactor vessel followed by separation via sedimentation; in order to intensify the process, there is a possibility of the introduction of a reactive or hybrid distillation unit, either heterogeneously (solid catalyst) or homogenously.

The use of a reactive distillation column to perform heterogeneous alkylation would have several benefits, the first being the ability to allow continuous operation in comparison to natural settling techniques. The intensification of the original reactor–separator design into a single column may also produce an economic incentive. Furthermore, there are several configurations (Dragomir and Jobson, 2005) which can be employed in order to achieve the exact separation of components required, as shown below.

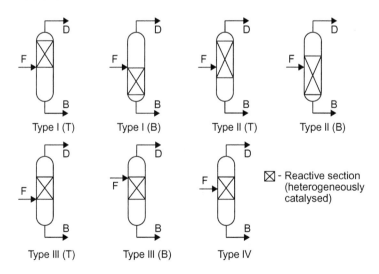

The various set-ups possible when considering reactive distillation for a single-feed heterogeneous system.

A seminar on reactive distillation from Professor Andrej Gorak has been summarised in Appendix A.

Concept 2 – A reactive contactor–separation column or rotating packed bed

A further intensification opportunity could arise if a gas–liquid contactor column is considered. It is a relatively simple principle which would see the simple separation of sulphuric acid and hydrocarbons, as sulphuric acid has a high boiling point (BP = 337°C). Difficulties may arise when considering the path of the gas through the column and thus the gas–liquid hold-up. This is because the reaction proceeds best at the point when the isobutane, olefin and sulphuric acid are in sufficient contact with one another, with higher conversions being achieved as a result of prolonged contact until all reactants have been used up.

The contactor column would work much in the same way as the reactive distillation unit, with various levels of packing or trays to promote gas–liquid contacting for the hydrocarbon feed and the acid-liquid catalyst. The specific design of this would utilise conventional absorption column design principles, resulting in counter-current gas–liquid contacting. Isobutane recycle could be introduced at multiple stages to ensure the correct reaction sequence is promoted, a diagram of which is detailed below:

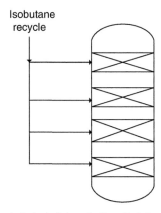

Multiple stage isobutane recycle to help intensify the alkylation process.

The concept could use either trays or packing, so long as the liquid–gas hold-up is long enough for the alkylation reaction to proceed. This is detailed as being up to several minutes within industry.

Similar principles would arise when considering the rotating packed bed, as this is essentially a further intensified separations column. Clearly, this leads to a more conventional intensified process, with greater gas–liquid contacting unit area than the standard contactor column in an inherently smaller unit. It may, however, also prove difficult to incorporate sufficient heat removal units into the design of a rotating packed bed as the centrifugal force would disrupt conventional fluid dynamics for liquid cooling.

Concept 3 – Implementation of a reactive compact heat exchanger

As the alkylation reaction is exothermic, a reactive heat exchanger unit (HEX-reactor) could be employed in order to effectively utilise the otherwise wasted heat. HEX-reactors are usually derived from existing PCRs and generally have the greatest potential when there is a fast exothermic reaction, which produces a certain amount of by-product.

As there are essentially three feeds to the alkylation reaction (isobutane recycle, olefin feed and sulphuric acid catalyst), the reactive heat exchanger will have to be versatile and robust in its design. Various different diffusion-bonded heat exchangers allows multiple inlets, while also increasing the safety aspects of handling hazardous fluids. This has a big part to play in the design as it allows isobutane to be either mixed with the olefin feed or added on a passage-by-passage basis, leading to a reduction in isobutane consumption and potentially energy usage elsewhere in the plant.

One existing reactive heat exchanger which is capable of handling these requirements is known as the ShimTech® reactor developed by CHART Energy & Chemicals below (Stryker, 2006):

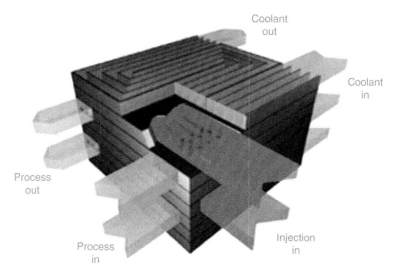

A three-dimensional representation of a ShimTech® reactive heat exchanger, detailing the use of three or more streams.

Potential design specifications of the ShimTech® exchanger would include:

- Stainless steel use to minimise corrosion of the strong acid catalyst used.
- Channels with dimensions greater than 1 mm in order to mitigate against potential fouling.
- Catalyst injection along the reaction pathway.
- Diffusion bonded to endure pressures above that of the reaction (85 psi or 6 bar).

- The size of the reactor cannot be detailed specifically, as the plant requirements are unknown, however, there is a scalability factor associated with the design.

This type of reactor has further benefits when considering the reaction rate, as it is determined by the space velocity. Highest yields are generally obtained at the lowest space velocity, which means a low olefin feed flow rate for a fixed acid flow, or a high acid catalyst rate for a fixed olefin flow. In the refining industry this is usually a value between 0.1 and 0.4 (Frankenburg, 1958).

As this type of HEX-reactor is versatile, other PI blockers such as acid concentration or temperature control become very manageable, resulting in an increase in efficiency. One potential design flaw may however arise when considering the mixing within the channels of the ShimTech® reactor. This could potentially be tackled by the use of custom-made channels.

Furthermore, the reactor intensification, the acid–hydrocarbon separator unit will also be intensified by recycling acid rich emulsion rather than flat acid (Caton et al., 2008). This decreases acid consumption, as well as reducing side polymerisation reactions by keeping alkyl sulphates (Step 2, first section) in contact with the olefins and isobutene. The separator unit can therefore be reduced to 25–35% of its original size, just by varying the operating conditions and increasing the recycle ratio of acid emulsion. It has also been proposed that a reduction in acid consumption can result in energy savings of up to 38% (US Department of Energy, 2006).

Step 9: Select the best concept

If the heterogeneous reactive separation concept is first considered, one can see potential pitfalls may occur. These include problems encountered when trying to achieve the alkylation with a solid catalyst. The main reason for this being that the catalyst activity dramatically reduces when switching from a liquid to a solid, as only 20–30% of the active sites on a solid catalyst are strong enough to perform the alkylation reaction. Coking of the reactor can further increase the poor activity of a solid catalyst, resulting in regular regeneration increasing processing time and therefore reducing the effectiveness of the intensification in terms of processing time. The only mitigation step to ensure coking (polymerisation reactions) does not take place is to increase catalyst strength; something which cannot be assured in the heterogeneous system.

Hybrid separation for a liquid–gas system would potentially allow simple separation principles to be employed to ensure high gas–liquid contacting; however, such columns do not allow adequate manipulation of reaction conditions when compared to the ShimTech® HEX-reactor option. Furthermore, the energy required to operate the rotating packed bed may become a substantial factor when considering the continuous utilisation within the alkylation process.

From the information detailed in Step 8, the best design concept was therefore deemed to be the use of a reactive heat exchanger, coupled with the application of acid emulsion recycle. The main reason for this being the versatility associated with such a unit in comparison to other ideas generated, resulting in the PI blockers being overcome. Costs for the specific requirements of the HEX-reactor are hard to obtain; however, an estimate would be in the region of £100,000.

Step 10: Implement laboratory scale PI protocols

Implementing a laboratory or pilot scale PI unit is essential to the successful introduction of the larger unit into an existing process as it helps to ensure the suggested PI will function as intended. Laboratory scale PI investigations can often prove difficult, as it is hard to create a small-scale experiment to replicate the intensified operation of a full-scale facility. In terms of the ShimTech® HEX-reactor relevant testing could be subcontracted to CHART Energy and Chemicals who would be charged with the fabrication of a full-scale unit.

One variable which would require verification within the laboratory is the mixing ability of such a piece of equipment. One such method to do this would be to mix three specific chemicals (1-Naphthol, 2-Naphthol and diazotised sulphanilic acid) via the mixing sensitive Bourne reaction scheme (Reynolds, 2002). If the mixing in the system is perfect the chemicals produce a faster reaction which will be favoured. However, if mixing is poor the fast reactions are mixing limited and by-products are formed. The by-products produced can be used to gauge how well the system is mixed. This could be completed in a section of the suggested reactor or in a replica, economics permitting.

Step 11: Compare intensified concept with conventional or existing plant

Using the concept of a HEX-reactor allows process intensification to be observed in the following ways:

- Increased safety associated with the operation of the unit, including mitigation against leakages of flammable hydrocarbon or corrosive catalyst via diffusion-bonded flanges.
- Enhanced heat transfer and effective heat removal, allowing for reduced acid consumption.
- Enhanced mixing and therefore increased acid hydrocarbon contacting.
- Reduced sulphuric acid catalyst consumption via efficient heat removal and control of the isobutane-to-acid ratio.
- Reduced isobutane use and therefore reduced energy across plant via a reduction in downstream processing.
- As previously stated, conventional plants may require several reactors to achieve the necessary continuous production of alkylate. Potential to intensify several typical alkylation reactors into one or two units (for continuous operation), resulting in intensification in terms of process size.
- A reduction in carbon dioxide emissions through energy savings.

Step 12: Make the final choice of plant

The main design concepts of the ShimTech® HEX-reactor would bring together the required residence time along with heat removal to produce a unit which eliminated hotspots and ensured adequate contact between sulphuric acid and hydrocarbon.

This allows the reduction of side reactions, including polymerisation, producing a pure product with high conversion.

In order to intensify the separation unit, the acid will be recycled as an emulsion rather than flat acid. This provides a reduction in size by approximately 2/3 and a reduction in energy of 38%. Overall, the proposed process intensified unit is estimated to be approximately 20–50% of the size of the original equipment, with further opportunities being available downstream of the alkylation reactor.

The suggested intensified equipment would also allow an almost immediate implementation, requiring rerouted process streams as opposed to entire plant turn-around or shutdown. This may increase its desirability to be taken up at existing alkylation facilities by refinery managers.

The checklist may now be addressed.

1) Is my plant/process currently running under optimum conditions?
Conventional alkylation systems are often considered to be running near optimum conditions when conversion is high and acid consumption is low. Most favourable operation sees the rate limiting factors overcome. Optimum cooling may not be achieved if water is used.

2) If not, why not?
A coolant with a higher affinity for heat removal would allow optimisation for the refrigeration process, ensuring more effective heat removal.

3) Is it cost-effective to reach optimum running conditions with my current plant/process?
A refrigerant such as, R-410A (mix of difluoromethane and pentafluoroethane) may result in more effective cooling and therefore increased efficiency, providing the heat can be effectively extracted from the coolant elsewhere in the plant. Alternatively, effluent refrigeration could be used in accordance with process integration to allow intensified use of inventory, removing the requirement for a specific refrigeration system.

4) If my plant/process is optimised, is there still a case for looking at PI?
Regardless of the level of optimisation, process intensification can still offer significant benefits in the following areas:
- Lower inventory of reactants/chemicals and therefore a more inherently safe operation.
- Increased level of safety within the operation conditions of the facility e.g. diffusion bonded heat exchangers less likely to leak, or reduced chemical inventory.
- The production of a smaller, more environmentally friendly plant.
- Increased throughput via a more effective process.

5) Do I know all the current operating parameters on my plant/process?
Ranges of values are given for operating parameters, as petrochemical companies are careful about releasing proprietary information surrounding chemical processes. General process conditions have been listed previously.

6) Can I obtain these data?

All data can be obtained upon contact with the refinery or processing facility under consideration for PI, through either historic data or through process monitoring.

7) Do I have information on the operating parameters of any replacement plant?

As the proposed intensification solution is that of an existing company, data should be reasonably accessible in terms of general operation and sizing. Due to the versatility of the unit and the different variables considered, the estimated operating parameters may differentiate from the actual operating conditions and it would therefore follow that a member of CHART Energy and Chemicals be present upon process validation and commissioning.

8) Will any replacement plant/process have significantly different requirements to the existing plant (throughput, be continuous or intermittent [batch], product type, operating temperatures, pressures, quality control)?

The proposed PI plant allows the operation of a continuous process in the same fashion as the old plant, except with the need for less equipment. The exact required number of new intensified equipment is not known; however, an estimation relating the size of the new intensified units to the old reactors can provide a guide.

As previously mentioned, the intensification of the plant also involves changing the recycle ratio of acid to the original reactor, via an acid–hydrocarbon emulsion.

9) Are these requirements realistic for a conventional plant?

The suggested PI implementation keeps all the main alkylation principles in mind and is therefore deemed to be achievable within existing petrochemical industry.

Changing the acid recycle ratio and composition is something which is realistic for an alkylation facility as detailed by STRATCO, a DuPont company (Caton et al., 2008; Caton, Racen, & Troutman, 2008).

10) If they are, why do I want to study a PI plant?

Although requirements for a conventional plant are realistic, the study of a PI plant allows scope for innovation alongside providing considerable incentives:

- A safer operation and plant.
- Purer product with less side reactions.
- A more visually environmentally friendly operation.

11) If they are not realistic for a conventional plant, how will a PI plant help?

As the suggested ShimTech® intensified HEX-reactor adds a large degree of flexibility, the intensified process will allow greater control of the reaction and alkyl product when compared with the conventional process.

12) Will a radical change of process/plant affect my product?

The product should theoretically only increase in yield, as well as reducing in the unwanted polymerisation side reactions which take place. However, a dramatic change in the processing route for high-octane compounds may result in some potentially adverse side effects when all equipment is configured together.

13) Is my knowledge of PI sufficient to answer the above?
Current knowledge of PI and chemical engineering as a discipline is sufficient to answer the above questions. Implementation would require verification from experts within the fields of the suggested PI and alkylation.

14) If not, are there others in my organisation who could assist me?
Within the organisation of Heriot-Watt University resides a visiting lecturer by the name of Professor David Reay, who is an industry expert on process intensification and a chair of the PIN network. No other members of my organisation are deemed to be an expert on the previous topics covered.

15) Should I consider employing outside consultants?
Outside consultants would be necessary if the suggested PI equipment were to be implemented within an existing refinery or alkylation unit.

16) Do I have sufficient time and resources within my own team?
Sufficient time would be set aside within the organisation to fulfil the project needs, including the relevant project management plan and equipment tie-in schedule. As previously mentioned, outside resources in terms of consultancy would be required to implement the intensified equipment into the existing process.

17) Do my timescales and potential budget allow for any R&D work?
Research and development would be applied to accurately predict the behaviour of the new equipment not only on the original process, but also on the downstream apparatus and processes. This is essential to the success of any project and therefore sufficient budget would be allocated.
The petrochemical industry also adheres to a high level of safety and thus new processes would need to be suitably developed in order to pass existing regulations.

18) If not, am I limited to equipment that is already fully developed?
Limitations should not be a problem if the process equipment is fully investigated using research and development techniques. It should be noted, however, that there is a limited budget and time available for any process and therefore some areas of the process could possibly be limited.

19) If this is the case, is such equipment available?
The equipment associated with the project is readily available via CHART Energy and Chemicals, as previously advised, in the form of a ShimTech® reactive heat exchanger. Whilst the equipment is available, it must be optimised and configured to ensure it is effective in its application.

20) Do I then have the information to carry out: A risk assessment (health and safety), a technical feasibility study and a rigorous financial appraisal?
The health and safety considerations associated with the new process equipment must be assessed via a risk assessment. As the chemicals being processed are very much the same, potential issues are thought to be scarce; however, a HAZOP study will be completed.

A technical feasibility study can be conducted by any chemical or otherwise competent engineer who is familiar with the process and has access to the relevant conditions/product specifications.

At cost engineer within a refinery can complete a financial appraisal of the proposed equipment in a short period of time. A quote would have to be obtained from the PI equipment supplier following a cost–benefit analysis on the existing plant.

21) Have I considered the availability of grants?

Potential grants available to which the intensified alkylation process applied are:

- European grant for innovation, with a budget of €50.5billion for the next 7 years.
- Engineering and Physical Science Research Council science and innovation awards, £3–5 million over 5 years.
- Government grants from £60k–250k for innovation, research and development providing certain conditions are met.
- Further grants relating to energy saving incentives may also be obtained.

22) Can I now make a comprehensive case for examining a PI solution to my plant/process?

The implementation of PI can now be thoroughly considered by the combination of all the major deciding factors including safety, cost, technical applicability, compatibility and process limitations.

References

Alkylation, SET Laboratories, Inc., 2011, October 20th. Retrieved October 20th, 2011, from SET Laboratories, Inc.: <http://setlaboratories.com/alkylation/tabid/101/Default.aspx#Sulfuric> Acid Alkylation Process.

Caton, J., Racen, B., Troutman, G., 2008, September. Introduction to Sulfuric Acid Alkylation Unit Design. DuPont STRATCO Clean Fuel Technol., 1–28.

de Jong, P., Bitter, J. H., & van der Eerden, P., 2004-2012. Process intensification in alkylation processes using zeolite catalysts with molecular traffic control. Retrieved November 13, 2011, from Netherlands Organisation for Scientific Research: <http://www.nwo.nl/nwohome.nsf/pages/NWOP_65CC9B_Eng>.

De Lasa, H.I., Doğu, G., 1991. Chemical Reactor Technology for Environmentally Safe Reactors and Products. Kluwer Academic Publishers, London, Ontario, Canada.

Dragomir, R.M., Jobson, M., 2005. Conceptual design of single-feed hybridreactivedistillation columns. Chem. Eng. Sci. 60 (16), 4377–4395.

Frankenburg, W.G., 1958. Advances in Catalysis and Related Subjects, vol. 10. Academic Press Inc., New York.

Johnson, D. W., 1994. Sulfuric Acid Release Report. Retrieved October 22nd, 2011, from Quest Consultants Inc.:<http://www.questconsult.com/>.

Kranz, K., 2003, May. Alkylation Chemistry. In: Ken Kranz, Mechanisms, Operating Variables, and olefin Interactions Managing R&D Chemist, STRATCO, DuPont Company

Lee, L.-M., Harriott, P., 1977. The Kinetics of Isobutane Alkylation in Sulfuric Acid. Ind. Eng. Chem. Proc. Des. Dev, 282–287.

Liolios, G., 2001. Acid Runaways in a Sulfuric Acid Alkylation Unit. Glenn Liolios, Vice President, STRATCO Inc.

Process Instruments Inc Acid Alkylation, 2011. Retrieved November 05, 2011, from Process Instruments: <http://www.process-instruments-inc.com/pages/acid.html>

Reynolds, I., 2002. Laboratory Protocol of PI. *8th PIN Meeting*, (p. BHR Group Limited). Cranfield University.

Stryker, L., 2006. Compact Heat Exchange Reactors. Retrieved November 13, 2011, from CHART Industries: <www.chart-ind.com/app_ec_reactortech.cfm>.

US Department of Energy, 2006. Energy Bandwidth for Petroleum Refining Processes. Office of Energy Efficiency and Renewable Energy Industrial Technologies Program, p. 17.

Appendix A

Reactive and Hybrid Separations: Modelling and Experimental Verification[6]

Prof Andrej Górak, Technische Universität Dortmund

Professor Andrej Górak is a member of the Department of Chemical and Biochemical Engineering at the Technische Universität, Dortmund. More specifically, he is heavily involved with the laboratory of fluid separations group, which is one of the leaders in the field of modelling and simulation of hybrid separation processes. Other major focuses at the university are on process intensification and reactive separations in which Górak has numerous publications. The university itself is very large, with the chemical engineering department holding 1,000 students, over 100 PhD students and 12 full-time professors.

The main focus of the seminar was to discuss the opportunities within reactive and hybrid separations. An overview of each of these was given with specific examples. Reactive separations involve: reactive distillation, reactive absorption, reactive extraction and membrane reactor. Hybrid separations involve: distillation and membrane; extractions and crystallisation; and distillation and crystallisation. Górak likened his line of work to an "innovative palaeontologist" referring to the changing of existing separation and reaction methods in order to achieve a better result.

Reactive Distillation

In general, there are several steps to chemical production; however, the idea proposed by Gorak and his team was to replace two stages with one intensified stage of reaction and separation.

Reactive separations occur at the same time and place within the apparatus. A reactor can simply be placed inside the column, typically resulting in a column with three sections, with the middle section being a reaction zone in which lies catalytic structured packing. The main principle being that no waste is produced, allowing for simultaneous reaction and separation. It is important to know exactly how to design the internal of the column, with relevant layers of separation packing and catalytic reaction packing. The combination is the key to successful operation (Multipak), and conversions of up to 100% are possible with the correct equilibrium shift. The column itself is 50 mm in diameter with a packing height of 5.4 m.

One major problem is that the reactions have to be accurate e.g. if by-products are produced it can create problems within the column and it may not function as intended. To solve this issue rigorous testing and validation takes place.

---

[6] Professor Górak gave a seminar to chemical engineering students at Heriot-Watt University in 2011, and this in part, stimulated this PI case study.

Hybrid separations

Górak also touched upon hybrid separations, explaining how they utilise membranes within a distillation column in order to provide pervaproation (PV) and vapour permeation (VP). One issue which presented itself when membranes and distillation were combined was the different combinations in which the separate processes took place. In order to get around the problem of testing many different combinations and set-ups, the membrane was introduced inside the column between packing. The optimum configuration was achieved by three PhD students spending a combined total of 9 years developing the technology.

The future of this line of work involves combining experiments and models to bring new ideas into the industry, as well as developing existing products and processes further to enhance unit operations. Distillation is currently the most developed unit operation, with over 40,000 columns worldwide, representing 70% of the industry energy costs. Reactive distillation is not as developed, reactive and hybrid (bio)separations belong to the emerging area of process intensification. Enhancement in this field will allow energy savings which can translate into large capital savings across the industry.

Appendix 1: Abbreviations Used

ACR	agitated cell reactor
ADL	Arthur D Little
ANL	Argonne National Laboratory (USA)
API	active pharmaceutical ingredient
ATR	autothermal reforming reactor
Bopd	barrels of oil per day
BUCT	Beijing University of Chemical Technology
CAPEX	capital expenditure
CHEAP	compact heat exchanger advancement programme
CHE	compact heat exchanger
CHP	combined heat and power
CIP	cleaning in place
CT	The Carbon Trust (UK)
CVD	chemical vapour deposition
DMFC	direct methanol fuel cell
DSMC	direct simulation Monte Carlo
EEBPP	The UK Government's Energy Efficiency Best Practice Programme
EFCE	European Federation of Chemical Engineering
EHD	electrohydrodynamics (an enhancement technique)
ELM	emulsion liquid membranes
EPSRC	Engineering and Physical Sciences Research Council
FBR	fluidised bed reactor
FPHE	formed plate heat exchanger
FPSO	floating production, storage and offloading
GLS	gas/liquid/solid
HEX	reactor heat exchanger reactor
HEXAG	Heat Exchanger Action Group
HF	hydrogen fluoride
HSE	Health and Safety Executive
IChemE	Institution of Chemical Engineers
IKE	Institut für Kernenergetic (Germany)
IPCC	Intergovernmental Panel on Climate Change
IP PCR	in passage printed circuit reactor
IRMPD	infrared multiphoton dissociation
ITP	Industrial Technologies Programme (USA)
JIT	just-in-time
LFM	lattice frame material
MAB	multiple adiabatic bed

MEMS	microelectromechanical systems
MHD	magnetohydrodynamics
MILC	metal-induced lateral crystallisation
MPI	main plant items
MRR	mean relative roughness
MVR	mechanical vapour recompression (Table 4.2)
NASA	National Aeronautics and Space Administration (USA)
OCM	oxidative coupling of methane
PCHE	printed circuit heat exchanger
PCR	printed circuit reactor
PEM	polymer electrolyte membrane
PHE	plate heat exchanger
PI	process intensification
PIN	process intensification network
POGT	partial oxidation gas turbine
ppm	parts per million
PVC	polyvinyl chloride
QA	quality assurance
RD	reactive distillation
RE	reactive extraction
RPB	rotating packed bed
RZB	rotating zigzag bed (Chapter 6)
SDR	spinning disc reactor
SEPA	Scottish Environmental Protection Agency
STR	stirred tank reactor
TEMA	Tubular Exchanger Manufacturers Association
TFT	thin film transistor
TICM	tubular inorganic catalytic membrane
TNT	trinitrotoluene
UN	United Nations
VCR	variable channel reactor

Appendix 2: Nomenclature

(note that most symbols are defined as they arise in the text).

A	species
A	pre-exponential function (in Arrhenius equation)
A	specific transfer area
C	cost (Chapter 2)
C_A	concentration of species
D	characteristic dimension (Chapter 2)
D	solute diffusivity (Equation 5.9)
d	diameter
E_A	activation energy
f	frequency of oscillation
F_o	Fourier number
h	film heat transfer coefficient (Chapter 3)
k	thermal conductivity
k	the rate constant (in Arrhenius equation)
k_L	local mass transfer coefficient (Chapter 5)
l	litre
M	mass flow rate
m	number of baffles (Chapter 5)
N	number of tanks in series (Chapter 5)
N_p	power number
N_u	Nusselt number
P	production capacity (Chapter 2)
P	particle motion (Chapter 5)
P	power (equation for N_p)
P_r	Prandtl number
Q	local heat flux (Equation 5.18)
r	radius
r_A	rate of heat loss of species
R	constant (section 5.3.7)
R	surface shear stress (Chapter 3)
R_e	Reynolds number
Re_n	Reynolds number based on net flow (Chapter 5)
s	film thickness
s, y	boundary conditions on spinning disc (Chapter 5)
Sc	Schmidt number
Sh	Sherwood number

t	time
t_e	film exposure time
tpa	tonnes per annum
T	temperature
T	surface tension/unit length (after equation 5.18)
U	film velocity
u	velocity (Chapter 3)
v	velocity
V	volume
V_m	maximum oscillation velocity
V_s	settling velocity

Definitions

Selectivity (%) = moles of product/(moles of feed in − moles of feed out) × 100
Conversion (%) = [moles (feed in) − moles (feed out)]/moles (feed in) × 100

Greek symbols

ε	power density (W/m^3)
ρ	density
μ	viscosity
ω	angular velocity (SDR)
θ	contact angle
ψ	velocity ratio
γ	indicates uniformity of suspension (section 5.3.7)

Suffices

n	(R_e)
	(R_o)
s	film thickness
w	wall

Appendix 3: Equipment Suppliers

The number of suppliers of equipment that can be used for intensified processes is rising. Suppliers include well-established manufacturers of compact heat exchangers. Several CHE suppliers listed also make HEX-reactors and/or microreactors. There are now a number of dedicated PI equipment suppliers, who also have a role in R&D. These include NiTech in the UK, IMM in Germany and many companies in the USA. Other companies have been in the PI field for a number of years, without promoting their equipment under the PI banner.

Note: The lists below are not exhaustive and have been compiled from information available to the authors at the time of writing. The listing of a supplier of goods or services does not constitute an endorsement by the authors of either its product or its competence. Neither does the omission of a supplier discriminate against its competence.

In many cases, brief data are given after the organisation details, summarising the PI-related activities of that organisation. Other data may of course be sought via the websites that are identified in the vast majority of cases. The names and addresses of other suppliers may be found in commercially available trade directories, (e.g. Compass Directory), University data sources, trade associations, or from Web searches.

Compact Heat Exchangers and Heat Exchanger Reactors
Brayton Energy

710 Rue de Vernon, Unité no.7, Gatineau, QC, J9J 3K5 Canada
Tel: (819) 557 1777; Fax: (819) 557 1778
www.braytonenergy.ca
Interesting compact gas-gas, gas-liquid and mesh heat exchangers.

Chart Energy and Chemicals

The Creative Industries Centre, Wolverhampton Science Park, Glaisher Drive, Wolverhampton WV10 9TG, UK
Tel: +44 (0) 1902 313870; Fax: +44 (0) 1902 429853
www.chartindustries.com
The Chart-kote and Marbond units are typical of the range offered by Chart.

Chart Industries, Inc. (US Company HQ).

One Infinity Corporate Centre Drive, Suite 300, Garfield Heights, Ohio 44125-5370 USA
Tel: +1 440 753 1490; Fax: +1 440 753 1491
www.chartindustries.com

Heatric Ltd

46 Holton Road, Holton Heath, Poole, Dorset BH16 6LT, UK
Tel: +44 (0) 1202 627000; Fax: +44 (0) 1202 632299
www.heatric.com
Heatric makes compact heat exchangers (the PCHE) and heat exchanger-reactors (the PCR). Heatric also make microreactors.

Laminova Production AB

PO Box 30, SE-662 21 Åmål, Sweden
Tel: +46 (0) 532 611 00; Fax: +46 (0) 532 128 63
www.laminova.se

Microinnova Engineering GmbH

Europapark 1, 8412 Allerheiligen bei Wildon, Austria
Tel: +43 (0) 3182 62626-0; Fax: +43 (0) 3182 62626-101
www.microinnova.com
(See also micro heat exchangers.)
The company also makes PI mini flow plant.

Porvair Ltd

Brampton House, 7 Regis Place, Bergen Way, King's Lynn, Norfolk PE30 2JN, UK
Tel: +44 (0) 1553 765500; Fax: +44 (0) 1553 765599
www.porvair.com
Porvair is a specialist filtration and environmental technology group, with operations in both the UK, US and China. The foam (ceramic or metal) can be used for compact heat exchangers and/or reactors. The company promotes the filtration more than the heat exchanger capabilities (2012).

SWEP International

Hjalmar Brantings vag 5, PO Box 105, 261 22 Landskrona, Sweden
Tel: +46 418 40 04 00; Fax: +46 418 292 95
www.swep.net
SWEP is a manufacturer of highly compact plate heat exchangers, including specialised units supplied for the Rotex heat pump.

Thermatron Engineering Inc

687 Lowell Street, Methuen, MA 01844, USA

Tel: +1 978 687 8844; Fax: +1 978 687 2477

www.thermatroneng.com

Thermatron specialises in the design and manufacture of compact liquid-to-air heat exchangers and above ambient cooling systems, utilising state-of-the-art technology such as fins and enhanced plates. Electronics thermal control is the main application area.

Within Lab

8A Petworth Street, London SW11 4QR, UK

Tel: +44 (0) 20 7617 7316

www.within-lab.com

The company makes highly compact heat exchangers using 3D printing methods (additive layer manufacturing).

Drying
GEA Process Engineering Inc.

9165 Rumsey Road, Columbia, MD 21045, USA

Tel: +1 410 997 8700; Fax: +1 410 997 5021

www.niroinc.com

A combination of vacuum and microwave drying can be beneficial (preferably at lower moisture contents for it to be economically viable) for the pharmaceutical and food industries.

Effluent Treatment
Hidrostal Process Engineering Ltd

The Galloway Centre, Hambridge Lane, Newbury, Berks RG14 5TL, UK

Tel: +44 (0) 1635 550440; Fax: +44 (0) 1635 550140

www.hidrostal.co.uk

(Venturi aeration.)

Hidrostal manufacture pumps and pumping systems, particularly for the water and waste industry.

Extractors
Beggs, Cousland and Company

Studio 3, Doges, Templeton on the Green, 62 Templeton Street, Glasgow G40 1DA, Scotland, UK

Tel: +44 (0) 141 556 5288; Fax: +44 (0) 141 554 7447
www.beggcousland.com
The company manufactures the Becoflex rotary demister/scrubber.

Extruders
B&P Process Equipment and Systems

Baker Perkins Ltd., Manor Drive, Paston Parkway, Peterborough,
PE4 7AP, UK
Tel: +44 1733 283000; Fax: +44 1733 283001
www.bakerperkinsgroup.com
B&P provides twin screw and single screw extruders to a wide variety of processing industries.

Thermo Electron Corporation

Emerald Way, Stone, Staffordshire ST15 0SR, UK
Tel: +44 (0) 1785 81 36 48
www.thermo.com/mc

Fluidised Bed Systems
Torftech Ltd

188 First Street, Greenham Business Park, Thatcham RG19 6HW, UK
Tel: +44 (0) 845 868 7241; Fax: +44 (0) 845 867 5199
www.torftech.com
Torftech is a Canadian company with major offices in the UK and Canada, offering reactor technologies for more rapid and accurate control of processes requiring gas–solid contact, often at higher temperatures than those experienced in conventional equipment.

Fuel Cells
INI Power Systems Inc

175 Southport Drive, Suite 100, Morrisville, NC 27560, USA
Tel: +1 919 677 7112
www.inipower.com
INI Power Systems produces direct methanol laminar flow fuel cells mainly for use in laptops, although other applications exist.

Heat Pipes
Los Alamos National Laboratory

PO Box 1663, Los Alamos, NM 87545, USA
Tel: (505) 667 5061; Fax: +1 505 665 4411
www.lanl.gov
LANL has an extensive history in developing heat pipes and their many applications within and out with chemical engineering applications. Current work is further developing the use of heat pipes in outer space although presently, applications are extremely vast; from cooling CPUs to nuclear power cells.

Thermacore International

780 Eden Rd., Lancaster, PA 17601, USA
Tel: +1 (717) 569 6551; Fax: +1 717 569 8424
www.thermacore.com

Thermacore Europe Ltd

12, Wansbeck Business Park, Ashington, Northumberland NE63 8QW, UK
Tel: +44 (0) 1670 859 500; Fax: +44 (0) 1670 859 539
www.Thermacore.co.uk
Thermacore has supplied heat pipes for use in process reactors.

Induction Heated Mixers
T. Giusti Ltd

Rixon Road, Wellingborough, Northamptonshire NN8 4BA, UK
www.giusti.co.uk

Instrumentation
ATA Scientific Pty

ATA Scientific, PO Box 2172, Taren Point, NSW 2229, Australia
Tel: +61 (02) 9541 3500
www.atascientific.com.au
The company supplies instrumentation for measuring rapid kinetics and is also involved in high gravity spinning mixers.

Intensified Process Plant Design, Development and Manufacture

Advanced Heat Engineering Ltd

Unit 2, Capenhurst Technology Park, Capenhurst, Chester CH1 6EH, UK
Tel: +44 (0) 151 347 2900
www.ctechinnovation.com
The company concentrates on processes where electric field can be used to enhance performance – e.g. microwaves and induction heating.

Alfa Laval Thermal

PO Box 73, S-221 00 Lund, Sweden
Tel: +46 46 36 65 00; Fax: +46 46 32 35 79
www.alfalaval.com
Heat exchanger reactors are now in the company's inventory (the Art Plate reactor), as well as heat exchangers and separators; and whole process plant design and build.

AM Technology Ltd

The Heath Business & Technical Park, Runcorn,Cheshire, WA7 4QX
Tel: +44 (0) 1928 51 54 54
www.amtechuk.com
Reactor control systems, innovative continuous reactor technologies including the variable geometry reactor, Coflore flow reactors and micro CSTRs.

B&P Process Equipment and Systems

Baker Perkins Inc., 3223 Kraft Ave. S.E. Grand Rapids, MI 49512-2027, USA
Tel: +1 616 784 3111; Fax: +1 616 784 0973
www.bakerperkinsgroup.com

Parr Instrument Company

211 Fifty Third Street, Moline, Illinois 61265-1770, USA
Tel: +1 (309) 762 7716; Fax: +1 (309) 762 9453
www.parrinst.com
The Parr Instrument company manufactures combustion bombs, calorimeters, chemical reactors (including tubular and CSTRs), pressure vessels, and related equipment developed specifically for laboratory use.

Xytel Corporation

4220 South Church Street Extension, Roebuck, South Carolina 2937, USA
Tel: +1 864 576 9777; Fax: +1 864 576 9799
www.xytelcorp.com
Turnkey projects on miniature process plants, demonstration units, etc.

Zeton B.V.

PO Box 9, NL-7500 AA Enschede, The Netherlands
Tel: +31 (0) 53 428 4100; Fax: +31 (0) 53 428 4199
www.zeton.com
PI pilot plant specialists.

Micro-Fluidics and Mems
Atotech Deutschland GmbH

Postfach 21 07 80, D-10507 Berlin, Germany
Tel: (+49) 30 349 85 0; Fax: (+49) 30 3485 777
www.atotech.com
Atotech manufactures for the electronics and micro sectors and has a wide range of
capabilities, including etching.

Dolomite Microfluidics

Unit 1, Anglian Business Park, Orchard Rd, Royston SG8 5TW, UK
Tel: +44 (0) 1763 242491; Fax: +44 (0) 1763 246125
www.dolomite-microfluidics.com
Dolomite Microfluidics applies microfluidic systems in instrument design.

STMicroelectronics

39, Chemin du Champ des Filles, C. P. 21, CH 1228 Plan-Les-Ouates, Geneva, Switzerland
Tel: +41 22 929 2929; Fax: +41 22 929 29
www.st.com
Systems on chips – including MEMS.

Tronics Microsystems

55 rue du Pré de l'Horme, 38926 CROLLES Cedex, France
Tel: +33 (0) 4 76 97 29 50; Fax: +33 (0) 4 76 97 29 51
www.tronicsgroup.com
Tronics Microsystems manufactures and delivers unique MEMS components.

Micro-Heat Exchangers
Karlsruhe Institute of Technology

Hermann-von-Helmholtz-Platz 1, 76344 Eggenstein-Leopoldshafen, Germany
Tel: +49 721 608-0
www.kit.edu

Developing micro heat-exchangers for use in lab, pilot and production scale processes. Improving heat transfer by combining aspects of different traditional designs.

Microinnova Engineering GmbH

Europapark 1, A-8412 Allerheiligen/Wildon, Austria
Tel: +43 (0) 3182 62626 0; Fax: +43 (0) 3182 62626 101
www.microinnova.com
Counter-flow micro heat exchanger: the novel Wt-series was developed as a heat exchanger for liquid/liquid, gas/liquid or gas/gas applications and can also serve for evaporation or condensation.

Microreactors (See Also Compact Heat Exchangers & Heat Exchanger-Reactors)

Clariant Competence Centre

Brüningstrasse 50, 65929 Frankfurt am Main Germany
Tel:+49 6196 757 60; Fax: +49 6196 757 8856
www.clariant.com

Corning S.A.S.

CETC Research Centre, Centre de Recherche, F-77210 Avon, France.
Tel: +33 1 6469 7521.
www.corning.com
Corning SAS manufacture glass microreactors.
See also www.micronit.com

Holst Centre/TNO

High Tech Campus 31, NL-5656 AE, Eindhoven, Netherlands
Tel: +31 40 40 20 400; Fax: +31 40 40 20 699
www.holstcentre.com AND www.tno.nl
The HELIX reactor is being developed here.

Micro Reactor Technologies Inc (Now part of Pall Corporation)

897 Independence Ave, Bldg. 3D, Mountain View, CA 94043, USA
Tel: +1 650 968 4527; Fax: +1 650 618 1878
www.pall.com

Oxford Catalysts

115E-H Milton Park, Abingdon, Oxfordshire, OX14 4RZ, UK
Tel: +44 (0) 1235 841 700; Fax: +44 (0) 1235 841 701
www.oxfordcatalysts.com
(See also Velocys – both are part of Oxford Catalysts Group)

Systanix, Inc.

7 Simon Ct, Piscataway, NJ 08854, USA
Tel: +1 732 690 8870; Fax: +1 732 289 6111
www.systanix.com
Systanix, Inc. has developed a scalable and adaptable system for chemical synthesis called SysFlo. With its patent-pending microreactor and catalyst technology, SysFlo's modular design enables 'highly efficient, on-demand chemical production while decreasing equipment footprint'.

ThalesNano Nanotechnology Inc

Graphisoft Park, Záhony u. 7., H-1031 Budapest, Hungary
Tel: +36 880 8500; Fax: +36 880 8501
www.thalesnano.com
ThalesNano Nanotechnology produces nanotechnology solutions for pharmaceutical and biochemical companies. Their products include the H-CUBE and the CATCART.

Velocys, Inc.

7950 Corporate Boulevard, Plain City, Ohio 43064, USA
Tel: +1 614 733 3300; Fax: +1 614 733 3301
www.velocys.com
Velocys produces a range of plant based upon the shims stacked to form microreactors and other unit operations. Now owned by Oxford Catalysts Group (see above).

Mixers

Charles Ross and Son Company

710 Old Willets Path, Hauppauge, NY 11788, USA.
Tel: +1 631 234 0500; Fax: +1 631 234 0691
www.mixers.com
The Batch Model High Shear Rotor-Stator mixer design consists of a single stage rotor that turns at high speed within a stationary stator. As the rotating blades pass the stator, they mechanically shear the contents. The batch model can be either permanently mounted to a vessel or suspended over a vessel on a portable lift. The mobile configuration offers the flexibilty to use a single mixer in multiple vessels. It also allows the user to vary the position of the stator to process a variety of materials.

Ehrfeld Mikrotechnik BTS GmbH

Mikroforum Ring 1, 55234 Wendelsheim, Germany
Tel: +49 (0) 6734 919300; Fax: +49 (0) 6734 919 305
www.ehrfeld.com
Modular microreaction system micromixers for continuous precipitation of micro- and nano-particles. (Part of Bayer).

T. Guisti Ltd

Rixon Road, Finedon Road Industrial Estate, Wellinborough, Northants NN8 4BA, UK
Tel: +44 (0) 1993 229933; Fax: +44 (0) 1993 272363
www.briggsplc.co.uk
Magnetic mixers are suitable for simple, aqueous or low viscosity mixing where hygiene is key. The unit is mounted, usually on the bottom of tank, air operated, single phase, DC or flame-proof. There is no penetration of the vessel wall and therefore no seal. The impellers are easily removed for clean-out-of place or can be sterilised in place by spray ball or flooding.

Microinnova Engineering GmbH

Microinnova Engineering GmbH, Europapark 1, A-8412 Allerheiligen/Wildon, Austria
Tel: +43 (0) 3182 62626-0; Fax: +43 (0) 3182 62626-101
office@microinnova.com
www.microinnova.com
The Falling Film Micro Reactor utilises a multitude of thin falling films that move by gravity force, for typical residence times of seconds up to about one minute. Its unique properties are the specific interface of $20\,000\,m^2/m^3$ and the good temperature control by an integrated heat exchanger. High mass and heat transfer were achieved performing direct fluorination of toluene with elemental fluorine in this device.

Mikroglas chemtech GmbH

Galileo-Galilei-Str. 28, 55129 Mainz, Germany
Tel: +49 6131/5555-0; Fax: +49 6131/55550
www.mikroglas.de
Specialist in glass microtechnology products.

Mixtech Ltd

Bredgar Road, Gillingham, Kent ME8 6PN, UK
Tel: +44 (0) 1634 386 683; Fax: +44 (0) 1634 386 684
www.mixtech.com
US company with many offices worldwide. In-line mixers are provided to the process industry in a wide variety of materials and sizes for low capital and operating cost mixing.

Pursuit Dynamics

Shackleton House, Kingfisher Way, Hinchingbrooke Business Park, Huntingdon, PE29 6HB, UK
Tel: +44 (0) 1480 422050; Fax: +44 (0) 1480 422059
www.pursuitdynamics.com
Highly efficient food mixing process.

Silverson Machines Limited

Waterside, Chesham, Bucks HP5 1PQ, UK
Tel: +44 (0) 1494 786331; Fax: +44 (0) 1494 791452
www.silverson.co.uk
Silverson manufactures high shear mixers.

Koch International bv

Molenbaan 9, PO Box 5177, 2900 ED Capelle a/d Ijessel, The Netherlands
Tel: +31 (0) 10 264 6526; Fax: +31 (0) 10 264 6525
www.koch-glitsch.com
Coalescers, mist eliminators etc. No longer do RPBs.

Sulzer Chemtech AG

Zürcherstrasse 14, 8401 Winterthur, Switzerland
Tel: +41 (0) 52 262 11 22; Fax: +41 (0) 52 262 01 01
www.sulzerchemtech.com

Mixer-Reactors

Corning Inc

One Riverfront Plaza, Corning, New York, 14831 USA
Tel: +1 607 974 9000
www.corning.com
Corning specialises in high quality glass and ceramics. As a company, they have successfully applied their manufacturing process to microreactor manufacture.

Corning S.A.S.

CETC Research Centre, Centre de recherche, F-77210 Avon, France
Tel: +33 1 64 69 7044
www.corning.com
Corning makes glass-based three layer reactor heat exchangers which can be stacked to increase throughput. Has the usual continuous process advantages. Also makes microreactors.

Dena Technology Ltd

Beevor Street, Barnsley, South Yorkshire, S71 1HN, UK
Tel: +44 (0) 1226 388 805
www.dena.co.uk
Nanotechnology and intensive processing of tyre waste.

Hitachi

Various sites.

www.hitachi.com

Hitachi has developed microreactors made by MEMS. Results show a bromination reaction yield up 40%. It uses a numbering up process, rather than scaling up, to move from lab to production more quickly. They are also working on small-scale direct methanol fuel cells for mobile phones and laptops, as conventional battery technology has reached saturation, while devices require ever more power.

Micronit Microfluidics

Colosseum 15, 7521 PV Enschede, The Netherlands

Tel: +31 53 850 6 850; Fax: +31 53 850 6 851

www.micronit.com

Micronit Microfluidics produces microchip applications including pumping equipment and micro reactors.

Syrris

UK head office: 27 Jarman Way, Royston, Herts SG8 5HW, UK

Tel: +44 (0) 1763 242 555; Fax: +44 (0) 1763 242 992

www.syrris.com

Zeton B.V.

P.O. Box 9, NL-7500 AA Enschede, The Netherlands

Tel: +31 (0) 53 428 41 00; Fax: +31 (0) 53 428 41 99

www.zeton.nl/index.htm

Zeton B.V. produces a wide range of reactors for different industries including microreactors.

Oscillatory Baffle Reactors (OBRs)
NiTech Solutions Ltd

www.nitechsolutions.co.uk

NiTech technologies are centered on the design and implementation of baffled chemical reactors and process reactors known as the Continuous Oscillatory Baffled Reactor (COBR™) and the Tubular Reactor (TBR™).

Equipment made by: Alconbury Weston Ltd, Unit D3, Fenton Trade Park, Dewsbury Road Fenton, Stoke-on-Trent ST4 2TE, UK

Porous Metals/Metal Foams
Porvair Ltd

700 Shepherd Street, Hendersonville, NC 28792, USA
Tel: +01 828 697 2411; Fax: +01 828 693 1868
www.porvair.com
See also Compact Heat Exchangers and Heat Exchanger Reactors for UK address.

Separators
Challenger Process Systems Co

21249 Hwy 110 South Troup, TX 75789, USA
Tel: +1 (903) 839 7291; +1 800 657 6576
www.challengerps.com

FL Smidth Krebs

5505 West Gilette Road, Tuscon, AZ 85743, USA
Tel: +1 520 744 8200; Fax: +1 520 744 8300
www.flsmidth.com
Krebs Engineers is a world leader in hydrocyclone separation solutions and was purchased by FL Smidth in 2007. Recognised for its knowledge/expertise in the use of hydrocyclones for the recovery and classification of solids and removal of oil from water.

FMC Technologies (was CDS)

Business Park, Ijsseloord 2, Delta 101, 6825 MN Arnhem, The Netherlands
Tel: +31 26 799 9100; Fax: +31 26 799 9119
www.fmctechnologies.com
FMC designs and manufactures separation equipment and technologies for the oil and gas industry.

GasTran Systems LLC

1768 East 25th Street, Cleveland, OH 44114, USA
Tel: +1 866 427 8726; Fax: +1 216 391 7004
www.gastransfer.com
GasTran Units use specially engineered materials to shear an incoming fluid stream into extremely fine droplets. This process dramatically increases the surface area of the fluid to facilitate proven chemical processes. GasTran process intensification technology is significantly more efficient than current alternatives because it is continuously shearing and coalescing the liquid, exposing surface area to the gas medium. Process

intensification means the highly efficient GasTran System replaces many traditional mass transfer methods such as tray or packed towers, membrane filtration systems, aeration systems, venturis, activated carbon a nd chemical treatments, while reducing operating costs.

Julius Montz GmbH

Postbox 530, D-40705 Hilden, Germany
Tel: +49 (0) 2103 8940; Fax: +49 (0) 2103 89477
www.montz.de
Montz supply dividing wall distillation columns.

Thomas Broadbent and Sons Ltd

Queen Street South, Huddersfield, West Yorkshire HD1 3EA, UK
Tel: +44 (0) 1484 477200; Fax: +44 (0) 1484 516142
www.broadbent.co.uk
The company manufactures batch and continuous centrifuges for separation, etc.

Twister BV

Einsteinlaan 20, 2289 CC Rijswijk (ZH), The Netherlands
Tel: +31 (0) 70 303 0006; Fax: +31 (0) 70 399 6859
www.twisterbv.com

Weir Minerals

Halifax Road, Todmorden, Lancashire OL14 5RT, UK
Tel: +44 (0) 7106 814251; Fax: +44 (0) 7106 815350
www.weirminerals.com
Weir minerals are a global supplier of pumps and hydrocyclones to the minerals industry. The company has developed the Cavex hydrocyclone which offers longer life, greater throughput and classification efficiency than previous hydrocyclone designs.

Wenzhou Jinzhou Group International Trading Corporation

Rm. 1112, Comprehensive Building, Jinzhou Industrial Zone, Caodai Guoxi Ouhai Wenzhou, China
Tel: +577 89615569/88058180
jzgroup.machinery@gmail
www.jzmachinery.com
Manufacture rotating packed beds for high gravity distillation duties.

Taylor-Couette Reactors
Kreido Laboratories

1140 Avenida Acaso, Camarillo, CA 93012, USA
Tel: +1 805 398 3499

Kreido Biofuels Inc

1070 Flynn Road, Camarillo, CA 93012, USA
Tel: +1 805 389 3499; Fax: +1 805 384 0989
info@kreido
www.kreido.com/fac_howbiodiesel.htm
Kreido Biofuels spent seven years researching and developing in fluid dynamics in order to develop a novel method of biodiesel manufacture. The result is their STT® technology. STT® (spinning tube in a tube reactor) is a chemical process intensification system which provides significant time and cost savings to the company.

R.C. Costello and Assoc., Inc.

1611 S. Pacific Coast Highway, Suite 210, Redondo Beach, CA 90277, USA
Tel: +1 310 792 5870; Fax: +1 310 792 5877
www.rccostello.com
The company is also involved in other PI technologies including reactive distillation and divided wall columns, and the shock wave power reactor.

Tube Inserts
Cal Gavin Ltd

Minerva Mill Innovation Centre, Station Road, Alcester, Warwickshire B49 5ET, UK
Tel: +44 (0) 1789 400401; Fax: +44 (0) 1789 400411
www.calgavin.com

Ormiston Wire

1 Fleming Way, Worton Road, Isleworth, Middlesex TW7 6EU, UK
Tel: +44 (0) 20 8569 7287; Fax: +44 (0) 20 8569 8601
www.ormiston-wire.co.uk

Ultrasound Systems
Hielscher Ultrasonics GmbH

Warthestr. 21
D-14513 Teltow, Germany

Tel: +49 3328 437 420; Fax: +49 3328 437 444
info@hielscher.com
www.heilscher.com
Industrial ultrasound processing plant.

Prosonix Ltd.

The Magdalen Centre, Robert Robinson Avenue, Oxford Science Park, Oxford OX4
4GA, UK
Tel: +44 (0) 1865 784250; Fax: +44 (0) 1865 784251
www.prosonix.co.uk
Their technology aims at controlling and actively manipulating nucleation and subsequent crystal growth behaviour. For instance, this technology has been applied to increase sodium oxalate impurity removal from the process liquor at the Aughinish Alumina refinery in Ireland. The existing process of recycling precipitated seed to act as a precipitation initiator was subject to the constraint that crystals rapidly become poisoned by other species present in the liquor and therefore could not be recycled in great quantity. The ultrasound technology is essentially applied very simply by 'bolting-on' to the pipe.

Appendix 4: R&D Organisations, Consultants and Miscellaneous Groups Active in PI

NOTE: As with Appendix 3, the list below is not exhaustive and has been compiled from information available to the authors at the time of writing. The listing of a supplier of goods or services does not constitute an endorsement by the authors of either its product or its competence. Neither does the omission of a supplier discriminate against its competence.

Aachenerverfahrenstechnik (AVT)

Institut für Verfahrenstechnik der RWT, Turmstr. 46, 52064 Aachen, Germany
Tel: +49 (0) 241 80 95470; Fax: +49 (0) 241 80 92252
www.avt.rwth-aachen.de
German Education Institute conducting research into specific PI areas. Topics include micro structured membrane reactors, electrophoresis, and phenol removal from water using coated membrane contactors.

Accentus plc

528.10 Unit 2, Rutherford Avenue, Harwell Science and Innovation Campus, Didcot, Oxfordshire OX11 0QJ, UK
Tel: +44 (0) 1235 434320; Fax: +44 (0) 1235 434329
www.accentus.co.uk
Accentus licenses intellectual property, including that for two PI units, a compact gas-to-liquids unit (www.compactgtl.com) and a vortex-intensifed gas scrubber, V-Tex.

Aee Intec
Operating Agent for IEA Solar Heating and Cooling Programme (Process Industries)

Christoph Brunner, AEE - Institute for Sustainable Technologies, A-8200 Gleisdorf, Feldgasse 19
Tel: +43 (0) 3112 5886 70; Fax: DW 18
c.brunner@aee.at
www.aee-intec.at/index.php?lang=en

Amalgamated Research Inc

2531 Orchard Drive East, Twin Falls, Idaho 83301, USA
Tel: +1 208 735 5400; Fax: +1 208 733 8604
www.arifractal.com
Amalgamated Research is a process research and development company and specialises in fluid handling and distribution. An example of the process intensification equipment offered is the patented fractal ion exchange system.

ANSYS Inc

Southpointe, 275 Technology Drive, Canonsburg, PA 15317, USA
Tel: +1 724 746 3304; Fax: +1 724 514 9494
www.ansys.com
ANSYS Inc designs and licences simulation software, widely used by engineers and designers in several areas. Modelling.

AspenTech, Inc

200 Wheeler Road, Burlington, Massachusetts 01803, USA
Tel: +1 781 221 6400; Fax: +1 781 221 6410
www.aspentech.com
AspenTech's focus has been on applying process engineering know-how, to modelling the manufacturing and supply chain processes that characterise the process industries.

Battelle, Pacific Northwest National Laboratory

PO Box 999 Richland, WA 99352, USA
Tel: +1 509 375 2121
www.pnnl.gov
Research into a range of microchemical and thermal systems such as micro fuel cells, compact heat exchangers, microscale absorbers, etc. The Velocys Company was a spin-out from PNNL.

Bayer Technology Services

D-51368 Leverkusen, Germany
www.bayertechnology.com
Research into microreactors and other miniaturisation technologies.

Beijing University of Chemical Technology

Research Centre of High Gravity Engineering and Technology
Beijing 100029

Professor Chen, Jianfeng – Head.
http://wwwold.buct.edu.cn/english/txt/SD/cce/faculty.htm
Carries out research on rotating packed beds etc. in a variety of applications.

BHR Group Ltd

Cranfield, Bedford MK43 0AL, UK
Tel: +44 (0) 1234 750422; Fax: +44 (0) 1234 750074
www.bhrgroup.com
A major UK operator in PI R&D. Activities: Intensified mixing and HEX-reactor
R&D. Pilot plants are available. The Flex-reactor was developed at BHRG. Now
active in intensification of biodiesel.

Cambridge University

Department of Chemical Engineering and Biotechnology, New Museums Site,
Pembroke Street, Cambridge, CB2 3RA, UK
Tel: +44 (0) 1223 334777; Fax: +44 (0) 1223 334796
www.ceb.cam.ac.uk and www.ceb.cam.ac.uk/groups.php?id=12
The Polymer Processing Research group at the University of Cambridge undertakes
research in a number of areas relevant to process intensification, namely microcap-
illary films and oscillatory flow mixing.

The Centre for Process Innovation

Wilton Centre, Wilton, Redcar TS10 4RF, UK
Tel: +44 (0) 1642 455 340; Fax: +44 (0) 1642 447 298
www.uk-cpi.com
The CPI works on chemical and bio-processing with facilities including micro-
reactors, a spinning disk reactor, and a continuous oscillatory baffle reactor demonstrator. They can also offer pilot plant design, process optimisation and related
engineering and consultation services.

Clarkson University

Process Intensification and Clean Technology Group, Department of Chemical
and Biomolecular Engineering, PO Box 5705, Clarkson University, Potsdam, NY
13699, USA
Tel: +1 315 268 6325; Fax: +1 315 268 6654
www.clarkson.edu/projects/pict/index.htm
The major centre for PI research at Clarkson University includes spinning disc
reactors and microreactors.

Commercial, Chemical and Development Company

1, Umayal Street, Chennai 600 010, Tamil Nadu, India
Tel: +91 44 26440167; Fax: +91 44 26440062

www.ccdcindia.com
Chemical engineering process design, including PI.

Cranfield University

College Road, Cranfield, Bedfordshire MK43 0AL, UK
Process Systems Engineering Group
Tel: +44 (0) 1234 754749
www.cranfield.ac.uk/soe/departments/ope/pse/research.html

Dalhousie University

Chemical Engineering, 1360 Barrington St. F Bldg., Halifax, Nova Scotia, Canada
B3J 1Z1
Tel: +001 (902) 494 3953; Fax: (902) 420 7639
chemicalengineering.dal.ca
Work on inline intensified mixing, multiphase reactors and other unit operations.

David Reay & Associates

PO Box 25, Whitley Bay, Tyne & Wear NE26 1QT, UK
Tel: +44 (0) 191 251 2985
www.drassociates.co.uk
PI and energy consulting engineers

Delft University of Technology

Stevinweg, 2628 CN Delft, The Netherlands
Tel: +31 15 2785404; Fax: +31 15 2781855
www.tudelft.nl/en
Active in many areas of PI, including distillation equipment, catalysis and general
PI research.

Département CREST FEMTO-ST/UMR CNRS 6174

Parc technologique, 2, avenue Jean Moulin, 90000 Belfort, France
Tel: 03 81 85 39 97; Fax: 03 81 85 39 98
www.femto-st.fr
Research includes the design of micro heat exchangers for micro refrigerators dedicated to the cooling of electronic components.

Ecole Supérieure de Chimie Physique Electronique de Lyon

Domaine Scientifique de la Doua, Bâtiment Hubert Curien 43, boulevard du 11
Novembre 1918, BP 82077 - 69616 Villeurbanne Cedex France
Tel: +33 (0) 472 43 17 00; Fax: +33 (0) 472 43 16 68

www.cpe.fr/?lang=en
Research into process engineering and biochemistry, and optimising the process from the initial materials to end products. Key points are zero pollution and defects, safety, increasing speed and reducing cost.

Energy Research Centre of the Netherlands (ECN)

PO Box 1, 1755 ZG Petten, The Netherlands
Tel: +31 224 56 4949
www.ecn.nl
Research topics at ECN in the field of process intensification are: membrane reactors in which the reaction is combined with separation of the reaction product from the reaction zone; membrane contactors, where reactants are fed in a controlled way to a reaction, leading to a higher quality of the desired product with lower amounts of by-products, and membrane emulsification. ECN is also active in areas related to the gas turbine reactor.

Europic

European Process Intensification Centre (EUROPIC)
p/a Leeghwaterstraat 44, 2628 CA Delft, The Netherlands
Tel: +31 15 27 87843
info@europic-centre.eu
www.europic-centre.eu
EUROPIC's mission is knowledge & technology transfer in the area of process intensification. Aiming to support and integrate the entire value chain EUROPIC actively connects science and business in innovative, application-driven research and creates interfaces between end users, engineering companies and technology providers.

Paying special attention to the quality and reliability of supplied information EUROPIC performs technology scouting, benchmarking and trend analyses, issues position papers and provides its members with worldwide consulting services as well as specialised courses and trainings.

Food Technology Centre

Wageningen UR, PO Box 17, 6700 AA Wageningen, The Netherlands
Tel: +31 (0) 317 475123; Fax: +31 (0) 317 883011
www.ftc.wur.nl and www.ftc.wur.nl/EN/werkvelden/menu4.asp
Dutch food technology specialist organisation developing entire systems. In their own words, 'Process intensification is a powerful tool to improve the efficiency and lower the cost of operations. Process steps that always used to be separated are now integrated to allow preservation, fractionation or production, sometimes all at the same time. FTC takes a leading position in technologies that can be used in hybrid processes.'

Hazard Evaluation Laboratory Ltd

9-10 Capital Business Park, Manor Way, Borehamwood, Hertfordshire WD6 1GW, UK
Tel: +44 (0) 20 8736 0640; Fax: +44 (0) 20 8736 0641
www.helgroup.com
Safety evaluation of reactors, etc. Reactor kinetics.

Health and Safety Laboratory

Hapur Hill, Buxton, Derbyshire SK 17 9JN, UK
Tel: +44 (0) 1298 218218; Fax: +44 (0) 1298 218822
www.hsl.gov.uk
The Health and Safety Laboratory (HSL) is Britain's leading industrial health and safety facility with over 30 years of research experience across all sectors. They have capabilities in a wide range of topics including: fire, explosion and process safety, including reactor safety.

Heriot-Watt University

Chemical Engineering, School of Engineering and Physical Sciences, Riccarton, Edinburgh EH14 4AS, Scotland
Tel: +44 (0) 131 449 5111
www.hw.ac.uk
Activities: biointensification; OBRs; highly compact heat exchangers.

Honda Research Institute USA Inc

800 California St., Suite 300, Mountain View, CA 94041, USA
Tel: 650 314 0400
www.honda-ri.com/HRI_Us

Honda Research Institute Europe GmbH

Carl-Legien-Straße 30, 63073 Offenbach/Main, Germany
Tel: +49 69 89011 750; Fax: +49 69 89011 749
www.honda-ri.org
Optimisation of micro-heat exchangers, CFD, analytical approach and multi-objective evolutionary algorithms.

Institut für Mikrotechnik Mainz GmbH (IMM)

Carl-Zeiss-Strasse 18-20, 55129 Mainz, Germany
Tel: +49 6131 990 388; Fax: +49 6131 990 205
www.imm-mainz.de
The Chemical Process Technology Department (CPT) is active in the field of chemical micro process technology and in similar, innovative continuous-flow processing technologies.

Institute for Membrane Technology (ITM-CNR)

Via Pietro Bucci, Cubo 17C, c/o University of Calabria, 87030 Rende (CS), Italy
Tel: (39) (0984) 492 050; Fax: (39) (0984) 402103
www.itm.cnr.it
This institute researches a wide variety of membranes; from the actual transport phenomena through many membrane types, to the many possible applications where membranes might be of use (such as organic or biological synthesis, to separation of components in a mixture) along with catalytic and reactor membranes.

IRCELYON, Institut de recherches sur la catalyse et l'environnement de Lyon, UMR5256

2 avenue Albert Einstein, 69626 Villeurbanne cedex, France
Tel: +33 (0) 472 445 300; Fax: +33 (0) 472 445 399
www.ircelyon.univ-lyon1.fr
Research on reactors and structured catalysts.

KenaTech Process Engineering

P.O. Box 1842, Medina, OH 44258-1842, USA
Tel: +1 (330) 725 7091; Fax: +1 (330) 725 7091
tak@kenatech.com
www.kenatech.com

Kvaerner Process Technology

The Technology Centre, Princeton Drive, Thornaby, Stockton-on-Tees TS17 6PY, UK
Tel: +44 (0) 1642 853800; Fax: +44 (0) 1642 853801
Activities: R&D, plant design, intensified processes and unit operations.

LAAS CNRS (Laboratoire d'Architecture et d'Analyse des Systèmes)

7 avenue du Colonel Roche, 31077 Toulouse Cedex 4, France
Tel: +335 61 33 62 00; Fax: +335 61 55 35 77
www.laas.fr/laas
Four thematic clusters define the research areas of the laboratory. These are the pole micro and nanosystème, pole critical systems, pole modelling, optimisation and control systems, and the pole robots and autonomous systems.

Laboratoire des Sciences du Génie Chimique, CNRS/ENSIC/INPL

LSGC, GROUPE ENSIC, BP 451, F-54001 Nancy Cedex France
Tel: +33 (0) 3 83 17 51 90; Fax: +33 (0) 3 83 32 29 75
www.ensic.inpl-nancy.fr

Research on various PI related topics for production of chemicals; biochemicals and materials; and treatment of wastes and polluted materials. Looking at optimising processes to save energy and improve safety and cleanliness of equipment.

Lappeenranta University of Technology

Laboratory of Process Systems Engineering, Department of Chemical Technology, Skinnarilankatu 34, Fin-53850 Lappeenranta, Finland
Tel: +358 5 621 2165; Fax: +358 5 621 2165
www.lut.fi/kete/laboratories/Process_Systems_Engineering/mainpage.htm#top
The Laboratory of Process Systems Engineering is active in teaching and research into processes engineering. Research topics within the group include methodologies in process research and design; process intensification; and modelling and simulation.

Loughborough University

Department of Chemical Engineering, Loughborough, Leicestershire LE11 3TU, UK
Tel: +44 (0) 1509 222 533; Fax: +44 (0) 1509 223 923
www.lboro.ac.uk/departments/cg
Has been involved in modelling and simulation of the flux responses of a gas–solid catalytic micro-reactor.

Louisiana Tech University

Institute for Micromanufacturing, 305 Wisteria Street, P.O. Box 3178, Ruston, LA 71272, USA
Tel: +1 318 257 2000
www.latech.edu
Micro heat exchangers fabricated by diamond machining.

Manchester University

Centre for Process Integration, School of Chemical Engineering and Analytical Science, The University of Manchester, Oxford Road, Manchester, M13 9PL, UK
Tel: +44 (0) 161 306 4380 or +44 (0) 161 306 8750; Fax: +44 (0) 161 236 7439
www.ceas.manchester.ac.uk/research/centres/centreforprocessintegration
The CPI is a leader in process design and integration, utilising cutting edge technology and producing highly regarded research. It has been heavily involved in bringing together process intensification and process integration.

Max Planck Institute for Dynamics of Technical Systems

Sandtorstr.1, 39106 Magdeburg, Germany
Tel: +49 391 61100; Fax: +49 391 610 500
www.mpi-magdeburg.mpg.de

Reactive distillation is explored in this laboratory as a means of combining separate reaction and distillation stages into a single unit.

Newcastle University

Process Intensification Group, (PIG), School of Chemical Engineering and Advanced Materials, Merz Court, Newcastle upon Tyne NE1 7RU, UK
pig.ncl.ac.uk
The Process Intensification Group at Newcastle University has developed novel approaches to equipment design and process synthesis with the aim of miniaturising process plants and making them environmentally friendly and flexible in terms of manufacturing capabilities and providing rapid response to market demand. The group's research activities include: spinning disc reactors, micro reactors, oscillatory flow reactors, compact heat exchangers, catalytic reactors, HiGee separation technology and nanoparticle technology.

Process Intensification and Miniaturisation Group (PIM)

Tel: +44 (0) 191 2227269; Fax: +44 (0) 191 2225292
www.ncl.ac.uk/pim/index.htm
Some of the main areas of research within this group at Newcastle University are cellular polymers, biomass waste gasification, nano-structured micro-porous materials and spinning disc reactors.

Post Mixing

7 Tippet Way, Pittsford NY 14534-4520, USA
Telephone: +1 (585) 507 4318
www.postmixing.com
Consultancy fronted by Tom Post, a fluids mixing expert working in a wide range of processing industries including petrochemical, biochemical, food, plastics and polymers. Specialising in a number of PI concepts, such as optimisation of processes and intensive mixing.

Process Intensification Consultants (India)

Tel: 91 40 23041352
www.processintensification.net
The research and products include spinning disc reactors, inline mixers, micro channel reactors and compact heat exchangers. The group has grown out of IIT Kanpur in India.

Process Systems Enterprise Limited

6th Floor East, 26-28 Hammersmith Grove, London W6 7HA, UK
Tel: +44 (0) 20 8563 0888; Fax: +44 (0) 20 8563 0999
www.psenterprise.com
Process modelling, including PI unit operations.

Rhodia Europe

Coeur Défense – Tour A – 37ème étage, 110, esplanade Charles de Gaulle, 92931 Paris La Défense Cedex, France
Tel: +33 1 53 56 64 64
www.rhodia.com
Rhodia is an international industrial group committed to sustainable development. Rhodia is the partner of the great players in the automotive, tyre, electronics, perfumes, hygiene, beauty and home maintenance fields.

Sheffield University

Chemical Engineering Department, Mappin Street, Sheffield, S1 3JD, UK
Tel: +44 (0) 114 222 7500; Fax: +44 (0) 114 222 7501
www.shef.ac.uk/cbe
Sheffield University's research includes mixing in microchannels and their applications, and micro distillation.

Tam Kang University

Department of Mechanical Engineering, Tamsui Campus, No.151, Yingzhuan Rd. Tamsui Dist., New Taipei City 25137 Taiwan, Republic of China
Tel: +886 2 26215656; Fax: +886 2 26223204
foreign.tku.edu.tw/TKUEnglish
Machining of thin metal foils with specially contoured diamond cutting tools allows the production of small and very smooth fluid microflow channels for micro heat exchanger applications. The plates are stacked and bonded with the vacuum diffusion process to form a cross-flow, plate-type heat exchanger. These fabrication techniques allow the production of small heat exchangers with a very high volumetric heat transfer coefficient and inherent low weight. The design and fabrication process for a copper-based, cross-flow micro heat exchanger has been developed. The micro heat exchanger provided a volumetric heat transfer coefficient of nearly $45\,MW/m^3\,K$.

TNO Environment, Energy and Process Innovation

PO Box 342, 7300 AH Apeldoorn, The Netherlands
Tel: +31 88 866 22 12; Fax +31 88 866 22 48
www.tno.nl
Activities: Intensified reactors; other intensified unit operations.

TWI Ltd

Granta Park, Great Abington, Cambridge CB21 6AL, UK
Tel: +44 (0) 1223 899000
www.twi.co.uk

TWI has expertise in surface sculpturing, to give surface features at the micro- or nanoscale that are appropriate to PI processes such as heat exchangers and micro-reactors.

Unilever Research and Development

Olivier van Noortlaan 120, 3133 AT, Vlaardingen, The Netherlands
Tel: +31 10 460 6933; Fax: +31 10 460 5800
www.unilever.com
One of the research areas involving PI is the use of power ultrasound as a means of textile laundering and cleaning. Ultrasound can increase the mass transfer rates by a factor of six in the laundering process. This is beneficial as traditional means use vast quantities of water and chemicals and are usually temperature limited (as not to damage the textiles), whilst the use of power ultrasound can reduce the amount of water and chemicals used as well as the processing time to achieve the same desired results.

Chemical Engineering Dept, University College London

Torrington Place, London WC1E 7JE, UK
www.ucl.ac.uk/chemeng
Activities: Design of catalytic microreactors; zeolite microreactors; modelling of catalytic plate reactors.

Universidade Estadual de Maringá

Av. Colombo, 5.790, Jd. Universitário, Maringá, Paraná, Brasil CEP 87020-900
Tel: +55 44 3261 4352
www.uem.br
Particulate materials that are cohesive when wet are usually difficult to dry in traditional fluidised beds, and therefore are rarely used in these instances. This research looks into how, by using a rotating-pulsed fluidised bed, it is possible to dry these materials, without the traditional problems, whilst preserving the quality of the desired product. By pulsating the gas, rather than a continuous flow of gas, energy savings of up to 50% can be achieved. The rotation ensures that there is a much more even distribution of wet product that needs to be dried, allowing even filtration through the particles.

University of Pittsburgh

Department of Chemical and Petroleum Engineering, Room 1249 Benedum Hall
Pittsburgh, PA 15261, USA
Tel: +1 (412) 624 9630; Fax: +1 (412) 624 9639
www.engr.pitt.edu
Professor Morsi is leading an extensive research effort to design and scale-up multiphase reactors such as bubble columns, slurry bubble-columns, high pressure/ temperature stirred vessels, and trickle-bed reactors.

The University of Queensland

Advanced Water Management Centre, Brisbane, Qld 4072, Australia
Tel: +61 7 3365 4730; Fax: +61 7 3365 4726
www.awmc.uq.edu.au
Looking into a wide variety of techniques for the recovery and purification of water
supplies.

University of Rostock

Leibniz-Institut für Katalyse e. V. (LIKAT Rostock), Albert-Einstein-Str. 29 a,
18059 Rostock, Germany
or
Berlin Büro, Volmerstr. 7 B, 12489 Berlin, Germany
Tel: +49 (30) 6392 4028
www.catalysis.de/Home.7.0.html
Process intensification and inorganic synthesis – including micro-reactor technology, membrane reactors sand catalyst coating.

University of Twente

Drienerlolaan 5, 7522 NB Enschede, PO Box 217, 7500 AE Enschede, The
Netherlands.
Tel: +31 53 4899 111; Fax: +31 53 4892 000
www.utwente.nl
One of the research areas focuses on micro-reactors with different catalysts fixed in
different compartments within the micro-reactor. This allows a much more efficient
use for the chemical pathway of the desired product to take place, as these reactions
can take place simultaneously, eliminating the need for many reactors in series, or
one reactor that has to be operated in a batch style, replacing the catalyst for each
stage, etc. Ideal flow with no build-up of product is essential in these reactors to
avoid the accumulation of intermediate product, as is being investigated by using
methods such as immiscible liquids, etc.

The University of Western Australia

Centre for Strategic Nano-Fabrication, 35 Stirling Highway, Crawley, WA 6009,
Australia.
Tel: +61 (08) 6488 3326; Fax: +61 (08) 6488 8683
www.strategicnano.uwa.edu.au
Research into synthesis of nanoparticles and nano-hybrids for use in spinning disc
processing technology. Nano-catalysis for the fine chemical industry.

University of York

Green Chemistry Centre of Excellence, Chemistry Department, Heslington, York
YO10 5DD, UK
Tel: +44 (0) 1904 434550; Fax: +44 (0) 1904 432705
www.york.ac.uk/chemistry/research/green
The Green Chemistry Centre of Excellence works at the cutting edge of chemical
research in the areas of clean synthesis, catalysis, novel materials and applications
for renewable resources.

URENCO (Capenhurst) Ltd

Capenhurst, Chester CHI 6ER, UK
Tel: +44 (0) 151 473 4000; Fax: +44 (0) 473 4040
www.urenco.co.uk
Urenco has developed ultra high-speed centrifuge technology, primarily for the
nuclear industry.

Yole Development

Yole Développement, Le Quartz, 75 cours Emile Zola, 69100 Lyon-Villeurbanne
France
Tel: +33 04 72 83 01 80; Fax: +33 04 72 83 01 83
www.yole.fr
Microtechnologies for process intensification.

Appendix 5: A Selection of Other Useful Contact Points, Including Networks and Websites

In this appendix, data are given on a number of organisations involved in supporting PI, including groupings of PI-active companies. There are also web addresses of useful patent databases.

A5.1 UK Assistance/Information/Support Programmes
The Carbon Trust

The Carbon Trust, 4th Floor, Dorset House, 27-45 Stamford Street, London SE1 9NT, UK
Tel: +44 (0)20 7170 7000
www.carbontrust.com
As part of its aim to help move to a low carbon economy in the UK, the Carbon Trust has an approach to accelerating research towards commercial value – the Directed Research initiative (including Research Acceleration). Its aim is to focus on overcoming the specific technical barriers that are holding back the next generation of low carbon technologies. A wide range of technologies is part of its brief. It has included PI.

Engineering and Physical Sciences Research Council (EPSRC)

Polaris House, North Star Avenue, Swindon SN2 1 ET, UK
Tel: +44 (0) 1793 444114; Fax: +44 (0) 1793 444009
www.epsrc.ac.uk
Supports R&D at universities.

A5.2 Centres of Excellence/Groupings
AXELERA (ARKEMA, SUEZ, CNRS, IFP and RHODIA)

Pôle de compétitivité à vocation mondiale Chimie-Environnement Lyon and Rhône-Alpes.
http://www.axelera.org/en/

577

Includes a PI project – the 'Factory of the Future' involving the partners in the technology platform.

Enki (Efficient Networking of Knowledge for Innovation)

www.enki2.com

Enki integrates a risk-analysis and decision-making method, not only in the exploratory phase, but continuously throughout the development phases by an evaluation of ideas and potential invention; risk management tool for innovation projects and innovation portfolio management. Enki also integrates sustainability thinking via innovative technologies as an opportunity. The challenge is not only to provide service to the consumer society, but to perform it with markedly lower reliance on materials, energy, labour and waste – sustainable development and PI are inherent in the organisation's approach.

EFCE (European Federation of Chemical Engineering)

www.efce.info

Since 1953, the European Federation of Chemical Engineering has promoted scientific collaboration and supported the work of engineers and scientists in 28 European countries. Moreover, from the very beginning, Eastern and Central European countries were included. It has promoted PI activities within the sector.

INERIS (Institut National de l'Environnement Industriel et des Risques)

http://www.ineris.fr/en

I studies and research to prevent the risk of economic activities and provide any benefit to facilitate the adoption by companies of that goal. Safety in processes, e.g. supercritical processing.

International Energy Agency – Heat Pump Implementing Agreement

Data accessible via the IEA Heat Pump Centre includes Annex 33 - Compact heat exchangers in heat pumping equipment.

Joint Research Centre, European Commission.

Institute for Systems, Informatics and Safety (ISIS)
http://poplar.sti.jrc.it/public/ian/aware/AWAREWEB.htm
ISIS specialise in safety work, venting, etc and co-ordinate the AWARE programme (Advanced Warning Against Runaway Events).

SenterNovem

SenterNovem is an agency of the Dutch Ministry of Economic Affairs, promoting sustainable development and innovation, both within the Netherlands and abroad.

The Fuel Cell and Hydrogen Association.

www.fchea.org
Integral participant of America's $1.2 billion hydrogen fuel initiative which, as part of its brief, is examining new, novel (low carbon) technologies for the generation of hydrogen. The UK Hydrogen association is based in North East England.

A5.3 Patent Sources

As well as protecting your own inventions, be they PI technologies or applications of PI, a search of patent databases often provides invaluable information on existing PI technologies. The web has made such searches rapid, cheap and easy to implement. The data below refer to some major patent databases.

Searches can be done on the basis of key words for the technology, patent numbers, (where known), companies or inventors. Care should be taken in ascertaining where free access stops and charges begin, e.g. for ordering the full patent specification.

www.ipo.gov.uk – The UK Patent Office (Intellectual Property Office) has its own website offering a variety of facilities. Access to patent abstracts and other services is available.

www.espacenet.com – This is the website of the European Patent Database that also lists world and other patents. Easy to use and access is free.

patft.uspto.gov – This is the United States Patent and Trademark Office home page and is the official site for searching the US patent database.

www.delphion.com – The Delphion Intellectual Property Network also lists US patents, searchable by key words, patent numbers, assignees, etc., (charges may be applicable).

A5.4 Networks

Networks and specialist groups within professional institutions are ideal media for the dissemination of data of all types on a technology. They thus provide ready opportunities for those wishing to have an introduction to a technology, and to meet people with experience of its application, access to funding opportunities and other technology transfer activities.

There are networks specifically directed at PI, but others have PI as a peripheral interest. It is recommended that those that are interested get in touch with the relevant contact person or visit the website (where available) to help assess the usefulness and relevance to them. The websites generally have good interactive links to other sites relevant to PI.

BioIndustry Association

Has a natural interest in process intensification.
www.bioindustry.org

BRITEST (Batch Route Innovative Technology Evaluation and Selection Techniques).

The BRITEST website contains some very useful links.
www.britest.co.uk

HEXAG (Heat Exchanger Action Group)

www.hexag.org

Process Intensification Network (PIN)

This network meets in the UK but has an international membership.
www.pinetwork.org

PIN-NL (Process Intensification Network Netherlands)

www.linkedin.com/groups/PINNL-Process-Intensification-Network-Netherlands-4321849/about *and*
http://www.agentschapnl.nl/en

Index

Note: Page numbers followed by '*f*' and '*t*' refer to figures and tables, respectively.

Printed and bound by CPI Group (UK) Ltd, Croydon, CR0 4YY

08/05/2025

01864786-0005